Land Use Cover Datasets and Validation Tools

David García-Álvarez •
María Teresa Camacho Olmedo •
Martin Paegelow • Jean François Mas
Editors

Land Use Cover Datasets and Validation Tools

Validation Practices with QGIS

 Springer

Editors
David García-Álvarez
Departamento de Geología
Geografía y Medio Ambiente
Universidad de Alcalá
Alcalá de Henares, Spain

Martin Paegelow
Département de Géographie
Aménagement et Environnement
Université de Toulouse Jean Jaurès
Toulouse, France

María Teresa Camacho Olmedo
Departamento de Análisis Geográfico
Regional y Geografía Física
Universidad de Granada
Granada, Spain

Jean François Mas
Laboratorio de Análisis Espacial
Centro de Investigaciones en Geografía Ambiental
Universidad Nacional Autónoma de México
Morelia, Mexico

ISBN 978-3-030-91000-6   ISBN 978-3-030-90998-7   (eBook)
https://doi.org/10.1007/978-3-030-90998-7

This Springer imprint is published by the registered company Springer Nature Switzerland AG
The registered company address is: Gewerbestrasse 11, 6330 Cham, Switzerland

Acknowledgments

This research was funded by the following projects: *INCERTIMAPS: Suitability and uncertainty of land use and land cover maps for the analysis and modelling of territorial dynamics* (PGC2018-100770-B-100) financed by the Spanish State Research Agency (SRA) and the European Regional Development Fund (ERDF) and *Herramientas para la Enseñanza de la Geomática con programas de Código Abierto* (PE117519) supported by the Programa de Apoyo a Proyectos para la Innovación y Mejoramiento de la Enseñanza (PAPIME), Universidad Nacional Autónoma de México (UNAM).

The editors would like to show their appreciation to all the contributing authors, without whom this book would not have been possible. They are also grateful to Alexis Vizcaino, our Springer editor, Sabina Nanu, for her thorough work ensuring the coherence and correct format of this book, and Nigel Walkington, for his diligent proofreading.

Finally, the editors would like to acknowledge the support provided by their respective research centres: Departamento de Geología, Geografía y Medio Ambiente, Universidad de Alcalá, Spain; Departamento de Análisis Geográfico Regional y Geografía Física, Universidad de Granada, Spain; Département de Géographie, Aménagement, Environnement, Université Toulouse Jean Jaurès, France; and Centro de Investigaciones en Geografía Ambiental (CIGA), Universidad Nacional Autónoma de México (UNAM), Mexico.

Contents

Part II   Data Access and Visualization

Visualization and Communication of LUC Data   69
Francisco Escobar

Sample Data for Thematic Accuracy Assessment in QGIS   85
Miguel Ángel Castillo-Santiago, Edith Mondragón-Vázquez,
and Roberto Domínguez-Vera

Part III   Tools to Validate Land Use Cover Maps: A Review

**Basic and Multiple-Resolution Cross-Tabulation to Validate Land
Use Cover Maps**   99
María Teresa Camacho Olmedo and David García-Álvarez

**Metrics Based on a Cross-Tabulation Matrix to Validate Land Use
Cover Maps**   127
Jean-François Mas, David García-Álvarez, Martin Paegelow,
Roberto Domínguez-Vera, and Miguel Ángel Castillo-Santiago

**Pontius Jr. Methods Based on a Cross-Tabulation Matrix to Validate Land
Use Cover Maps**   153
Martin Paegelow, Jean-François Mas, Marta Gallardo, María Teresa Camacho
Olmedo, and David García-Álvarez

Editors and Contributors

About the Editors

David García-Álvarez, Ph.D. is a postdoctoral researcher at the Department of Geology, Geography and Environment, at the University of Alcalá, Spain. He has previously worked at the University of Granada and the Joint Research Centre of the European Commission. His research interests include Land Use Cover change mapping and modelling, uncertainty of spatial data, shared economies in the accommodation sector and the analysis of regional and local territorial dynamics. For more detailed information please visit his personal Website at https://sites.google.com/view/dagaral/.

María Teresa Camacho Olmedo, Ph.D. is a Professor at the Department of Geographical Regional Analysis and Physical Geography, at the University of Granada, Spain. She leads the project entitled Suitability and Uncertainty of Land Use and Cover Maps for the Analysis and Modelling of Territorial Dynamics (INCERTIMAPS), funded by the Spanish State Research Agency (SRA) and the European Regional Development Fund (ERDF) (ref. PGC2018-100770-B-100) (2019–2021), after successfully completing various other competitive projects since 2003. Her current research includes land use and land cover modelling, simulation models and scenarios and territorial dynamics. She has published numerous scientific papers and books. For more detailed information please visit her Website at http://geofireg.ugr.es/pages/profesorado/Camacho_olmedo.

Martin Paegelow, Ph.D. is a Professor at the Department of Geography, Land Planning and Environment at the University of Toulouse Jean Jaurès, France, and member of GEODE (Geography of Environment) UMR 5602 CNRS laboratory. In 1991, he obtained a Ph.D. in geography and was accredited to supervise research in 2004. He is a specialist in environmental geography and geomatics. His principal research areas are environmental management and geomatic solutions with a special focus on geomatic land change modelling and its validation. He is also strongly committed to the teaching side of his job and since 2004 has been the co-director of the Master's Degree in Geomatics in Toulouse. For more detailed information please visit his Website at http://w3.geode.univ-tlse2.fr/permanents/paegelow.php .

Jean François Mas, Ph.D. is a tenured Professor in the Center of Research in Environmental Geography (Centro de Investigaciones en Geografía Ambiental, CIGA) at the National Autonomous University of Mexico (Universidad Nacional Autónoma de México, UNAM). He specializes in the fields of remote sensing, geographical information science and spatial modelling. His research interests include land use/land cover change monitoring and

modelling, accuracy assessment of spatial data, forest inventory and vegetation cartography. He has published more than 70 peer-reviewed scientific publications, advised 9 Ph.D. dissertations, supervised 15 master's degree students and participated in 35 research projects. For more detailed information please visit his Website at http://www.ciga.unam.mx/index.php/mas
.

Contributors

María Teresa Camacho Olmedo Departamento de Análisis Geográfico Regional y Geografía Física, Universidad de Granada, Granada, Spain

Miguel Ángel Castillo-Santiago Departamento de Observación y Estudio de la Tierra, la Atmósfera y el Océano, El Colegio de la Frontera Sur, San Cristóbal de las Casas, Mexico

Roberto Domínguez-Vera Departamento de Observación y Estudio de la Tierra, la Atmósfera y el Océano, El Colegio de la Frontera Sur, San Cristóbal de las Casas, Mexico

Francisco Escobar Departamento de Geología, Geografía y Medio Ambiente, Universidad de Alcalá, Alcalá de Henares, Spain

Marta Gallardo Departamento de Geografía, Universidad Nacional de Educación a Distancia, Madrid, Spain

David García-Álvarez Departamento de Geología, Geografía y Medio Ambiente, Universidad de Alcalá, Alcalá de Henares, Spain

Francisco José Jurado Pérez Departamento de Análisis Geográfico Regional y Geografía Física, Universidad de Granada, Granada, Spain

Javier Lara Hinojosa Departamento de Análisis Geográfico Regional y Geografía Física, Universidad de Granada, Granada, Spain

Jean-François Mas Laboratorio de Análisis Espacial, Centro de Investigaciones en Geografía Ambiental, Universidad Nacional Autónoma de México, Morelia, Mexico

Ramón Molinero-Parejo Departamento de Geología, Geografía y Medio Ambiente, Universidad de Alcalá, Alcalá de Henares, Spain

Edith Mondragón-Vázquez Departamento de Observación y Estudio de la Tierra, la Atmósfera y el Océano, El Colegio de la Frontera Sur, San Cristóbal de las Casas, Mexico

Sabina Florina Nanu Departamento de Análisis Geográfico Regional y Geografía Física, Universidad de Granada, Granada, Spain

Martin Paegelow Département de Géographie, Aménagement, Environnement, Université Toulouse Jean Jaurès, Toulouse, France

Jaime Quintero Villaraso Departamento de Análisis Geográfico Regional y Geografía Física, Universidad de Granada, Granada, Spain

About This Book

David García-Álvarez, María Teresa Camacho Olmedo,
Martin Paegelow, and Jean-François Mas

Abstract

This chapter offers an introduction to the book and is specifically recommended for all readers intending to do the practical exercises it contains. It also provides readers with all the information they require to make the most of the book's contents. In this chapter, we explain the aim, structure and intended audience for this book. We also give the readers a few tips and guidelines about how to make best use of it. This is followed by a description of the software and the data used to do the practical exercises. In the last section of this chapter, we offer a detailed explanation about how we conducted the review of the LUC datasets carried out for Chap. "Land Use Cover Datasets: A Review" and Part IV of the book.

Keywords

Introduction • Tutorial • QGIS • R • Datasets

The original version of this chapter was revised: The URL link on page 11 in section "5.5 Data" has been updated. The correction to this chapter is available at https://doi.org/10.1007/978-3-030-90998-7_23

D. García-Álvarez (✉)
Departamento de Geología, Geografía y Medio Ambiente, Universidad de Alcalá, Alcalá de Henares, Spain
e-mail: David.garcia@uah.es

M. T. Camacho Olmedo
Departamento de Análisis Geográfico Regional y Geografía Física, Universidad de Granada, Granada, Spain

M. Paegelow
Département de Géographie, Aménagement et Environnement, Université de Toulouse Jean Jaurès, Toulouse, France

J.-F. Mas
Laboratorio de Análisis Espacial, Centro de Investigaciones en Geografía Ambiental, Universidad Nacional Autónoma de México, Morelia, Mexico

1 Introduction

This chapter sets out the aims of this book and explains the methods and approaches applied in its production. It also aspires to be a guide, offering readers instructions as to how best to use the book. We therefore strongly encourage all readers to read this chapter carefully, so as to gain a clearer understanding of all the different aspects analysed in this book. This chapter also provides essential information for those wishing to do the practical exercises in this book.

We begin by presenting the aims of the book and we offer a few tips explaining how each group of users can make best use of this book according to their particular requirements. Then, we provide information about the software and the data required to carry out the practical exercises presented in Parts II and III (Sect. 5). In the last section, we offer a detailed explanation of the review of LUC datasets carried out in Chap. "Land Use Cover Datasets: A Review" and Part IV (Sect. 6).

The book is the fruit of two research projects which seek to provide a clearer understanding of the uncertainties associated with Land Use Cover maps and with the results of Land Use Cover Change modelling exercises (INCERTIMAPS Project: Suitability and uncertainty of land use and land cover maps for the analysis and modelling of territorial dynamics) and the promotion of Open Access software for teaching spatial science (PE117519: Herramientas para la Enseñanza de la Geomática con programas de Código Abierto). See complete information about these projects in the section Acknowledgements.

2 What is the Main Aim of This Book?

The aim of this book is to provide an up-to-date state of the art on Land Use Cover (LUC) datasets and validation tools. The book summarizes the available information and makes it accessible to any interested user, including some of the latest developments in the field.

The book was conceived as a practical tool to inform readers about currently available LUC datasets at global and supra-national scales and to help them understand more about the validation of LUC data and LUCC modelling exercises, so enabling them to validate their own data and models. To this end, the book combines brief theoretical explanations with practical information and exercises.

Part I of the book briefly covers the theoretical foundations of LUC mapping, LUCC modelling and the analysis and assessment of their associated uncertainties. Parts II and III were conceived as practical guides to enable any reader to use any of the tools and data. Part II covers the visualization of LUC data and the production of reference datasets to validate LUC maps. Part III describes the use of common validation tools and the interpretation of their results. All the practical exercises are accompanied by an explanation of the basic theory behind them, so as to enable users to understand the analyses and the principles on which the techniques are based. Finally, Part IV of the book characterizes the most relevant available LUC data. It also provides all the necessary information as to how to download and use the datasets.

As the book aims to reach the widest possible audience, the theory is briefly explained in simple, understandable terms. Practical exercises are implemented in QGIS, an open-source Geographical Information System, which can be downloaded for free.

3 Who is the Book Aimed At?

The book is aimed at anyone interested in Land Use Cover (LUC) mapping, Land Use Cover Change modelling and Land Use Cover Change analysis. Although to make full use of the book, some background in the field is recommended, it aims to be accessible and useful to all kinds of user, regardless of their level of expertise. Nonetheless, a basic knowledge of spatial analysis and GIS analysis is required to understand a lot of the information provided.

The book will be particularly useful for researchers working in the fields of LUC mapping and LUCC modelling and especially for those interested in validation methods and the available sources of LUC data. Those interested in the application of open-source software in LUC may also find this book very useful, as it is the only book working with open-source software that focuses on these topics from a holistic perspective. For the QGIS community, the book provides the relevant information and tools to enable users to take full advantage of the software and expand the fields in which it can be effectively applied.

4 How to Use This Book?

The book can be used in different ways, depending on the type of user and their particular background and interests. With this in mind, it has been conceived as a flexible tool that can be used for a wide variety of purposes.

Beginners in this field are referred to Chap. "Land Use Cover Mapping, Modelling and Validation. A Background", as are other users interested in gaining an overall picture of LUC mapping, LUCC modelling and the essential concepts required for uncertainty and validation analyses. This short, yet comprehensive chapter sets out the basic theoretical principles on which the rest of the book is based and is therefore recommended reading for all users.

For LUC data visualization and creation, readers are referred to Part II of this book. It provides an overview of the different options available for symbolizing LUC data and LUC change in GIS. It also addresses some of the problems associated with the spatial visualization of LUC information. This part of the book also includes a tutorial on the creation of a set of sample points for LUC data validation with QGIS.

Users interested in the validation of LUC datasets and Land Use Cover Change (LUCC) modelling exercises should refer to Chap. "Validation of Land Use Cover Maps: A Guideline". This provides guidelines for validating different LUC products: single LUC maps, LUC map series, and outputs from LUCC modelling exercises. The different tools and methods referred to in these guidelines are then described in detail and applied in practice in the example exercises in Part III of the book.

Users interested in doing the example QGIS exercises appearing in this book should refer to Sect. 5 of this chapter, which presents all the data and the cases studied in this book. It also offers essential information about the particular version of QGIS that we use and about how to integrate R software into QGIS, a necessary step when carrying out some of the exercises set out in the book.

Those interested in LUC data sources should refer to Chap. "Land Use Cover Datasets: A Review", which offers an introduction to LUC mapping at global and supra-national scales, including a review of the different datasets available. Part IV of the book offers in-depth descriptions of most of the datasets that are available for download, detailing their specific characteristics and how they can be accessed. The methodology followed in the review of the datasets is described in Sect. 6 of this chapter.

5 QGIS Exercises: Software, Study Areas and Data

5.1 GIS Software

Of all the Geographical Information Systems (GIS) currently available, in this book we use QGIS, a well-known, open-source GIS software that is widely used and recognized. It provides a unified interface to many other relevant open-source GIS software programmes, such as SAGA, GDAL, GRASS or LasTools (Menke et al. 2016). It also allows integration with R, a powerful open-source software for statistical analysis.

We opted for the QGIS 3.10.13 "A Coruña" version of QGIS for the practical exercises included in this book. This is because it was the newest long-term release version of QGIS available when we began writing the book.

Users could try other versions of QGIS when doing the exercises included in this book. However, they should bear in mind that the exercises have been created and tested using the version indicated above and that certain issues and errors may arise when using any other version of QGIS. Earlier versions of QGIS prior to QGIS 3 are strongly discouraged, as important changes were made in the software between versions 2 and 3 and many features of QGIS 3 do not work in earlier versions of the software.

The latest version of QGIS is available at the QGIS website (www.qgis.org). Users who require a specific version of this software should visit: https://qgis.org/downloads/. Full documentation relating to the software can also be found at the official website: https://www.qgis.org/en/docs/index.html, where inexperienced QGIS users will find a brief introduction to the software interface and the main tools.

Several user manuals are also available to help beginners make the most of the software. These include the books published by Packt (Graser et al. 2017; Cutts and Graser 2018) and the series of manuals coordinated by Baghdadi et al. (2018a, b, c, d), which contain both generic and thematic GIS exercises.

5.2 QGIS Plugins

QGIS works with plugins written in the C++ and Python programming languages. These plugins are an easy way to expand the capabilities of the software, which is why many of the features of the software are currently implemented through these plugins.

There are two types of plugins: core and external plugins (QGIS Project 2020). The core plugins are maintained by the QGIS Development Team and automatically form part of the distributed software. The external plugins are developed by a community of users and are available at the QGIS Python Plugins Repository (https://plugins.qgis.org/plugins/).

The external plugins may be up-to-date or outdated and are usually available for specific QGIS versions. The official plugin repository includes information about all these questions. External plugins that are still in the early stages of development and have not been widely used are marked by QGIS as experimental plugins and are not directly available through the software.

Several QGIS plugins are used in the exercises presented in this book (Table 1). In all cases, we used the most up-to-date versions of these plugins as of when we began writing. Some of the plugins may have been updated since then, which could lead to certain differences in the interface and the results. This is something that readers should be aware of when using the plugins.

The Semi-Automatic Classification Plugin is one of the most important QGIS plugins and is used in many of the exercises in this book. It was developed and updated by Luca Congedo (2016) and provides a comprehensive interface and set of tools for classifying remote sensing imagery. This includes many tools for validating image classifications, which are also used in this book. For more information on the plugin and how to use it, users must refer to the plugin manual (Luca Congedo 2016) and official website (see Table 1).

LecoS (Landscape ecology Statistics) is a plugin developed by Jung (2016) to calculate the spatial metrics usually employed in the field of landscape ecology. Although other methods can be implemented in QGIS to calculate these metrics, the LecoS plugin is the best-known QGIS tool for this purpose. All the relevant information about the plugin is available at the official website (see Table 1).

The R Processing Provider allows the R software capabilities to be integrated into QGIS. Full documentation on the plugin is available at the official website (see Table 1). Users can also find extra information on the plugin and the way the R language can be integrated into QGIS in the official documentation on QGIS.[1] To find out more about how to integrate R into QGIS, users should consult Sect. 5.3 of this chapter.

QuickMapServices is a very used QGIS plugin that allows to import to the QGIS interface many different web-map services of different kinds (XYZ tiles, TMS, WMS, WMTS, ESRI ArcGIS Services). More information on the plugin is available in the official website (see Table 1)

---

[1] https://docs.qgis.org/3.4/en/docs/user_manual/appendices/qgis_r_syntax.html#. https://docs.qgis.org/3.4/en/docs/user_manual/appendices/qgis_r_syntax.html#syntax-summary-for-qgis-r-scripts.

Table 1 QGIS plugins employed in the practical exercises of the book

Plugin	URL
Processing R Provider	https://north-road.github.io/qgis-processing-r/
Semi-automatic Classification	https://fromgistors.blogspot.com/p/semi-automatic-classification-plugin.html
LecoS—Landscape Ecology Statistics	https://conservationecology.wordpress.com/qgis-plugins-and-scripts/lecos-land-cover-statistics/
MapAccurAssess	https://doi.org/10.5281/zenodo.5419130
QuickMapServices	https://nextgis.com/blog/quickmapservices/
Google Earth Engine Data Catalog	https://github.com/sandroklippel/qgis_gee_data_catalog/wiki

and the manual recommended by the plugin's authors, in Russian.[2]

The Google Earth Engine Data Catalog plugin provides direct access in QGIS to the data catalog that takes part of the Google Earth Engine platform. Users will need a Google account to make use of this plugin. However, not much information is available about the plugin. If needing more information, users are referred to its official website (see Table 1).

We also use MapAccurAssess, a plugin specifically developed for the exercises of this book by Domínguez Vera (2021). Although not available yet in the official QGIS plugin repository, it can be downloaded from the official repository of information accompanying this book (see Table 1). The plugin provides a tool for assessing the accuracy of classified Land Use Cover images, taking into account the recommendations made by Olofsson et al. (2013). For more information about the plugin, users are referred to the plugin manual, in Spanish (Domínguez Vera, 2021). It is also available in the official repository for this book.

To install any of these plugins in QGIS, access the "Manage and install plugins…" tool in the plugins menu to find the plugin you require. Once selected, click on the "Install Plugin" option (Fig. 1). In the "Settings" tab of the tool, users can also make experimental and deprecated plugins available in QGIS. To install MapAccurAssess, use the "Install from ZIP" tab, select the downloaded file and then click "Install Plugin" (Fig. 2).

5.3 Integrating R into QGIS

Some of the exercises presented in this book use R, a free, open-source statistical software. QGIS enables the R environment to be integrated into the software, making it easier

for any QGIS user to take full advantage of the tools available through R.

QGIS does not have the required tools to compute all the validation tools and methods that have been reviewed in this book. We have therefore had to implement some of them in QGIS through the R processing environment. Users wishing to find out more about R and its integration into QGIS, with practical exercises about how to use both software packages in combination, should consult the manual by Islam (2018).

To integrate R into QGIS, users must begin by downloading the R software. R and any of its associated data can be downloaded from a comprehensive file network, from which users must select the mirror closest to their location at https://cran.r-project.org/mirrors.html.

Once downloaded and installed, users must also install a series of packages in R to execute the different tools and methods included in the book (Table 2). This step cannot be carried out through the QGIS interface. Users must open R and manually install the different packages. To do this, select *Packages > Install Package(s)…* from the menu (Fig. 3). In the window that opens, select the mirror from which to download the packages (Fig. 4). Finally, select the package to be installed (Fig. 5). Installation of the package may take a little while to complete. Installation is complete when the R console allows the user to write new code (Fig. 6).

Table 2 lists the packages required to do the different exercises appearing in this book. In the table, next to each package name, we offer a link to the website with all the information about the package: description, download link, reference manual, etc.

After installing R and the required packages, we need to install the QGIS plugin that allows us to integrate the two software packages. This is the "Processing R provider" plugin. Instructions to this end can be found in Sect. 5.2 of this chapter. After installing the plugin, users must download the scripts we have developed to integrate the R tools and capabilities into QGIS. These scripts are listed in Table 3 and are available at https://doi.org/10.5281/zenodo.5418985 in the official repository for this book.

[2] https://gis–lab-info.translate.goog/qa/quickmapservices.html?_x_tr_sl=ru&_x_tr_tl=en&_x_tr_hl=en.

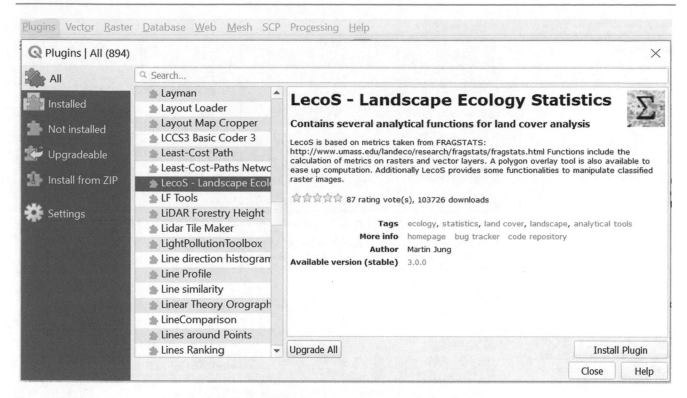

Fig. 1 QGIS plugins. Standard plugin installation workflow

Fig. 2 QGIS plugin. Plugin installation from a zip file

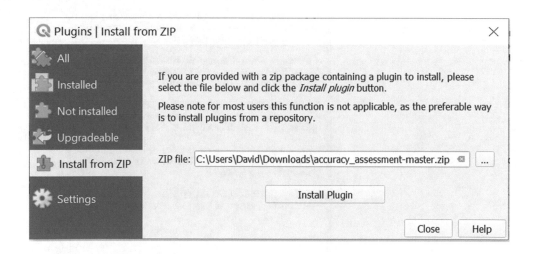

Table 2 List of R packages required to use the R scripts provided in this book

Package	URL
intensity.analysis	https://cran.r-project.org/web/packages/intensity.analysis/index.html
raster	https://cran.r-project.org/web/packages/raster/index.html
Rgdal	https://cran.r-project.org/web/packages/rgdal/index.html
ROCR	https://cran.r-project.org/web/packages/ROCR/index.html
sabre	https://cran.r-project.org/web/packages/sabre/index.html
sf	https://cran.r-project.org/web/packages/sf/index.html
sp	https://cran.r-project.org/web/packages/sp/index.html

Fig. 3 Integrating R in QGIS. Installing the required pachakes in R: first step

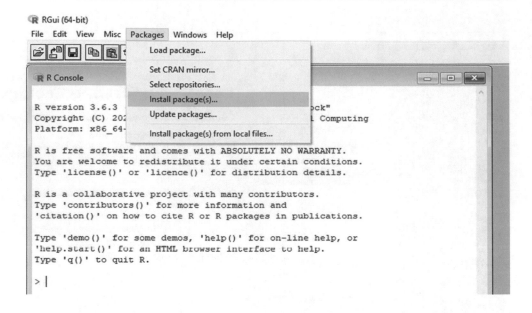

Fig. 4 Integrating R in QGIS. Installing the required pachakes in R: second step (mirror selection)

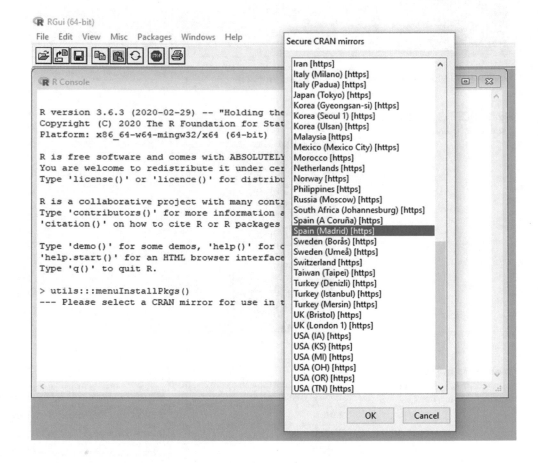

Once downloaded, the script files must be pasted into the R scripts folder of QGIS. The path to this folder can be found in the "Options" menu of QGIS. To access it, go to *Settings > Options....* and then select the "Processing" submenu (Fig. 7). In the "Providers" tab, there is a specific tab for "R". After opening this tab, a list appears including the "R scripts folder" path, which indicates where users must save the scripts that come with the book.

Fig. 5 Integrating R in QGIS. Installing the required pachakes in R: third step (package(s) selection)

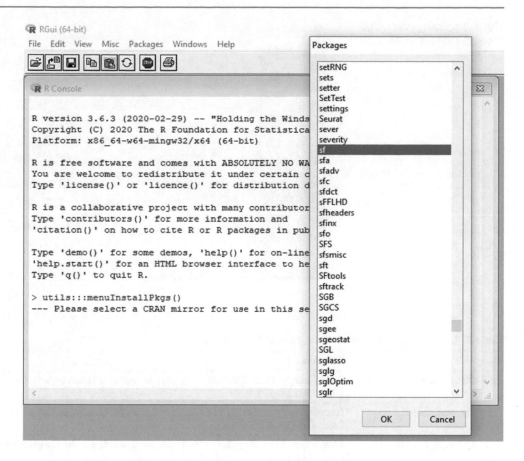

Fig. 6 Integrating R in QGIS. Installing the required pachakes in R: end of the workflow

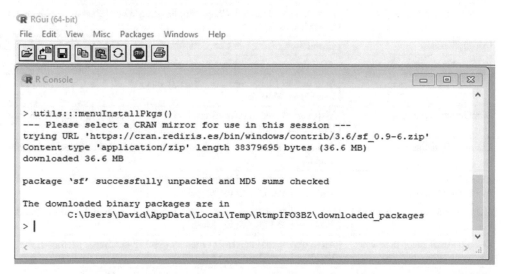

Fig. 7 Integrating R in QGIS.
R configuration in QGIS

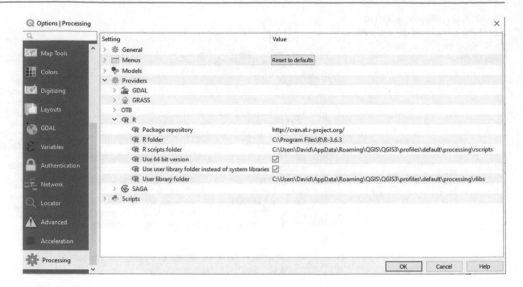

Table 3 List of the R scripts developed for use in this book

Script file	Function
LUCCBudget.rsx	LUCC budget (Sect. 2 in Chap. "Pontius Jr. Methods Based on a Cross-Tabulation Matrix to Validate Land Use Cover Maps")
Intensity_analysis.rsx	Intensity analysis (Sect. 6 in Chap. "Pontius Jr. Methods Based on a Cross-Tabulation Matrix to Validate Land Use Cover Maps")
Stable_change_flow_matrix.rsx	Flow matrix (Sect. 7 in Chap. "Pontius Jr. Methods Based on a Cross-Tabulation Matrix to Validate Land Use Cover Maps")
Flow_matrix_graf.rsx	Flow matrix (Sect. 7 in Chap. "Pontius Jr. Methods Based on a Cross-Tabulation Matrix to Validate Land Use Cover Maps")
Correlation.rsx	Correlation (Sect. 1 in Chap. "Validation of Soft Maps Produced by a Land Use Cover Change Model")
ROCAnalysis.rsx	ROC analysis (Sect. 2 in Chap. "Validation of Soft Maps Produced by a Land Use Cover Change Model")
MapCurves_raster.rsx	Map curves (recommended for raster data) (Sect. 1 in Chap. "Advanced Pattern Analysis to Validate Land Use Cover Maps")
MapCurves_vector.rsx	Map curves (recommended for vector data) (Sect. 1 in Chap. "Advanced Pattern Analysis to Validate Land Use Cover Maps")
Change_Statistics.rsx	Change statistics (Sect. 1 in Chap. "Metrics Based on a Cross-Tabulation Matrix to Validate Land Use Cover Maps")
Individual Areal Inconsistency.rsx	Areal and spatial agreement metrics (Sect. 2 in Chap. "Metrics Based on a Cross-Tabulation Matrix to Validate Land Use Cover Maps")
Individual Spatial Agreement.rsx	Areal and spatial agreement metrics (Sect. 2 in Chap. "Metrics Based on a Cross-Tabulation Matrix to Validate Land Use Cover Maps")
Overall Areal Inconsistency.rsx	Areal and spatial agreement metrics (Sect. 2 in Chap. "Metrics Based on a Cross-Tabulation Matrix to Validate Land Use Cover Maps")
Overall Spatial Agreement.rsx	Areal and spatial agreement metrics (Sect. 2 in Chap. "Metrics Based on a Cross-Tabulation Matrix to Validate Land Use Cover Maps")
Overall Spatial Inconsistency.rsx	Areal and spatial agreement metrics (Sect. 2 in Chap. "Metrics Based on a Cross-Tabulation Matrix to Validate Land Use Cover Maps")
Local accuracy assessment statistics.rsx	Overall, user and producer's accuracies through GWR (Sect. 1 in Chap. "Geographically Weighted Methods to Validate Land Use Cover Maps")

Fig. 8 Location map of the
Asturias Central Area

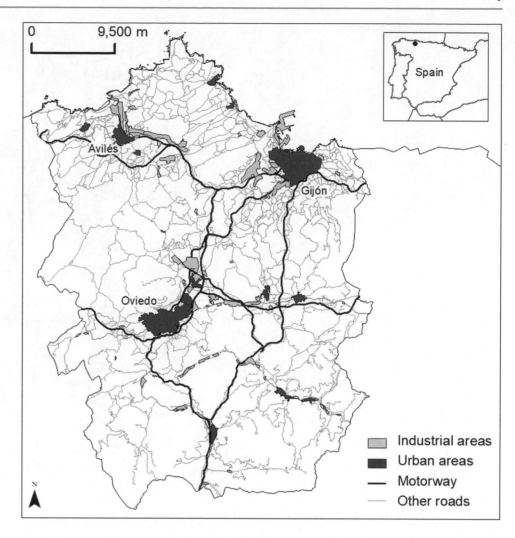

5.4 Study Areas

The exercises provided in this book are applied to three
specific study areas: the Ariège Valley (France), the Asturias
Central Area (Spain) and the Marqués de Comillas municipality (Mexico). We now offer a brief introduction to these
study areas, so as to give readers the contextual information
they require for a clearer understanding of the results of the
exercises.

5.4.1 The Asturias Central Area (Spain)

The Asturias Central Area is a rural-industrial-urban area
located in the heart of Asturias, in Northern Spain (Fig. 8). It
hosts around 80% of the Asturian population and most of its
economic activity (Rodríguez Gutiérrez et al. 2009). It is
made up of a polycentric set of cities of different sizes that
play a complementary socioeconomic role. The cities are
surrounded by a network of villages and plenty of rural
space, where a traditional rural economy and lifestyle is
mixed with peri-urban dynamics (Rodríguez Gutiérrez et al.
2013).

The cities at the top of the urban hierarchy are Oviedo,
Gijón and Avilés, which concentrate most of the urban LUC
dynamics in recent decades (Gobierno del Principado de
Asturias 2016). The area within the triangle formed by the
three cities has also been the subject of important LUC
dynamics, with the emergence of new industrial and residential developments, attracted by the accessibility that the
area's extensive transport network provides (Méndez García
and Ortega Montequín 2013). The south of the Asturias
Central Area is dominated by small industrial cities, mainly
Mieres and Langreo, located in long, narrow valleys where
there is almost no new space for development (Prada Trigo
2011). These were formerly mining/industrial towns which
are now in decline.

5.4.2 Ariège Valley (France)

The Ariège Valley area consists of the central part of the
valley formed by the River Ariège, which is situated is in the
department of the same name about 70 km south of Toulouse (Fig. 9). It covers an area of 1113 km² and has a
population of about 80,000 inhabitants. The Ariège Valley is

Fig. 9 Map showing the location
of the Ariège Valley

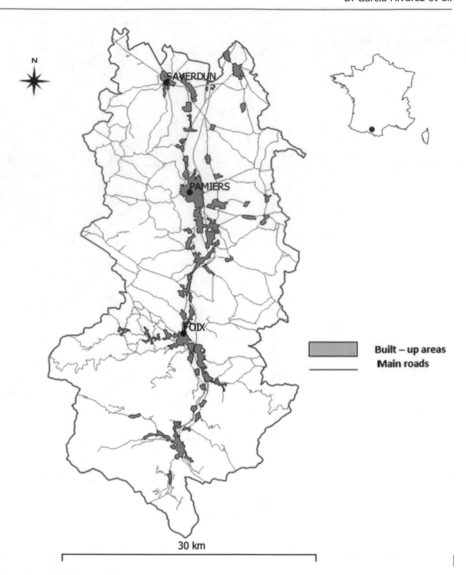

a rural area with agriculture in the northern part and wooded land in the south, approaching the Pyrenees. The largest town is Pamiers, in the centre of the valley, with about 15,000 inhabitants, while the departmental capital, Foix, has a population of 9700. Saverdun, in the north of the valley, has 4900 inhabitants.

In the past, the Ariège Valley was a centre for industrial and mining activities while today it is mainly rural. Tourism is increasingly common. The most notable LUC dynamics are reforestation and the increase in built-up areas, which are mainly concentrated along the river.

5.4.3 Marqués de Comillas (Mexico)

Marques de Comillas is a physiographical region of the Lacandon rainforest in Chiapas, Mexico (Fig. 10). Bounded by two rivers, the Usumacinta and the Lacantun, it comprises approximately 15% (2032 km^2) of the Lacandon region. The climate is hot and humid, with an average annual temperature of 24.3 °C and average annual precipitation of 2960 mm, most of which falls from May to December (García-Amaro 2004).

A colonization programme by the Mexican Government in the 1970s encouraged the establishment of farming communities in forest-covered areas, promoting agriculture, agroforestry (cacao) and cattle ranching, which is currently the most important business activity. Over the last 40 years, Marqués de Comillas has suffered a dramatic loss in forest cover; in the mid-1980s, forests occupied 83% of the region, while today, this has fallen to just 29%, less than half of which are well-preserved forests. The landscapes are now made up above all of mosaics of agricultural lands, cattle pastures and human settlements.

Fig. 10 Location of Marqués de Comillas

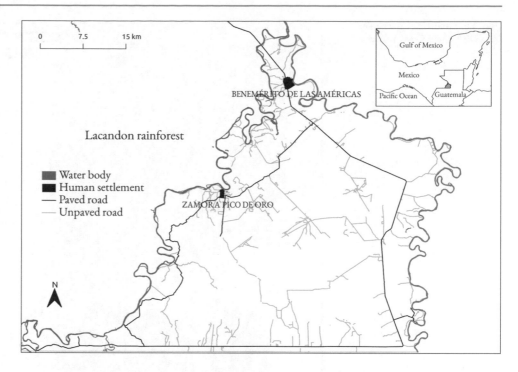

5.5 Data

All the data used in the example exercises provided in this book can be found online and downloaded at https://doi.org/10.5281/zenodo.5418318 in the official repository for this book. This data consists of LUC maps for the three different study areas (Ariège Valley, Asturias Central Area and the Marqués de Comillas municipality) and the data from LUC modelling exercises for the first two. The data for Ariège Valley comes from the work carried out Nabila Bounoua and Jéromine Le Campion, students of the Master in Geomatics SIGMA at the University of Toulouse Jean Jaurès.

Detailed information on the LUCC modelling exercises developed for the Asturias Central Area and Ariège Valley can be found in studies by García-Álvarez et al. (2019) and Bounoua and Le Campion (2019). The LUC maps for these two areas were obtained from two different datasets: CORINE Land Cover, SIOSE. The LUC map for the Marqués de Comillas municipality was obtained through the classification of satellite imagery.

We will now briefly describe the LUC datasets and maps that form part of the database for each study area. At the end of this section, there is a table with all the files used in this book.

CORINE Land Cover (CLC) is a pan-European dataset of LUC information available for five different dates from 1990 to 2018. It provides detailed, coherent LUC information for most of the countries in Europe. It is usually carried out by photointerpretation in vector format at a scale of 1:100,000, with a Minimum Mapping Unit (MMU) of 5–25 ha and a Minimum Mapping Width (MMW) of 100 m. Detailed

information about this dataset can be found in Chap. " General Land Use Cover datasets for Europe" of this book.

A simpler version of CLC is used in the Ariège Valley (Fig. 11) and the Asturias Central Area (Fig. 12) case studies. In the latter, CLC is available in both vector and raster format. Although CLC is officially distributed in raster format at a spatial resolution of 100 m, the CLC rasters for the study areas in this book are provided at a different spatial resolution: 50 m for Asturias and 15 m for Ariège. These rasters were obtained after rasterizing the CLC vector layers.

SIOSE (Sistema de Información sobre Ocupación del Suelo de España) is a Spanish dataset in vector format that provides very detailed LUC information. It was obtained by photointerpretation of aerial imagery at 1:25,000, with a MMU of 0.5–2 ha and a MMW of 10 m. It follows a specific data model aimed at objects, which means that all the land uses and covers in a polygon are described by a specific code. This means that instead of being assigned to a specific LUC category, each polygon is described by a code detailing its LUC composition.

Some of the maps in the Asturias Central Area case study were obtained after simplification of the SIOSE database. The maps were obtained after the classification of each SIOSE polygon into a single category and after the rasterization at 50 m of the original vector dataset (Fig. 12). More information on how this operation was performed can be found in García-Álvarez (2018). Extra information about the characteristics of SIOSE can be found in Valcárcel et al. (2008) and García-Álvarez and Camacho Olmedo (2017).

The Marques de Comillas LUC map (Fig. 13) is part of a database on Land Cover and Land Cover/Land Use Changes

Fig. 11 Land Use Cover map
(CORINE Land Cover) Ariège
Valley

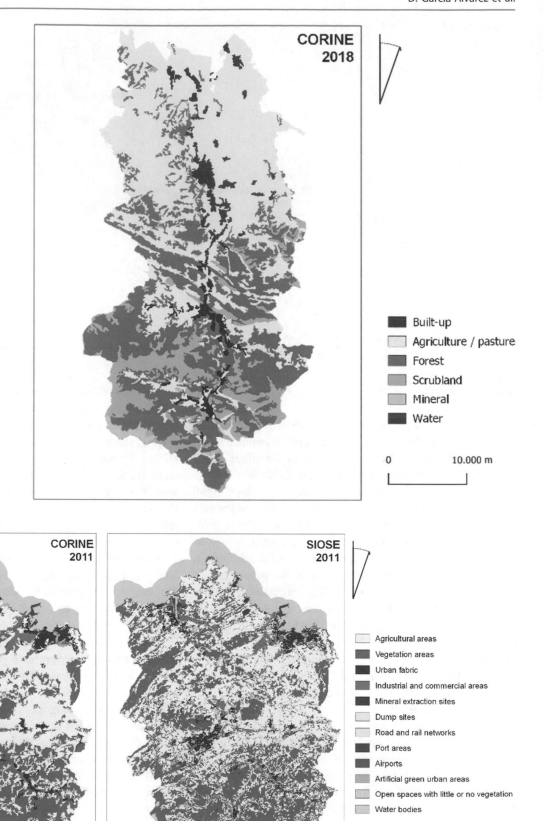

Fig. 12 Land Use Cover maps (CORINE Land Cover, SIOSE) Asturias Central Area

Fig. 13 Land Use Cover Map
Marqués de Comillas

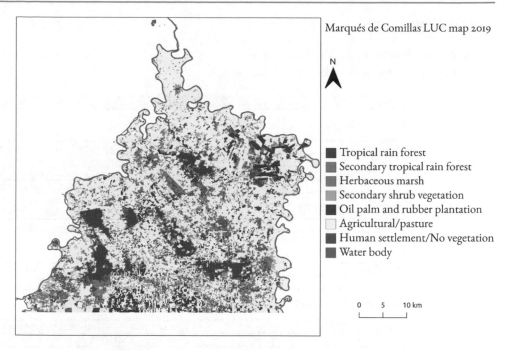

Marqués de Comillas LUC map 2019

N

■ Tropical rain forest
■ Secondary tropical rain forest
■ Herbaceous marsh
■ Secondary shrub vegetation
■ Oil palm and rubber plantation
□ Agricultural/pasture
■ Human settlement/No vegetation
■ Water body

0   5   10 km

in the State of Chiapas in Mexico. The original database covers 7.5 million ha, of which the Marqués de Comillas map covers a small section of approximately 200,000 ha. The maps were computed via a supervised classification of 2019 Sentinel-2 imagery. They were subsequently photo-interpreted to correct errors from the supervised stage as well as to include information on agricultural land uses. The map contains eight thematic categories describing levels of forest conservation, and other land uses; the approximate scale is 1:40,000, with an MMU of one ha. More information can be found at the following link: https://bosqueschiapasdemo.ecosur.ourecosystem.com/.

In the following tables, we list the files from the different datasets and LUC modelling exercises described above that have been used in different exercises in this book. More datasets are available online, including extra LUC maps and model drivers not considered in the exercises in this book.

The tables include information about the name of the file available for download and the descriptive name used to refer to these files in the book. For each dataset, we also provide the projection of the dataset and the file describing the legend of the maps. A document listing all these characteristics for the layers only available online is also provided when downloading the data.

Ariège Valley (Val d'Ariège)

Projection: WGS84/UTM 31N (EPSG: 32631)
Associated files: BD_Val_Ariege (Word document file): explanation and legend

File name	Name in the book
CORINE Land Cover	
CLC_2000	CORINE Land Cover Map Val d'Ariège 2000
CLC_2012	CORINE Land Cover Map Val d'Ariège 2012
CLC_2018	CORINE Land Cover Map Val d'Ariège 2018
Model Drivers	
Roads_dist	Distance to roads
Simulation output	
CLC_predict_2018	Simulation LCM Val d'Ariège 2018
CLC_predict_2018_soft_UTM	Soft prediction LCM Val d'Ariège 2018
00_12_18_transition_2_to_1	Transition potential map from agricultural to artificial areas
00_12_18_transition_3_to_1	Transition potential map from forests to artificial areas
Markov18_class1_utm	Markovian probability map for artificial areas Ariège Valley

Asturias Central Area

Projection: WGS84/UTM 30N (EPSG: 32630)
Associated files: Legend Asturias maps (spreadsheet)

File name	Name in the book

CORINE Land Cover

File name	Name in the book
C05.shp	CORINE Land Use Vector Map Asturias Central Area 2005
C05.rst	CORINE Land Use Map Asturias Central Area 2005
C05_v00.rst	CORINE Land Use Map Asturias Central Area 2005 v.0
C11.shp	CORINE Land Use Vector Map Asturias Central Area 2011
C11.rst	CORINE Land Use Map Asturias Central Area 2011
Changes_CORINE_05_11.rst	CORINE Land Use Changes Asturias Central Area 2005-2011
C18.rst	CORINE Land Use Map Asturias Central Area 2018

SIOSE

File name	Name in the book
S11.shp	SIOSE Land Use Vector Map Asturias Central Area 2011
S11.rst	SIOSE Land Use Map Asturias Central Area 2011

CORINE Land Use Cover Change model and simulation

File name	Name in the book
Simulation_C11.rst	Simulation CORINE Asturias Central Area 2011
Simulation_C11_SIOSEDemands.rst	Simulation CORINE 2 Asturias Central Area 2011
Simulated_changes_CORINE_05_11.rst	Simulated CORINE changes Asturias Central Area 2005-2011
C_Suitability _factor_Urban_Fabric.rst	Urban fabric suitability map – CORINE model

Marqués de Comillas

Projection: WGS84/UTM 15N (EPSG: 32615)
Associated files: Marques_LUC_datasets (Word document file): dataset description and legend

File name	Name in the book
LandCoverMarques2019.tif	Marqués de Comillas Land Use Land Cover Map 2019
RandomSample_Buffer.shp	Photo-interpreted reference dataset – Marqués de Comillas 2019
RandomSample.shp	Centroids of sample sites – Marqués de Comillas
random_sample_points.shp	Marqués de Camilla random sample points from Mexico (2019)
limit.shp	Boundaries of Marqués de Comillas study area

6 Review of Land Use Cover Datasets

Chapter "Land Use Cover Datasets: A Review" and Part IV of the book contain a review of the Land Use Cover datasets available at global and supra-national scales. Due to the limited extent and scope of this book, we did not review national and regional LUC datasets, which are far too numerous for our purposes.

The datasets we reviewed are classified into two groups, depending on the information they provide. The first group is made up of the datasets that provide information about the different land uses or covers without focusing on any one of them in particular, i.e. general LUC datasets. The second

group consists of the LUC datasets that map a specific land use or cover in detail (e.g. vegetation, croplands, built-up areas…). These are referred to as thematic LUC datasets. Some datasets are difficult to assign to one of the two groups, as they map a wide range of LUC categories while also providing specific detail on just one of them. The authors decided which group to assign them to on a case-by-case basis.

The datasets were also classified according to their extent, differentiating between global and supra-national LUC datasets. The first group of datasets maps land uses or covers all over the Earth, while the second maps them for a specific area covering more than one country. The maps in the second group may cover a whole continent or focus on just a few countries.

Table 4 List of repositories and web portals distributing LUC information at global and supra-national scales

Web portal	Description
Copernicus Land Monitoring Service	Web portal for the thematic land monitoring service provided by the Copernicus programme. It offers information on land monitoring at global, pan-European and local scales for the European Union
FAO GeoNetwork	Catalog of spatial datasets developed and maintained by the Food and Agriculture Organization (FAO). It has a specific section on LUC information
Geo-Wiki	Platform developed to collect LUC information via crowdsourcing. Its viewer hosts different LUC datasets, either from external projects or developed through the Geo-Wiki project and similar crowdsourcing approaches
Google Earth Engine	Catalogue of spatial datasets to be used as part of Google Earth Engine. There is a specific section on LUC datasets
Land Processes Distributed Active Archive Center (LP DAAC)	Archive of spatial data managed in partnership by the USGS and NASA, which distributes most of the information produced by these institutions
Wekeo	Copernicus Data and Information Access Service (DIAS), which provides a cloud-based platform to access and process Copernicus data. It includes a catalogue of data with many LUC products produced within the context of the Copernicus programme
FROM-GLC	Web portal developed and maintained by experts from Tsinghua University, which distributes all the LUC datasets produced by the team associated with this university since the FROM-GLC project

When making the review, we consulted the most relevant web portals and repositories of LUC data (Table 4). A few selected papers, reports and other relevant documents reviewing or comparing LUC datasets were also consulted (Manakos and Braun 2014; Mora et al. 2014; Grekousis et al. 2015; Tsendbazar et al. 2015; Diogo and Koomen 2016; Klotz et al. 2016; Pérez-Hoyos et al. 2017; Fritz et al. 2019).

Very old or outdated maps, which were produced according to traditional cartographic methods, are not included in this review. Nor are other old maps that combine LUC information with other data about climate or biogeographic variables, such as the maps produced by Matthews (1983) and Olson et al. (1983). Traditional maps obtained through photointerpretation of aerial imagery and field survey, which offer information about certain specific land covers such as vegetation and agricultural areas, are not included in the review either. Although they may be interesting sources for historical LUC change analysis, they are usually only available for national or more detailed areas and normally have not been digitalized.

There are plenty of other spatial datasets that provide important information for studying specific land covers. For vegetation covers, maps of live biomass are a good example (Kindermann et al. 2008; Thurner et al. 2014). These datasets were not included in our review because they are not specific sources of LUC information focusing exclusively on land cover. However, there is an enormous amount of data like this that may be useful for the study and characterization of LUC. This data comes in many different forms and from a range of different sources.

Part IV of the book characterizes in detail all the reviewed LUC datasets that are currently available for download and may be relevant for a wide community of users. Datasets produced at very coarse scales or which are already very outdated are not described in Part IV, as they are of limited utility for most members of the LUC community. LUC datasets currently unavailable for download are not characterized in Part IV either. We tried to obtain, either online or by contacting the authors, all the global or supra-national datasets to which we found references. Some of them, however, are no longer available. These datasets have not been reviewed.

The LUC datasets described in Part IV were characterized according to the following elements: information about the project or context within which they were produced; information about their method of production; description of the data available for download; and practical information for using the dataset in an effective way. For each dataset we also provide all the technical references in which it is described as well as other references of interest in which it is used or analysed. A table summarizing the main characteristics of the dataset (extent, temporal availability, spatial resolution, updates, accuracy…) is also provided.

References

Baghdadi N, Mallet C, Zribi M (2018a) QGIS and generic tools. Wiley, Hoboken

Baghdadi N, Mallet C, Zribi M (2018b) QGIS and applications in territorial planning. Wiley, Hoboken

Baghdadi N, Mallet C, Zribi M (2018c) QGIS and applications in water and risks. Wiley, Hoboken

Baghdadi N, Mallet C, Zribi M (2018d) QGIS and applications in agriculture and forest. Wiley, Hoboken

Bounoua N, Le Campion J (2019) L'évolution de l'occupation des sols et notamment de l'urbanisation sur la Vallée de l'Ariège. Rapport de Master II Géomatique SIGMA

Cutts A, Graser A (2018) Learn QGIS: your step-by-step guide to the fundamental of QGIS 3.4, 4th ed. Packt, Birmingham

Diogo V, Koomen E (2016) Land cover and land use indicators: review of available data. OCDE Green Growth Papers. https://www.oecd-ilibrary.org/environment/land-cover-and-land-use-indicators_5jlr2z86r5xw-en. Accessed 26 June 2021

Domínguez Vera RdJ (2021) Documentación para el plugin MapAccurAssess. https://doi.org/10.5281/zenodo.5419130. Accessed 26 June 2021

Fritz S, See L, Bayas JCL et al (2019) A comparison of global agricultural monitoring systems and current gaps. Agric Syst 168:258–272. https://doi.org/10.1016/j.agsy.2018.05.010

García-Álvarez D, Camacho Olmedo MT (2017) Changes in the methodology used in the production of the Spanish CORINE: uncertainty analysis of the new maps. Int J Appl Earth Obs Geoinf 63:55–67. https://doi.org/10.1016/j.jag.2017.07.001

García-Álvarez D, Camacho Olmedo MT, Paegelow M (2019) Sensitivity of a common land use cover change (LUCC) model to the minimum mapping unit (MMU) and minimum mapping width (MMW) of input maps. Comput Environ Urban Syst 78. https://doi.org/10.1016/j.compenvurbsys.2019.101389

García-Álvarez D (2018) The influence of scale in LULC modelling. A comparison between two different LULC maps (SIOSE and CORINE). In: Camacho Olmedo MT, Paegelow M, Mas J-F, Escobar F (eds) Geomatic approaches for modeling land change scenarios. Springer, Cham, Switzerland, pp 187–213

García-Amaro E (2004) 1928–1999: modificaciones al sistemas de clasificación climática de Köppen. Edit Instituto de Geografía, UNAM, México

Gobierno del Principado de Asturias (2016) Directrices subregionales de ordenación del Área Central de Asturias. Avance: objetivos y criterios

Graser A, Mearns B, Mandel A, et al (2017) QGIS: becoming a GIS Power User. Learning path. Packt, Birmingham, Mumbai

Grekousis G, Mountrakis G, Kavouras M (2015) An overview of 21 global and 43 regional land-cover mapping products. Int J Remote Sens 36:5309–5335. https://doi.org/10.1080/01431161.2015.1093195

Islam S (2018) Hands-on geospatial analysis with R and QGIS. Packt, Birmingham

Jung M (2016) LecoS—a python plugin for automated landscape ecology analysis. Ecol Inform 31:18–21. https://doi.org/10.1016/j.ecoinf.2015.11.006

Kindermann GE, McCallum I, Fritz S, Obersteiner M (2008) A global forest growing stock, biomass and carbon map based on FAO statistics. Silva Fenn 42:387–396. https://doi.org/10.14214/sf.244

Klotz M, Kemper T, Geiß C et al (2016) How good is the map? A multi-scale cross-comparison framework for global settlement layers: evidence from Central Europe. Remote Sens Environ 178:191–212. https://doi.org/10.1016/j.rse.2016.03.001

Luca Congedo (2016) Semi-automatic classification Plugin documentation release 4.8.0.1. https://buildmedia.readthedocs.org/media/pdf/semiautomaticclassificationmanual-v4/latest/semiautomaticclassificationmanual-v4.pdf. Accessed 26 June 2021

Manakos I, Braun M (2014) Land use and land cover mapping in Europe. In: Practices and trends. Springer, Dordrecht, Heidelberg, New York, London

Matthews E (1983) Global vegetation and land use: new high-resolution data bases for climate studies. J Clim Appl Meteorol 22:474–487. https://doi.org/10.1175/1520-0450(1983)0222.0.CO;2

Méndez García B, Ortega Montequín M (2013) Ciudad difusa y territorio: el caso del Área Central Asturiana. Ciudad Rev Del Inst Urbanística La Univ Valladolid 16:131–144

Menke K, Smith R, Pirelli L, Van Hocscn J (2016) Mastering QGIS. Packt, Birmingham

Mora B, Tsendbazar N-E, Herold M, Arino O (2014) Global land cover mapping: current status and future trends. In: Manakos I, Braun M (eds) Land use and land cover mapping in Europe. Practices and trends. Springer, Dordrecht, Heidelberg, New York, London, pp 11–30

Olofsson P, Foody GM, Stehman SV, Woodcock CE (2013) Making better use of accuracy data in land change studies: estimating accuracy and area and quantifying uncertainty using stratified estimation. Remote Sens Environ 129:122–131. https://doi.org/10.1016/j.rse.2012.10.031

Olson JS, Watts JA, Allison LJ (1983) Carbon in live vegetation of major world ecosystems. https://cdiac.ess-dive.lbl.gov/epubs/ndp/ndp017/ndp017appA.pdf. Accessed 26 June 2021

Pérez-Hoyos A, Rembold F, Kerdiles H, Gallego J (2017) Comparison of global land cover datasets for cropland monitoring. Remote Sens 9:1118. https://doi.org/10.3390/rs9111118

Prada Trigo J (2011) Desarrollo, patrimonio y políticas de revitalización en ciudades intermedias de especialización minero-industrial. In: El caso de Langreo (Asturias). Consejo Económico y Social del Principado de Asturias, Oviedo

QGIS Project, 2020QGIS Project (2020) QGIS user guide. Release 3.4

Rodríguez Gutiérrez F, Menéndez Fernández R, Fernández Prieto JA (2013) Las villas en el sistema territorial asturiano. Eria 90:31–54

Rodríguez Gutiérrez F, Menéndez Fernández R, Blanco Fernández J (2009) El área metropolitana de Asturias. In: Ciudad Astur: el nacimiento de una estrella urbana en Europa. Trea, Oviedo

Thurner M, Beer C, Santoro M et al (2014) Carbon stock and density of northern boreal and temperate forests. Glob Ecol Biogeogr 23:297–310. https://doi.org/10.1111/geb.12125

Tsendbazar NE, de Bruin S, Herold M (2015) Assessing global land cover reference datasets for different user communities. ISPRS J Photogramm Remote Sens 103:93–114. https://doi.org/10.1016/j.isprsjprs.2014.02.008

Valcárcel N, Villa G, Arozarena A, et al (2008) SIOSE, a successful test bench towards harmonization and integration of land cover/use information as environmental reference data. In: Chen J, Jiang J, Peled A (eds) The international archives of the photogrammetry, remote sensing and spatial information sciences, vol XXXVII, part B8. ISPRS, Beijing

Land Use Cover Mapping, Modelling and Validation. A Background

David García-Álvarez, María Teresa Camacho Olmedo,
Jean-François Mas, and Martin Paegelow

Abstract

In this chapter, we offer a brief introduction to the main concepts associated with Land Use Cover (LUC) mapping, Land Use Cover Change (LUCC) modelling and the uncertainty and validation of LUC and LUCC data and model outputs. The chapter summarizes the theoretical fundamentals required to understand the rest of the book. First, we define Land Use and Land Cover concepts that have been extensively discussed and debated in the literature (Sect. 2). Second, we review the history of LUC mapping, from the first manually produced maps to the advent of aerial and satellite imagery and the production of new datasets with much greater detail and accuracy (Sect. 3). Third, we address the usefulness of LUC data and LUCC analysis for society (Sect. 4), contextualizing all these studies and efforts within the framework of Land Change Science (Sect. 5). Fourth, we offer a brief introduction to LUCC modelling, its purpose, uses and the different stages that make up a LUCC modelling exercise (Sect. 6). We also offer a brief introduction to the different types of LUCC models currently available. Finally, we present the concepts of uncertainty and validation and offer a brief introduction to the topic (Sect. 7). The chapter also includes a short list of recommendations for further reading for those who wish to explore the theory presented here in more depth.

Keywords

Land Use • Land Cover • Land Use Cover Change • Land Use Cover mapping • Land Change Science • Land Use Cover Change modelling • Uncertainty • Validation

1 Introduction

Land Use and Land Cover (LUC) data is an important source of information for a wide range of users from different backgrounds and scientific disciplines. It provides an overview of the different covers on the Earth's surface (e.g. vegetation, agricultural fields, rocks, water, artificial surfaces…) and how they evolve over time. It also traces how these covers are used (land use) and how this use changes.

LUC data can be very useful in an array of different fields. It is especially valuable for understanding the impact that many natural and human-induced processes, such as climate change, deforestation and urbanization, can have on the Earth's surface. As a result, LUC research has been receiving increasing attention over recent decades, and the number of fields making use of this data is on the rise.

Researchers have been proposing new methods and techniques for producing LUC maps. This has increased the number of LUC datasets available at global, continental, regional and local scales. This has also led to an increase in the number of users who decide to make their own LUC maps. The validation of LUC data has also been the subject of specific research and new methods, strategies and techniques have been proposed for validating and analysing LUC maps.

Despite all these advances, many users are still unaware of the wide range of datasets available, while others lack a clear understanding of the methods or techniques that can be

D. García-Álvarez (✉)
Departamento de Geología, Geografíay Medio Ambiente,
Universidad de Alcalá, Alcalá de Henares, Spain
e-mail: David.garcia@uah.es

M. T. Camacho Olmedo
Departamento de Análisis Geográfico Regional y Geografía
Física, Universidad de Granada, Granada, Spain

J.-F. Mas
Laboratorio de Análisis Espacial, Centro de Investigaciones en
Geografía Ambiental, Universidad Nacional Autónoma de
México, Morelia, Mexico

M. Paegelow
Département de Géographie, Aménagement et Environnement,
Université de Toulouse Jean Jaurès, Toulouse, France

used to validate LUC data. Thus, in addition to producing more LUC datasets, more information is required. Users must be able to find out more about the most appropriate datasets for their field of study, and the general uncertainties and limitations of each one. They should also be informed about the methods that can be used to assess the specific utility and uncertainties of this data for their line of research.

2 Land Use versus Land Cover

Although Land Use and Land Cover are often combined, for example, in references to LUC maps and information, they in fact have quite separate meanings. Many authors have proposed complementary definitions (Di Gregorio and Jansen 1998; Campbell and Wynne 2011; Giri 2016a; Wulder et al. 2018) and the European directive INSPIRE, which establishes an Infrastructure for Spatial Information in the European Community, also includes a definition of each term (see text box below). On the basis of these various sources, we have opted for the following definitions.

Directive INSPIRE (2007/2/EC)

Land Cover: Physical and biological cover of the earth's surface including artificial surfaces, agricultural areas, forests, (semi-)natural areas, wetlands, water bodies.

Land Use: Territory characterised according to its current and future planned functional dimension or socio-economic purpose (e.g. residential, industrial, commercial, agricultural, forestry, recreational).

Land cover refers to the Earth's biophysical covers. Areas without a specific cover, such as areas of bare rock or bare soil, are also regarded as land covers. By contrast, land use refers to the activities that humans carry out on the Earth's surface or on a specific land cover.

A land cover can have one or multiple uses, or even none. An artificial surface could be used to host people (e.g. residential area), production (e.g. industrial area) or leisure activities (e.g. sports facilities). In maps at coarser scales, this artificial surface can host all these uses together. For example, an urban area is an artificial cover which has multiple uses. Bare rock, on the other hand, often hosts no land use of any kind.

A specific land use can also be associated with multiple land covers at the same time. An airport is a land use that is usually associated with several artificial covers, such as buildings, roads and runways, and also with vegetation covers, like grassland.

Whereas land covers are usually visible in aerial or satellite images, land uses are more difficult to distinguish. For instance, a building could have multiple uses: apartments, offices, industrial plants, sports facilities, etc. Sometimes the land use can be deduced from contextual information in the image, but, in most cases, additional information is required. This makes map production more difficult and expensive. As a result, most maps only provide information about land covers. In other cases, they focus on the land use of certain specific covers, such as artificial or agricultural areas, so providing both Land Use and Land Cover (LUC) data. This is why in LUC science, we generally talk about Land Use and Land Cover information, as the two aspects tend to be combined within the same datasets.

3 Land Use and Land Cover Mapping: A History

Some information on Land Use and Land Cover was available prior to the advent of remote sensing instruments (Campbell 1983). However, it was the appearance of aerial and, above all, satellite images that promoted the production of systematic LUC maps at regional, continental and global scales (Loveland 2016).

Before the emergence of aeroplanes and satellites, the main method for map production was ground survey (Wallis 1981; Fuller et al. 1994; Crone 2000). This was a time-consuming, laborious process that made systematic mapping of vast territories a difficult task. However, various important projects to map national territories were carried out in the eighteenth and nineteenth centuries without the use of aerial imagery (Collier 2009a). Most of these projects involved topographic or cadastral maps, like the first French topographic survey finished in 1793, the French Napoleonic cadastre which began in 1807 or the Austrian cadastral survey launched in 1762 (Collier 2009a; Rochel et al. 2017). There are also striking examples of systematic exercises to map LUC information, such as the Land Utilization Survey of Great Britain, conducted from 1931 to 1938 (Campbell 1983). Nonetheless, the general rule was for land use information to be presented as part of other maps with more general purposes (e.g. topographic, cadastral maps) or a very thematic approach (e.g. agricultural uses and production) (Campbell 1983).

With the advent of aerial imagery and, later, satellite imagery, mappers obtained a view of the Earth's surface from the top of the atmosphere or from space. Mapping became easier and cheaper (Fuller et al. 1994). Instead of going out to the field to collect information, mappers could photointerpret and extract most of the features on the Earth's surface from the imagery, including land uses and covers.

Information collected in the field was still required to validate what was photointerpreted and to include some extra information that was not discernible in the image (Steiner 1965; Campbell 1983). However, these tasks were less time-consuming and demanding than the original ground survey activities.

Aerial images became increasingly common from the beginning of the twentieth century, with the development of the aeroplane industry within the context of the two World Wars (Collier 2009b). Most nations started or boosted ambitious national mapping programmes for strategic or economic purposes. Many national topographic or cadastral mapping projects were completed during this period (Collier 2009a). Some pioneer land use mapping projects were also launched at that time, such as the Michigan Land Economic Survey in the early 1920s and the Rural Land Classification Survey conducted by the Tennessee Valley Authority, which began in the 1930s (Steiner 1965). There was even a plan to create the first global land use map, with the foundation of a World Land Use Commission in 1949 and the mapping of different test areas in the 1950s and 1960s (Campbell 1983). However, mapping was still costly and very time-consuming. Although much easier than before, photointerpretation was a manual task carried out using rudimentary tools that required a great deal of time and effort (Steiner 1965; Campbell 1983).

The launch of the first satellite into space in 1957 proved a turning point in the history of LUC mapping (Emery and Camps 2017). Satellites provide a periodic imagery coverage of the Earth's surface. Once satellites started to provide images of the Earth, a homogeneous, cheap mosaic of the entire surface of the Earth soon became available (Morain 1998; Chuvieco 2016).

Satellites record the reflectance of the Earth's surface in different regions of the electromagnetic spectrum. The reflectance curve for each land cover can be independently characterized and defined (Chuvieco 2016; Emery and Camps 2017). In this way, satellite imagery gives mappers the information they need to draw the land covers on the Earth's surface automatically, so reducing the need for photointerpretation or human intervention in the process (Campbell and Wynne 2011; Chuvieco 2016). Nonetheless, the mapping of LUC covers from imagery reflectance has various important issues that can result in uncertainty and errors. One land cover can present several different spectral responses due to variations in vegetation density and phenology. Different land covers can also present a similar spectral response. This problem, known as spectral confusion, is critical in diverse and complex landscapes and can lead to large numbers of classification errors.

Despite these limitations, the availability of satellite imagery and the ease with which land cover information could be obtained from them boosted the production of land cover maps, which until then had been relatively rare (Comber 2008). Whereas most of the LUC information available in the pre-satellite era had been focused above all on land use, from then onwards, maps focusing on land cover or on a mixture of land cover and land use became predominant (Fisher and Unwin 2005; Comber 2008).

Manual photointerpretation was still common in the early years of satellite remote sensing (Campbell 1983). It benefited from computer-assisted procedures, such as on-screen digitalization. However, it was progressively replaced by digital procedures with the development of powerful computers and the improvement of classification and image treatment methods (Loveland 2016). Nonetheless, even today manual photointerpretation still plays an important role in the production of LUC maps. Recent examples of Land Use Cover mapping over large areas using visual interpretation include maps of Europe (CORINE Land Cover; see Feranec et al. (2007)), Africa (AFRICOVER; see Di Gregorio and Latham (2003); Fritz et al. (2015)) and China (Zhang et al. 2014).

As LUC mapping became easier, cheaper and quicker, many institutions, scientists and other users began producing LUC datasets at all the different scales (Grekousis et al. 2015; Loveland 2016). Initial efforts were mainly focused on regional and national scales (Loveland 2016). However, the appearance of the first satellites with sensors providing free imagery covering the whole Earth at coarse resolutions allowed the first global LUC datasets to be developed (Congalton et al. 2014; Mora et al. 2014; Grekousis et al. 2015).

The AVHRR sensor on board the NOAA weather satellites launched in 1978 (Campbell and Wynne 2011), and the VEGETATION sensor, installed in the SPOT satellite in 1998 (Gutman et al. 2012a), provided the first sources of satellite imagery for global mapping exercises (Congalton et al. 2014; Gong et al. 2016). Landsat, which was first launched in 1972, provided the first source of satellite imagery at medium spatial resolutions, which could be used for LUC mapping at regional and local scales (Belward and Skøien 2015).

Since then, LUC mapping practice has been developed in parallel with the launch of new satellites and the increasing improvement in their spatial and spectral resolutions (Belward and Skøien 2015). This process has also been spurred by the appearance and consolidation of public and private initiatives focusing on Earth Observation and LUC monitoring (Herold et al. 2016; Wulder et al. 2018). Although many such organizations now exist, perhaps the most important are the United States Geological Survey (USGS) and the European Space Agency (ESA).

The key role played by the USGS is undeniable. It authored the first research laying down the foundations of modern LUC mapping (Anderson et al. 1976; Gutman et al.

2008) and is also responsible for some of the most important Earth-monitoring projects today (Barber 2019; Szantoi et al. 2020). The ESA has also played an important role, especially recently after the launch of the Copernicus programme with the support of the European Commission (Szantoi et al. 2020). The constellation of Sentinel satellites and the Copernicus land monitoring products, produced by the European Environmental Agency (EEA) and the Joint Research Centre (JRC), have enabled important advances in the production of detailed, high-quality LUC information that is updated periodically (Manakos and Braun 2014; Grekousis et al. 2015; Herold et al. 2016).

Users now have more information available than ever (Belward and Skøien 2015; Grekousis et al. 2015; Giri 2016a). Many LUC products have been developed and are ready to use, with abundant, detailed documentation about their characteristics (Grekousis et al. 2015; Diogo and Koomen 2016). There are numerous sources of satellite imagery, some of which are pre-treated and are available free of charge (Belward and Skøien 2015). Many methods have been developed for image processing and LUC mapping, such as classification algorithms (Bruzzone and Demir 2014; Yu et al. 2014; Khatami et al. 2016). Many methods and techniques have also been proposed for assessing the validly of LUC information (Strahler et al. 2006; Stehman and Foody 2019). Most of these methods and techniques are available on widely used software and are readily accessible to any user (Bastin et al. 2013; Mas et al. 2014b; Brovelli et al. 2018). All this has encouraged research into the production of LUC information and has widely extended its use, which has also led to an increase in published research on the topic, especially in the last 25 years (Yu et al. 2014).

4 Uses of LUC Data

The importance and utility of Land Use and Land Cover information is beyond doubt. LUC data is a valuable source of information for scientists (Bontemps et al. 2012; Manakos and Braun 2014). It gives them a better understanding of the interactions between societies and the environment (Lu et al. 2004), an aspect of special interest for many social sciences such as geography or economics (Geoghegan 1998; Green et al. 2005). LUC data can also be used to monitor a range of different natural and environmental processes (e.g. hydrological, meteorological…), a question of great interest for many natural sciences (Rindfuss et al. 2004).

Policymakers also need LUC data for proper resource management and to help them deal with many of the challenges facing society today (Szantoi et al. 2020). It allows them to understand where land resources are located and how and when they change (Strand 2013; Thackway et al. 2013).

Campbell (1983) reviewed some of the applications of LUC data in policymaking in the USA at different scales. He found that "almost all governmental units have a continuing requirement to create and implement laws and policies that directly or indirectly involve existing or future land use". Local administrations need land use information for spatial planning. Regional and national governments may require LUC information for water management, flood control or in the design and assessment of environmental policies. At the international level, LUC data provides important evidence on which to base decisions regarding many of the global challenges facing society today.

Most of the current global agendas refer to policy objectives involving Land Use and Land Cover. They play a direct role in 7 out of 17 UN Sustainable Development Goals (SDGs), and in the UN Framework Convention on Climate Change (UNFCCC), the Convention on Biological Diversity, the UN Convention to Combat Desertification (UNCCD) and the Ramsar Convention on Wetlands (Szantoi et al. 2020). LUC data is required to monitor many of the targets or actions proposed in these agreements, so emphasizing the need for global LUC maps (Diogo and Koomen 2016).

The Group on Earth Observations (GEO) has defined eight Social Benefit Areas (SBAs) in which Earth observations, including LUC data, provide useful evidence in support of policymaking.[1] They are biodiversity and ecosystem sustainability, disaster resilience, energy and mineral resource management, food security and sustainable agriculture, infrastructure and transportation management, public health surveillance, sustainable urban development and water resources management. Specifically, LUC data can help, among other things, to characterize the land for disease control; monitor fires; assess the potential of land for biofuel production and wind or hydropower generation; and assess the role of LUC changes in the dynamics of hydrological systems and vegetation (Giri 2016b).

Among scientists, LUC maps are frequently used as a basis for modelling exercises (Tsendbazar et al. 2015; Herold et al. 2016). At a global scale, climate change models require global LUC maps (Sophie et al. 2011). At regional and local scales, land use and cover change models have emerged as valuable tools for policy support (Van Delden et al. 2011; White et al. 2015). These models are built on LUC datasets (Sohl and Sleeter 2012).

LUC information is also used for many other research activities, most of them related to the different policy fields mentioned above. In recent years, it has been applied, for example, in studies analysing habitat distribution and ecosystem services (Jacob et al. 2003; Brown 2013), spatial

---

[1] https://earthobservations.org/geo_wwd.php#.

patterns of biodiversity (Zimmermann et al. 2010; Tuanmu and Jetz 2014), and ecosystem status and biogeochemical cycling (Johnson and Patil 1998; Lawrence et al. 2012), etc. A wide variety of processes are also studied using LUC data. Bielecka (2019) review some of the most common processes analysed through the CORINE Land Cover database. These include agricultural abandonment, urbanization, afforestation, deforestation, landscape fragmentation, etc.

5 Land Change Science

Although LUC information is employed for manifold purposes, the field taking most advantage of this data is Land Use and Land Cover Change (LUCC) analysis (Feranec et al. 2007; Verburg et al. 2009; Bielecka 2019). LUCC analysis is the study of the changes in the land uses and covers on the Earth's surface, and their causes and consequences (Moran et al. 2012). LUCC is not usually studied as an end in itself, and the focus is normally on understanding its impact on a range of other natural or human-induced processes (Gutman et al. 2012a). Many of them have already been mentioned when explaining the general utility of LUC data.

LUC change analyses are widely used in climate change studies (Sophie et al. 2011), the study of hydrological systems (Carlson and Traci Arthur 2000; Cuo et al. 2009), weather conditions (Marshall et al. 2004), soil erosion (Cebecauer and Hofierka 2008), loss of biodiversity (Cebecauer and Hofierka 2008), as well as in research into ecosystem services (Hu et al. 2008) or animal habitats (Lawler et al. 2004). The utility of LUC data increases when historical information is available, as it allows us to track LUC changes over time (Verburg et al. 2011; García-Álvarez and Camacho Olmedo 2017).

The importance of LUCC studies has led to the emergence of a specialist field called Land Change Science (Gutman et al. 2012a; Turner 2017), which is also referred to as Land Use Science or Land System Science (Müller and Munroe 2014). This is defined as a "transdisciplinary field" that "seeks to understand the dynamics of land cover and land use as a coupled human–environment system to address theory, concepts, models, and applications relevant to environmental and societal problems, including the intersection of the two" (Turner et al. 2007). One of its hallmarks is the integration of natural and social sciences via a holistic approach (Rindfuss et al. 2004; Gutman et al. 2012a). Land Change Science now has its own specialists, who work at the confluence between these fields of knowledge (Moran et al. 2012; Müller and Munroe 2014).

Land Change scientists are responsible for monitoring LUC change, understanding it and modelling for the future, so obtaining knowledge and evidence that may be useful for policymaking (Turner et al. 2007). Land Change is part of the wider field of research addressing Global Environmental Change, for which historical series of LUC data are required (Turner et al. 2007; Janetos 2012). This is why Land Change Science has emerged in parallel to the growth in remote sensing observation and the appearance of the first time series of Earth observation data (Moran et al. 2012; Turner 2017).

Many international programmes and organizations have stressed the importance of LUCC and Land Change Science (Giri 2016b). Turner (2017) claims that the science first originated in the joint programme on LUCC funded by the International Geosphere Biosphere Program (IGBP) and the International Human Dimensions Programme (IHDP). Other programmes that have emphasized the importance of LUCC studies include the U.S. Climate Change Science Program, the Global Land Project and the Group on Earth Observations (GEO) and the United States Global Change Research Program (USGCRP) (Gutman et al. 2012b; Moran et al. 2012). Some of these programmes are specifically focused on LUCC as a specialist interest, lying at the heart of their activities. These include the Land Cover and Land Use Change (LCLUC) programme run by NASA and the Global Observation of Forest and Land Cover Dynamics (GOFC-GOLD) programme (Gutman et al. 2012b).

6 Land Use and Land Cover Change Modelling

As previously noted, Land Change Science is not only a question of analysing and understanding LUC changes, but it also seeks to model them in the near future (Gutman et al. 2012a; Turner 2017). Once we have understood what has changed, where it has changed, why it has changed (drivers or causes), how it has changed and what the consequences are, we can then take a step further and try to understand how different change trends can affect human-natural ecosystems. This is especially useful for policymaking. By evaluating different change scenarios, we can understand what the future may look like and what we can do to put the policy objectives we are seeking into practice (Oxley et al. 2002; Soares-Filho et al. 2006; Escobar et al. 2018).

Land Use and Land Cover Change Modelling (LUCCM) is about understanding the LUC dynamics at work within a given Earth system and modelling their future evolution (Verburg et al. 2004; Paegelow and Camacho Olmedo 2008). To understand these dynamics, we need to study how the system has changed in the past and analyse the processes that gave rise to these changes (Plata Rocha 2010; Toro Balbotín 2014). By studying these processes in detail, we can identify the drivers behind the changes taking place (Bürgi et al. 2005; Kolb et al. 2013). Once we know what changes are occurring and why, we can conceptualize this information and translate it into modelling terms.

Models allow us to play around with the system we are studying so as to predict how different policies affect LUC and the changes they may cause (Van Delden et al. 2011). Models also help us understand how these changes may evolve in the future under different socio-economic conditions (Antoni et al. 2018). At a more modest level, LUCC models also enable us to study and analyse these systems in detail, so as to obtain a more in-depth understanding of them (Hewitt et al. 2014).

LUC maps are the main input for LUCC models (Sohl and Sleeter 2012; Grinblat et al. 2016), forming the base on which all processes are conceptualized (García-Álvarez et al. 2019b). LUC maps conceptualize the landscape to be modelled: they present the LUC categories into which the landscape is divided and determine the spatial detail of the model (Conway 2009; García-Álvarez et al. 2019a). They are also often used as a reference for studying LUC changes in the past (Burnicki et al. 2010) and for validating LUCC models (Van Vliet et al. 2016).

Many types of LUCC models are available today (National Research Council 2014). Although there is no standard, globally accepted classification, we can broadly distinguish between process and pattern-based LUCC models (Brown et al. 2013). The latter assume that changes in the landscape pattern are the result of the processes and dynamics taking place, and that each pattern is a consequence of a specific process (Mas et al. 2014a). These models simulate the pattern and its changes. They are therefore heavily reliant on time series of LUC maps and the changes they show.

Process-based models simulate the processes taking place, rather than the pattern (O'Sullivan and Perry 2013). There are different kinds of process-based models, with agent-based LUCC models gaining increasing popularity. These models simulate the behaviour of the agents or actors that take part in the system being modelled and their interactions (Crooks and Heppenstall 2012). These agents cause the processes taking place on the ground and the changes in the landscape pattern. Although important, LUC maps do not play the same key role in these models as they do in pattern-based models, as most of the parameters used in process-based models are inferred from other sources (Mas et al. 2014a).

LUCC models can also be classified according to the scale of analysis, their stochastic or deterministic nature, the type of scenarios they can produce and the techniques and methods they apply (García-Álvarez 2018a). For example, some models include Markov chains to estimate the quantity of simulated change in the future (Sang et al. 2011; Eastman and Toledano 2018). These are usually calculated on the basis of the changes that took place between two LUC maps in the past (Sinha and Kimar 2013; Mas et al. 2014a), so increasing the importance of LUC data in the modelling exercise.

Modelling exercises normally consist of four main phases: calibration, simulation, validation and the proposal of scenarios (Camacho Olmedo et al. 2018), although other phase-based structures have also been proposed. In almost all cases, researchers differentiate between the calibration and the validation phase (Pontius Jr. et al. 2004; Gallardo 2014; Van Vliet et al. 2016). Nonetheless, some studies omit the validation stage, choosing solely to explore the modelled system and its behaviour.

Calibration refers to the setting-up and parametrization of the model (Clarke 2004; Mas et al. 2018). The users define the objectives of the exercise, and the data and model to be used. They then parametrize the model in line with their understanding of the simulated system. After the initial results are obtained, the model is adjusted to obtain the best possible results (Van Vliet et al. 2016). Once the model is fully calibrated and a simulation has been obtained, this must be validated by comparing it with reference data that were not used earlier on in the modelling exercise (Pontius Jr. and Malanson 2005; Paegelow and Camacho Olmedo 2008).

The methods and techniques used for calibration are similar to, if not the same as, those used in the validation phase (Mas et al. 2018). In the calibration phase, the results obtained from the model are compared with reference data so as to obtain a model that properly simulates the system being studied (Van Vliet et al. 2016). The model is then validated with independent data sources, not used in the calibration phase (Pontius Jr. and Malanson 2005; Van Vliet et al. 2011). Thus, whereas calibration fits the model to the reference data, validation makes sure that there is a good fit over time and not just for the date of the reference map. In this way, it ensures that the processes that explain the changes in the system being studied were correctly modelled.

7 Uncertainty and Validation

The increased availability of satellite and aerial imagery and the development of new methods and techniques for image processing and classification has enabled the production of an increasing number of LUC maps and time series of LUC maps at all scales (Yu et al. 2014; Grekousis et al. 2015; Giri 2016a). The same trend can be observed in the application of LUCC models, which has become very common as a result of easy access to LUC maps and LUCC modelling software (Sohl and Sleeter 2012; Ferchichi et al. 2017).

With the increasing production and use of LUC maps and LUCC models, more attention has been paid to the uncertainty and limitations of these data and analyses (Yeh and Li 2006; Krüger 2016; Loveland 2016; Ferchichi et al. 2017; García-Álvarez et al. 2019b). Uncertainty can be defined as "the lack or the degree of certainty about any data or geospatial analysis due to the difference between reality and its representation through geospatial data or tools" (García-Álvarez et al. 2019b). Understanding how different these maps and exercises are from real landscapes and processes and, therefore, how reliable they are is essential. This is the only way of knowing how accurate the information we obtain from these maps and analyses is and to what extent it can be used as a basis for taking policy decisions.

It is important to realize that all spatial data and analyses contain some degree of uncertainty (Longley et al. 2011). They are an abstraction and simplification of real landscapes and processes (Comber et al. 2005; Devillers and Jeansoulin 2006). This means that the maps and models are themselves just conceptualizations of different processes and features of the Earth. When we conceptualize a landscape on a map, what we are actually doing is simplifying it to obtain elements with which we can work and experiment.

In the case of LUC maps, the complexity and variety of real landscapes is normally translated into a given set of categories (Di Gregorio and Jansen 1998; Herold and Di Gregorio 2012). Land Use and Land Covers do not always fit into a precise, clear-cut classification, as they show heterogeneous, mixed patterns that cannot be easily classified within a specific category (Di Gregorio and Jansen 1998; Villa et al. 2008). This makes it difficult to clearly define a particular land use and to distinguish it on the ground from all other land uses, establishing boundaries between them (Fassnacht et al. 2006). Some degree of uncertainty is therefore inevitable in the classification process.

Mapping the full complexity of the Earth remains beyond human capacity, and even beyond existing computer capabilities (Unwin 1995; Murayama 2012). The smaller or coarser the scale, the greater the need for abstraction or simplification (Lloyd 2014). At whatever scale we work, we are capable of assimilating similar amounts of information. This means that at larger or finer scales we can add details, while at smaller or coarser scales we can only show the essentials.

To understand the uncertainty and limitations of our data and analyses, we usually carry out uncertainty assessments (Van Asselt 2000; Jcgm 2008; Abreu and Ralha 2017; García-Álvarez et al. 2019b). In general, when we assess our data and analyses against reference data to evaluate the reliability of the information they provide, we are said to be validating the data or models (Fonte et al. 2015; Van Vliet et al. 2016). Validation can therefore be defined as the process by which we assess how certain or reliable a piece of

data or result is. This is done by comparing it against other data or information that we use as a reference and consider to be true.

Although validation is already a common practice and there are many methods, strategies and reference data available for validating LUC maps and LUCC models, there is still a lot of room for improvement. In the case of LUCC maps, when Olofsson et al. (2013) carried out their review, up to 15% of the papers addressing land change with LUC maps did not include any proof of data validation. They also found that most of the reviewed papers did not include all the relevant information about the accuracy of the measured changes. The review carried out by Yu et al. (2014) produced even less hopeful results: of 6771 papers including some type of LUC mapping exercise, only 1585 reported overall accuracy measures. Morales-Barquero et al. (2019) found that only 32% of the papers they reviewed provided a reproducible accuracy assessment and recommended that more statistically rigorous accuracy assessment practices be encouraged.

In LUCCM, several authors emphasized the importance of analysing the uncertainty of the results, even when general validation exercises are carried out (Li and Wu 2006; Krüger 2016). In fact, Van Asselt (2000) criticized the widespread use of validation exercises in modelling as a tool "to sell the model as being scientifically credible", without proper discussion and analysis of the uncertainties and limitations of the modelling exercise. Sohl et al. (2016) consider the lack of information regarding uncertainty and the failure to quantify it as one of the reasons hampering the adoption of LUCC models in decision-making.

The uncertainty of most of the available LUC datasets has been assessed in a large range of research studies (Grekousis et al. 2015; Tsendbazar 2016). However, these studies do not usually address all possible sources of uncertainty. Some limitations have been reported regarding the validation of specific areas and categories, which are heterogenous and, therefore, more difficult to map (Leyk et al. 2005; Fassnacht et al. 2006). The mapping accuracy of these categories and areas is not usually well characterized, as validation exercises only assess the general uncertainty or validity of the whole dataset (Prestele et al. 2016). Moreover, the validity of a specific dataset will depend on how it is used (Castilla and Hay 2007). An LUC map considered invalid for a specific type of study could be a reliable source of information for another study at another scale and with different aims. Maps like these are often described as "fit for use" or "fit for purpose" (Chrisman 2010). In addition, users often process the datasets in some way, so introducing sources of uncertainty that need to be evaluated (Nienkemper and Menz 2016). When using a series of LUC maps, additional uncertainties may arise. As Olofsson et al. (2013) noted, even when two independent maps are both very accurate, it is possible that the accuracy of the change map obtained by post-classification comparison will be low due to error propagation.

Many users develop their own maps, given the increasing availability of free imagery and tools with which to process and classify the images easily (Belward and Skøien 2015; Yuan et al. 2020). They need to validate the maps that they produce both for general purposes and for the specific use for which they were designed (Chuvieco 2016). The LUCCM community also need to validate the results of their modelling exercises (Paegelow and Camacho Olmedo 2008). To correctly interpret these results, they also need to understand the uncertainty of the LUC databases on which LUCC models are built (Prestele et al. 2016; García-Álvarez 2018b), given that input data and, specifically, input LUC maps, are considered one of the main sources of uncertainty in LUCCM (Verburg et al. 2013; Houet et al. 2015).

8 Conclusions

Many frequent users of LUC data and LUCC models are unaware of the latest developments in validation and uncertainty analysis of LUC data. It is also possible that they have limited knowledge of many of the datasets currently available for carrying out LUC exercises.

Many of the recent advances in this field remain within closed scientific communities and are not disseminated among the wider LUC community outside the research arena. This book seeks to respond to their needs. It provides an overview of the state of the art on LUC datasets, including time series of LUC maps, and the tools and methods available for LUC map validation. It also presents and explains frequently used tools and guidelines for validating the results produced by LUCC models. As many of the tools and techniques reviewed here are used in both LUC mapping and LUCC modelling validation exercises, in this book we address these two analyses together.

A full validation exercise, characterizing all the uncertainties of a given dataset or model, is a complex task that requires a high level of expertise and a wide range of tools and strategies, each one addressing different sources of uncertainty. This is beyond the scope of this book. Here we focus on the quantitative validation of LUC maps and LUCC model results. For detailed information about qualitative analyses of uncertainty, we refer readers to more specialized bibliography, depending on the specific objectives of their research. Readers wishing to find out more about other important aspects of uncertainty and validation practice, such as uncertainty communication, are also referred to specific literature on this topic.

Further Reading

Giri C (ed) (2012) Remote sensing of land use and land cover. Principles and applications. CRC Press.

This is one of the main reference books on Land Use Cover mapping, focusing specifically on LUC mapping and analysis. It offers an overview of the main concepts associated with LUC mapping and remote sensing and provides an introduction to this field, tracing its history. It also addresses the main methodological issues in relation to LUC mapping using remote sensing techniques, such as validation practices, land cover change detection and image classification methods. In the third part, the book includes examples of regional LUC mapping and LUCC monitoring for different parts of the world.

Manakos I, Braun M (2014) Land Use and Land Cover Mapping in Europe: Practices & Trends. Springer, Dordrecht, Heidelberg, New York, London.

Focused on Europe, this book is part of the reference bibliography for LUC mapping and LUCC monitoring. It provides a state of the art of LUC mapping globally, for Europe and at a national level for some of the European countries. Several chapters focus on remote sensing practices and methods for LUC mapping and LUCC detection. The book also has several introductory chapters on the role of remote sensing in the production of LUC information. Other chapters focus on the LUCC monitoring of processes relevant for policymaking.

Camacho Olmedo MT, Paegelow M, Mas J-F, Escobar F (2018) Geomatic Approaches for Modeling Land Change Scenarios. Springer, Cham, Switzerland.

This book provides an up-to-date review of LUCCM practice. The first part describes each of the LUCCM phases: calibration, simulation, validation and proposal of scenarios. Each chapter also presents common methods and strategies, implemented in different modelling software, for setting up and running a LUCC modelling exercise. The book also includes a series of technical notes for many of these tools and techniques, as well as short presentations of standard LUCC modelling software that is currently available. Common applications of LUCC models for thematic analyses and methodological studies are also described.

García-Álvarez D, Van Delden H, Camacho Olmedo MT, Paegelow M (2019) Uncertainty Challenge in Geospatial Analysis: An Approximation from the Land Use Cover Change Modelling Perspective. In: Koutsopoulos K, de

Miguel González R, Donert K (eds) Geospatial Challenges in the 21st Century. Springer, pp 289–314.

This book chapter offers a synthetic overview of uncertainty in LUCCM. It includes a theoretical explanation of what uncertainty is and analyses its different dimensions. It also presents the different sources of uncertainty in LUCCM and reviews different strategies and methods for managing it.

Gutman G, C. Janetos A, Cochrane COJ, et al. (2012) Land Change Science. Observing, Monitoring and Understanding Trajectories of Change on the Earth's Surface. Springer Netherlands, Dordrecht.

Although outdated (it was initially edited in 2004), this book provides an introduction to Land Change Science and Land Use Cover Change analysis. The experience acquired with the International Land Use and Land Cover (LUCC) Research Programme of the NASA is the leitmotif of the book. It provides an overview of Land Change Science, defining its main concepts and presenting the main international initiatives in LUCC research. It also offers an overview of the main processes of change analysed within the LUCC framework and its utility for policymaking and other fields. The book has various chapters focusing on methodological issues, some of which refer to LUCCM.

Belward AS, Skøien JO (2015) Who launched what, when and why; trends in global land-cover observation capacity from civilian earth observation satellites. ISPRS J Photogramm Remote Sens 103:115–128. https://doi.org/10.1016/j.isprsjprs.2014.03.009

This paper offers an overview of the history of civilian earth observation satellite missions that produce information that can be used in LUC mapping. It describes various different space missions and reflects on how useful they have been for the LUC community.

References

Abreu CG, Ralha CG (2017) Uncertainty Assessment in Agent-Based Simulation: An Exploratory Study. In: Negotiation and Argumentation in Multi-Agent Systems. pp 36–50

Anderson JR, Hardy EE, Roach JT, Witmer RE (1976) A Land Use and Land Cover Classification System for Use with Remote Sensor Data

Antoni JP, Judge V, Vuidel G, Klein O (2018) Constraint Cellular Automata for Urban Development Simulation: An Application to the Strasbourg-Kehl Cross-Border Area. pp 293–306

Barber C (2019) Monitoring Land Change with USGS LCMAP Science Products. In: American Geophysical Union, Fall Meeting 2019

Bastin L, Buchanan G, Beresford A et al (2013) Open-source mapping and services for Web-based land-cover validation. Ecol Inform 14:9–16. https://doi.org/10.1016/j.ecoinf.2012.11.013

Belward AS, Skøien JO (2015) Who launched what, when and why; trends in global land-cover observation capacity from civilian earth observation satellites. ISPRS J Photogramm Remote Sens 103:115–128. https://doi.org/10.1016/j.isprsjprs.2014.03.009

Elzbieta B, Agnieszka J (2019) Intellectual Structure of CORINE Land Cover Research Applications in Web of Science: A Europe-Wide Review. Remote Sens 11:2017. https://www.mdpi.com/2072-4292/11/17/2017

Bontemps S, Herold M, Kooistra L et al (2012) Revisiting land cover observation to address the needs of the climate modeling community. Biogeosciences 9:2145–2157. https://doi.org/10.5194/bg-9-2145-2012

Brovelli MA, Minghini M, Molinari ME et al (2018) Capacity building for high-resolution land cover intercomparison and validation: What is available and what is needed. Int Arch Photogramm Remote Sens Spat Inf Sci - ISPRS Arch 42:15–22. https://doi.org/10.5194/isprs-archives-XLII-4-W8-15-2018

Brown DG, Verburg PH, Pontius RG Jr, Lange MD (2013) Opportunities to improve impact, integration, and evaluation of land change models. Curr Opin Environ Sustain 5:452–457. https://doi.org/10.1016/j.cosust.2013.07.012

Brown G (2013) The relationship between social values for ecosystem services and global land cover: An empirical analysis. Ecosyst Serv 5:58–68. https://doi.org/10.1016/j.ecoser.2013.06.004

Bruzzone L, Demir B (2014) A Review of Modern Approaches to Classification of Remote Sensing Data. In: Manakos I, Braun M (eds) Land Use and Land Cover Mapping in Europe. Springer, Practices & Trends, pp 127–143

Bürgi M, Hersperger AM, Schneeberger N (2005) Driving forces of landscape change - current and new directions. Landsc Ecol 19:857–868. https://doi.org/10.1007/s10980-005-0245-3

Burnicki AC, Brown DG, Goovaerts P et al (2010) Propagating error in land-cover-change analyses: impact of temporal dependence under increased thematic complexity. Int J Geogr Inf Sci 24:1043–1060. https://doi.org/10.1080/13658810903279008

Camacho Olmedo MT, Paegelow M, Mas JF, Escobar F (2018) Geomatic Approaches for Modeling Land Change Scenarios. An Introduction. Springer, Cham, Switzerland

Campbell JB (1983) Mapping the land: aerial imagery for land use information

Campbell JB, Wynne RH (2011) Introduction to remote sensing, 5th edn. The Guilford Press, New York, London

Carlson TN, Traci Arthur S (2000) The impact of land use - Land cover changes due to urbanization on surface microclimate and hydrology: A satellite perspective. Glob Planet Change 25:49–65. https://doi.org/10.1016/S0921-8181(00)00021-7

Castilla G, Hay GJ (2007) Uncertainties in land use data. Hydrol Earth Syst Sci 11:1857–1868. https://doi.org/10.5194/hess-11-1857-2007

Cebecauer T, Hofierka J (2008) The consequences of land-cover changes on soil erosion distribution in Slovakia. Geomorphology 98:187–198. https://doi.org/10.1016/j.geomorph.2006.12.035

Chrisman N (2010) Development in the Treatment of Spatial Data Quality. Fundam Spat Data Qual:21–30https://doi.org/10.1002/9780470612156.ch1

Chuvieco E (2016) Fundamentals of Satellite Remote Sensing. An environmental approach, 2 edition. CRC Press, Boca Raton

Clarke KC (2004) The limits of simplicity: toward geocomputational honesty in urban modeling. In: Atkinson P, Foody G, Darby S, Wu F (eds) Geodynamics. CRC Press, Boca Raton, pp 215–232

Collier, 2009.aCollier P (2009a) Mapping, Topographic. Int Encycl Hum Geogr 409–420https://doi.org/10.1016/B978-008044910-4.00073-0

Collier P (2009b) Photogrammetry/Aerial Photography. In: International Encyclopedia of Human Geography. pp 151–156

Comber A, Fisher P, Wadsworth R (2005) What is land cover? Environ Plan B Plan Des 32:199–209. https://doi.org/10.1068/b31135

Comber AJ (2008) The separation of land cover from land use using data primitives. J Land Use Sci 3:215–229. https://doi.org/10.1080/17474230802465173

Congalton RG, Gu J, Yadav K et al (2014) Global land cover mapping: A review and uncertainty analysis. Remote Sens 6:12070–12093. https://doi.org/10.3390/rs61212070

Conway TM (2009) The impact of class resolution in land use change models. Comput Environ Urban Syst 33:269–277. https://doi.org/10.1016/j.compenvurbsys.2009.02.001

Crone GR (2000) Historia de los mapas. Fondo de Cultura Económica, Madrid

Crooks AT, Heppenstall AJ (2012) Introduction to agent-based modelling. In: Heppenstall A, Crooks AT, See LM, Batty M (eds) Agent-Based Models of Geographical Systems. Springer, Dordrecht, Heidelberg, London, New York, pp 85–105

Cuo L, Lettenmaier DP, Alberti M, Richey JE (2009) Effects of a century of land cover and climate change on the hydrology of the Puget Sound basin. Hydrol Process 23:907–933. https://doi.org/10.1002/hyp.7228

Devillers R, Jeansoulin R (2006) Spatial Data Quality: Concepts. In: Fundamentals of Spatial Data Quality. ISTE, London, UK, pp 31–42

Di Gregorio A, Jansen LJ (1998) Land Cover Classification System (LCCS): Classification Concepts and User Manual. FAO, Rome

Di Gregorio A, Latham J (2003) Africover Land Cover Classification and Mapping Project. In: Encyclopedia of Life Support Systems (EOLSS). pp 236–254

Diogo V, Koomen E (2016) Land Cover and Land Use Indicators: Review of available data

Eastman JR, Toledano J (2018) A Short Presentation of CA_MARKOV. In: Camacho Olmedo MT, Paegelow M, Mas J-F, Escobar F (eds) Geomatic Approaches for Modeling Land Change Scenarios. Springer, Cham, Switzerland, pp 481–484

Emery W, Camps A (2017) Introduction to Satellite Remote Sensing. Elsevier, Atmosphere, Ocean, Cryosphere and Land Applications

Escobar F, Van Delden H, Hewitt R (2018) LUCC Scenarios. In: Camacho Olmedo MT, Paegelow M, Mas J-F, Escobar F (eds) Geomatic Approaches for Modeling Land Change Scenarios. Springer, Cham, Switzerland, pp 81–97

Fassnacht KS, Cohen WB, Spies TA (2006) Key issues in making and using satellite-based maps in ecology: A primer. For Ecol Manage 222:167–181. https://doi.org/10.1016/j.foreco.2005.09.026

Feranec J, Hazeu G, Christensen S, Jaffrain G (2007) Corine land cover change detection in Europe (case studies of the Netherlands and Slovakia). Land Use Policy 24:234–247. https://doi.org/10.1016/j.landusepol.2006.02.002

Ferchichi A, Boulila W, Farah IR (2017) Reducing uncertainties in land cover change models using sensitivity analysis. Knowl Inf Syst. https://doi.org/10.1007/s10115-017-1102-9

Fisher P, Unwin D (2005) Land Use and land cover: Contradiction or Complement. In: Fisher P (ed) Re-Presenting GIS. Wiley, pp 85–98

Fonte CC, Bastin L, See L et al (2015) Usability of VGI for validation of land cover maps. Int J Geogr Inf Sci 29:1269–1291. https://doi.org/10.1080/13658816.2015.1018266

Fritz S, See L, Mccallum I et al (2015) Mapping global cropland and field size. Glob Chang Biol 21:1980–1992. https://doi.org/10.1111/gcb.12838

Fuller R, Groom GB, Jones AR (1994) The land cover map of Great Britain: an automated classification of Landsat TM data. Photogramm Eng Remote Sensing 60:553–562

Gallardo M (2014) Cambios de usos del suelo y simulación de escenarios en la Comunidad de Madrid. Universidad Complutense de Madrid https://eprints.ucm.es/id/eprint/25253/

García-Álvarez D (2018a) Aproximación al estudio de la incertidumbre en la modelización del Cambio de Usos y Coberturas del Suelo (LUCC). Universidad de Granadan https://digibug.ugr.es/handle/10481/52908

García-Álvarez D (2018b) The influence of scale in LULC modelling. A comparison between two different LULC maps (SIOSE and CORINE). In: Camacho Olmedo MT, Paegelow M, Mas J-F, Escobar F (eds) Geomatic Approaches for Modeling Land Change Scenarios. Springer, Cham, Switzerland, pp 187–213

García-Álvarez D, Camacho Olmedo MT (2017) Changes in the methodology used in the production of the Spanish CORINE: Uncertainty analysis of the new maps. Int J Appl Earth Obs Geoinf 63:55–67. https://doi.org/10.1016/j.jag.2017.07.001

García-Álvarez D, Camacho Olmedo MT, Paegelow M (2019a) Sensitivity of a common Land Use Cover Change (LUCC) model to the Minimum Mapping Unit (MMU) and Minimum Mapping Width (MMW) of input maps. Comput Environ Urban Syst 78:. https://doi.org/10.1016/j.compenvurbsys.2019.101389

García-Álvarez D, Van Delden H, Camacho Olmedo MT, Paegelow M (2019) Uncertainty Challenge in Geospatial Analysis: An Approximation from the Land Use Cover Change Modelling Perspective. In: Koutsopoulos K, de Miguel GR, Donert K (eds) Geospatial Challenges in the 21st Century. Springer, pp 289–314

Geoghegan J (1998) "Socializing the Pixel" and Pixelizing the Social" in Land-Use and Land-Cover Change. In: National Research Council (ed) People and Pixels: Linking Remote Sensing and Social Science. National Academies Press, pp 51–69

Giri CP (2016a) Remote sensing of land use and land cover: Principles and applications

Giri CP (2016b) Brief overview of remote sensing of land cover. In: Remote Sensing of Land Use and Land Cover: Principles and Applications. pp 3–12

Gong P, Yu L, Li C et al (2016) A new research paradigm for global land cover mapping. Ann GIS 22:87–102. https://doi.org/10.1080/19475683.2016.1164247

Green GM, Schweik CM, Randolph JC (2005) Linking Disciplines across Space and Time: Useful Concepts and Approaches for Land-Cover Change Studies. In: Moran EF, Ostrom E (eds) Seeing the Forest and the Trees: Human-environment Interactions in Forest Ecosystems. MIT Press

Grekousis G, Mountrakis G, Kavouras M (2015) An overview of 21 global and 43 regional land-cover mapping products. Int J Remote Sens 36:5309–5335. https://doi.org/10.1080/01431161.2015.1093195

Grinblat Y, Gilichinsky M, Benenson I (2016) Cellular Automata Modeling of Land-Use/Land-Cover Dynamics: Questioning the Reliability of Data Sources and Classification Methods. Ann Am Assoc Geogr 106:1299–1320. https://doi.org/10.1080/24694452.2016.1213154

Gutman G, Byrnes R, Masek J et al (2008) Towards monitoring land-cover and land-use changes at a global scale: The global land survey 2005. Photogramm Eng Remote Sensing 74:6–10

Gutman G, C. Janetos A, Cochrane COJ, et al (2012a) Land Change Science. Observing, Monitoring and Understanding Trajectories of Change on the Earth's Surface. Springer Netherlands, Dordrecht

Gutman G, Justice C, Sheffner E, Loveland T (2012b) The NASA Land Cover and Land Use Change Program. In: Gutman G, C. Janetos A, Cochrane COJ, et al (eds) Land Change Science. Remote Sensing and Digital Image Processing. Springer, pp 17–29

Herold M, Di Gregorio A (2012) Evaluating land-cover legends using the un land cover classification system. In: P. Giri C (ed) Remote

Sensing of Land Use and Land Cover. Principles and Applications. CRC Press, pp 65–89

Herold M, See L, Tsendbazar NE, Fritz S (2016) Towards an integrated global land cover monitoring and mapping system. Remote Sens 8:1–11. https://doi.org/10.3390/rs8121036

Hewitt R, Van Delden H, Escobar F (2014) Participatory land use modelling, pathways to an integrated approach. Environ Model Softw 52:149–165. https://doi.org/10.1016/j.envsoft.2013.10.019

Houet T, Vacquié L, Sheeren D (2015) Evaluating the spatial uncertainty of future land abandonment in a mountain valley (Vicdessos, Pyrenees - France): Insights from model parameterization and experiments. J Mt Sci 12:1095–1112. https://doi.org/10.1007/s11629-014-3404-7

Hu H, Liu W, Cao M (2008) Impact of land use and land cover changes on ecosystem services in Menglun, Xishuangbanna, Southwest China. Environ Monit Assess 146:147–156. https://doi.org/10.1007/s10661-007-0067-7

Jacob BG, Regens JL, Mbogo CM et al (2003) Occurrence and Distribution of Anopheles (Diptera: Culicidae) Larval Habitats on Land Cover Change Sites in Urban Kisumu and Urban Malindi, Kenya. J Med Entomol 40:777–784. https://doi.org/10.1603/0022-2585-40.6.777

Janetos AC (2012) Research Directions in Land-Cover and Land-Use Change. In: Gutman G, C. Janetos A, Cochrane COJ, et al. (eds) Land Change Science. Remote Sensing and Digital Image Processing. Springer, pp 449–457

Jcgm JCFGIM (2008) Evaluation of measurement data — Guide to the expression of uncertainty in measurement. Int Organ Stand Geneva ISBN 50:134. https://doi.org/10.1373/clinchem.2003.030528

Johnson GD, Patil GP (1998) Quantitative multiresolution characterization of landscape patterns for assessing the status of ecosystem health in watershed management areas. Ecosyst Heal 4:177–187. https://doi.org/10.1046/j.1526-0992.1998.00091.x

Khatami R, Mountrakis G, Stehman SV (2016) A meta-analysis of remote sensing research on supervised pixel-based land-cover image classification processes: General guidelines for practitioners and future research. Remote Sens Environ 177:89–100. https://doi.org/10.1016/j.rse.2016.02.028

Kolb M, Mas J-F, Galicia L (2013) Evaluating drivers of land-use change and transition potential models in a complex landscape in Southern Mexico. Int J Geogr Inf Sci 27:1804–1827. https://doi.org/10.1080/13658816.2013.770517

Krüger C (2016) Uncertainties in land change modeling

Lawler JJ, O'Connor RJ, Hunsaker CT et al (2004) The effects of habitat resolution on models of avian diversity and distributions: A comparison of two land-cover classifications. Landsc Ecol 19:517–532. https://doi.org/10.1023/B:LAND.0000036151.28327.01

Lawrence PJ, Feddema JJ, Bonan GB et al (2012) Simulating the biogeochemical and biogeophysical impacts of transient land cover change and wood harvest in the Community Climate System Model (CCSM4) from 1850 to 2100. J Clim 25:3071–3095. https://doi.org/10.1175/JCLI-D-11-00256.1

Leyk S, Boesch R, Weibel R (2005) A conceptual framework for uncertainty investigation in map-based land cover change modelling. Trans GIS 9:291–322. https://doi.org/10.1111/j.1467-9671.2005.00220.x

Li H, Wu J (2006) Uncertainty analysis in ecological studies: an overview. In: Wu J, Jones KB, Li H, Loucks OL (eds) Scaling and Uncertainty Analysis in Ecology: Methods and Applications. Springer, Dordrecht, pp 45–66

Lloyd CD (2014) Exploring Spatial Scale in Geography. Wiley, Chichester

Longley PA, Goodchild MF, Maguire DJ, Rhind DW (2011) Geographical Information Systems and Science, 2nd edn. Wiley, Chichester

Loveland TR (2016) History of land-cover mapping. In: P. Giri C (ed) Remote Sensing of Land Use and Land Cover: Principles and Applications. CRC Press, pp 13–22

Lu D, Mausel P, Brondízio E, Moran E (2004) Change detection techniques. Int J Remote Sens 25:2365–2401. https://doi.org/10.1080/0143116031000139863

Manakos I, Braun M (2014) Land Use and Land Cover Mapping in Europe. Practices & Trends. Springer, Dordrecht, Heidelberg, New York, London

Marshall CH, Pielke RA, Steyaert LT, Willard DA (2004) The impact of anthropogenic land-cover change on the Florida Peninsula Sea Breezes and warm season sensible weather. Mon Weather Rev 132:28–52. https://doi.org/10.1175/1520-0493(2004)132%3c0028:TIOALC%3e2.0.CO;2

Mas J-F, Kolb M, Paegelow M et al (2014) Inductive pattern-based land use/cover change models: A comparison of four software packages. Environ Model Softw 51:94–111. https://doi.org/10.1016/j.envsoft.2013.09.010

Mas J-F, Pérez-Vega A, Ghilardi A et al (2014) A Suite of Tools for Assessing Thematic Map Accuracy. Geogr J 2014:1–10. https://doi.org/10.1155/2014/372349

Mas J-F, Paegelow M, Camacho Olmedo MT (2018) LUCC modeling approaches to calibration. In: Camacho Olmedo MT, Paegelow M, Mas J-F, Escobar F (eds) Geomatic Approaches for Modeling Land Change Scenarios. Springer, Cham, Switzerland, pp 1–15

Mora B, Tsendbazar N-E, Herold M, Arino O (2014) Global Land Cover Mapping: Current Status and Future Trends. In: Manakos I, Braun M (eds) Land Use and Land Cover Mapping in Europe. Practices & Trends. Springer, Dordrecht, Heidelberg, New York, London, pp 11–30

Morain SA (1998) A brief history of remote sensing applications, with emphasis on Landsat. In: National Research Council (ed) People and Pixels: Linking Remote Sensing and Social Science. The National Academies, Washington, DC, pp 28–50

Morales-Barquero L, Lyons M, Phinn S, Roelfsema C (2019) Trends in Remote Sensing Accuracy Assessment Approaches in the Context of Natural Resources. Remote Sens 11:2305. https://doi.org/10.3390/rs11192305

Moran EF, Skole DL, Turner BL (2012) The Development of the International Land-Use and Land-Cover Change (LUCC) Research Program and Its Links to NASA's Land-Cover and Land-Use Change (LCLUC) Initiative. In: Gutman G, Janetos AC, Justice CO et al (eds) Land Change Science. Springer, Netherlands, pp 1–15

Müller D, Munroe DK (2014) Current and future challenges in land-use science. J Land Use Sci 9:133–142. https://doi.org/10.1080/1747423X.2014.883731

Murayama Y (2012) Introduction: Geospatial Analysis. In: Progress in Geospatial Analysis. pp 1–9

National Research Council (2014) Advancing Land Change Modeling: Opportunities and Research Requirements. National Academies Press, Washington, D.C.

Nienkemper P, Menz G (2016) Thematic resolution in conservation monitoring - Assessment of the impact of classification detail on landscape analysis using the example of a biosphere reserve. Erdkunde 70:237–253. https://doi.org/10.3112/erdkunde.2016.03.03

O'Sullivan D, Perry GLW (2013) Spatial Simulation: Exploring Pattern and Process. Wiley, Chichester

Olofsson P, Foody GM, Stehman SV, Woodcock CE (2013) Making better use of accuracy data in land change studies: Estimating accuracy and area and quantifying uncertainty using stratified estimation. Remote Sens Environ 129:122–131. https://doi.org/10.1016/j.rse.2012.10.031

Oxley T, Jeffrey P, Lemon M (2002) Policy Relevant Modelling: Relationships Between Water, Land Use, and Farmer Decision Processes. Integr Assess 3:30–49. https://doi.org/10.1076/iaij.3.1.30.7413

Paegelow M, Camacho Olmedo MT (2008) Advances in geomatic simulations for environmental dynamics. In: Paegelow M, Camacho Olmedo MT (eds) Modelling Environmental Dynamics. Springer, Berlin, Heidelberg, pp 3–54

Plata Rocha W (2010) Descripción, análisis y simulación del crecimiento urbano mediante tecnologías de la información geográfica. El caso de la Comunidad de Madrid. Universidad de Alcalá, Departamento de Geografía

Pontius RG Jr, Huffaker D, Denman K (2004) Useful techniques of validation for spatially explicit land-change models. Ecol Modell 179:445–461. https://doi.org/10.1016/j.ecolmodel.2004.05.010

Pontius RG Jr, Malanson J (2005) Comparison of the structure and accuracy of two land change models. Int J Geogr Inf Sci 19:243–265. https://doi.org/10.1080/13658810410001713434

Prestele R, Alexander P, Rounsevell MDA et al (2016) Hotspots of uncertainty in land-use and land-cover change projections: a global-scale model comparison. Glob Chang Biol 22:3967–3983. https://doi.org/10.1111/gcb.13337

Rindfuss RR, Walsh SJ, Turner BL et al (2004) Developing a science of land change: challenges and methodological issues. Proc Natl Acad Sci U S A 101:13976–13981. https://doi.org/10.1073/pnas.0401545101

Rochel X, Abadie J, Avon C, et al (2017) Quelles sources cartographiques pour la définition des usages anciens du sol en France ? Rev For Française:353.https://doi.org/10.4267/2042/67866

Sang L, Zhang C, Yang J et al (2011) Simulation of land use spatial pattern of towns and villages based on CA–Markov model. Math Comput Model 54:938–943

Sinha P, Kimar L (2013) Markov Land Cover Change Modeling Using Pairs of Time-Series Satellite Images. Photogramm Eng Remote Sens 79:1–15. https://doi.org/10.14358/PERS.79.11.1037

Soares-Filho BS, Nepstad DC, Curran LM et al (2006) Modelling conservation in the Amazon basin. Nature 440:520–523

Sohl TL, Sleeter BM (2012) Role of Remote Sensing for Land-Use and Land-Cover Change Modeling. In: P. Giri C (ed) Remote Sensing and Land Cover: Principles and Applications. CRC Press, pp 225–239

Sohl TL, Wimberly MC, Radeloff VC et al (2016) Divergent projections of future land use in the United States arising from different models and scenarios. Ecol Modell 337:281–297. https://doi.org/10.1016/j.ecolmodel.2016.07.016

Sophie B, Pierre D, Eric VB, et al (2011) Producing global land cover maps consistent over time to respond the needs of the climate modelling community. 2011 6th Int Work Anal Multi-Temporal Remote Sens Images, Multi-Temp 2011 - Proc 161–164. https://doi.org/10.1109/Multi-Temp.2011.6005073

Stehman SV., Foody GM (2019) Key issues in rigorous accuracy assessment of land cover products. Remote Sens Environ 231:111199.https://doi.org/10.1016/j.rse.2019.05.018

Steiner D (1965) Use of air photographs for interpreting and mapping rural land use in the United States. Photogrammetria 20:65–80. https://doi.org/10.1016/0031-8663(65)90035-9

Strahler AH, Boschetti L, Foody GM, et al (2006) Global Land Cover Validation: Recommendations for Evaluation and Accuracy Assessment of Global Land Cover Maps

Strand GH (2013) The Norwegian area frame survey of land cover and outfield land resources. Nor Geogr Tidsskr 67:24–35. https://doi.org/10.1080/00291951.2012.760001

Szantoi Z, Geller GN, Tsendbazar NE et al (2020) Addressing the need for improved land cover map products for policy support. Environ Sci Policy 112:28–35. https://doi.org/10.1016/j.envsci.2020.04.005

Thackway R, Lymburner L, Guerschman JP (2013) Dynamic land cover information: Bridging the gap between remote sensing and natural resource management. Ecol Soc 18.https://doi.org/10.5751/ES-05229-180102

Toro Balbotín D (2014) Analyse de la détérioration de la forêt de la cordillère de la Costa dans le sud chilien: géomatique et modélisation prospective appliquée sur une forêt patrimoniale de la province d'Osorno (41° 15' - 41° 00' latitude Sud). Université Toulouse 2 Le Mirail. Laboratorie GEODE

Tsendbazar N (2016) Global land cover map validation, comparison and integration for different user communities. Wageningen University

Tsendbazar NE, de Bruin S, Herold M (2015) Assessing global land cover reference datasets for different user communities. ISPRS J Photogramm Remote Sens 103:93–114. https://doi.org/10.1016/j.isprsjprs.2014.02.008

Tuanmu MN, Jetz W (2014) A global 1-km consensus land-cover product for biodiversity and ecosystem modelling. Glob Ecol Biogeogr 23:1031–1045. https://doi.org/10.1111/geb.12182

Turner BL (2017) Land Change Science. International Encyclopedia of Geography: People, the Earth. Environment and Technology. John Wiley & Sons Ltd., Oxford, UK, pp 1–6

Turner BL, Lambin EF, Reenberg A (2007) The emergence of land change science for global environmental change and sustainability. Proc Natl Acad Sci 104:20666–20671. https://doi.org/10.1073/pnas.0704119104

Unwin DJ (1995) Geographical information systems and the problem of "error and uncertainty." Prog Hum Geogr 19:549–558. https://doi.org/10.1177/030913259501900408

Van Asselt MBA (2000) Perspectives on Uncertainty and Risk - the PRIMA Approach to Decision Support. Kluwer Academic Publishers, Boston, Dordrecht, London

Van Delden H, Van Vliet J, Rutledge DT, Kirkby MJ (2011) Comparison of scale and scaling issues in integrated land-use models for policy support. Agric Ecosyst Environ 142:18–28. https://doi.org/10.1016/j.agee.2011.03.005

Van Vliet J, Bregt AK, Brown DG et al (2016) A review of current calibration and validation practices in land-change modeling. Environ Model Softw 82:174–182. https://doi.org/10.1016/j.envsoft.2016.04.017

Van Vliet J, Bregt AK, Hagen-Zanker A (2011) Revisiting Kappa to account for change in the accuracy assessment of land-use change models. Ecol Modell 222:1367–1375. https://doi.org/10.1016/j.ecolmodel.2011.01.017

Verburg PH, Neumann K, Nol L (2011) Challenges in using land use and land cover data for global change studies. Glob Chang Biol 17:974–989. https://doi.org/10.1111/j.1365-2486.2010.02307.x

Verburg PH, Schot P, Dijst M, Veldkamp A (2004) Land use change modelling: current practice and research priorities. GeoJournal 61:309–324. https://doi.org/10.1007/s10708-004-4946-y

Verburg PH, Tabeau A, Hatna E (2013) Assessing spatial uncertainties of land allocation using a scenario approach and sensitivity analysis: A study for land use in Europe. J Environ Manage 127:S132–S144. https://doi.org/10.1016/j.jenvman.2012.08.038

Verburg PH, van de Steeg J, Veldkamp A, Willemen L (2009) From land cover change to land function dynamics: A major challenge to improve land characterization. J Environ Manage 90:1327–1335. https://doi.org/10.1016/j.jenvman.2008.08.005

Villa G, Valcarcel N, Caballlero ME, et al (2008) Land Cover Classifications: An Obsolete Paradigm. In: Chen J, Jiang J, Nayak S (eds) The International Archives of the Photogrammetry, Remote

Land Use Cover Mapping, Modelling and Validation. A Background

33

Sensing and Spatial Information Sciences. ISPRS, Beijing, pp 609–614

Wallis H (1981) The History of Land Use Mapping. Cartogr J 18:45–48. https://doi.org/10.1179/caj.1981.18.1.45

White R, Engelen G, Uljee I (2015) Modeling cities and regions as complex systems: from theory to planning applications. The MIT Press, Cambridge, Massachusetts, London, England

Wulder MA, Coops NC, Roy DP et al (2018) Land cover 2.0. Int J Remote Sens 39:4254–4284. https://doi.org/10.1080/01431161.2018.1452075

Yeh AG-O, Li X (2006) Errors and uncertainties in urban cellular automata. Comput Environ Urban Syst 30:10–28. https://doi.org/10.1016/j.compenvurbsys.2004.05.007

Yu L, Liang L, Wang J et al (2014) Meta-discoveries from a synthesis of satellite-based land-cover mapping research. Int J Remote Sens 35:4573–4588. https://doi.org/10.1080/01431161.2014.930206

Yuan K, O'Neil P, Torrejon D (2020) Landsat's past paves the way for data democratization in earth science. In: Data Democracy. pp 147–161

Zhang Z, Wang X, Zhao X et al (2014) A 2010 update of National Land Use/Cover Database of China at 1:100000 scale using medium spatial resolution satellite images. Remote Sens Environ 149:142–154. https://doi.org/10.1016/j.rse.2014.04.004

Zimmermann P, Tasser E, Leitinger G, Tappeiner U (2010) Effects of land-use and land-cover pattern on landscape-scale biodiversity in the European Alps. Agric Ecosyst Environ 139:13–22. https://doi.org/10.1016/j.agee.2010.06.010

Validation of Land Use Cover Maps: A Guideline

María Teresa Camacho Olmedo, David García-Álvarez, Marta Gallardo, Jean-François Mas, Martin Paegelow, Miguel Ángel Castillo-Santiago, and Ramón Molinero-Parejo

Abstract

This chapter offers a general overview of the available tools and strategies for validating Land Use Cover (LUC) data—specifically LUC maps—and Land Use Cover Change Modelling (LUCCM) exercises. We give readers some guidelines according to the type of maps they want to validate: single LUC maps (Sect. 3), time series of LUC maps (Sect. 4) or the results of LUCCM exercises (Sect. 5). Despite the fact that some of the available methods are applicable to all these maps, each type of validation exercise has its own particularities which must be taken into account. Each section of this chapter starts with a brief introduction about the specific type of maps (single, time series or modelling exercises) and the reference data needed to validate them. We also present the validation methods/functions and the corresponding exercises developed in Part III of this book. To this end, we address, in this order, the tools for validating Land Use Cover data based on basic and Multiple-Resolution Cross-Tabulation (see chapter "Basic and Multiple-Resolution Cross Tabulation to Validate Land Use Cover Maps"), metrics based on the Cross-Tabulation matrix (see chapter "Metrics Based on a Cross-Tabulation Matrix to Validate Land Use Cover Maps"), Pontius Jr. methods based on the Cross-Tabulation matrix (see chapter "Pontius Jr. Methods Based on a Cross-Tabulation Matrix to Validate Land Use Cover Maps"), validation practices with soft maps produced by Land Use Cover models (see chapter "Validation of Soft Maps Produced by a Land Use Cover Change Model"), spatial metrics (see chapter "Spatial Metrics to Validate Land Use Cover Maps"), advanced pattern analysis (see chapter "Advanced Pattern Analysis to Validate Land Use Cover Maps") and geographically weighted methods (see chapter "Geographically Weighted Methods to Validate Land Use Cover Maps").

Keywords

Land Use Cover • Land Use Cover Change Modelling exercises • Validation

M. T. Camacho Olmedo (✉)
Departamento de Análisis Geográfico Regional y Geografía Física, Universidad de Granada, Granada, Spain
e-mail: camacho@ugr.es

D. García-Álvarez · R. Molinero-Parejo
Departamento de Geología, Geografía y Medio Ambiente, Universidad de Alcalá, Alcalá de Henares, Spain

M. Gallardo
Departamento de Geografía, Universidad Nacional de Educación a Distancia, Madrid, Spain

J.-F. Mas
Laboratorio de Análisis Espacial, Centro de Investigaciones en Geografía Ambiental, Universidad Nacional Autónoma de México, Morelia, Mexico

M. Paegelow
Département de Géographie, Aménagement et Environnement, Université de Toulouse Jean Jaurès, Toulouse, France

M. Á. Castillo-Santiago
Departamento de Observación y Estudio de la Tierra, la Atmósfera y el Océano, El Colegio de la Frontera Sur, San Cristóbal de las Casas, Mexico

1 Introduction

Validation is a required step prior to the effective use of any Land Use Cover (LUC) dataset or of the results of a Land Use Cover Change Modelling (LUCCM) exercise. We need to understand to what extent these datasets and results are uncertain in order to be able to assess the limits that these uncertainties may impose on the conclusions of our analyses and studies.

There are many methods, tools and strategies currently available for validating LUC data and LUCCM exercises. However, comprehensive guidelines providing users with clear instructions and recommendations about how to carry out this validation are scarce. Olofsson et al. (2013, 2014) review the validation of land change maps and offer a series of recommendations as to how to perform a credible scientific validation, accepting that other recommendations or good practice guidelines could be equally valid and perhaps even more so. Paegelow et al. (2014, 2018) propose a variety of validation techniques and error analysis which can be used to validate different LUCCM exercises.

In this chapter, we aim to provide readers with a general overview of the available tools and strategies for validating LUC data—specifically LUC maps—and LUCCM exercises. We give readers different guidelines according to the type of maps they want to validate: single LUC maps (Sect. 3), time series of LUC maps (Sect. 4) and results of LUCCM exercises (Sect. 5). Although some of the available methods and tools can be applied to all these maps, each type of validation exercise has its own specific aspects that users must bear in mind. For example, the results of LUCCM exercises include soft and hard LUC maps. The hard outputs of a model—hard maps—are very similar to input LUC maps, while the soft outputs—soft maps—are continuous and ranked. We therefore also present some validation methods that focus specifically on soft maps.

Before presenting these validation methods and functions, it is important to make clear that visual inspection is an essential part of any validation exercise. It can provide a great deal of information about the uncertainties of the data being evaluated, which are not detected by the quantitative methods reviewed in this book. Visual inspection should be conducted during all validation exercises, at the beginning, at the end and throughout the entire process.

2 Validation Methods/Functions and Exercises Presented in Part III of This Book

This chapter is intended as a presentation of Part III of this book. Figure 1 shows the validation methods/functions and the corresponding exercises presented in the chapters and sections of Part III. With this in mind, in this chapter we address, in this order: the available tools for validating Land Use Cover data related with basic and Multiple Resolution Cross-Tabulation (see chapter "Basic and Multiple-Resolution Cross Tabulation to Validate Land Use Cover Maps"), metrics derived from the Cross-Tabulation matrix (see chapter "Metrics Based on a Cross-Tabulation Matrix to Validate Land Use Cover Maps"), methods proposed by Pontius Jr. based on the Cross-Tabulation matrix (see chapter "Pontius Jr. Methods Based on a Cross-Tabulation Matrix to Validate Land Use Cover Maps"), validation practices with soft maps produced by Land Use Cover Change models (see chapter "Validation of Soft Maps

Produced by a Land Use Cover Change Model"), spatial metrics (see chapter "Spatial Metrics to Validate Land Use Cover Maps"), advanced pattern analysis (see chapter "Advanced Pattern Analysis to Validate Land Use Cover Maps") and geographically weighted methods (see chapter "Geographically Weighted Methods to Validate Land Use Cover Maps").

The exercises presented in Part III have been applied using the Quantum GIS (QGIS) software and R scripts. To homogenize the exercises across the different chapters, they have the same standard objectives: *to validate a map (t₁) against reference data/map (t₁)* (single LUC map); *to validate a series of maps with two or more time points (t₀, t₁, t₂...)* (LUC maps series/ LUC changes); and, for results from LUCCM exercise, *to validate soft maps produced by the model against a reference map of changes (t₀ – t₁)* (soft LUC maps), *to validate a simulation (T₁) against a reference map (t₁)* (single LUC map - hard LUC maps) and *to validate simulated changes (t₀ – T₁) against a reference map of changes (t₀ – t₁)* (LUC maps series / LUC changes – hard LUC maps). However, in certain specific cases, additions have been made to these standard titles. In addition to the applications of each method/function implemented in the practical exercises in this book, the cells shaded in grey in Fig. 1 indicate that the method has other potential applications that are not described here.

3 Validation of Single Land Use Cover Maps

The validation of single LUC maps is the most widespread practice of all those addressed in this book. Foody (2002) concludes that there is no single universally acceptable measure of accuracy but rather a variety of indices, each sensitive to different features. Creating a single, all-purpose measure of classification accuracy would therefore seem an almost impossible goal. However, accuracy assessment must follow certain guidelines and principles in order to guarantee scientifically defensible assessment of map accuracy (Stehman 1999; Stehman and Czaplewski 1998).

Users have been validating their maps since the advent of digital remote sensing and the first classifications of digital imagery, as a means of assessing to what extent the classified images resemble the real LUC on the ground. Now, several decades later, the validation of single LUC maps is a very common practice, and although new methods and tools have been developed over the years, the original ones remain popular. These are based above all on the comparison of the assessed LUC map with reference datasets through cross-tabulation (Foody 2002; Strahler et al. 2006). In recent years, the use of pattern analysis and other validation methods has become increasingly common.

Validation methods/functions		Single LUC map	LUC maps series/ LUC changes	LUCC modelling exercises		
				Soft LUC maps	Hard LUC maps	
					Single LUC map	LUC maps series/ LUC changes
Book chapters	Chapter sections	To validate a map (t1) against reference data/map (t1)[1]	To validate a series of maps with two or more time points (t0, t1, t2...)	To validate soft maps produced by the model against a reference map of changes (t0 – t1)	To validate a simulation (T1) against a reference map (t1)	To validate simulated changes (t0 – T1) against a reference map of changes (t0 – t1)
Basic and Multiple-Resolution Cross Tabulation	1. Basic Cross Tabulation	Exercise 1	Exercise 4	Exercise 2	Exercise 3	
	2. Multiple-Resolution Cross Tabulation	Exercise 1		Exercise 2	Exercise 3	
Metrics based on a Cross Tabulation Matrix	1. Change statistics		Exercise 1			
	2. Areal and spatial agreement metrics	Exercise 1			Exercise 2	Exercise 3
	3. Kappa Indices	Exercise 1			Exercises 2, 3	
	4. Agreement between maps at global and stratum level					Exercise 1
	5. Accuracy assessment statistics	Exercise 1				
Pontius Jr. methods based on a Cross Tabulation matrix	1. Null Model					Exercise 1
	2. LUCC Budget		Exercise 1			
	3. Quantity and allocation disagreement		Exercise 1			
	4. Figure of Merit (FoM) and complementary Producer's and User's accuracy					Exercises 1, 2, 3
	5. Incidents and States		Exercise 1			
	6. Intensity analysis		Exercise 1			
	7. Flow matrix		Exercise 1			
Validation of soft maps produced by a Land Use Cover Change model	1. Correlation			Exercise 1		
	2. Receiver Operating Characteristic (ROC)			Exercise 1		
	3. Difference in Potential (DiP)			Exercises 1, 2		
	4. Total uncertainty, quantity uncertainty, allocation uncertainty			Exercise 1[2]		
Spatial metrics	1. Spatial metrics	Exercise 1	Exercises 4, 5, 6		Exercise 2	Exercise 3
Advanced pattern analysis	1. Map Curves	Exercise 1	Exercise 4		Exercise 2	Exercise 3
	2. Change on pattern borders		Exercise 1			
	3. Allocation distance error				Exercises 1, 2	
Geographically weighted methods	1. Overall, user's and producer's accuracy through GWR	Exercise 1				

[1] Titles of exercises included in the corresponding sections.

[2] This validation technique is not calculated against a reference map of changes.

Fig. 1 Validation methods/functions and corresponding exercises presented in Part III of this book for single LUC maps, LUC maps series/LUC changes and LUCCM exercises. The grey cells highlight the possible applications of each method/function

The reference datasets for validating single LUC maps may be obtained from different sources of LUC data. These can be classified into two main groups: ground samples and reference LUC maps. However, in the validation exercises, other reference spatial data can also be used, such as the raw imagery used in the classification process or the soft maps obtained as a result.

The ground samples collected through field surveys provide highly accurate, detailed data. However, this information is very expensive to obtain and fieldwork is not an option when working with large study areas. This is why most reference LUC samples are obtained by photointerpretation or classification of satellite imagery. The data obtained via photointerpretation must be of higher quality that the data being validated. This usually involves careful interpretation of a set of samples using imagery with a higher spatial resolution than the images used to create the map. Another option is photointerpretation of the same imagery used to obtain the dataset, applying a different workflow and methods or techniques that guarantee better quality.

Those using these methods to obtain LUC samples for validation purposes should provide information about their accuracy or uncertainty. When obtaining reference data by field surveys or photointerpretation, users must take particular care when selecting the sampling strategy they will apply during the collection of this information, as it can have an important impact on the results of the validation exercise

and on their validity (see chapter "Visualization and Communication of LUC Data").

LUC maps can also be validated against other LUC maps. In these cases, the reference LUC map must have a higher spatial resolution and greater detail that the map being assessed. They must also be of proven quality, i.e. maps or datasets with verified accuracy and uncertainty. Although less precise, validation exercises carried out by comparing the evaluated map with other LUC maps are quick and very cheap, hence their popularity. This also allows a wider set of methods and techniques to be used compared to the possibilities offered by reference datasets other than maps.

Users can also validate their LUC maps against additional sources of information other than reference datasets, in order to characterize the maps in more detail and gain a clearer picture of their uncertainty. Such sources include raw imagery, which is often used in the classification process, or the soft maps obtained from it, which are used to assess the characteristics of the pixels that make up each class. Raw imagery can be used to evaluate the reflectance value for all the pixels belonging to a particular class and how close it is to the reference reflectance value used in the classification process. When available, users can also compare each category pixel with soft maps showing the percentage of each pixel belonging to each of the LUC categories under consideration. Similar insights into the accuracy of LUC maps can be obtained by comparing them with continuous LUC data (reference data), such as the Vegetation Continuous Fields (VCF) products.

If we focus on validation tools (Fig. 1), the agreement between the reference data/map (t_1) and the LUC map under evaluation (t_1)—the two maps should have the same date t_1—can be assessed using the **cross-tabulation matrix**[1] (see Sect. 1 in chapter "Basic and Multiple-Resolution Cross Tabulation to Validate Land Use Cover Maps"). This is also referred to in the literature as the confusion or error matrix, or as the contingency table. Cross tabulation is usually the first step in any validation exercise, as the raw matrix provides plenty of information regarding the spatial agreement between the LUC map being validated and the reference dataset.

In some cases, the level of agreement may vary at different levels of spatial detail. For example, when spatially aggregated and simplified, the LUC map being evaluated may show more agreement with the reference dataset. The choice of spatial resolution is therefore a source of uncertainty. To account for this uncertainty, we can cross-tabulate the assessed and reference datasets at **multiple spatial resolutions** (see Sect. 2 in chapter "Basic and Multiple-Resolution Cross Tabulation to Validate Land Use Cover Maps"), i.e. the original resolution and other coarser ones.

Different metrics are calculated from the confusion matrix (see chapters "Metrics Based on a Cross-Tabulation Matrix to Validate Land Use Cover Maps" and "Pontius Jr. Methods Based on a Cross-Tabulation Matrix to Validate Land Use Cover Maps"). These metrics summarize the agreement between reference and validated datasets in a single value and are therefore very easy to interpret. As a result, they have been widely used in LUC validation.

The most common metrics are the **accuracy assessment statistics** (see Sect. 5 in chapter "Metrics Based on a Cross-Tabulation Matrix to Validate Land Use Cover Maps") and the **Kappa Indices** (see Sect. 3 in chapter "Metrics Based on a Cross-Tabulation Matrix to Validate Land Use Cover Maps"). The accuracy assessment statistics are standard metrics that provide information about the similarity between two georeferenced data. They are obtained from the cross-tabulation matrix and enable the extraction of specific information contained in the matrix. They include, among others, the overall, producer's and user's accuracy metrics. They are usually supplied with the cross-tabulation matrix, providing extra information in addition to that provided by the matrix itself (e.g. category area adjusted by the error level, confidence intervals…).

Of all these metrics, the most commonly used in validation exercises is probably Overall accuracy. There has been great debate in the literature about the threshold above which the Overall accuracy of a map can be considered acceptable. The 85% threshold proposed by Anderson (1971) was the common reference for many years and continues to be applied by a lot of users nowadays (Wulder et al. 2006; Foody 2008). However, there is no specific accuracy threshold regarded as valid for all study cases and datasets. The acceptable level of accuracy will depend on the intended application of the dataset and the characteristics of the area being mapped. As regards different scales and spatial resolution, we cannot compare the accuracy of global or supra-national LUC maps with that of regional and local ones, which are not subject to the same level of simplification or abstraction as the global or supra-national maps.

The overall accuracy metric does not provide information about the accuracy at which each category on the LUC map is mapped. Important differences are often identified in terms of the relative accuracy of the different categories. Mixed LUC categories do not usually show the same accuracy as spectrally pure categories. At high levels of thematic detail, very similar LUC categories can be easily confused and will, therefore, have lower levels of accuracy. Users must take these differences at the category level into account and report the accuracy values for each category. The general approach for **agreement between maps at global and stratum level** may be useful to this end (see Sect. 4 in chapter "Metrics Based on a Cross-Tabulation Matrix to Validate Land Use Cover Maps"). Some authors talk specifically about Overall and Individual Spatial Agreement, proposing different metrics for these purposes (Yang et al. 2017; Islam et al. 2019) (see **Areal and spatial agreement metrics** in Sect. 2 in chapter "Metrics Based on a Cross-Tabulation Matrix to Validate Land Use Cover Maps").

It is also important to remember that the accuracy of a LUC map is not usually the same across the entire mapped area and considerable spatial variations are possible. The bigger the area being mapped, the more likely it is for there to be spatial differences in accuracy levels across the mapped area. The cross-tabulation matrix does not provide information about these spatial differences. When mapping large study areas made up of different, clearly distinguishable regions, each region can be validated independently, producing a specific cross-tabulation matrix in each case. The global analysis would cover the entire map, while specific areas of the map (e.g. a region, a municipality…) could also be analysed at the stratum level.

Overall Accuracy is highly correlated with the Kappa Index (Olofsson et al. 2014), which explains why both metrics provide similar information. One difference is that Kappa takes into account the agreement expected by chance,

---

[1] The methods/functions presented in the corresponding chapters in Part III of this book are highlighted in **bold**.

a factor that is not considered in Overall Accuracy. The Kappa Index (see Sect. 3 in chapter "Metrics Based on a Cross-Tabulation Matrix to Validate Land Use Cover Maps") has been criticized by a range of authors, who claim that it can sometimes be misleading (Pontius and Millones 2011; Olofsson et al. 2014). Moreover, standard indices such as overall, producer's and user's accuracy have the advantage that they can be interpreted as measures of the probability of encountering pixels, patches, etc. that have been allocated to the correct category (Stehman 1997).

The methods mentioned above do not employ fuzzy logic and, instead, apply a binary logic when calculating agreement, i.e. the two elements agree or don't agree. Partial agreements are not considered. However, there are some tools for calculating map agreement that incorporate fuzzy logic, such as the Fuzzy Kappa or the Fuzzy Kappa Simulation (Woodcock and Gopal 2000).

Other metrics, similar to Kappa, have also been proposed. Usually they aim to outperform Kappa and correct some of its associated problems. These include, among others, the F-Score (Pérez-Hoyos et al. 2020), Scott's pi statistic (Gwet 2002) and Krippendorff's α-coefficient (Kerr et al. 2015). These metrics are not widely used and they provide similar information to Kappa, which is why we do not recommend that they be used in a standard LUC validation exercise.

Extensive research by Pontius Jr. has given rise to other metrics based on the cross-tabulation matrix which can be used to validate a single LUC map against a reference map (see chapter "Pontius Jr. Methods Based on a Cross-Tabulation Matrix to Validate Land Use Cover Maps"). **Quantity & allocation disagreement** (see Sect. 3 in chapter "Pontius Jr. Methods Based on a Cross-Tabulation Matrix to Validate Land Use Cover Maps") (Pontius and Millones 2011) compares the agreement between maps regarding the proportions allocated to the different categories and regarding the way they are allocated, i.e. differences in the quantities allocated to each category and differences in their location. These metrics complement the cross-tabulation table, so enabling users to take full advantage of the information it provides. Quantity and Allocation disagreement is a very good method for validating a single map against a reference map (García-Álvarez and Camacho Olmedo 2017).

Users can also specifically assess the pattern of the map they want to validate to find out how much its pattern coincides with that of the reference map. Pattern agreement can be assessed using **Spatial metrics** (see Sect. 1 in chapter "Spatial Metrics to Validate Land Use Cover Maps") and the **Map Curves** method (see Sect. 1 in chapter "Advanced Pattern Analysis to Validate Land Use Cover Maps"). Spatial metrics allow us to characterize different aspects of the map's pattern in detail, such as its fragmentation, the proportion allocated to each category, the complexity of the patches… (Botequilha et al. 2006; Forman 1995). Initially

developed within the field of landscape ecology, these metrics are also widely used for characterizing the pattern of categorical maps. For its part, Mapcurves (Hargrove et al. 2006) provides a single value summarizing the pattern agreement between two maps. In both cases, we should always compare maps drawn at the same spatial and thematic resolution, as any changes in resolution would severely alter the pattern of the map, so rendering the comparison uninformative.

Geographic weighting methods (GWR) (see chapter "Geographically Weighted Methods to Validate Land Use Cover Maps") can also be used to study the spatial distribution of LUC accuracy measures. The overall, user's and producer's accuracy metrics mentioned above are derived from the cross-tabulation matrix and are therefore not spatial metrics, i.e. they provide overall information for the entire area, without assessing the spatial distribution of error and accuracy. The application of **Overall, user's and producer's accuracy metrics through GWR** (see Sect. 1 in chapter "Geographically Weighted Methods to Validate Land Use Cover Maps") can help the user to assess the suitability of the LUC data and to observe local variations in accuracy and error on the map (Comber 2013). In some cases, local assessments may be necessary because they can uncover possible clusters of errors in the LUC data. By adapting logistic Geographically Weighted Regression (GWR) (Brunsdon et al. 1996), the spatial variations in Boolean LUC (classified data) and fuzzy LUC (reference data) can be modelled, providing maps that show the distribution of the overall, user's and producer's accuracy metrics.

4 Validation of Land Use Cover Maps Series/Land Use Cover Changes

There is no common practice or set of methods for validating or evaluating the uncertainty of a LUC map series with two or more time points (t_0, t_1, t_2…). Most of the exercises for the validation of LUC data only refer to single LUC maps, without focusing specifically on the LUC change studied through a series of LUC maps.

One of the facets that users most demand from LUC data is the ability to study and display LUC changes over time. We therefore need methods and tools to assess the uncertainty of the changes that are measured from LUC maps. It is worth noting that the individual accuracy of two LUC maps involved in a post-classification comparison offers few clues as to the accuracy of change, because the relation between the errors in the two maps is unknown. As pointed out by Olofsson et al. (2013), even when both maps are highly accurate, it is possible that the change map accuracy will be low and the estimated change area strongly biased.

One of the main limitations when it comes to validating LUC changes and LUC map series is the lack of reference data. We could obtain reference datasets via photointerpretation or field surveys. However, it is difficult to guess where the LUC changes will take place, as they may happen at different places and with different intensities and patterns over space and time. In addition, there is a clear lack of LUC map series showing accurate, validated LUC change that could be used as reference data. Another option would be to validate the LUC changes against other types of reference data. This could be done for example by comparing the LUC changes measured over a time series of LUC maps against the difference in reflectance between two satellite images for the same time period. This is because when LUC change takes place, there is a significant change in the reflectance value registered by the satellite capturing the images.

Nevertheless, as commented earlier, the most common situation is that there are no reference datasets available. In these cases, the uncertainty of the LUC map series must be assessed by evaluating the consistency and the logic of the measured LUC change. The tools and techniques recommended here provide a great deal of information to the user. However, the final interpretation of the measured LUC change will be subjective, based on the user's expertise and understanding of the study area. In this situation, visual inspection can be very useful for quickly understanding many of the uncertainties in the time series of LUC maps that cannot be measured using quantitative metrics. This is why we recommend visual inspection as a first essential step prior to the validation of any LUC map or LUC modelling exercise.

Users must be aware that LUC change usually represents a very small portion of the mapped area. For a specific, not very large landscape, we would only expect a few features to change over a short period of time. In addition, the same area would not normally be expected to be affected by various successive changes. On the contrary, when an area changes, the new land use or cover tends to remain unchanged over time. In addition, there are some LUC transitions that make less sense than others. For example, one would not expect an artificial area to change to vegetation or agricultural land. These general assumptions may be adapted in line with the particular characteristics of the study area and also within the context of each element being analysed.

The same validation techniques reviewed above for single LUC maps (Sect. 3) can also be applied when comparing measured and reference changes or just for evaluating the consistency and logic of measured LUC change. However, some tools are specific to time series (Fig. 1).

The **cross-tabulation matrix** (see Sect. 1 in chapter "Basic and Multiple-Resolution Cross Tabulation to Validate Land Use Cover Maps") is the tool that provides most information about the change happening between two LUC maps. For a time series, we can compare each pair of LUC maps to find out the changes that take place at each date and the area they cover, for the map as a whole and at category level. We can summarize the main processes of change in our study area, such as, for example, the artificialization or deforestation rates for each time period. This gives us an overview of the change that has taken place over our map series and makes it easier to interpret some of the inconsistencies in measured change. Some authors also propose making a summary of all the transitions taking place, associating some of them with a default degree of uncertainty (Gómez et al. 2016; Hao and Gen-Suo 2014). For example, a transition from artificial surfaces to agricultural areas is not expected and could therefore be assigned a high degree of uncertainty.

Multi-resolution cross-tabulation (see Sect. 2 in chapter "Basic and Multiple-Resolution Cross Tabulation to Validate Land Use Cover Maps") offers a means of checking whether some of the errors, inconsistencies or uncertainties we detect at the original resolution are not detected at coarser resolutions. When this happens, the errors and inconsistencies probably arise due to the level of detail at which the dataset was created.

The cross-tabulation matrix is an excellent source of information, which we can easily summarize using other tools and metrics. As commented in Sect. 3, **Areal and spatial agreement metrics** (see Sect. 2 in chapter "Metrics Based on a Cross-Tabulation Matrix to Validate Land Use Cover Maps") and **Kappa Indices** (see Sect. 3 in chapter "Metrics Based on a Cross-Tabulation Matrix to Validate Land Use Cover Maps") are used to assess the agreement between two maps. Despite their limitations, these metrics can be used to chart, in a generic way, the persistence or changes between two dates. If two maps in a series undergo the normal rate of change that we associate with any landscape, the differences between them should be slight, which means that the Kappa and agreement metrics should reflect high levels of coincidence between the maps being compared.

The **Agreement between maps at global and stratum level** (see Sect. 4 in chapter "Metrics Based on a Cross-Tabulation Matrix to Validate Land Use Cover Maps") analysis could provide additional specific information about the agreement in a time series of LUC maps at whole map level, or for a given stratum, i.e. a smaller area or a specific LUC category. **Accuracy assessment statistics** can also be calculated for a LUC map series, either globally (see Sect. 5 in chapter "Metrics Based on a Cross-Tabulation Matrix to Validate Land Use Cover Maps") or locally (Sect. 1 in chapter "Geographically Weighted Methods to Validate Land Use Cover Maps"). For example, when the LUC map series is obtained using a base map that is progressively updated, the first stage is to validate the base map of the series using the same procedure described earlier for validating single LUC maps. Once this has been done, we can validate the changes against a reference dataset of changes through cross-tabulation, obtaining from the resulting table the overall, producer's and user's accuracy metrics. Pouliot and Latifovic (2013) coined the term Update Accuracy (UA) to refer to the accuracy of the measured changes. They refer to the accuracy of the base map as the Base Map Accuracy (BMA). They also propose a metric called Time Series Accuracy (TSA) as the mean accuracy of all the LUC maps that make up the series, individually validated through a specific reference LUC dataset for each case.

Change statistics (see Sect. 1 in chapter "Metrics Based on a Cross-Tabulation Matrix to Validate Land Use Cover Maps") (FAO 1995; Puyravaud 2003) are widely used to assess land use and cover changes. These indices measure, for example, relative change or rates of change and allow us to compare the change between regions of different sizes. These indices can be complemented by the change matrix obtained from cross-tabulation. They are calculated from the map series itself, rather than from the cross-tabulation matrix.

Robert Gilmore Pontius Jr. has made major contributions to the family of validation techniques based on the cross-tabulation matrix (chapter "Pontius Jr. Methods Based on a Cross-Tabulation Matrix to Validate Land Use Cover Maps"). The **LUCC budget** (see Sect. 2 in chapter "Pontius Jr. Methods Based on a Cross-Tabulation Matrix to Validate Land Use Cover Maps") (Pontius et al. 2004) provides more information about the changes that take place between pairs of maps. It differentiates between net and gross changes, therefore, allowing us to gain a clearer understanding of the transitions and swaps between categories, providing useful additional information to identify category confusion over time. Category confusion arises when the same area is mapped as different, albeit similar, categories at different points in time, when no change has actually taken place.

Quantity and allocation disagreement (see Sect. 3 in chapter "Pontius Jr. Methods Based on a Cross-Tabulation Matrix to Validate Land Use Cover Maps") show, at overall and category level, differences between pairs of maps in terms of category proportions due to the different allocation of the categories. Few changes are expected in a time series of maps. This means that quantity and allocation disagreement should be low and should centre on the most dynamic categories.

The number of **incidents and states** (see Sect. 5 in chapter "Pontius Jr. Methods Based on a Cross-Tabulation Matrix to Validate Land Use Cover Maps") (Pontius et al. 2017) also provides information that can help identify errors. This technique allows us to identify those areas that are more dynamic than expected, i.e. those that change a lot over a short period of time, always transitioning between the same categories. **Intensity analysis** (see Sect. 6 in chapter "Pontius Jr. Methods Based on a Cross-Tabulation Matrix to Validate Land Use Cover Maps") (Aldwaik and Pontius 2012) compares the rates of LUC change between periods, categories, and transitions. Based on the assumption that a category or area is expected to change at similar levels of intensity over time, this analysis enables us to identify those categories that do not comply with this assumption. The **Flow matrix** (see Sect. 7 in chapter "Pontius Jr. Methods Based on a Cross-Tabulation Matrix to Validate Land Use Cover Maps") (Runfola and Pontius 2013) measures the instability of annual land use change over different time intervals, so as to identify anomalies relative to the amount of change over the whole time series.

Spatial metrics (see Sect. 1 in chapter "") and **Map curves** (see Sect. 1 in chapter "Advanced Pattern Analysis to Validate Land Use Cover Maps") enable us to characterize the pattern of each LUC map in the series. We do not expect the pattern of the map to vary significantly over the time period being analysed. This means that only smooth changes should be observed when comparing the spatial metrics for each of the periods analysed.

Spatial metrics that specifically measure the areas that change between pairs of maps may also be useful. In the case of a pair of maps or a time series, the detection of **change on pattern borders** (see Sect. 2 in chapter "Advanced Pattern Analysis to Validate Land Use Cover Maps") (Paegelow et al. 2014) enables us to identify data errors resulting from different data sources, different classifiers or spectral responses. For example, the noise or error shown by a time series of LUC maps often arises due to border areas between categories being interpreted differently each year. Users can specifically analyse the changes that take place in these border patches, often elongated and less than 1 or 2 pixels wide, so helping them to identify potential errors. These patches can also be characterized through the calculation of spatial metrics.

5 Validation of Land Use Cover Change Modelling Exercises

Validating a LUCC modelling exercise is a complex task. In this case, we are not validating a single LUC map or a series of LUC maps, but a model application made up of multiple inputs, which interact to deliver new results. When validating LUCC modelling exercises, users tend to focus exclusively on the validation of the model's hard maps, i.e. maps with a categorical legend similar to the input LUC maps (Camacho Olmedo et al. 2018). These hard maps are the main final output of any modelling exercise, but not the only one. To properly validate a LUCC modelling exercise we should focus not only on the scenario generated by the model, but also on the other outputs and inputs.

Given the nature of this book, we will be dealing exclusively with the validation of LUC maps associated with LUCC modelling exercises: input LUC maps, output soft LUC maps and output hard LUC maps. Users must bear in mind that other sources of data can be used in LUCC modelling exercises and can be validated via complementary methods.

Modellers can begin a modelling exercise by evaluating the uncertainty of the input LUC maps used in the model and their changes according to the guidelines set out in Sects. 3 and 4 above. This is because the quality of the input LUC maps can have a significant effect on the performance of the model. When setting up LUCC models, it is essential to understand the changes that take place in the set of input and reference maps. An assessment of the uncertainty of these LUC changes is therefore vital for determining and characterizing the uncertainty of the LUCC modelling exercise.

In the following subsections, we present the validation tools for output LUC maps, i.e. the products obtained by the model, differentiating between soft and hard LUC maps.

5.1 Soft LUC Maps

Soft LUC maps, also referred to as suitability, change potential or change probability maps, are produced by the model to express the propensity to change over space, that is, the potential of each pixel to become a specific category in the future (Camacho Olmedo et al. 2018). Modellers can assess the internal behaviour and coherence of the model they are building by comparing the model's soft maps with the maps of simulated changes. They can also find out to what extent the changes simulated by the model coincide with the areas of highest potential in the respective maps for each modelled category. In addition, they can compare the soft maps obtained by different models and assess their level of agreement.

Soft LUC maps are usually validated against a reference map of changes ($t_0 - t_1$), and there are various methods for carrying out this analysis (see chapter "Validation of Soft Maps Produced by a Land Use Cover Change Model"). The Pearson and Spearman **correlation** (see Sect. 1 in chapter "Validation of Soft Maps Produced by a Land Use Cover Change Model") is appropriate for a quick assessment of the soft map, by computing it against the map of observed change (Bonham-Carter 1994; Camacho Olmedo et al. 2013). The **Receiver Operating Characteristic (ROC)** (see Sect. 2 in chapter "Validation of Soft Maps Produced by a Land Use Cover Change Model") (Pontius and Parmentier 2014) is used to assess soft maps by comparing them with the observed binary event map. A highly predictive model

produces a soft map in which the highly ranked values coincide with the actual event. In soft maps, the **Difference in Potential** (DiP) proposed by Eastman et al. (2005) (see Sect. 3 in in chapter "Validation of Soft Maps Produced by a Land Use Cover Change Model") compares the relative weight of values allocated to changed areas, in other words the difference between the mean potential in the areas of change and the mean potential in the areas of no change (Pérez-Vega et al. 2012).

In short, the previous three methods evaluate the relationship between the observed changed area and the soft LUC map, assuming that a good model output allocates the highest change probability values to the areas that did actually change, and the lowest change probability values to the areas that did not change. Unlike the previous methods, the **total uncertainty, quantity uncertainty and allocation uncertainty** indices (see Sect. 4 in chapter "Validation of Soft Maps Produced by a Land Use Cover Change Model") (Krüger and Lakes 2016) are not calculated against a reference map of changes, and instead estimate uncertainty by adding together misses and false alarms based on soft prediction score levels.

In addition to these specific indices for soft LUC maps, validation can also be conducted after reclassifying the original soft maps, so transforming continuous, ranked maps (soft) into categorical maps (hard) (see Sects. 1 and 2 in chapter "Basic and Multiple-Resolution Cross Tabulation to Validate Land Use Cover Maps"). This preliminary step enables most of the validation tools presented in this chapter to be applied for this purpose.

5.2 Hard LUC Maps

The second output obtained by the model is the hard LUC map. Also known as prospective LUC maps, these are simulated LUC maps with an identical categorical legend to the input LUC maps (Camacho Olmedo et al. 2018). The hard maps must be validated in order to understand more about the behaviour of the model and how well it simulates changes. These maps provide a clearer picture of the characteristics of the simulated changes and how they resemble our reference data.

5.2.1 Single LUC Maps

The simulation (T_1) can only be validated against a single LUC map (t_1) if both maps correspond to the same year. This will also enable users to apply the panoply of tools presented in Sect. 3. The **Accuracy assessment statistics,** computed either globally (see Sect. 5 in chapter "Metrics Based on a Cross-Tabulation Matrix to Validate Land Use Cover Maps") or locally (see Sect. 1 in chapter "Geographically Weighted Methods to Validate Land Use Cover Maps") could also be applied to validate the simulation against other LUC data such as ground points.

In addition to this generic list of tools, some metrics are specifically used for validating the hard LUC maps obtained from LUCCM exercises. **Allocation distance error** (see Sect. 3 in chapter "Advanced Pattern Analysis to Validate Land Use Cover Maps") (Paegelow et al. 2014) measures the relevance of simulation errors by computing the distance between a false positive (commission) and the closest object in the reference map, considering the minimum distance or the centroids of the area in question.

5.2.2 LUC Maps Series/LUC Changes

The most appropriate, most complete validation procedure for hard maps must include three different maps: the simulation (T_1), a reference LUC map for the same year (t_1) and the base map over which the simulation is executed (t_0). In other words, if our modelling exercise starts in the year 2010, we will need a base map for 2010 to establish the initial landscape on which the simulation will be calculated. Then, if we run a simulation for the year 2020, we will also need a reference map for 2020 in order to be able to compare how well our model simulates change. By comparing the simulation and the reference map we can understand to what extent the simulation matches the reference data. The changes that take place on the reference map and the simulation can be extracted by comparing them with the base map. The changes extracted from the two maps can then be compared so as to find out how well the simulated changes agree with the changes that took place on the reference maps.

There are many tools for validating and understanding the errors and uncertainties of simulated changes. In fact, all the methods and strategies explained in Sect. 4 can be applied in LUCC modelling. In this case, however, the main purpose is to achieve the best possible fit between the results of the model and the reference data.

The majority of metrics are obtained from the **cross-tabulation matrix** (see Sect. 1 in chapter "Basic and Multiple-Resolution Cross Tabulation to Validate Land Use Cover Maps"). The cross-tabulation matrix offers a detailed picture of the changes that were simulated (by cross-tabulating the simulation with the base map), the changes we used as a reference (by cross-tabulating the reference map with the base map) and the agreement and disagreement between the simulation and the reference map (by cross-tabulating the simulation with the reference map). The cross-tabulation matrix can also be used to summarize simulated and reference change in a series covering the main processes of change (artificialization, deforestation…). This enables us to quickly identify the changes that have taken place in our simulation and to spot potential change patterns that do not make sense.

Cross tabulation can be carried out at **multiple resolutions** (see Sect. 2 in chapter "Basic and Multiple-Resolution Cross Tabulation to Validate Land Use Cover Maps") (the original and coarser ones), to find out at which resolution there is the greatest agreement. Sometimes, the simulation and the reference landscape do not agree on the details but show high consistency at coarser scales. This implies that the model is unable to simulate the precise location of the changes, but it does simulate the main patterns of change correctly.

Different metrics have been proposed for summarizing the agreement between the simulation and the reference maps that the cross-tabulation matrix shows in raw (see chapter "Metrics Based on a Cross-Tabulation Matrix to Validate Land Use Cover Maps"). The **Areal and spatial agreement metrics** (see Sect. 2 in chapter "Metrics Based on a Cross-Tabulation Matrix to Validate Land Use Cover Maps") could be applied to summarize the agreement between two maps of changes, the simulated and the reference change maps, overall or per category. **Kappa** (see Sect. 3 in chapter "Metrics Based on a Cross-Tabulation Matrix to Validate Land Use Cover Maps") also summarizes the overall agreement between two maps. However, it has been widely criticized because it assesses the similarity between the simulation and the reference map, but does not distinguish between the areas that change between the two dates and those that do not. Therefore, in maps that simulate permanence correctly, the Kappa metric will be high. Accordingly, we only recommend Kappa for assessing how well permanence is simulated, and it should not be used for a detailed assessment of the accuracy of simulated changes. The Kappa Simulation proposed by Van Vliet et al. (2011) takes the standard Kappa flaws regarding LUCC modelling into account. It focuses on the agreement between the changes in the simulation and the changes in the reference map with regard to the initial map used as a base for the simulation.

The **Agreement between maps at global and stratum level** (see Sect. 4 in chapter "Metrics Based on a Cross-Tabulation Matrix to Validate Land Use Cover Maps") analysis can assess for a specific LUC transition, for example, whether the agreement between an observed (reference map) and a simulated transition varies or not for several distance classes resulting from a driver (e.g. distance to roads). Other metrics, such as **change statistics** (see Sect. 1 in chapter "Metrics Based on a Cross-Tabulation Matrix to Validate Land Use Cover Maps"), are widely used for characterizing the simulated changes, providing extra information that may be helpful for their validation.

Pontius proposes several metrics for validating simulated change (see chapter "Pontius Jr. Methods Based on a Cross Tabulation Matrix to Validate Land Use Cover Maps"). Some of them can also be used to validate time series of

LUC maps and were therefore described in Sect. 2. The **LUCC budget** (see Sect. 2 in chapter "Pontius Jr. Methods Based on a Cross Tabulation Matrix to Validate Land Use Cover Maps") technique helps users to understand the changes that take place between the simulation and the base map and between the reference and the base maps. This tool calculates the gross and net changes, overall and per category, as well as the category swaps, in both the simulated and the reference landscapes. This enables us to assess in detail whether the changes we simulated are similar to the changes that take place on the reference maps and follow the same trends.

Quantity & allocation disagreement (see Sect. 3 in chapter "Pontius Jr. Methods Based on a Cross Tabulation Matrix to Validate Land Use Cover Maps") differentiates, at an overall level and per category, between the (dis)agreement between two maps in terms of the proportion of the map occupied by each category (quantities) and the (dis) agreement due to the allocation of the categories in the same/different places on the map (allocation). It is therefore useful for assessing how much of the disagreement is due to the way the model simulates quantities and how much is due to its incorrect allocation of categories. By making the analysis at the category level, it also allows us to assess where (i.e. in which categories) the errors and uncertainties arise.

If a chronological series of simulations (more than two-time points) is available, **Incidents and States** (see Sect. 5 in chapter "Pontius Jr. Methods Based on a Cross Tabulation Matrix to Validate Land Use Cover Maps" may also be employed. This metric helps identify pixels that follow illogical transition patterns, with changes at successive time intervals between the same pair of categories (e.g. from agricultural to urban fabric and then back to agricultural).

Intensity analysis (see Sect. 6 in chapter "Pontius Jr. Methods Based on a Cross Tabulation Matrix to Validate Land Use Cover Maps") compares the different intensities of change per category in simulations and reference maps over at least three points in time. In this way we can assess whether our model correctly simulated the change trend displayed by the reference data. The **flow matrix** (see Sect. 7 in chapter "Pontius Jr. Methods Based on a Cross Tabulation Matrix to Validate Land Use Cover Maps") could also be applied to validate simulated changes in a generic way, assessing the stability and instability of the real and simulated changes over time.

The **Null model** (Pontius and Malanson 2005) (see Sect. 1 in chapter "Pontius Jr. Methods Based on a Cross Tabulation Matrix to Validate Land Use Cover Maps") compares the agreement between the base map for the simulation and the reference map versus the agreement between the simulation and the reference map. If the former

is higher than the latter, our modelling exercise could be judged to have performed poorly, in that the accuracy of the obtained simulation is lower than that for a reference map in which no change takes place. This assertion may be clarified by using other validation tools to obtain a clearer understanding of the logic and pattern of the simulated change. The null model is also a valuable tool for evaluating how well the model simulates permanence.

The **Figure of Merit** (Pontius et al. 2008) **and complementary Producer's and User's accuracy**, (see Sect. 4 in chapter "Pontius Jr. Methods Based on a Cross Tabulation Matrix to Validate Land Use Cover Maps") also measure the agreement between simulated changes and changes in the reference map. The Figure of Merit technique is recommended when trying to assess the model's ability to correctly simulate change. The different components of the Figure of Merit can be used to discover whether the model estimates more or less change than the reference map. It is also highly recommended for evaluating the congruence of model outputs and model robustness. This is a form of validation that evaluates the agreement between simulations obtained using different models or using the same model parametrized in different ways (Paegelow et al. 2014; Camacho Olmedo et al. 2015).

None of the above tools assesses the accuracy of the pattern of LUC change in the simulation. This aspect is important because even if the quantities simulated are wrong and the categories are not allocated in the same positions as in the reference maps, the pattern of LUC change may have been simulated correctly. Pattern can be validated using **Spatial metrics** (see Sect. 1 in chapter "Spatial Metrics to Validate Land Use Cover Maps") and the **Map Curves** (see Sect. 1 in chapter "Advanced Pattern Analysis to Validate Land Use Cover Maps") method, which compare the pattern of the simulation with the pattern of the reference landscape.

Spatial metrics characterize many different elements of the landscape: fragmentation, shape complexity, category proportions, diversity…. They can be calculated specifically for the simulated and reference changes, so allowing users to identify the specific pattern characteristics of the features that changed during the simulation period. In this way we can understand the size and shape of the simulated changes, inferring from this information how logical or uncertain they may be.

The MapCurves method gives a summary figure for the pattern agreement between two maps, and is therefore much easier to interpret. However, it does not provide all the complex detail that can be revealed by applying the different spatial metrics.

We can also analyse the changes that take place on the borders of existing patches and the changes that result in the appearance of new patches. This distinction may be useful for identifying errors or inconsistencies. The detection of

change on pattern borders (see Sect. 2 in chapter "Advanced Pattern Analysis to Validate Land Use Cover Maps") enables us to evaluate and identify errors in the simulations, which may be due to different parameters being applied in the model allocation procedure, such as, for example, the use of a contiguity filter. The **Allocation distance error** (see Sect. 3 in chapter "Advanced Pattern Analysis to Validate Land Use Cover Maps") calculates the distance between wrongly simulated patches and reference patches, so as to gain a better picture of how well the patches are simulated. In this sense, a model that wrongly allocates change close to areas that actually change on the ground would be considered to have performed better than a model that allocates them further away.

References

Aldwaik SZ, Pontius RG (2012) Intensity analysis to unify measurements of size and stationarity of land changes by interval, category and transition. Landsc Urban Plan 106:103–114

Anderson JR (1971) Land-use classification schemes used in selected recent geographic applications of remote sensing. Photogramm Eng Remote Sens 37:379–387

Bonham-Carter GF (1994) Tools for map analysis: map pairs. In: Bonham-Carter GF (ed) Geographic information systems for geoscientists, pp 221–266. Pergamon. ISBN 9780080418674, https://doi.org/10.1016/B978-0-08-041867-4.50013-8

Botequilha Leitao A, Miller J, Ahern J, McGarigal K (2006) Measuring landscapes: a planner's handbook. Island Press, Washington, Covelo, London

Brunsdon C, Fotheringham AS, Charlton ME (1996) Geographically weighted regression: a method for exploring spatial nonstationarity. Geogr Anal 28(4):281–298. https://doi.org/10.1111/j.1538-4632.1996.tb00936.x

Camacho Olmedo MT, Paegelow M, Mas J-F (2013) Interest in intermediate soft-classified maps in land change model validation: suitability versus transition potential. Int J Geogr Inf Sci 27 (12):2343–2361

Camacho Olmedo MT, Pontius RG Jr, Paegelow M, Mas JF (2015) Comparison of simulation models in terms of quantity and allocation of land change. Environ Model Softw 69(2015):214–221. https://doi.org/10.1016/j.envsoft.2015.03.003

Camacho Olmedo MT., Mas JF, Paegelow M (2018) The Simulation Stage in LUCC modeling. In: Camacho Olmedo M, Paegelow M, Mas JF, Escobar F (eds) Geomatic approaches for modeling land change scenarios. Lecture Notes in Geoinformation and Cartography. Springer, Cham, pp 27–51. Publisher Name Springer, Cham Print ISBN 978-3-319-60800-6 Online ISBN 978-3-319-60801-3 eBook Packages Earth and Environmental Science. https://doi.org/10.1007/978-3-319-60801-3_3

Comber AJ (2013) Geographically weighted methods for estimating local surfaces of overall, user and producer accuracies. Remote Sens Lett 4(4):373–380. https://doi.org/10.1080/2150704X.2012.736694

Eastman JR, Van Fossen ME, Solarzano LA (2005) Transition potential modelling for land cover change. In: Maguire D, Goodchild M, Batty M (eds), GIS, Spatial analysis and modeling. ESRI Press, Redlands, California

FAO (1995) Forest resources assessment 1990. Global synthesis. FAO, Rome

Foody GM (2002) Status of land cover classification accuracy assessment. Remote Sensing Environ 80(1):185–201. https://doi.org/10.1016/S0034-4257(01)00295-4

Foody GM (2008) Harshness in image classification accuracy assessment. Int J Remote Sens 29(11):3137–3158

Forman RTT (1995) Land mosaics: the ecology of landscapes and regions. Cambridge University Press, Cambridge, United Kingdom

García-Álvarez D, Camacho Olmedo MT (2017) Changes in the methodology used in the production of the Spanish CORINE: Uncertainty analysis of the new maps. Int J Appl Earth Obs Geoinf 63:55–67. https://doi.org/10.1016/j.jag.2017.07.001

Gwet K (2002) Kappa statistic is not satisfactory for assessing the extent of agreement between raters. Series: statistical methods for inter-rater reliability assessment

Gómez C, White JC, Wulder MA (2016) Optical remotely sensed time series data for land cover classification: A review. ISPRS J Photogramm Remote Sens 116:55–72. https://doi.org/10.1016/j.isprsjprs.2016.03.008

Hao G, Gen-Suo J (2014) Assessing MODIS land cover products over China with probability of interannual change. Atmosph Ocean Sci Lett 7(6):564–570. https://doi.org/10.1080/16742834.2014.11447225

Hargrove WW, Hoffman FM, Hessburg PF (2006) Mapcurves: a quantitative method for comparing categorical maps. J Geogr Syst 8:187–208. https://doi.org/10.1007/s10109-006-0025-x

Islam S, Zhang M, Yang H, Ma M (2019) Assessing inconsistency in global land cover products and synthesis of studies on land use and land cover dynamics during 2001 to 2017 in the southeastern region of Bangladesh. J Appl Remote Sens 13(04):1. https://doi.org/10.1117/1.JRS.13.048501

Kerr GHG, Fischer C, Reulke R (2015) Reliability assessment for remote sensing data: beyond Cohen's kappa. In: International geoscience and remote sensing symposium (IGARSS), pp 4995–4998. https://doi.org/10.1109/IGARSS.2015.7326954

Krüger C, Lakes T (2016) Revealing uncertainties in land change modeling using probabilities. Trans GIS 20(4):526–546. https://doi.org/10.1111/tgis.12161

Olofsson P, Foody GM, Stehman SV, Woodcock CE (2013) Making better use of accuracy data in land change studies: estimating accuracy and area and quantifying uncertainty using stratified estimation. Remote Sens Environ 129(2013):122–131. https://doi.org/10.1016/j.rse.2012.10.031

Olofsson P, Foody GM, Herold M et al (2014) Good practices for estimating area and assessing accuracy of land change. Remote Sens Environ 148:42–57. https://doi.org/10.1016/j.rse.2014.02.015

Paegelow M, Camacho Olmedo MT, Mas JF and Houet T (2014) Benchmarking of LUCC modelling tools by various validation techniques and error analysis. Cybergeo Eur J Geogr [En ligne] Systèmes, Modélisation, Géostatistiques, document 701, mis en ligne le 22 décembre 2014. ISSN: 1278-3366. CNRS-UMR Géographie-cités 8504. https://doi.org/10.4000/cybergeo.26610

Paegelow M, Camacho Olmedo MT, Mas JF (2018) Techniques for the validation of LUCC modeling outputs. In: Camacho Olmedo M, Paegelow M, Mas JF, Escobar F (eds) Geomatic approaches for modeling land change scenarios. Lecture Notes in Geoinformation and Cartography. Springer, Cham, pp 53–80. Publisher Name Springer, Cham Print ISBN 978-3-319-60800-6 Online ISBN 978-3-319-60801-3 eBook Packages Earth and Environmental Science

Pérez-Hoyos A, Udías A, Rembold F (2020) Integrating multiple land cover maps through a multi-criteria analysis to improve agricultural monitoring in Africa. Int J Appl Earth ObsGeoinf 88:102064. https://doi.org/10.1016/j.jag.2020.102064

Pérez-Vega A, Mas JF, Ligmann-Zielinska A (2012) Comparing two approaches to land use/cover change modeling and their implications for the assessment of biodiversity loss in a deciduous tropical forest. Environ Model Softw 29(1):11–23

Pontius RG Jr, Malanson J (2005) Comparison of the structure and accuracy of two land change models. Int J Geogr Inf Sci 19:243–265. https://doi.org/10.1080/13658810410001713434

Pontius RG Jr, Millones M (2011) Death to Kappa: birth of quantity disagreement and allocation disagreement for accuracy assessment. Int J Remote Sens 32:4407–4429. https://doi.org/10.1080/01431161.2011.552923

Pontius RG Jr, Parmentier B (2014) Recommendations for using the relative operating characteristic (ROC). Landscape Ecol 29(3):367–382

Pontius RG Jr, Shusas E, McEachern M (2004) Detecting important categorical land changes while accounting for persistence. Agr Ecosyst Environ 101:251–268

Pontius Jr RG, Boersma W, Castella JC, Clarke K, de Nijs T, Dietzel C, Duan Z, Fotsing E, Goldstein N, Kok K, Koomen E, Lippitt CD, McConnell W, MohdSood A, Pijanowski B, Pithadia S, Sweeney S, Trung TN, Veldkamp AT, Verburg PH (2008) Comparing input, output, and validation maps for several models of land change. Ann Reg Sci 42(1):11e47

Pontius RG Jr, Krithivasan R, Sauls L et al (2017) Methods to summarize change among land categories across time intervals. J Land Use Sci 12:218–230. https://doi.org/10.1080/1747423X.2017.1338768

Pouliot D, Latifovic R (2013) Accuracy assessment of annual land cover time series derived from change-based updating. In: Proceedings of the 7th International workshop on the analysis of multi-temporal remote sensing images: "our dynamic environment" (MultiTemp 2013), pp 1–3. https://doi.org/10.1109/Multi-Temp.2013.6866005

Puyravaud J-P (2003) Standardizing the calculation of the annual rate of deforestation. For Ecol Manage 177(1–3):593–596

Runfola DSM, Pontius RG (2013) Measuring the temporal instability of land change using the Flow matrix. Int J Geogr Inf Sci 27:1696–1716. https://doi.org/10.1080/13658816.2013.792344

Stehman SV, Czaplewski RL (1998) Design and analysis for thematic map accuracy assessment: fundamental principles. Remote Sens Environ 64:331–344

Stehman SV (1997) Selecting and interpreting measures of thematic classification accuracy. Remote Sens Environ 62(1):77–89. https://doi.org/10.1016/S0034-4257(97)00083-7

Stehman SV (1999) Basic probability sampling designs for thematic map accuracy assessment. Int J Remote Sens 20(12):2423–2441. https://doi.org/10.1080/014311699212100

Strahler AH, Boschetti L, Foody GM, Friedl MA, Hansen MC, Herold M, Mayaux P, Morisette JT, Stehman SV, Woodcock CE (2006) Global land cover validation: recommendations for evaluation and accuracy assessment of global land cover maps. Office for official publications of the European communities. GOFC-GOLD Report No 25. Luxemburg

Van Vliet J, Bregt AK, Hagen-Zanker A (2011) Revisiting Kappa to account for change in the accuracy assessment of land-use change models. Ecol Model 222(8):1367–1375. https://doi.org/10.1016/j.ecolmodel.2011.01.017

Woodcock CE, Gopal S (2000) Fuzzy set theory and thematic maps: accuracy assessment and area estimation. Int J Geogr Inf Sci 14(2):153–172. https://doi.org/10.1080/136588100240895

Wulder MA, Franklin SE, White JC, Linke J, Magnussen S (2006) An accuracy assessment framework for large-area land cover classification products derived from medium-resolution satellite data. Int J Remote Sens 27(4):663–683. https://doi.org/10.1080/01431160500185284

Yang Y, Xiao P, Feng X, Li H (2017) Accuracy assessment of seven global land cover datasets over China. ISPRS J Photogramm Remote Sens 125:156–173. https://doi.org/10.1016/j.isprsjprs.2017.01.016

Land Use Cover Datasets: A Review

David García-Álvarez and Sabina Florina Nanu

Abstract

This chapter presents a review of Land Use Cover (LUC) datasets at global and supranational scales. To this end, we differentiate between LUC maps (Sect. 3) and reference LUC datasets (Sect. 4). The former map how different land uses or covers are distributed across the Earth's surface. The latter provides a sample of LUC data for specific points on Earth and are normally used in LUC mapping and modelling calibration and validation exercises. We also include a brief presentation of the main producers of LUC datasets (Sect. 2). The LUC maps reviewed here are classified according to different criteria. First, we differentiate between general LUC maps (Sect. 3.2), which provide information about all land uses and covers on Earth, and thematic LUC maps (Sect. 3.3), which focus on the mapping of a specific land use or cover. Second, we classify general and thematic LUC maps according to their extent, distinguishing between global and supra-national LUC maps. The general maps are classified according to the continent for which they provide information, either fully or partially, while the thematic maps are classified according to the type of land use or cover they focus on. Most of the datasets reviewed in this chapter are characterized in detail in Part IV of this book, to which this chapter acts as an introduction. This chapter includes a series of tables with all the datasets, indicating those for which a detailed description is provided in Part IV.

Keywords

Land Use • Land Cover • General maps • Thematic maps • Reference datasets

1 Introduction

Nowadays, there are many sources of Land Use Cover (LUC) data. The availability of LUC data has been increasing since the end of the last century, in line with the development of remote sensing techniques and easier access to aerial and satellite imagery. LUC data is available at all spatial scales, from local to global. Access to spatial information, including LUC datasets, has also improved in the last decade with the development of the open access culture.

Most of the LUC data being produced today refers to LUC maps, which are either single, one-off maps or form part of a time series. These maps provide layers of spatial data with LUC information for each part of the area being mapped at one (single maps) or several points in time (series of maps). Other spatial sources of LUC information include reference datasets used to validate LUC maps or train remote sensing classifiers. Although datasets of this kind have been produced since the beginning of the satellite remote sensing era, they have only recently become widely available for general purposes.

In this chapter, we review the main producers of LUC maps and the most relevant LUC datasets currently available —both LUC maps and data packages with reference data. Although this aspires to be a comprehensive review, some LUC products may be missing. We focus on the datasets that are available for download and can be used in practice. When relevant, we also mention others that are currently unavailable for download.

Many older LUC maps are not included, because they were drawn at very coarse resolution using old-fashioned production methods and therefore cannot meet the demands of modern users. Because of the scope and extent of the

D. García-Álvarez (✉)
Departamento de Geología, Geografía y Medio Ambiente, Universidad de Alcalá, Alcalá de Henares, Spain
e-mail: David.garcia@uah.es

Sabina Florina Nanu
Departamento de Análisis Geográfico Regional y Geografía Física, Universidad de Granada, Granada, Spain

book, we focus exclusively on datasets at global and supra-national levels. A detailed description of the approach followed when carrying out this review appears in chapter "About This Book" of this book.

The most important datasets reviewed in this chapter are described in detail in Part IV of this book (chapters "Global General Land Use Cover Datasets with a Single Date "–"Supra-national Thematic Land Use Cover Datasets"), where users can find a detailed description of each dataset, including classification schemes, production methods and download options.

2   The Producers of LUC Data

We have classified LUC data producers into four main groups (Fig. 1): (i) Individual users and small actors; (ii) Research projects; (iii) Governmental and other organizations; and (iv) citizens producing LUC information through Volunteering Geographic Information (VGI) initiatives. The type of LUC data produced by each group varies.

At local and detailed scales, many organizations and users create their own LUC datasets. The fact that they have easy access to aerial/satellite imagery and to software for processing, photointerpreting and classifying these images has facilitated this process. This allows users to obtain very specific datasets that match their particular requirements. The datasets created for small projects and for specific purposes are not usually disseminated and remain the property of the communities or users that produce them. When these datasets are made available, they are often provided without the necessary technical information and general metadata.

At regional, national, supra-national and global scales, an increasing number of LUC databases are being produced for a broad range of users. Often these databases are specially designed for specific communities, such as the climate change research community. In other cases, they provide more general LUC information for a wide range of research fields and as support for policy decisions.

There are two main producers of LUC datasets. Firstly, nationally or internationally funded research projects, which

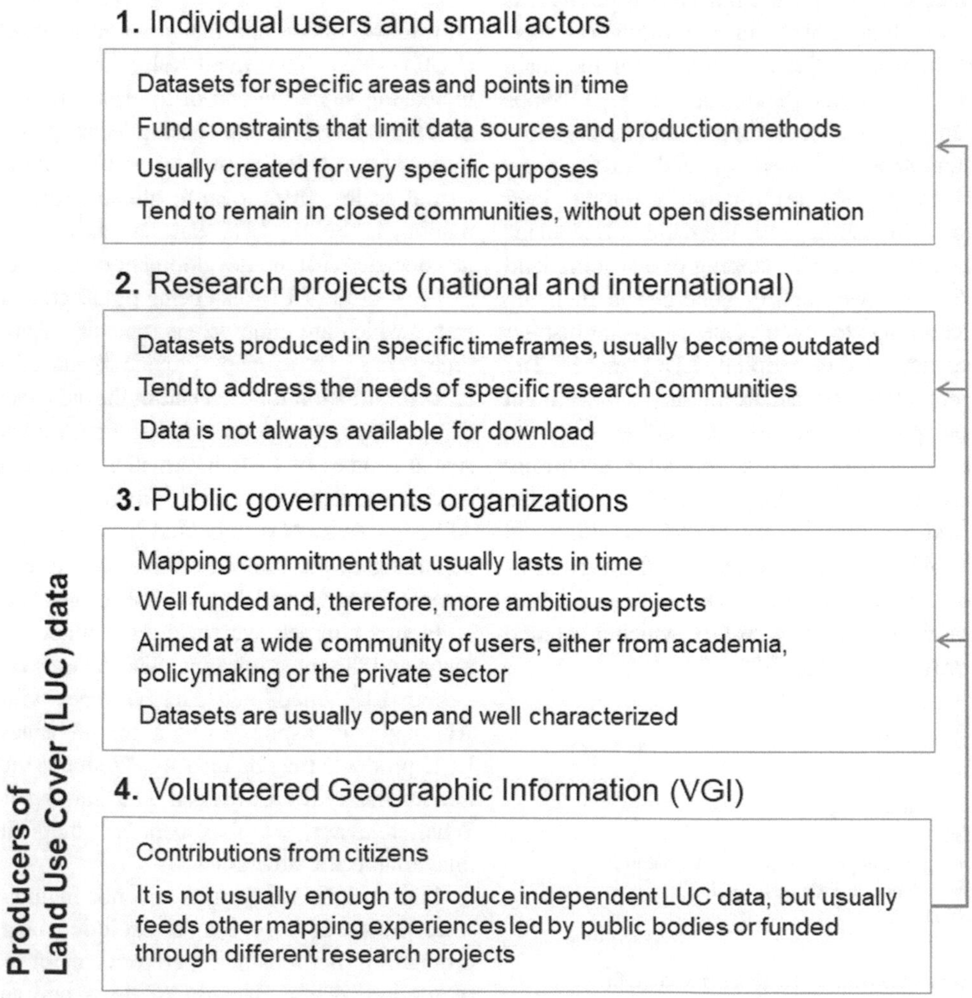

Producers of Land Use Cover (LUC) data

1. Individual users and small actors

Datasets for specific areas and points in time

Fund constraints that limit data sources and production methods

Usually created for very specific purposes

Tend to remain in closed communities, without open dissemination

2. Research projects (national and international)

Datasets produced in specific timeframes, usually become outdated

Tend to address the needs of specific research communities

Data is not always available for download

3. Public governments organizations

Mapping commitment that usually lasts in time

Well funded and, therefore, more ambitious projects

Aimed at a wide community of users, either from academia, policymaking or the private sector

Datasets are usually open and well characterized

4. Volunteered Geographic Information (VGI)

Contributions from citizens

It is not usually enough to produce independent LUC data, but usually feeds other mapping experiences led by public bodies or funded through different research projects

Fig. 1 Classification and characterization of LUC data producers

produce the datasets in collaboration with different universities and research institutions. The limited timeframe of these projects often affects the continuity of the mapping work they perform, and the datasets are not usually improved or updated once the project has come to an end. Dissemination of the data may also be affected by the end of funding. The Global Land Cover Facility, a reference initiative in the field of LUC research, which recently went offline,[1] is a perfect example of this problem.

Depending on the specific objectives of the projects and the institutions involved, these datasets may or may not be available for download. The quality of metadata and auxiliary information can also vary a lot from one project to the next. In some cases, a lot of technical and auxiliary information is provided, while in others users can only access the dataset itself and the research paper in which it is presented.

Governmental and other organizations are the other big producers of LUC data. In these cases, the objective is to provide information about the areas for which the organization is responsible or the areas affected by its policies and/or decisions. This data is a useful source of information for the policymaking process and is usually part of wider cartographic efforts by national and regional governments, and sometimes by international organizations, to provide geographic information of reference.

As these projects are part of official mapping work conducted by nations, regions and other large organizations, they are usually backed by significant long-term funding. These databases are therefore more likely to be updated or improved in the future. Another advantage is that they usually provide highly detailed, accurate information. They are also quite flexible. As a result, these databases are widely used by the whole scientific community, public and private sector professionals and many other users.

In recent years, there has been an increase in the data produced by members of the public through crowdsourcing or similar practices. This kind of information is known as Volunteered Geographic Information (VGI) and is part of a movement called 'citizen science', in which private citizens participate in scientific research, either by gathering or validating data or by assisting in any of the other phases of the scientific process.

Approaches of this kind allow local knowledge and expertise to be incorporated into data production. Highly detailed, up-to-date datasets can be produced easily and cheaply. Nevertheless, important issues can arise in terms of data quality and uncertainty, due to possible inconsistencies in the methods and procedures followed by the contributors, their different levels of expertise, etc.

---

[1] https://spatialreserves.wordpress.com/2019/01/07/global-land-cover-facility-goes-offline/.

3 Land Use Cover Maps

Reviewing all the LUC maps currently available is a daunting task, which perhaps explains why it has rarely been attempted. To our knowledge, the only researchers to carry out an extensive review of LUC maps at global and regional scales were Grekousis et al. (2015). They focused on general LUC products synthetizing all the land uses and land covers on Earth, so overlooking the increasing trend towards thematic LUC datasets that provide detailed mapping of a specific land use or land cover (e.g. forest, crop areas…).

The dividing line between general and thematic LUC products is not always clear. Some LUC maps, for example, provide general information on several different land covers (e.g. artificial, vegetation, water) while providing a detailed study of just one of them, thereby adopting a thematic approach. Although, in our review, we classify LUC maps as either general or thematic, readers should be aware of these possible inconsistencies.

Both types of LUC maps, general and thematic, can also be classified according to the extent they cover, differentiating between global, supranational, national, regional and local LUC maps. However, a comprehensive review of national, regional and local maps would be a huge task that is beyond the scope of this book. We will therefore be focusing exclusively on global and supranational LUC maps.

LUC maps for national and, especially, for regional and local areas, are usually only available for developed countries, or even highly developed countries, which can afford to invest in the production of spatial information and in research programmes. The most developed nations of the European Union, Australia and the United States usually have detailed LUC datasets, not only at a national level but also for specific regions. In China, the government has invested heavily in research, so enabling the production of national and regional LUC products. China is, together with the USA, the country producing most research on LUC mapping today (Yu et al. 2014).

3.1 Platforms and Repositories

A few online platforms and repositories provide an overview of the LUC datasets available. The Geo-Wiki platform (www.geo-wiki.org) is one of the most recent. It was initially developed to collect reference LUC information through crowdsourcing and to create a hybrid LUC map. It now hosts both general and thematic LUC maps. The Google Earth Engine Platform, which was also recently launched, includes a repository of spatial datasets, with a specific section devoted to Land Cover data (https://developers.google.com/earth-engine/datasets/tags/landcover).

The FAO Geonetwork repository (www.fao.org/geonetwork/) makes a great deal of spatial datasets available to users. The repository includes a specific section on LUC data. It hosts LUC maps at all scales and is a valuable source of LUC information for developing countries. The Land Processes Distributed Active Archive Center (LP DAAC) (https://lpdaac.usgs.gov/) holds most of the LUC datasets produced by NASA and the United States Geological Survey (USGS), in addition to other important global datasets.

The Copernicus Land Monitoring System website (https://land.copernicus.eu/) is the main source of LUC products created through the Copernicus programme, and is of particular interest for those working with European LUC information. All Copernicus layers are also available through the WEkEO Copernicus DIAS service (https://wekeo.eu/), a cloud-based platform that provides access to Copernicus datasets and to various tools for processing them, including all the land monitoring data.

3.2 General Land Use Cover Maps

3.2.1 Global LUC Maps

The production of global LUC datasets started at the end of the twentieth century. By then, coarse-resolution satellite imagery was available for producing consistent global LUC datasets at a low cost. A previous attempt had been made to create a global LUC map through photointerpretation of aerial imagery (Campbell 1983). Some authors also mention the maps developed by Matthews (1983), Olson et al. (1983) and Wilson and Henderson-Sellers (1985), when reviewing the first global LUC datasets. However, these datasets are quite thematic, focusing particularly on vegetation. They were created by combining existing maps with data obtained in the field and via interpretation of aerial imagery (Giri 2005).

The first global general LUC map of which we have record dates from 1994 (Table 1) (Defries and Townshend 1994). It was a global LUC map obtained after classification of AVHRR imagery data at a very coarse resolution: one degree (≈111 km at the Equator). This project was led by the Laboratory for Global Remote Sensing of the University of Maryland.

The next global LUC maps were also produced by the team from Maryland. These were an improvement on their original map. Two maps were produced at spatial resolutions of 8 km and 1 km, respectively (DeFries et al. 1995; Hansen et al. 2000). For years, they were distributed through the Global Land Cover Facility. However, since this repository went online, only the map at 1 km has been available. The other two maps are now outdated, both due to their very coarse resolution, of little use for most of today's applications, and because of the methods employed in their production.

A lot of new maps have been produced since these first global general LUC maps appeared, especially since 2010. Tables 1 and 2 provide a synthetic overview of these efforts. When available, the tables include a reference to the section of this book where these datasets are described in detail. For the datasets providing a time series of maps, we also specify to what extent LUC changes can be studied over the series of maps without important sources of uncertainty.

As in the case of the pioneering maps from the University of Maryland, all the datasets reviewed here have been

Table 1 List of available global general LUC maps with a single date

LUC map	Spatial resolution	Timeframe	Number of classes	Description note
Mathews Global Vegetation/Land Use	≈111km	1983	32	-
UMD LC Classification	1 km	1992/93	14	Sect. 1 in chapter "Global General Land Use Cover Datasets with a Single Date"
GLCC 2.0 Global	1 km	1992/93	17 (IGBP)	Sect. 2 in chapter "Global General Land Use Cover Datasets with a Single Date"
GLC2000	1 km	1999/2000	22	Sect. 3 in chapter "Global General Land Use Cover Datasets with a Single Date"
GMRCA LULC	10 km	2000	10	-
Geo-Wiki Hybrid	300 m	2000/05	10	Sect. 4 in chapter "Global General Land Use Cover Datasets with a Single Date"
LADA LUC map	≈8.3 km	2007	40	Sect. 5 in chapter "Global General Land Use Cover Datasets with a Single Date"
GLC-SHARE	1 km	2014 and before	11	Sect. 6 in chapter "Global General Land Use Cover Datasets with a Single Date"
OSM Landuse/Landcover	10 m	2017 and before	14	Sect. 7 in chapter "Global General Land Use Cover Datasets with a Single Date"

Table 2 List of available global general LUC datasets with a time series of maps

LUC map	Spatial resolution	Timeframe	Number of classes	Does it support change detection?	Description note
GLASS-GLC	5 km	1982–2015	8	Yes	Sect. 1 in chapter "Global General Land Use Cover Datasets with a Time Series of Maps"
LC-CCI	300 m	1992–2018	37	Yes	Sect. 2 in chapter "Global General Land Use Cover Datasets with a Time Series of Maps"
GLC30	30 m	2000, 2010, 2020	10	Yes	Sect. 3 in chapter "Global General Land Use Cover Datasets with a Time Series of Maps"
GLC250	250 m	2001, 2010	25	Not recommended	Sect. 4 in chapter "Global General Land Use Cover Datasets with a Time Series of Maps"
MCD12Q1	500 m	2001–2020	18	Not recommended	Sect. 5 in chapter "Global General Land Use Cover Datasets with a Time Series of Maps"
GLCNMO	1 km 500 m	2003 (1 km) 2008 (500 m) 2013 (500 m)	20	No	Sect. 6 in chapter "Global General Land Use Cover Datasets with a Time Series of Maps"
GlobCover	300 m	2005, 2009	23	No	Sect. 7 in chapter "Global General Land Use Cover Datasets with a Time Series of Maps"
FROM-GLC	30 m 10 m	2010 (30 m) 2015 (30 m) 2017 (30, 10 m)	11	Not recommended	Sect. 8 in chapter "Global General Land Use Cover Datasets with a Time Series of Maps"
CGLS-LC100	100 m	2015–2019	23	Yes	Sect. 9 in chapter "Global General Land Use Cover Datasets with a Time Series of Maps"

developed by research groups from different universities across the world, above all from China, Europe and the USA. The Joint Research Centre (JRC) of the European Commission and the USGS of USA have also been actively involved in many of these projects.

Most of these datasets are intended for use in climate change modelling, for which coherent global LUC maps at coarse resolutions are required. However, these databases are becoming increasingly popular and are used for many other purposes, a lot of them related with land change. This has been one of the drivers promoting the creation of new maps, with better quality and higher detail.

Below, we characterize the global LUC datasets produced in the last decades according to their method of production, level of accuracy and spatial, temporal and thematic resolutions. Over this period, map production methods have becoming increasingly complex in order to create more accurate maps that provide better spatial, temporal and thematic information.

The Production Methods

Nowadays, global LUC maps are created using improved and innovative production methods, involving advanced classifiers, such as those based on machine learning, as well as a lot of auxiliary data. In many cases, specific LUC categories are mapped through several specific procedures due to their particular patterns, reflectance behaviour, etc. Additional post-classification treatments have also become common in a bid to avoid some of the uncertainties and errors associated with the production of these maps.

In recent years, due to the increasing availability of LUC datasets, more and more global LUC maps are being produced by data fusion, in which new maps are created by combining existing datasets using a range of different algorithms and approaches. The aim of these projects is to create datasets with higher levels of accuracy and, therefore, less uncertainty. To this end, they usually combine the most accurate or highest quality LUC information from each dataset.

FAO-GLCShare is perhaps the best-known example of an attempt to build a new global LUC map from data fusion. It was created in 2014 by merging high-quality detailed national and regional LUC databases (Latham et al. 2014). In many cases, the new maps were obtained from the fusion of existing LUC datasets at global scales. Geo-Wiki Hybrid (See et al. 2015) is one of the most famous examples of maps created using this approach.

LUC maps obtained from data fusion do not have a single specific date of reference for the mapped area. When first produced, they are considered as up-to-date LUC databases. However, if they are not updated frequently, they eventually become obsolete and can no longer be regarded as useful sources for LUC change analysis.

The maps obtained through crowdsourcing, i.e. by aggregating a large number of individual inputs supplied by a community of people, could undergo the same problems.

Although still relatively rare, they could play an important role in the future. OSM-LULC, released in 2017 (Schultz et al. 2017), is the only example of a global general LUC map made with crowdsourced data.

These projects are usually updated on a regular basis. However, problems of coverage arise. In OSM-LULC, most of the world (except for specific test areas in Europe) is only partially mapped. Moreover, as they rely on volunteers to provide the information they require, the mapping and updating work is dependent on the volunteers' availability and willingness to participate. These may vary greatly from one country to the next and also over time. This is an inevitable source of uncertainty.

The recent advent of the Google Earth Engine (GEE) platform has encouraged the production of new global LUC maps, some general and others thematic. GEE provides a powerful cloud computing service, giving users the chance to process and classify tons of satellite imagery. This is particularly important when users do not have the necessary computer power to do this themselves. The availability of cloud-computing services will lead to an increase, in the near future, in the number of highly detailed LUC products being created using complex computer production methods. Many of these will be produced at global scales.

Accuracy

The development and application of new methods and techniques to produce LUC maps has not improved the accuracy of these datasets. Although some global LUC maps are more accurate than others, there is no correlation between time, the introduction of new methods and techniques and the achievement of higher levels of accuracy (Yu et al. 2014).

Global LUC datasets usually have accuracy levels of over 60%. In the best cases, they are around 80%. They are therefore still subject to high degrees of uncertainty. This is to be expected given the high level of abstraction they require. The entire surface of the Earth is being mapped according to the same method and must fit into the same legend. This means there is little room for local or regional specificities, which inevitably introduces a degree of uncertainty.

Spatial Resolution

LUC mapping has evolved over time, with the result that global LUC maps are produced at an increasing number of spatial resolutions. Initially, the AVHRR and VEGETATION sensors, with a spatial resolution of 1 km, were the main source of imagery for global LUC mapping. Later, imagery from MODIS (500 m) and MERIS (300 m) became the standard source of information. In recent years, it has become increasingly common to use the huge stock of Landsat imagery to produce global LUC maps at 30 m. Some projects have gone even further, producing global LUC maps at even finer resolutions. One example is the 2017 edition of FROM-GLC (10 m) (Chen et al. 2019), which was based on Sentinel-2 imagery.

Sentinel satellites will be providing free, long-term, high-quality imagery over the coming years. This may boost the production of global LUC maps at increasingly high levels of detail.

Temporal Resolution

The temporal resolution of LUC maps has also increased over time, especially in recent years. Historical time series of LUC maps are becoming more common (Table 2). When MODIS Land Cover (MCD12Q1) was launched in 2002, it was the first global LUC dataset to provide a series of LUC maps for different years (Friedl et al. 2002). It was later joined by GLCNMO, GlobCover, FROM-GLC and GLC30, which all provided new series of LUC maps for at least two different points in time.

However, in most of these series, LUC change cannot be reliably detected by cross-tabulating the different maps that make up the dataset. Different methods of production for each year, changes in the source of imagery, differences in the reflectance of the images, etc., introduce a lot of noise in the comparison. This makes it impossible to obtain meaningful results from LUC change analyses.

The latest version of the MODIS Land Cover (Collection 6) incorporated important changes in the product algorithm and workflow to account for these sources of uncertainty (Sulla-Menashe et al. 2019). However, change detection is still not supported and is therefore not recommended.

New time series of LUC maps have been produced recently with the specific purpose of enabling change detection. These include the LC-CCI (ESA 2017) and GLASS-GLC maps (Liu et al. 2020). They provide a long record of LUC information: with yearly maps for the period 1992–2018 in the case of the LC-CCI, and for the period 1982–2015 in the case of GLASS-GLC. The latter dataset has the longest, most frequent time series currently available. However, it uses a very coarse spatial resolution (5 km) and change detection using the GLASS-GLC map series is limited by various sources of uncertainty (Liu et al. 2020).

Classification Schemes

Unlike the spatial and temporal resolutions, there are no important variations over time in the thematic resolution of most global LUC products. In fact, standard LUC classification systems are now widely used so as to ensure that the different databases are comparable. One of the most common is the International Geosphere-Biosphere Programme

(IGBP) legend, which was used in one of the first LUC global maps ever released: the IGBP-Dis. Maps based on the IGBP legend usually distinguish around 17 categories.

The Land Cover Classification System (LCCS) proposed by the FAO in 1998 (Di Gregorio and Jansen 1998) has become the standard LUC classification method today. It is a flexible classification system that can be adapted to LUC maps at different scales and for different areas of the world. It first distinguishes between 8 broad land cover categories, each of which is later disaggregated into a varying number of subcategories based on a series of classifiers, which define the attributes or characteristics of each land cover. This enables users to adapt the classification detail to the required level of analysis. The resulting categories are mutually exclusive, as they are defined by different sets of classifiers. LCCS-based legends are hierarchical and comparable, so facilitating the comparison and analysis of global LUC maps by checking for agreements and differences.

3.2.2 Supra-national LUC Maps

A lot of international institutions and organizations need comprehensive and coherent worldwide data to support their activities. Global datasets are also required by research communities that study the whole Earth as a system. For their part, national governments and organizations require large amounts of data to support policymaking at a national level. Many other institutions, associations, professionals and researchers need very detailed data that is only available at regional and local scales.

Within this context, supra-national datasets do not provide much detail and work at a different scale to that at which most institutions and organizations implement their policies. They therefore do not meet the requirements of the research and policy-making communities working at global scales. This means that there is less interest and consequently less funding for datasets at these scales, hence the relative lack of supra-national LUC maps.

Supra-national LUC maps have been developed by the European institutions to assist policymaking and environmental monitoring in Europe. In other continents, supra-national LUC maps are usually developed within the context of different projects funded by international institutions, such as the FAO and various different US and European institutions. The latter include the European Space Agency (ESA) and the Joint Research Centre (JRC) of the European Commission, which have been actively involved in the production of supra-national LUC maps for many developing areas with important biodiversity values.

Europe

Europe is the continent with the widest range of supra-national LUC maps. The European Union (EU) has

certain powers over the European environment and is therefore interested in monitoring any changes in land use. To this end, the EU has invested in the production of EU-wide reference data as a reliable source of information on which to base their policy decisions. As a consequence, plenty of detailed, high-quality datasets are now available providing LUC information for the European continent (Table 3). The quality and detail of these datasets reveal the large amount of resources that the EU has invested in land monitoring, especially in recent years via the Copernicus programme.

Of all the European LUC datasets, CORINE Land Cover (CLC) is by far the best known. It is one of the oldest and most successful programmes on land monitoring, offering very high levels of accuracy and detail. All these qualities have made CLC a reference in LUC mapping worldwide. It is the only cross-country initiative working at similar scales that provides detailed, temporally rich LUC data, which can be used effectively for change detection. CLC is one of the best examples of decentralized, coordinated LUC mapping. CLC is produced at a national level, which allows European countries to develop their own national datasets while taking advantage of the work and the resources invested to create CLC.

A few non-European countries have mapped the land uses and covers in their entire nations or in certain specific areas following the CLC model. Some of them have done so with the help of the European institutions and other European research groups. These include Palestine, Morocco, Tunisia, San Salvador, Guatemala, Honduras, Haiti, Dominican Republic, Colombia, Burkina Faso and Gabon (Jaffrain 2011). Nevertheless, these maps are one-off, single-date LUC maps which do not provide the monitoring capacity provided by CLC in Europe.

Through the Copernicus programme, the EU has also developed coherent and consistent LUC mapping products aimed at monitoring the LUC dynamics of specific areas (e.g. coastal and metropolitan areas, riparian zones, Natura 2000 network...). These are very detailed products in both spatial and thematic terms, which have been designed to meet the needs of their potential community of users or to provide information in support of a range of different policies. Their production is centralized, so avoiding the inconsistencies that might result from a coordinated, decentralized production method. Although they were only recently launched, the EU has assured their long-term continuity, so providing consistent time series of data.

Two other series of LUC maps, which are complementary to CLC, are also available for Europe. Annual Land Cover is a recently launched product that provides annual LUC maps, so overcoming the temporal resolution limitations of CLC, which is only updated once every 6 years. Annual Land

Table 3 List of available general LUC datasets for Europe

LUC map	Extent	Spatial resolution/Scale	Timeframe	Number of classes	Does it support change detection?	Description note
HILDA	Europe (EU)	1 km	1900–2010 (every 10 years)	6	Yes	Sect. 1 in chapter "Global General Land Use Cover Datasets for Europe"
CLC	Europe (EU)	1:100,000 MMU: 25 ha	1990, 2000, 2006, 2012, 2018	44	Yes, through layer of changes	Sect. 2 in chapter "Global General Land Use Cover Datasets for Europe"
PELCOM	Europe (EU)	1 km	1997	16	One-date map	Sect. 3 in chapter "Global General Land Use Cover Datasets for Europe"
Annual Land Cover	Europe (EU)	30 m	2000–2019	33	No	Sect. 4 in chapter "Global General Land Use Cover Datasets for Europe"
GlobCorine	Europe	300 m	2005, 2009	17	No	Sect. 5 in chapter "Global General Land Use Cover Datasets for Europe"
Urban Atlas	Functional urban areas of Europe (EU)	1:10,000 MMU: 0.25-1 ha	2006, 2012, 2018	29	Yes, through layer of changes	Sect. 6 in chapter "Global General Land Use Cover Datasets for Europe"
N2K	Natura 2000 reserves of Europe (EU)	1:5000–1:10,000 MMU: 0.5 ha	2006, 2012, 2018	11	Yes	Sect. 7 in chapter "Global General Land Use Cover Datasets for Europe"
Riparian Zones	Riparian areas of Europe (EU)	1:10,000 MMU: 0.5 ha	2012, 2018	56	One-date map	Sect. 8 in chapter "Global General Land Use Cover Datasets for Europe"
Coastal Zones	Coastal regions of Europe (EU)	1:10,000 MMU: 0.5 ha	2012, 2018	71	Yes, through layer of changes	Sect. 9 in chapter "Global General Land Use Cover Datasets for Europe"
S2GLC 2017	Europe (EU)	10 m	2017	13	One-date map	Sect. 10 in chapter "Global General Land Use Cover Datasets for Europe"

Cover is produced as part of a project funded by the European Commission, which aims to create harmonized spatial datasets for Europe. However, it is not recommended for change detection, as there is a lot of inter-annual variability between LUC covers.

HILDA is another LUC dataset providing a long time series of LUC maps for Europe. Although it has a coarser resolution, it provides the longest time series of maps reviewed here: 1900–2010. It was produced by a research project team, who combined various different datasets and applied complex modelling techniques (Fuchs et al. 2013).

Africa

A large number of supra-national LUC maps have also been found for Africa (Table 4). Most of the datasets cover specific regions of the continent, such as Eastern, Western or Southern Africa. Areas that are particularly relevant for environmental research, such as the Congo Basin, have also been mapped.

Only a few projects tried to offer an overview of the LUC covers for the entire African continent. The FAO mapped the covers for many African countries as part of the AFRICOVER project, but did not encompass the whole continent. The first comprehensive, Africa-specific, general LUC dataset only appeared quite recently. It was produced by EU research and earth-observation organizations. No similar initiatives have been found for America, Asia and Oceania. They are also quite rare for Europe as a whole, where continental LUC data usually covers the EU and associated countries.

There are three datasets providing a time series of LUC maps for different African countries. However, only one of these (West Africa Land Use Land Cover) was obtained by applying a common mapping approach which provides LUC information for all mapped areas at the same dates. In the

Table 4 List of available general LUC datasets for Africa

LUC map	Extent	Spatial resolution/Scale	Timeframe	Number of classes	Does it support change detection?	Description note
West Africa Land Use Land Cover	West Africa	2 km	1975, 2000, 2013	26	Yes	Sect. 1 in chapter "General Land Use Cover Datasets for Africa"
SERVIR-ESA	Eastern and Southern Africa	30 m	Different dates depending on the country (1990–2015)	7	Yes	Sect. 2 in chapter "General Land Use Cover Datasets for Africa"
SADC Land Cover Database	Southern African Development Community	1:250,000	Different dates depending on the country (1990/99)	13	One-date map	Sect. 3 in chapter "General Land Use Cover Datasets for Africa"
AFRICOVER	Burundi, DR Congo, Egypt, Eritrea, Kenya, Rwuanda, Sudan, Tanzania, Uganda, Lybia, Malawi	1:200,000	Different dates depending on the country (1994/01)	8	One-date map	Sect. 4 in chapter "General Land Use Cover Datasets for Africa"
CCI LAND COVER—S2 PROTOTYPE	Africa	20 m	2016	10	One-date map	Sect. 5 in chapter "General Land Use Cover Datasets for Africa"
Congo Basin Vegetation Types	Congo Basin region	300 m	2000/07	20	One-date map	Sect. 6 in chapter "General Land Use Cover Datasets for Africa"

other two, the time series is made up of national or regional LUC maps produced for different years of reference, so hampering cross-country LUC change analyses.

The Americas
In the Americas, there is a clear distinction between the datasets covering North America and those covering South America and the Caribbean (Table 5). For North America, the North American Land Change Monitoring System (NALCMS) is of particular interest. It provides LUC maps for Canada, Mexico and the USA at three points in time. It is the only LUC supra-national American dataset with a time series of LUC maps. The NALCMS maps are created by merging datasets produced individually for each participating country following a similar approach.

Three different maps have been produced for South America, including in some cases the Caribbean. These were the result of various different research projects and activities and two of them (SERENA and South America 30 m) are no longer accessible for use.

South America 30 m, developed by Giri and Long (2014), provides the most up-to-date, detailed data. The SERENA map was designed to ensure its consistency with the NALCMS map (Blanco et al. 2013) so that together they could offer an overview of both North and South America. However, they had different spatial resolutions and were produced for different years of reference.

Asia and Antarctica
We only found one supra-national dataset for Asia, which covered the LUC of the Himalayan region (Table 6). It is possible that other supra-national datasets are available, although language barriers would prevent us from reviewing them properly. In any case, China is the most advanced country in Asia in terms of LUC mapping, and its research is focused above all on global and national mapping projects.

No supra-national maps are available for Oceania, due to its particular characteristics in which continental areas and islands are usually separate individual nations. These countries have no shared continental or inland regions for which a supra-national LUC dataset might be useful. As a result, no datasets of this kind have been produced.

Finally, a specific LUC map for Antarctica was produced recently by Chinese researchers (Hui et al. 2017). It is a vector LUC dataset for the reference year 2000, which differentiates between three land cover types. It is available online for any interested user.[2]

---

[2] https://zenodo.org/record/826032.

Table 5 List of available general LUC datasets for America

LUC map	Extent	Spatial resolution	Timeframe	Number of categories	Does it support change detection?	Description note
LBA-ECO LC-08	South America	1 km	1987/91	41	One-date map	Sect. 1 in chapter "General Land Use Cover Datasets for America and Asia"
NALCMS	North America	30 m 250 m	2005 (250 m) 2010 (250, 30 m) 2015 (30 m)	19	Partially	Sect. 2 in chapter "General Land Use Cover Datasets for America and Asia"
SERENA	South America	500 m	2008	22	One-date map	–
MERISAM2009	South America	300 m	2008/10	11	One-date map	Sect. 3 in chapter "General Land Use Cover Datasets for America and Asia"
South America 30 m	South America	30 m	2010	5	One-date map	–

Table 6 List of available general LUC datasets for Asia and Antarctica

LUC map	Scale	Timeframe	Number of categories	Does it support change detection?	Description note
The Himalaya Regional Land Cover database	1:350,000	2000	35	Yes, through layer of changes (1970/80–2007)	Sect. 4 in chapter "General Land Use Cover Datasets for America and Asia"
AntarcticaLC2000	1:100,000	2000	3	One-date map	–

3.3 Thematic Land Use Cover Datasets

Thematic Land Use Cover (LUC) datasets map parts of the Earth's surface as a specific land cover, considering not just its extent but also its intensity of distribution. They normally focus on land covers and provide very little information about land use. Thematic LUC maps are usually produced using automatic remote sensing techniques that find accurate land use characterization difficult.

Thematic LUC maps usually represent land covers in greater detail than general LUC maps. Some provide information about the proportion of the study area occupied by a particular land cover on the ground. In other cases, they delineate the extent of a specific cover with great detail and accuracy. Other thematic LUC maps share certain features with general LUC maps, in that they map the Earth according to a set of predefined categories, which are usually subclasses of a specific type of cover (e.g. vegetation). Many maps charting vegetation in its various different forms can therefore be regarded as thematic sources of LUC information in that they characterize a specific cover.

Some maps may provide thematic information about specific land covers together with other relevant data. This was especially true in the twentieth century, when many different maps combining biogeographic and climate information were produced for the climate and other research communities. These maps were usually produced by merging different techniques and datasets. Examples include the maps produced by Matthews (1983) and Olson et al. (1983). As these maps are now outdated and were not focused exclusively on land cover, we decided not to include them in this review.

Prior to the advent of satellite remote sensing, there were also a large number of traditional maps obtained through photointerpretation of aerial imagery and field surveys that provided information on certain specific land covers. These maps charted vegetation above all and, to a lesser extent, agricultural areas. These can be useful sources of information for historical LUC change analysis. However, as they are usually only available for national or more detailed areas and in many cases have not been digitalized, they are not reviewed here either.

There are also plenty of other spatial datasets that provide useful information for studying specific land covers. One example for vegetation covers are maps of live biomass (Kindermann et al. 2008; Thurner et al. 2014). Accordingly, there is a huge supply of information that can be used to study and characterize land covers, which comes in datasets of many different kinds. In this review, however, we will only be analysing datasets with a pure land cover approach.

The fact that thematic LUC maps focus on a single, specific cover normally means they are more accurate than general LUC maps. They are often more detailed too. This makes them especially useful for uncertainty analysis and validation exercises. As a general rule, they are a good source of reference data for studying land covers in a particular study area. However, they may not be as easy to use or to process as general LUC maps. If they provide too much information, users will have to process it to meet the specific needs of their studies.

The progress made in recent decades in the production of general LUC maps has also been achieved in thematic LUC mapping, with increasing levels of detail and more innovative, more complex methods. Some of the newest products have been produced using the cloud-computing capabilities of Google Earth Engine, which seems likely to play a key role in thematic LUC mapping in the future, and will allow more thematic datasets to be produced. Until now, the Landsat archive has been the most detailed source of imagery for LUC thematic mapping, although the imagery provided by the Sentinel constellation of satellites will soon enable users to expand the catalogue of thematic LUC datasets at highly detailed spatial resolutions of less than 30 m.

3.3.1 Global Thematic LUC Maps Focusing on Vegetation Covers

One of the most common features mapped by thematic LUC products is natural vegetation and tree and forest covers in particular. In fact, forest monitoring is one of the main applications of Landsat data, as reviewed by Hansen and Loveland (2012). This is because of widespread scientific interest in the study of vegetation dynamics and the fact that remote sensing techniques have made it much easier to characterize vegetation covers.

LUC maps focusing on vegetation covers usually offer coherent time series of LUC data that support change detection (Table 7). The most popular include the Vegetation Continuous Fields (VCF) datasets produced by NASA. These were first produced at the beginning of the 2000s and were obtained from AVHRR data at 1 km (Hansen et al. 2017). Since then, more VCF datasets have been produced at increasing levels of spatial detail, based above all on imagery from MODIS and Landsat (Hansen et al. 2003; Sexton et al. 2013). The temporal resolution of these products has also improved, with FCover providing information every 10 days for the period 1999–2020.

VCF datasets provide information about the vegetation cover fraction for each pixel in the analysed area. FCover is the only dataset that provides information on the percentage of vegetation cover, whereas all the others focus on tree or forest covers. Whereas FCover considers all kinds of natural vegetation, MEaSUREs VCF (VCF5KYR), MODIS VCF (MOD44B), Landsat VCF (GFCC) and the Hansen Forest Map focus exclusively on tree covers. In addition, GFCC and Hansen Forest Map include specific layers of forest change. Forests are mapped as such when a minimum fraction of their area is covered by trees. Therefore, changes in tree cover changes do not necessarily mean forest changes.

Two recent projects have explored the potential of radar data for mapping forest extent (Shimada et al. 2014; Martone et al. 2018). One of the advantages of radar data compared to optical sensors is that it is unaffected by weather and daylight conditions. This is particularly useful when mapping certain specific forest areas, such as those located in the tropics.

3.3.2 Global Thematic LUC Maps Focusing on Agricultural Covers

Agricultural areas are also widely mapped with specific LUC products (Table 8). Thematic agricultural LUC datasets usually show the extent of croplands and pasturelands or the cover fraction per unit of analysis, i.e. per pixel. In some cases, very detailed information on different types of crops is provided. These detailed LUC datasets are obtained from a wealth of detailed auxiliary information, as it is very difficult to accurately differentiate crop covers using standard remote sensing techniques.

Unlike other LUC thematic products, those mapping agricultural areas do not usually offer a time series, which means they cannot be used for land change analysis. Mapping agricultural areas is quite complex and this has hindered the production of coherent time series of agricultural LUC maps. One exception to this general trend was the dataset by Ramankutty and Foley (1999), who used historical sources of LUC data to model cropland cover on Earth from 1992 back to 1700. Another exception was the Harvested Area and Yield for 4 Crops maps, which provided information for three different dates.

3.3.3 Global Thematic LUC Maps Focusing on Artificial Covers

Built-up areas are becoming a common subject for thematic LUC products. As with the datasets focusing on vegetation covers, they provide time series of data which support change detection (Table 9). However, many of these maps are binary maps that only differentiate between urban/impervious and non-urban/non-impervious surfaces. They do not provide information about specific land uses so limiting their utility. However, people working with artificial surfaces are more interested in land use than in land cover, as artificial areas can be used for many different purposes, each of which has a different impact on the Earth.

Table 7 List of thematic LUC datasets characterizing vegetation covers

LUC map	Spatial resolution	Thematic information	Timeframe	Does it support change detection?	Description note
VCF5KYR	≈5.6 km	Percentage of tree cover, non-tree vegetation cover and bare ground	1982–1993 2001–2016	Possible	–
The World's Forests 2000	1 km	3 forest classes	1995/96	One-date map	Sect. 1 in chapter "Global Thematic Land Use Cover Datasets Characterizing Vegetation Covers"
Global mangrove distribution	30 m	Mangrove extent	1997/00	One-date map	–
FCover	300 m 1 km	Percentage of vegetation cover	Every 10 days from: 1999–2020 (1 km) and 2014 to the present (300 m)	Yes, through specific layers of change	Sect. 2 in chapter "Global Thematic Land Use Cover Datasets Characterizing Vegetation Covers"
Hybrid Forest Mask 2000	1 km	Percentage of forest cover	2000	One-date map	Sect. 3 in chapter "Global Thematic Land Use Cover Datasets Characterizing Vegetation Covers"
SYNMAP	1 km	26 vegetation classes	2000	One-date map	Sect. 4 in chapter "Global Thematic Land Use Cover Datasets Characterizing Vegetation Covers"
GFCC	30 m	Percentage of tree cover and forest gains/losses	2000, 2005, 2010, 2015 (tree cover) 1990–2000/2000–2005 (forest change)	Yes	Sect. 5 in chapter "Global Thematic Land Use Cover Datasets Characterizing Vegetation Covers"
Hansen Forest Map	30 m	Percentage of tree cover and forest gains/losses	2000–2019	Yes, through specific layers of forest gains and losses	Sect. 6 in chapter "Global Thematic Land Use Cover Datasets Characterizing Vegetation Covers"
MOD44B	250 m	Percentage of tree cover	2000–2019	Yes	Sect. 7 in chapter "Global Thematic Land Use Cover Datasets Characterizing Vegetation Covers"
PTC Global version	500 m 1 km	Percentage of tree cover	2003 (1 km) 2008 (500 m)	Possible	Sect. 8 in chapter "Global Thematic Land Use Cover Datasets Characterizing Vegetation Covers"
FNF	25 m	Forest extent	2007–2010 2015–2017	Possible	Sect. 9 in chapter "Global Thematic Land Use Cover Datasets Characterizing Vegetation Covers"
Forests of the World 2010	250 m	Percentage of tree cover	2010	One-date map	Sect. 10 in chapter "Global Thematic Land Use Cover Datasets Characterizing Vegetation Covers"
TanDEM-X Forest/Non-Forest Map	50 m	Forest/Non forest	2011/16	One-date map	Sect. 11 in chapter "Global Thematic Land Use Cover Datasets Characterizing Vegetation Covers"

Table 8 List of thematic LUC datasets characterizing agricultural covers

LUC map	Spatial resolution	Thematic information	Temporal frame	Does it support change detection?	Description note
Historic Croplands Dataset	0.5 degrees	Cropland proportion	1700–1992	Yes	–
1992 Croplands Dataset	10 km (5 min)	Cropland proportion	1992	One-date map	–
Harvested Area and Yield for 4 Crops (1995–2005)	10 km (5 min)	Map proportion for 4 crops	1995 2000 2005	Not for assessments at the cell level	–
GMRCA	10 km	66 categories grouped into 9 Rainfed cropland	2000	One-date map	–
GIAM	10 km	28 categories Irrigated cropland	2000	One-date map	–
Cropland and Pasture Area in 2000	10 km (5 min)	Cropland proportion Pastureland proportion	2000	One-date map	–
Harvested Area and Yield for 175 Crops	10 km (5 min)	Map proportion for 175 crops	2000	One-date map	–
Global Agricultural Lands	10 km	Cropland proportion Pastureland proportion	2000	One-date map	–
Global Cropland Extent	250 m	Cropland extent	2000/08	One-date map	Sect. 1 in chapter "Global Thematic Land Use Cover Datasets Global Thematic Land Use Cover Datasets"
IIASA-IFPRI Cropland Map	1 km	Percentage of cropland cover	2005	One-date map	Sect. 2 in chapter "Global Thematic Land Use Cover Datasets Global Thematic Land Use Cover Datasets"
GRIPC	500 m	3 cropland classes	2005	One-date map	Sect. 3 in chapter "Global Thematic Land Use Cover Datasets Global Thematic Land Use Cover Datasets"
FROM-GC	30 m	Cropland extent	2010	One-date map	–
GFSAD1KCD	1 km	8 cropland classes	2010	One-date map	Sect. 4 in chapter "Global Thematic Land Use Cover Datasets Global Thematic Land Use Cover Datasets"
GFSAD1KCM	1 km	5 cropland classes	2010	One-date map	Sect. 4 in chapter "Global Thematic Land Use Cover Datasets Global Thematic Land Use Cover Datasets"
Global Synergy Cropland Map	500 m	Percentage of cropland cover	2010	One-date map	Sect. 5 in chapter "Global Thematic Land Use Cover Datasets Global Thematic Land Use Cover Datasets"
UCL	250 m	Percentage of cropland cover	2014	One-date map	Sect. 6 in chapter "Global Thematic Land Use Cover Datasets Global Thematic Land Use Cover Datasets"
GFSAD30	30 m	Cropland extent	2015	One-date map	Sect. 7 in chapter "Global Thematic Land Use Cover Datasets Global Thematic Land Use Cover Datasets"
LADA Dominant crops	8.3 km	Up to 534 categories	Data fusion	One-date map	–
ASAP Land Cover Masks	1 km	Percentage of cropland/rangeland covers	2019	One-date map	Sect. 8 in chapter "Global Thematic Land Use Cover Datasets Global Thematic Land Use Cover Datasets"

Table 9 List of thematic LUC datasets characterizing artificial covers

LUC map	Spatial resolution	Thematic information	Timeframe	Does it support change detection?	Description note
Global Urban Land	30 m	Artificial areas extent	1980, 1990, 1995, 2000, 2005, 2010, 2015	Yes	Sect. 1 in "Global Thematic Land Use Cover Datasets Characterizing Artificial Covers"
GHSL	10 m (2018) 20 m (2016) 30 m, 250 m, 1 km (1975–2014)	Built-up areas extent Percentage of built-up areas (2014)	1975, 1990, 2000, 2014, 2016, 2018	Yes, except for the 2016 layer	Sect. 2 in "Global Thematic Land Use Cover Datasets Characterizing Artificial Covers"
GAIA	30 m	Artificial areas extent	1985–2018	Yes	Sect. 3 in "Global Thematic Land Use Cover Datasets Characterizing Artificial Covers"
GUB	30 m	Urban boundaries	1990, 1995, 2000, 2005, 2010, 2015, 2018	Yes	Sect. 3 in "Global Thematic Land Use Cover Datasets Characterizing Artificial Covers"
Global Urban Expansion 1992–2016	1 km	Urban areas extent	1992, 1996, 2000, 2006, 2010, 2016	Yes	Sect. 4 in "Global Thematic Land Use Cover Datasets Characterizing Artificial Covers"
ISA	1 km	Impervious area density	2000/01, 2010	Unknown	Sect. 5 in "Global Thematic Land Use Cover Datasets Characterizing Artificial Covers"
URB_MAP	500 m	Urban extent	2001/05	One-date product	–
HBASE	30 m 250 m 1 km	Urban areas extent	2010	One-date product	Sect. 6 in "Global Thematic Land Use Cover Datasets Characterizing Artificial Covers"
GMIS	30 m	Percentage of impervious areas	2010	One-date product	Sect. 6 in "Global Thematic Land Use Cover Datasets Characterizing Artificial Covers"
GUF	≈12 m ≈84 m	Built-up areas extent	2011	One-date product	Sect. 7 in "Global Thematic Land Use Cover Datasets Characterizing Artificial Covers"
WSF	10 m, 100 m, 250 m, 500 m, 1 km, 10 km	Settlement areas extent	1985–2015, 2014/15, 2019	In the future	Sect. 8 in "Global Thematic Land Use Cover Datasets Characterizing Artificial Covers"
GISM	30 m	Impervious areas extent	2015	One-date product	Sect. 9 in "Global Thematic Land Use Cover Datasets Characterizing Artificial Covers"

3.3.4 Global Thematic LUC Maps Focusing on Water and Other Covers

Some thematic LUC products focus specifically on water covers, two of which provide information on their change over time (Table 10). Other products offer a hybrid between general and thematic LUC datasets. These include the Global 1-km Consensus Land Cover, which provides a LUC thematic map for 12 different land covers (Tuanmu and Jetz 2014). It has 12 layers, each of which contains information about the fraction of the pixel occupied by the cover being mapped. A thematic LUC dataset with a similar approach

was obtained for 13 different covers as part of the ClimAfrica project for the period 1901–2017 (Churkina et al. 2009). Like other similar datasets already reviewed, it was obtained by a model based on different sources of historical LUC information.

3.3.5 Supra-national Thematic LUC Maps

We have only reviewed a few experiences of supra-national thematic LUC mapping (Table 11). The majority of them map vegetation covers, focusing especially on areas of special biodiversity or environmental value.

Table 10 List of thematic LUC datasets characterizing water and other covers

LUC map	Spatial resolution	Thematic information	Timeframe	Does it support change detection?
Historical land use based on Synmap landcover	0.5 degrees	13 themes (Map proportion for each)	1901–2007	Yes
Global Surface Water	30 m	Water occurrence 1–100	1984–2019	Yes, through specific product
CC WB	150 m	Water/no water	2000/12	One-date product
Daily Global Surface Water Change Database	500 m	Water 3 categories	2001–2016 (Daily)	Yes
Global 1-km Consensus Land Cover	1 km	12 themes (Map proportion for each)	Data fusion	One-date product

Table 11 List of thematic supra-national LUC datasets

LUC map	Extent	Spatial resolution/Scale	Thematic information	Timeframe	Does it support change detection?	Description note
TREES Vegetation Map of Tropical South America	Amazon basin	1 km	Vegetation 14 categories	1992	One-date map	–
Circumpolar Arctic Region Vegetation	Arctic region	1:7,500,000	Vegetation 20 classes	1993/95	One-date map	–
Insular Southeast Asia—Forest Cover Map	Insular Southeast Asia	1 km	5 forest classes	1998/00	One-date map	Sect. 1 in "Supra-national Thematic Land Use Cover Datasets"
Continental Southeast Asia—Forest Cover Map	Continental Southeast Asia	1 km	8 forest/wood classes	1998/00	One-date map	Sect. 2 in "Supra-national Thematic Land Use Cover Datasets"
Central Africa—Vegetation map	Cameroon Central African Republic Republic of Congo Equatorial Guinea Gabon DR Congo	1-5 km	Vegetation	1987/93	One-date map	–
Congo Basin Monitoring Map	Congo River Basin	57 m	Forest extent Forest probability Forest cover clearing	1990/00	Information on forest clearing	Sect. 3 in "Supra-national Thematic Land Use Cover Datasets"
FACET	DR Congo Congo Gabon	60 m	Forest 3 cover categories + gains and losses	2000 2005 2010	Yes	–
MARS Crop Mask Over Africa	Africa	250 m	Cropland extent	One date, different depending on the mapped area	One-date map	Sect. 4 in "Supra-national Thematic Land Use Cover Datasets"
HRL Impervious	Europe (EU)	10 m (after 2018) 20 m, 100 m (before 2018)	Extent and percentage of impervious areas	2006, 2009, 2012, 2015, 2018	Yes, through layer of changes	Sect. 5 in "Supra-national Thematic Land Use Cover Datasets"

(continued)

Table 11 (continued)

LUC map	Extent	Spatial resolution/Scale	Thematic information	Timeframe	Does it support change detection?	Description note
HRL Forests	Europe (EU)	10 m (after 2018) 20 m, 100 m (before 2018)	Percentage of tree cover areas, leaf type and forest type	2012, 2015, 2018	Yes, through layer of changes	Sect. 5 in "Supra-national Thematic Land Use Cover Datasets"
HRL Grasslands	Europe (EU)	10 m (after 2018) 20 m, 100 m (before 2018)	Extent of grassland areas	2015, 2018	Yes, through layer of changes	Sect. 5 in "Supra-national Thematic Land Use Cover Datasets"
HRL Water	Europe (EU)	10 m (after 2018) 20 m, 100 m (before 2018)	5 water-wet classes	2015, 2018	Unknown	Sect. 5 in "Supra-national Thematic Land Use Cover Datasets"
HRL Small Woody Features	Europe (EU)	5 m	Extent of Small Woody Features	2015	Not at the moment	Sect. 5 in "Supra-national Thematic Land Use Cover Datasets"
ESM	Europe (EU)	2 m (2015) 2.5 m (2012) 10 m (2012)	Built-up extent (2015) Residential areas extent (2012) 13 built-up categories (2012) Percentage of built-up areas (2012)	2012, 2015	No	Sect. 6 in "Supra-national Thematic Land Use Cover Datasets"

They are usually produced by international institutions, such as the European Commission, or research groups from internationally renowned universities. They are interested in monitoring and understanding the land dynamism of high biodiversity areas of worldwide importance.

The European Commission, through the Copernicus programme, is behind some of the few supra-national thematic LUC datasets that focus on other covers such as artificial surfaces or agricultural areas.

4 Reference Land Use Cover Data

Reference data is required to train supervised remote sensing classifiers and to validate LUC maps. Reference LUC datasets consist of a series of geographically distributed sample points with LUC information. Each point contains information about the specific land use or cover in the pixel or polygon of the Earth's surface represented by the point.

The reference datasets are subject to the same spatial abstraction required in LUC maps. Reference points are associated with a specific pixel or polygon. The level of abstraction required varies depending on the size of these points. The uncertainty of the reference information will also

vary accordingly. The fact that a single land use or cover is assigned to a whole pixel or polygon, even though they may contain other land uses or covers, can also produce uncertainty. In addition, there is always a degree of subjectivity in the decision to assign a pixel or polygon to a particular category, especially in borderline cases that are not clear-cut. This can create an additional source of uncertainty.

Relatively few general LUC reference datasets are currently available. This is because many reference datasets were created ad hoc every time a new LUC map was validated or reference data was required to train a remote sensing classifier, and it was therefore unnecessary to have a ready supply of general LUC reference datasets. These datasets are also affected by some degree of thematic generalization, as is any LUC map. LUC information must conform to a specific classification system or legend. Given the ad hoc nature of many reference datasets, the classification or legend used to classify the land uses and covers was normally also case-specific. However, the recent emergence of standard LUC reference datasets aimed at a wide range of users and research fields has extended the use of standard legends and classification systems, such as the FAO LCCS, when drawing up these datasets.

One of the most renowned LUC reference datasets is the Land Use Cover Area frame Sample (LUCAS), produced by EUROSTAT every 3 years since 2006. It is made up of more than 330,000 survey points across the EU.[3] An increasing number of countries have taken part in every new version of the survey. Of all the LUC reference datasets available, this is the most comprehensive. For each point, experts collect information about land uses, land covers and other relevant environmental parameters. LUCAS also includes four photographs for each surveyed point. It is the only LUC reference dataset reviewed that provides a coherent time series of data for different years.

In recent years, various reference datasets used to validate and train classifiers of global LUC maps have been made available online, so enabling them to be used for other purposes rather than just in the production of one specific map. The work done by the team from the GOFC-GOLD Land Cover Office is of special note. They collected and improved the reference datasets from six different LUC products (GLC2000, GlobCover 2005, STEP, VIIRS, GLCNMO and the urban dataset from the University of Tokyo). Samples of these datasets (with up to 70% of all the available reference points) are freely available for download on the project website.[4]

There is a growing trend to gather reference data through crowdsourcing and volunteering initiatives. Information gathered in this way is often referred to as Volunteered Geographic Information (VGI) and is part of citizen science. Members of the public create reference LUC information that will later be used to train classifiers and validate final maps. The information is gathered by local volunteers across the world, so taking advantage of local expertise. It is also a good source of cheap reference information. However, production methods of this kind have many related limitations and uncertainties.

The most famous of these initiatives is Geo-Wiki, which is frequently used to collect LUC information for calibration and validation practices. Geo-Wiki provides a user-friendly online tool that makes it very ease to visualize LUC maps and to collect the reference LUC data required to validate them. Many international research projects working on LUC mapping and citizen science have based their research on Geo-Wiki. One of the most important is the H2020 Land-Sense Citizen Observatory.[5] It produced a global LUC reference dataset over four campaigns (Fritz et al. 2017).

Sahariah et al. (2017) also produced a global LUC reference dataset for cropland land covers using Geo-Wiki and crowdsourcing. Both datasets are available online for any user interested in the PANGEA repository.[6]

The Australian Terrestrial Ecosystem Research Network (TERN) has developed a specific Geo-Wiki application to validate Australian LUC maps: AusCover.[7] Also associated with Geo-Wiki, the LACO-wiki platform provides another tool for the collection of LUC reference datasets.[8] Users can easily validate their own LUC maps on this platform, which includes a repository of reference data created or hosted by the community. It is a very comprehensive, user-friendly tool for LUC reference data production and LUC map validation, which has outperformed the capabilities of Geo-Wiki for this specific task.

Many other tools and platforms have been developed in recent years with similar purposes: Collect Earth, GLFC LT, VIEW-IT… (Bey et al. 2016). However, although these platforms offer the tools required to create LUC reference datasets through crowdsourcing, many of these datasets are not made available online. Even in the platforms based on crowdsourced information, the LUC reference data remains very case-specific and is not disseminated, so preventing its reuse in other situations.

Although they cannot be considered LUC data as such, volunteered geo-referenced photographs may be useful for obtaining reference LUC datasets. They provide a fixed picture of a landscape at a given point in time. By analysing the picture, users can identify the dominant land cover or land use, so obtaining LUC reference data.

Several initiatives for collecting volunteered photographs of specific geographic locations are already ongoing. Flickr is one of the most famous, although its purposes and objectives have little to do with science or scientific methods. The Degree Confluent Project (DCF)[9] aims to collect photographs and descriptions of each integer degree intersection of latitude and longitude on Earth. Geograph collects representative photographs of every single square km in England, Ireland[10] and Germany.[11] The Field Photo Library[12] collects geo-referenced photos across the earth. Google Maps also hosts pictures and is now regarded as a successor to Panoramio, a service similar to Flickr.

---

[3] https://ec.europa.eu/eurostat/web/lucas.
[4] http://www.gofcgold.wur.nl/sites/gofcgold_refdataportal.php.
[5] https://landsense.eu/.

[6] https://doi.pangaea.de/.
[7] https://application.geo-wiki.org/branches/auscover/.
[8] https://old.laco-wiki.net/en/Welcome.
[9] http://confluence.org/index.php.
[10] www.geograph.org.uk/
[11] https://geo-en.hlipp.de/
[12] http://www.eomf.ou.edu/photos/

Further Reading

Fonte CC, Bastin L, See L, et al. (2015) Usability of VGI for validation of land cover maps. Int J Geogr Inf Sci 29:1269–1291. https://doi.org/10.1080/13658816.2015.1018266

This paper reviews the main platforms and sources available for volunteer-based collection of LUC reference data and other information that may be useful for producing datasets of this kind. It also discusses the pros and cons of this approach for obtaining reference LUC data.

Grekousis G, Mountrakis G, Kavouras M (2015) An overview of 21 global and 43 regional land-cover mapping products. Int J Remote Sens 36:5309–5335. https://doi.org/10.1080/01431161.2015.1093195

Comprehensive review of general LUC datasets available at global and continental scales. It also reflects on the progress made and the challenges that lie ahead, proposing a series of recommendations for future LUC mapping practice.

Herold M, See L, Tsendbazar NE, Fritz S (2016) Towards an integrated global land cover monitoring and mapping system. Remote Sens 8:1–11. https://doi.org/10.3390/rs8121036

This paper summarizes the state of the art on global LUC mapping. It identifies the areas where most progress has been made in the field, referring in particular to the products with greater spatial detail and more frequent temporal information; the increasing importance of validation; the progressive implementation of the FAO Land Cover Classification System (LCCS) framework as the standard LUC classification method; and the increasing interest in citizen engagement. The paper also mentions some of the specific fields that have recently been the focus of scientific attention: data fusion; uncertainty analysis by data comparison; and quantification of LUC change. Finally, the authors reflect on the work that remains to be done and the challenges that lie ahead.

Mora B, Tsendbazar N-E, Herold M, Arino O (2014) Global Land Cover Mapping: Current Status and Future Trends. In: Manakos I, Braun M (eds) Land Use and Land Cover Mapping in Europe. Practices & Trends. Springer, Dordrecht, Heidelberg, New York, London, pp 11–30.

Book chapter offering a short but very comprehensive state of the art on global LUC mapping. It reviews the LUC datasets available in 2014 and summarizes the progress that had been made until then. It also points out the main issues with regard to global LUC mapping practice and objectives for the future. Many of these objectives have now been accomplished.

P. Giri C (ed) (2012) Remote sensing of land use and land cover. Principles and applications. CRC Press.

One of the reference books on Land Use Cover mapping and analysis. It provides an introduction to the field, tracing its history and an overview of the main concepts relating to LUC mapping and remote sensing. It also addresses the main methodological issues in relation to LUC mapping using remote sensing techniques, such as validation practices, land cover change detection and image classification methods. In Part III, the book includes examples of regional LUC mapping and LUCC monitoring.

See L, Fritz S, Perger C, et al. (2015) Harnessing the power of volunteers, the internet and Google Earth to collect and validate global spatial information using Geo-Wiki. Technol Forecast Soc Change 98:324–335. https://doi.org/10.1016/j.techfore.2015.03.002

Good description of the Geo-Wiki platform, its history, evolution and current capabilities. It also reviews some of the LUC reference datasets based on information collected through the platform.

Tsendbazar NE, de Bruin S, Herold M (2015) Assessing global land cover reference datasets for different user communities. ISPRS J Photogramm Remote Sens 103:93–114. https://doi.org/10.1016/j.isprsjprs.2014.02.008

The paper compares and analyses 12 LUC reference datasets in detail. These datasets are used in the production and validation of global LUC maps. This is one of the most comprehensive reviews of the LUC reference datasets currently available. It also assesses the potential reuse of these datasets, focusing on the data requirements imposed by different user communities. The authors try to identify the particular features that LUC reference datasets must have to enable them to be used by a wide range of users.

Wulder MA, Coops NC, Roy DP, et al. (2018) Land cover 2.0. Int J Remote Sens 39:4254–4284. https://doi.org/10.1080/01431161.2018.1452075

A long but detailed reflection on the progress that has been made and the changes in Land Cover mapping since the appearance of remote sensing.

References

Bey A, Díaz ASP, Maniatis D et al (2016) Collect earth: land use and land cover assessment through augmented visual interpretation. Remote Sens 8. https://doi.org/10.3390/rs8100807

Blanco PD, Colditz RR, López Saldaña G et al (2013) A land cover map of Latin America and the Caribbean in the framework of the SERENA project. Remote Sens Environ 132:13–31. https://doi.org/10.1016/j.rse.2012.12.025

Campbell JB (1983) Mapping the land: aerial imagery for land use information

Chen B, Xu B, Zhu Z et al (2019) Stable classification with limited sample: Transferring a 30-m resolution sample set collected in 2015 to mapping 10-m resolution global land cover in 2017. Sci Bull

Churkina G, Brovkin V, Von Bloh W et al (2009) Synergy of rising nitrogen depositions and atmospheric CO2 on land carbon uptake moderately offsets global warming. Global Biogeochem Cycles 23. https://doi.org/10.1029/2008GB003291

DeFries RS, Hansen MC, Townshend JRG, Sohlberg R (1995) Global land cover classification at 8 km spatial resolution: the use of training data derived from Landsat imagery in decision tree classifiers. Remote Sens Environ 19:3141–3168

DeFries RS, Townshend JRG (1994) NDVI-derived land cover classifications at a global scale. Int J Remote Sens 15:3567–3586. https://doi.org/10.1080/01431169408954345

Di Gregorio A, Jansen LJ (1998) Land Cover Classification System (LCCS): classification concepts and user manual. FAO, Rome

ESA (2017) Land cover CCI. Product user guide. Version 2.0

Friedl MA, McIver DK, Hodges JCF et al (2002) Global land cover mapping from MODIS: algorithms and early results. Remote Sens Environ 83:287–302. https://doi.org/10.1016/S0034-4257(02)00078-0

Fritz S, See L, Perger C et al (2017) A global dataset of crowdsourced land cover and land use reference data. Sci Data 4:1–8. https://doi.org/10.1038/sdata.2017.75

Fuchs R, Herold M, Verburg PH, Clevers JGPW (2013) A high-resolution and harmonized model approach for reconstructing and analysing historic land changes in Europe. Biogeosciences 10:1543–1559. https://doi.org/10.5194/bg-10-1543-2013

Giri C (2005) Global land cover mapping and characterization: Present situation and future research priorities. Geocarto Int 20:35–42. https://doi.org/10.1080/10106040508542334

Giri C, Long J (2014) Land cover characterization and mapping of South America for the year 2010 using landsat 30 m satellite data. Remote Sens 6:9494–9510. https://doi.org/10.3390/rs6109494

Grekousis G, Mountrakis G, Kavouras M (2015) An overview of 21 global and 43 regional land-cover mapping products. Int J Remote Sens 36:5309–5335. https://doi.org/10.1080/01431161.2015.1093195

Hansen M, DiMiceli C, Sohlberg R (2017) User guide for the MEaSURES Vegetation continuous fields product, version 1

Hansen MC, DeFries RS, Townshend JRG et al (2003) Development of 500 meter vegetation continuous field maps using MODIS data. Int Geosci Remote Sens Symp 1:264–266. https://doi.org/10.1109/igarss.2003.1293745

Hansen MC, Defries RS, Townshend JRG, Sohlberg R (2000) Global land cover classification at 1 km spatial resolution using a classification tree approach. Int J Remote Sens 21:1331–1364. https://doi.org/10.1080/014311600210209

Hansen MC, Loveland TR (2012) A review of large area monitoring of land cover change using Landsat data. Remote Sens Environ 122:66–74. https://doi.org/10.1016/j.rse.2011.08.024

Hui FM, Kang J, Liu Y et al (2017) AntarcticaLC2000: the new Antarctic land cover database for the year 2000. Sci China Earth Sci 60:686–696. https://doi.org/10.1007/s11430-016-0029-2

Jaffrain G (2011) CORINE Land Cover Outside of Europe. Nomenclature adaptation to other bio-geographical regions. Studies & project from 1990 to 2010. Final report

Kindermann GE, McCallum I, Fritz S, Obersteiner M (2008) A global forest growing stock, biomass and carbon map based on FAO statistics. Silva Fenn 42:387–396. https://doi.org/10.14214/sf.244

Latham J, Cumani R, Rosati I, Bloise M (2014) Global land cover SHARE (GLC-SHARE) database beta-release version 1.0-2014

Liu H, Gong P, Wang J et al (2020) Annual dynamics of global land cover and its long-term changes from 1982 to 2015. Earth Syst Sci Data 12:1217–1243. https://doi.org/10.5194/essd-12-1217-2020

Martone M, Rizzoli P, Wecklich C et al (2018) The global forest/non-forest map from TanDEM-X interferometric SAR data. Remote Sens Environ 205:352–373. https://doi.org/10.1016/j.rse.2017.12.002

Matthews E (1983) Global vegetation and land use: new high-resolution data bases for climate studies. J Clim Appl Meteorol 22:474–487. https://doi.org/10.1175/1520-0450(1983)0222.0.CO;2

Olson JS, Watts JA, Allison LJ (1983) Carbon in live vegetation of major world ecosystems.

Ramankutty N, Foley JA (1999) Estimating historical changes in global land cover: croplands from 1700 to 1992. Global Biogeochem Cycles 13:997–1027. https://doi.org/10.1029/1999GB900046

Sahariah P, Schlesinger P, Panging K et al (2017) A global reference database of crowdsourced cropland data collected using the Geo-Wiki platform. Sci Data 4:170136

Schultz M, Voss J, Auer M et al (2017) Open land cover from OpenStreetMap and remote sensing. Int J Appl Earth Obs Geoinf 63:206–213. https://doi.org/10.1016/j.jag.2017.07.014

See L, Schepaschenko D, Lesiv M et al (2015) Building a hybrid land cover map with crowdsourcing and geographically weighted regression. ISPRS J Photogram Remote Sens 103:48–56. https://doi.org/10.1016/j.isprsjprs.2014.06.016

Sexton JO, Song XP, Feng M et al (2013) Global, 30-m resolution continuous fields of tree cover: landsat-based rescaling of MODIS vegetation continuous fields with lidar-based estimates of error. Int J Digit Earth 6:427–448. https://doi.org/10.1080/17538947.2013.786146

Shimada M, Itoh T, Motooka T et al (2014) New global forest/non-forest maps from ALOS PALSAR data (2007–2010). Remote Sens Environ 155:13–31. https://doi.org/10.1016/j.rse.2014.04.014

Sulla-Menashe D, Gray JM, Abercrombie SP, Friedl MA (2019) Hierarchical mapping of annual global land cover 2001 to present: the MODIS collection 6 land cover product. Remote Sens Environ 222:183–194. https://doi.org/10.1016/j.rse.2018.12.013

Thurner M, Beer C, Santoro M et al (2014) Carbon stock and density of northern boreal and temperate forests. Glob Ecol Biogeogr 23:297–310. https://doi.org/10.1111/geb.12125

Tuanmu MN, Jetz W (2014) A global 1-km consensus land-cover product for biodiversity and ecosystem modelling. Glob Ecol Biogeogr 23:1031–1045. https://doi.org/10.1111/geb.12182

Wilson MF, Henderson-Sellers A (1985) A global archive of land cover and soils data for use in general circulation climate models. J Climatol 5:119–143. https://doi.org/10.1002/joc.3370050202

Yu L, Liang L, Wang J et al (2014) Meta-discoveries from a synthesis of satellite-based land-cover mapping research. Int J Remote Sens 35:4573–4588. https://doi.org/10.1080/01431161.2014.930206

Visualization and Communication of LUC Data

Francisco Escobar

Abstract

The increasing number of disciplines and public and private sectors interested in land use/land cover (LUC) information has boosted the demand for and the production of related cartographic products. However, the communicating power of the final maps may be impaired, if any of the cartographic transformations performed during the mapping process does not adapt well to the particular subject or area being mapped. This chapter takes the reader on a guided tour through the map production process, offering an overview of the cartographic language, the rules and practices that contribute to the success of the map as a communication tool and the most common forms in which LUC maps appear. Recent developments in geovisualization tools applied to LUC are also discussed.

Keywords

Cartography • Communication • Land Use Cover • Mapping

1 Introduction

The main purpose of cartography is to communicate geospatial information. The map serves as a channel through which a message is transmitted from the sender—the mapmaker—to the receiver—the map user (Robinson 1953, 1969; Muller 1975; Koláčny 1977; Ratajski 1978; Morrison 1976).

Like any other communication tool, cartography possesses its own language. The term "language" has been used by a number of authors in this field and can be defined as a system of signs enabling communication (Cauvin et al. 2010a). For communication to be successful, these signs should be capable of conveying to the reader the concepts that the author wishes to transmit. Given that maps also seek to convey information through signs, cartography must be considered part of semiotics. Indeed, as early as 1952, Robinson developed this idea by introducing a whole system of specific symbols for mapmaking (Robinson 1952).

Subsequently, various studies explored this concept in greater depth, culminating in 1967 with the seminal piece by Jacques Bertin "Semiology of Graphics", a genuine world reference on this subject. This was followed in 1978 by Ratajski, who outlined that, in modern thematic cartography, the ultimate goal of semiotics is to build an accurate, unambiguous cartographic language.

In cartography, semiotics unfolds as two different categories of signs; on the one hand it refers to **geometric signs**, the spatial dimensions (zero, one, two or three) and the geometric nature of map features (points, lines, polygons and volumes), and on the other, to **visual variables**, defined as the possible elementary variations in perceptible marks (Bertin 1967). This definition was frequently cited, and eventually revised, by other cartographers (Cauvin et al. 2010a; Robinson 1953; Robinson et al. 1984; Monmonier 1993; Slocum et al. 2005).

In this chapter we will be focusing on both kinds of signs and their role in the cartographic representation of land use/land cover (LUC).

Recent technological advances in the GIS industry have popularized cartography, giving rise to what some people refer to as a "geospatial society" in which maps are increasingly ubiquitous and used in all kinds of applications. This has brought new opportunities for cartography as a science but it also poses new challenges, one of which is that many new mapmakers lack the necessary cartographic skills to produce effective maps. Unfortunately, there are numerous examples in the literature that illustrate the fact that GIS has made it easy to produce large numbers of wrong or

F. Escobar (✉)
Departamento de Geología, Geografía y Medio Ambiente, Universidad de Alcalá, Alcalá de Henares, Spain
e-mail: francisco.escobar@uah.es

D. García-Álvarez et al. (eds.), *Land Use Cover Datasets and Validation Tools*,
https://doi.org/10.1007/978-3-030-90998-7_5

confusing maps more quickly than ever before. In the case of LUC mapping, no matter how sophisticated and expensive the technology for the collection and processing of the information may be, inexpert mapmakers often fail to communicate the relevant information correctly.

In order to help overcome these issues, this chapter aims to provide the basic ground rules for the correct representation and interpretation of LUC maps.

2 Geometric Signs

The geographic entities we find in the landscape are portrayed on maps as cartographic objects of varying geometric nature. Different land use areas are no exception and are usually depicted as polygons. The process for representing this information on a 2-dimensional piece of paper or on a screen is anything but simple as it involves, at least, the following transformations; (1) projecting the irregular and curved surface of the Earth on a plane, (2) selecting land use patches of sufficient size as to be visible (and readable) on the map, and (3) aggregating the information at the right administrative level when analysing LUC distribution over statistical spatial units. These three transformations have important implications for LUC mapping, which we will now go on to explain.

2.1 Cartographic Projection and LUC Mapping

The representation of our curved planet on a 2-dimensional map requires the application of mathematical models, known as cartographic or mapping projections, to project the Earth's surface on a plane (Slocum et al. 2005). Deformations occur during the projection process, which provide differentiating criteria to enable us to classify these projections into three big families; conformal, equidistant and equivalent, the last of which is also referred to as equal area.

- Conformal projections are used in navigation charts, as their main characteristic is the preservation of angles. Parallels and meridians intersect in a perpendicular manner, so forming four 90° angles at each intersection and an orthogonal network as a whole. However, these maps show important distortions in terms of the proportionality of areas and distances.
- Equidistant projections preserve the distances between specific pairs of points and distort areas and angles. These kinds of projections are mainly used in engineering and construction works.
- Equivalent or equal area projections preserve the proportionality of areas and by doing so distort the shapes and distances.

The bigger the area represented, the greater the impact of our choice of projection. This is noticeable in world maps where familiarity with the shapes of countries and continents make it easy for the reader to understand the deformations in each case. However, in smaller areas whose shape is not usually familiar to the general population, the map reader will find it difficult to notice the deformations. Of course, given the limited portion of the Earth's surface portrayed, the effects of the deformations are not as obvious as in world maps, but they do exist and can have an impact on LUC mapping. Since the choice of the projection results in significantly different maps, as Fig. 1 shows, the mapmaker must decide which projection system suits their map best. A bad choice could result not only in an unwanted distorted map, but also in a map that estimates metrics incorrectly. LUC analysts want metrics that inform the reader about different aspects of LUC, among them land use category distribution patterns and clusters, and especially the size of individual or groups of patches. This means that LUC maps must preserve the proportionality of areas. Conformal and equidistant projections are unsuitable for this purpose and equivalent projections must therefore be used.

2.2 The Minimum Mapping Unit in LUC Maps

The minimum mapping unit, or MMU, defines the size of the smallest cartographic object that will appear on the map (Cauvin et al. 2010a), in this way determining the resolution and by extension the most appropriate scale for the map.

Today, the predominance of digital maps over paper-based maps and their capacity to zoom in and out mean that the MMU is not as obvious as in the past. However, all maps are affected by the mapmaker's choices regarding their final scale, and the MMU has to be set in such a way as to facilitate the useability and readability of the map. In digital maps, the zoom feature may incorporate 'intelligent' functions, which allow it to display certain map elements, features and labels, solely at the appropriate level of zoom. The result is that when the user zooms out, the smaller features are hidden and when they zoom in again, more and more small features become visible. For the intelligent zoom to work properly, the mapmaker must establish a different MMU at each zoom level, in this way deciding which elements will be visible at each different scale, an important decision in the mapmaking process.

CORINE Land Cover is a well-known European project, which established an MMU of 25 hectares for areal entities and a minimum width of 100 m for linear features (European Environment Agency 2017). This means that in a printed map at the recommended working scale of 1:100,000 the MMU will occupy 0.5 cm^2 or 25 mm^2.

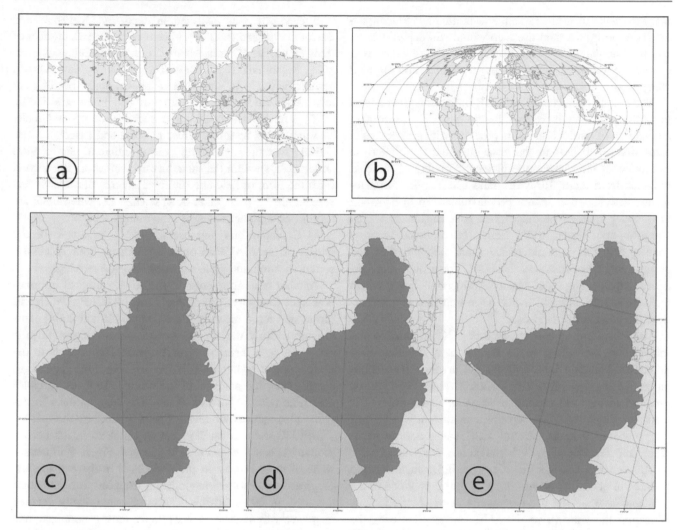

Fig. 1 Impact of the cartographic projection on map appearance at global and local (Guadiamar River Basin) scales. **a** Mercator projection (conformal), **b** Mollweide projection (equal area), **c** ETRS89 / UTM zone 29N (conformal), **d** Mollweide (equal area), **e** Europe Equidistant Conic (equidistant). For demonstration purposes only, the differences between (**d**) and (**e**) have been accentuated by applying a World and a European projection system respectively

The MMU also plays an important role in the data collection phase. Regardless of whether data is collected by field work or by interpretation of aerial or satellite imagery, the features that are smaller than the MMU will not appear on the map.

Some authors work almost exclusively with raster structures for which the pixel is the basic unit. As a result, they tend to conceive the MMU in terms of pixel size. From this perspective, it is generally accepted that the smallest observable feature in the final map, i.e. the MMU, should comprise at least four contiguous pixels (NOAA 2011).

When it comes to determining the MMU of LUC maps, it is important to differentiate between databases and maps. Patches that might be a suitable size for data analysis could be completely inappropriate for map publishing. Single pixels or small groups of pixels forming small areas below

the MMU threshold might be considered in data analysis, but would not appear on the map.

Three intrinsic characteristics of LUC mapping must be taken into consideration when deciding the most appropriate MMU: (i) Confusion between use and coverage, (ii) Definition of land use categories and associated land size, and (iii) High sensitivity of LUC maps to the interrelations between MMU and scale. The scale at which LUC information is expressed also has an enormous impact on the communication capacity of the resulting map (Wu and Harbin 2006; García-Álvarez et al. 2019).

In what is a common confusion between land use and land cover, different MMUs can result in maps showing different categories. For instance, at a relatively coarse resolution, a MMU of 1 km^2 would lead to an airport being depicted as such in both a land use map and a land cover map. However, if we increase the resolution by reducing the MMU to 50 m, the land use map would still depict it as an airport, but the land cover map would classify the areas covered by runways, buildings, or green areas into different categories.

The second characteristic of LUC information that affects the MMU is directly related to the first. The increasing availability of Earth observation products with greater spatial resolution could lead to the false idea that the higher the resolution of the images, the better the quality of the data obtained from them. However, land use, i.e. the "arrangements, activities and inputs people undertake in a certain land cover type to produce, change or maintain it" (Di Gregorio and Jansen 2000) cannot be observed in areas smaller than that required to carry out said activities and arrangements. For instance, the MMU for a LUC map category representing low-density residential development must be at least as small as the basic unit (house with garden) for this kind of land use.

The third intrinsic characteristic of LUC information that impacts on the MMU is its nature as a covering phenomenon. Mapping LUC information involves the delimitation of areas showing homogeneous coverage. This poses a problem in the data collection phase of small-scale LUC maps, in which the MMU covers a significantly large area that probably includes several LUC categories. In these cases, the identification of homogenous areas becomes a much more complex task. In order to assign a single value to the area in question, the cartographer must apply one of the available criteria. The most frequently used criteria include allocating the area: (i) to the LUC category covering the largest proportion of the area or (ii) to the predominant LUC category in the surrounding area. Related issues arise when attempting to downscale or upscale previously existing geospatial information. This increases the uncertainty of the map (García-Álvarez et al. 2019) and could give rise to the Modifiable Areal Unit Problem (MAUP) and the Category Aggregation Problem (CAP).

2.3 The Modifiable Areal Unit Problem (MAUP) and the Category Aggregation Problem (CAP)

LUC can be mapped and conceptualized in different ways; from the most typical LUC maps in which the areas are classified into homogeneous categories, to choropleth maps which summarize, at selected administrative levels, different statistical values for the LUC they contain. In all cases, LUC information is expressed via polygon-based geometry but the MAUP is most noticeable in choropleth maps.

The MAUP was analysed in depth by Openshaw and Taylor (1979) and its effects have been tested in a number of research studies (García-Álvarez et al. 2019; Cebrecos et al. 2018; Rajabifard et al. 2000). The MAUP appears when a specific variable is observed in spatial units of different levels within a hierarchical structure (Eagleson et al. 2002, 2003). The MAUP causes two effects—zoning and scale. The first refers to the different patterns and associated statistical measures resulting from different aggregation arrangements within the same hierarchical level. The second takes the form of new and different patterns of the analysed variable that appear when downscaling, i.e. when units are aggregated together to make larger units.

LUC mapmakers and users need to be aware of the impact of the MAUP in order to facilitate both successful communication and well-informed decision-making.

Another issue in relation to the downscaling of information is the Category Aggregation Problem (CAP), which was formulated more recently (Pontius and Malizia 2004). This problem refers to the important consequences of grouping the categories in a thematic legend together. This leads to the disappearance of certain subcategories from the legend, so complicating the analysis of the changes in these variables over time (García-Álvarez 2018). The aggregation of categories also reduces the level of detail offered by the map.

In LUC these constraints are key aspects in the correct production and analysis of related maps. Figure 2 illustrates some of these issues. At the scale used in these maps, the progressive categorical aggregation from left to right shows the need for larger MMUs. The most categorically detailed map is very difficult to read, while the most generalized map provides insufficient information. Setting the MMU therefore entails a trade-off between the scale, the level of analysis sought, and the number of categories. This means that both components (thematic and spatial) of the geographic information must be considered simultaneously when setting the MMU in LUC mapping.

3 Visual Variables

The expression 'visual variable' was used by J. Bertin (1967) to designate the components of a system of signs. Later on, Slocum et al. (2005) defined it as the variations and perceived differences in the signs used to represent a thematic phenomenon. Other terms adopted by cartographers when referring to visual variables are symbol, graphical variable, graphical primitive or mark. Bertin identified six visual variables: shape, orientation, colour, value, grain and size, which have since formed the basis of studies of cartographic semiotics (Slocum et al. 2005).

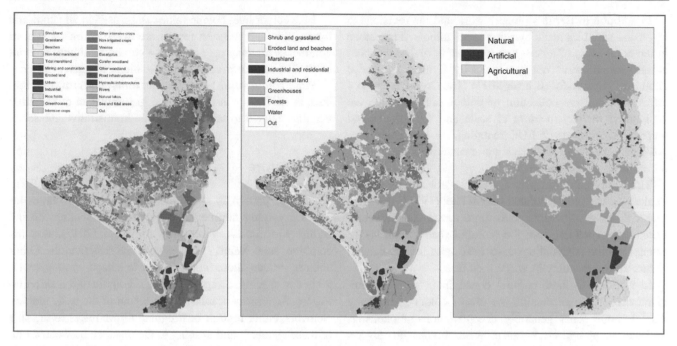

Fig. 2 Examples of LUC map information and issues arising from changes in the MMU and the aggregation of categories

3.1 Shape

Shape is the first variation distinguishable on any map. It helps identify the different types of objects appearing on a map, which are described by different contours. These contours may be regular and abstract (geometric signs) or figurative (pictograms). Shape corresponds to a nominal level of measurement and only allows us to convey either associations between objects with the same shape or differences between elements represented by different shapes. Shape is neither ordinal nor quantitative and cannot therefore be used for thematic phenomena with ordinal or quantitative levels of measurement (Cauvin et al. 2010a).

In LUC mapping as in any other kind of polygon-based mapping, shape can only affect filling patterns, not the shape of the polygons themselves. The only exception to this rule are cartograms, in which both the size and the shape of polygon objects vary in line with quantitative thematic values. In maps showing point and line features, shape is frequently used to highlight different associations between categorical objects.

3.2 Orientation

The orientation of a sign refers to its position relative to a reference framework and it is expressed in degrees (between 0 and 360). As with shape, orientation can only represent the attributes on a nominal level of measurement and can only affect point-based elements (Cauvin et al. 2010a). For line,

polygon or volume geometries, the orientation would only affect the filling patterns (textures) chosen. It is used much less frequently than other visual variables, especially in LUC mapping.

3.3 Colour Hue

Colour hue (often referred to simply as colour) is the most complex visual variable and its use in maps has been extensively analysed by cartographers (Bertin 1967; Robinson et al. 1984; Monmonier 1993; Slocum et al. 2005; Cauvin et al. 2010a). Colour varies depending on the light source, the reflective characteristics of the observed object and the human eye. The visible world is in fact composed of colourless matter but electromagnetic waves with different wavelengths are perceived as different colours by most people.

As a visual variable on a map, unlike shape and orientation, colour can be used not only in points, but also in lines and polygons. As regards its properties in relation to thematic information, colour is selective, separative and associative. Colour hues are neither ordered nor quantitative, which means they cannot be used to represent attributes measured at quantitative scales, and are therefore only suitable for representing phenomena measured at nominal scales. However, under certain conditions and when arranged in the appropriate order, colours can also be used to express order and opposition. For instance, yellow, orange and red can represent low, medium, and high data values, respectively (White 2017).

In addition to Bertin's pioneer work and the revisions to his visual variables made by subsequent authors, a milestone in the application of colour hue schemes in digital mapping is the ColorBrewer Tool developed by Cynthia Brewer at Penn State University (Brewer 2021). The ColorBrewer tool offers an extensive collection of colour ramps, which are well-suited for any measure of scale and for colour-blind map users. In terms of LUC mapping, an interesting proposal for colouring LUC maps with coarse pixel data can be found in Raposo et al. (2016).

The use of colour in mapping is also affected by its cultural connotations. As pointed out by Hall (1971), signs and gestures have different, sometimes even contradictory meanings depending on the cultural background. One example is the connotations associated with red, as danger, versus green, as safety in western cultures.

In addition to these cultural constraints, for map communication to be successful, the use of colour in mapping must honour some generally accepted conventionalisms. In LUC mapping, for instance, water bodies are always represented in light blue, while residential areas are normally depicted in red.

A very useful, well-known colour scheme for LUC mapping was established by the European Environmental Agency in the Corine Land Cover project (EEA 2017). Its 44 categories are represented by colours whose different hues are assigned to different groups of categories. In this way, artificial areas are represented in reds and purples, agricultural uses in yellow, forests in green, open spaces in grey and green, and wetlands and water bodies in blue.

3.4 Colour Value

White (2017) defined colour value as the lightness or darkness of a colour from pure black to pure white. Its variation constitutes "a continuous progression which the eye perceives in the grayscale stretching from black to white" (Bertin 1967) in a given area. Cauvin et al. (2010a) noted that the term progression conveys the basic property of this visual variable—order. It can be expressed as the ratio of the respective quantities of black and white.

As this is an excellent way of expressing order, it highlights the differences in a hierarchical system. Even though it is frequently used to represent quantities, the human capacity to associate different colour values with different quantities is very limited. Today, however, digital mapping allows black to be allocated in amounts that vary in proportion with the thematic value, so making it possible to use value ramps that overcome this limitation.

Like colour hue, colour value can be used in all geometric forms, although the best results are obtained on an area or volume, as the map user requires a certain minimum amount of surface area to perceive the variations of grey.

Since colour value is not suitable for representing nominal data, in LUC maps it is only used to summarize quantitative variables related to land use within administrative areas.

3.5 Texture

Texture or pattern is a complex visual variable that comprises a varying number of components depending on the author you consult. According to White (2017), textures combine size, value, hue, shape and orientation. Other authors reduce these components to shape, arrangement, grain and spacing. Shape is the basic graphic unit making up texture. Arrangement refers to the layout of the basic graphic elements, either regular or irregular. Grain refers to the size of these elements and spacing to the distance between them. The use of textures for data measured at different levels is also controversial. While White recommends that textures only be used for nominal and ordered attributes of areas and lines, other authors (Cauvin et al. 2010a) claim that they can also be used for quantitative data.

Nowadays, textures are not used as often in mapping as they once were. In the past, when colour printing was significantly more expensive, textures were frequently used to fill out areas containing nominal, ordinal or quantitative information. Today textures have largely been replaced by colour. However, they are sometimes used in combination with other covering visual variables such as colour hue or colour value, so as to increase the amount of information provided by the map.

Textures can be useful in LUC mapping when the basic LUC information is combined with other relevant information. In the case shown in Fig. 3, the area occupied by the Sierra de Guadarrama National Park in Spain has been texturized to differentiate it from the rest of the mapped area.

3.6 Size

Size is, together with colour value, the most frequently used visual variable for representing quantitative data. Size can be defined as the variation in the area or the volume of a sign. It is rarely used in LUC mapping as these maps are normally based on categorical data. Although in theory, size expresses quantity, order and selection (Bertin 1967), its use in representing qualitative information can lead to confusion.

Fig. 3 Use of textures in LUC mapping

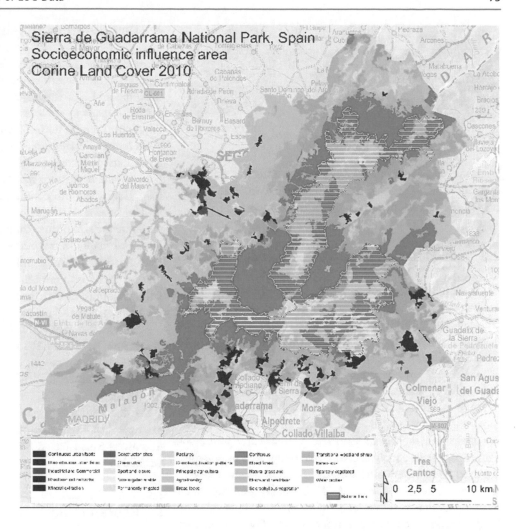

Thus, size is only recommended for representing ordinal or quantitative data.

As regards the geometries of the map, size can only be fully applied to points. In the case of lines, since the distance between two points is fixed, size can only be applied as a variation in line width. As for polygons, any variation in their size based on a quantitative attribute other than surface area would result in the loss of their cartographic projection properties. Given this constraint, when the nature of the attribute is such that its representation with size is recommended, polygons may be represented by a point, usually at their centroid, which varies in size according to the value attached to the polygon attribute.

LUC map products using this visual variable are therefore limited to those summarising quantities such as the proportion of land occupied by each land use category, the proportion of land undergoing a land use change between two dates, or other related quantitative variables. In all cases, these quantities are summarized on a superimposed spatial structure, usually administrative units.

3.7 Visual Variables and Geometric Dimension

In the previous paragraphs, we have seen how some visual variables adapt better than others to the varying geometric forms in which geographical information is presented.

Figure 4 summarizes the recommended use of the visual variables with different geometries. Green cells show optimal combinations, red cells show inapplicable combinations and yellow cells show the combinations that are subject to certain conditions. Points accept all visual variables with the exception of textures, although some points may be big enough to accommodate texture pattern. Given that lines are defined as the shortest distance between two points, they can only accept colour hue and colour value. However, a thick line can have different shapes and textures. As regards size, according to the above definition, lines can only vary in width, not in length. Polygons are more restrictive, in that they will only accept colour hue, colour value and texture. Any change in their shape, orientation or size would result in a distortion of the cartographic base which makes them

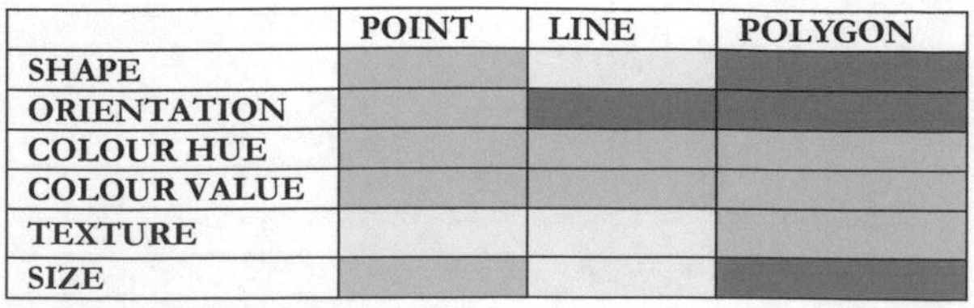

	POINT	LINE	POLYGON
SHAPE			
ORIENTATION			
COLOUR HUE			
COLOUR VALUE			
TEXTURE			
SIZE			

Fig. 4 Visual variables and geometric dimension

unusable. However, these three visual variables could be applied to polygons when they (the visual variables) form part of the texture pattern that fills these polygons.

3.8 Visual Variables and Measurement Level

In the above descriptions of the visual variables, we also outlined the meanings with which they are associated, and consequently the most suitable level of measurement for them. In general terms, the visual variables that can be ordered (colour value, size, texture and colour hue if properly ordered) are best suited for attributes measured at ordinal level. Visual variables indicating quantity (size and to some extent colour value and texture) can be applied to represent attributes at interval or ratio measurement levels. For their part, the visual variables with selective and associative properties, such as colour hue, shape and orientation, are used to represent attributes measured at nominal level.

Orientation is a special case. It usually has the same meaning as shape, but under certain conditions it can also be used to represent ordered attribute series. For instance, an arrow symbol pointed at any angle in the 360° of a circle could be associated with an ordered attribute depicting every point in a hierarchy based on the angle of the arrow.

As regards textures, their complex nature makes them suitable for any measurement level. Changes in the shape, orientation and colour hue of the pattern of elements that make up the texture would apply to attributes at nominal scale while size and colour value variations would be used to represent attributes at quantitative and ordinal measurement scales. Figure 5 summarises the recommended application of visual variables to represent attributes with different measurement levels.

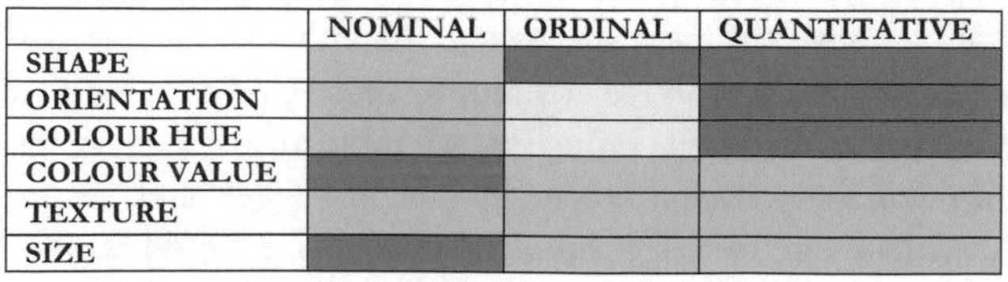

	NOMINAL	ORDINAL	QUANTITATIVE
SHAPE			
ORIENTATION			
COLOUR HUE			
COLOUR VALUE			
TEXTURE			
SIZE			

Fig. 5 Visual variables and associated level of measurement

4 Representing Nominal LUC Data

Most common LUC maps depict an area or region, highlighting with different colours the homogeneous patches of the different LUC categories it contains. As described above, for these maps to serve as successful communication tools, they must comply with a series of cartographic rules.

In terms of cartographic projection, the proportionality of areas must be preserved. If not, it would be impossible to compare the respective size of the different categories on the map. Equivalent projection must therefore be used.

The final size of the map will determine the scale and therefore the size of the Minimum Mapping Unit. In the case of digital maps, we recommend that an intelligent zoom be used so that the map only displays features equal to or greater than the minimum size. As a fixed image, the final LUC map must also strike a balance between the MMU and the number of LUC categories.

The visual variable best suited for categorical data is colour hue. Its use in LUC mapping must adhere to generally accepted conventions such as the use of blue colours to represent water bodies, reds and purples for built-up areas and so on.

In line with these recommendations, Fig. 6 presents an example LUC map for the Guadiamar River Basin area in Southwest Spain based on Corine Land Cover data for the year 2000.

5 Representing LUC Quantitative Data

As pointed out above, the cartographic representation of LUC quantitative data requires additional layers, such as administrative units, for the computation of these quantities at a meaningful spatial level. Some sort of selection must be undertaken in order for the resulting maps to be readable. Figure 7 shows examples of the percentage of land occupied by natural, agricultural and artificial land use categories respectively.

As with any map representing quantitative attributes, special attention must be paid to the number of intervals and their limits. An excessive number of intervals would make it difficult to differentiate between the associated symbols, regardless of whether they are based on size or colour value. By contrast, if too few intervals are used, this will reduce the level of detail of the information provided by the map. Brooks and Carruthers (1953) suggested that the number of classes should be less than or equal to five times the decimal logarithm of the number of observations. Other authors suggested that the number of classes should be equal to 3.3 times the decimal logarithm of the number of observations plus 1 (Huntsberger 1961). In both cases the number

Fig. 6 Example of LUC map. Guadiamar River Basin, Spain

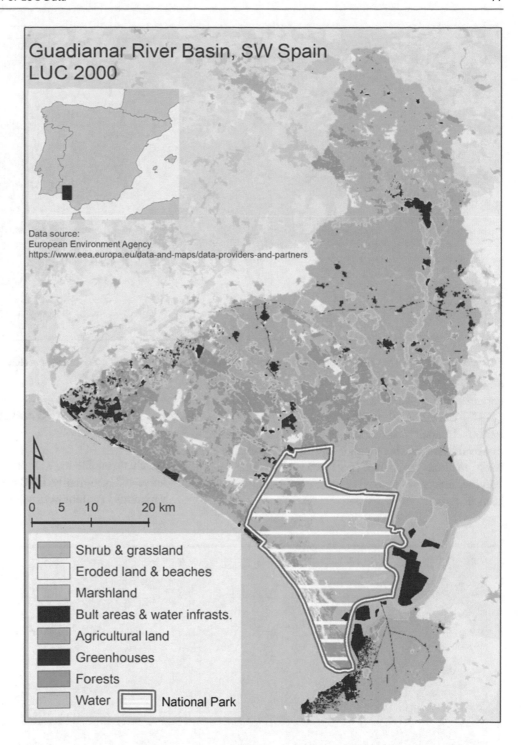

Guadiamar River Basin, SW Spain
LUC 2000

Data source:
European Environment Agency
https://www.eea.europa.eu/data-and-maps/data-providers-and-partners

0   5   10   20 km

Shrub & grassland
Eroded land & beaches
Marshland
Bult areas & water infrasts.
Agricultural land
Greenhouses
Forests
Water   National Park

of classes increases quickly in line with the number of observations, making it difficult for the map reader to differentiate between the symbols. The average maximum number of different colour values that humans can perceive in a map is seven (Olson 1975) and, according to Robinson (1998), the optimum number is five.

The limits established for each of the intervals have a strong impact on the final appearance and usefulness of the map. There are a large number of possible methods for establishing these limits, but not all of them adapt to all sorts of data. The distribution of the thematic variable must be taken into account, as some methods are only suited to

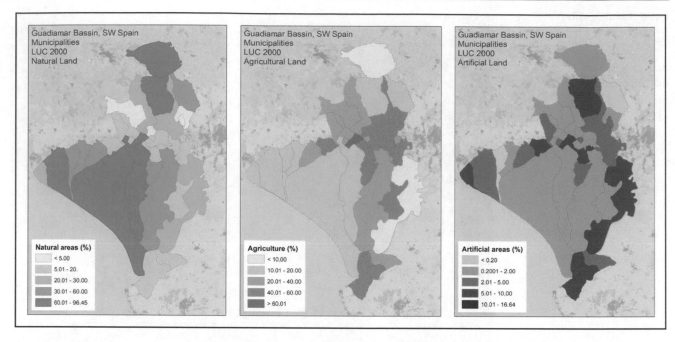

Fig. 7 Quantitative maps showing the area occupied by different land use categories in the Guadiamar River Basin

certain specific distributions. Following work by Monmonier (1982), Cauvin et al. (2010b) explained the details of the various different methods and analysed their advantages and disadvantages. In this chapter, we will be focusing on the main methods available in standard GIS software. The varying impact of three of the most common methods can be seen in Fig. 8.

6 Representing Qualitative and Quantitative LUC Data

Pie charts enable the simultaneous communication of qualitative and quantitative LUC data. The pie symbol can display variations in colour hue, colour value, size and texture. It can

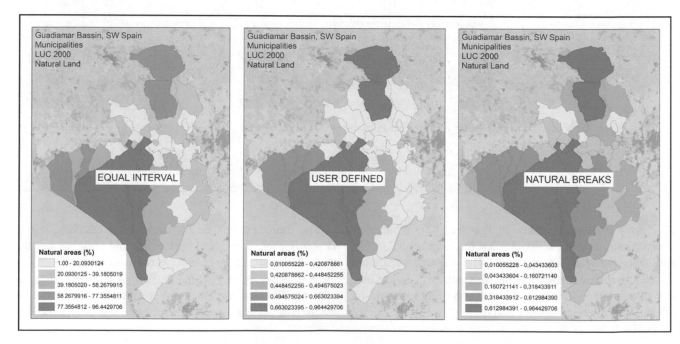

Fig. 8 Impact of the classification method in quantitative maps

represent nominal data by means of colour hue variations, while ordinal and quantitative data can be represented with size or colour value. Figure 9 shows the land occupied by natural, artificial and agricultural uses in the municipalities in the Guadiamar River Basin. Symbol size is proportional to the total area of the municipalities and the pie sections correspond to each of the categories coloured with a different hue.

7   Representing LUC Changes

One of the key areas in LUC studies is the analysis of the cover changes that have taken place in the past or are predicted to occur in the future, according to different scenarios (White and Engelen 1993; Camacho Olmedo et al. 2018; Hewitt et al. 2014; Guzman et al. 2020). The methods

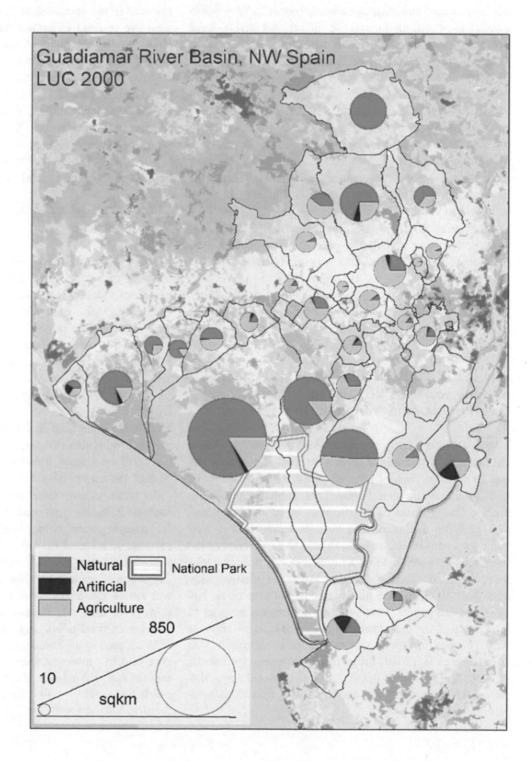

Fig. 9 Pie chart map representing the proportions of LUC categories in the municipalities in the Guadiamar River Basin

applied to undertake this analysis are usually based on the comparison of two input LUC maps with different dates.

The cartographic representation of the LUC change that has taken place between these two dates is often expressed in terms of the amount of land gained or lost by each land use category. This is a quantitative attribute and is therefore subject to the constraints summarized in Sect. 5.

As regards the representation of categories as nominal data, an excessively large number of land use categories in the input maps and their associated, theoretically possible transitions would in turn result in an excessively large number of new categories. This means that some kind of selection process must be performed. The options include: (i) reducing the map to the binary categories of "stable" and "changed"; (ii) selecting just one land use category to represent the areas gained or lost by it; and (iii) selecting the areas gained or lost by one specific land use category, in order to represent the land use categories from which or to which these areas have changed.

In order to make the comparison, the two input maps must be overlaid. During this process, it is highly likely that new areas of varying size will appear on the output map. The issues relating to the MMU discussed in Sect. 2.2. apply to the representation or possible generalization of these new polygons. Figure 10 presents a composite of two input maps with LUC information for 1956 and 1999 respectively, an output map showing areas that have undergone LUC changes between these dates and a second output map showing the main transitions that have taken place between LUC categories.

8 New Forms of Visualizing and Communicating LUC Data

Throughout the examples presented so far, we have made clear that LUC representation is a far from simple task and that LUC maps convey even the most relevant aspects of LUC information with difficulty. These limitations can have serious consequences when it comes to taking policy and land planning decisions. The abstract representation, normally by means of colour hues, of land use categories or transitions between them does not necessarily make it easier for users to understand the real landscape changes they represent. Policy makers may not be expert map users, and will therefore require more intuitive information in order to fully comprehend the impacts of predicted land use changes on landscapes, economy, society and the environment. Van Lammeren et al. (2010) found that users complained about an excessive amount of detail on A4-size printed maps, that the colours were too close, and that it was difficult to compare the maps.

In an attempt to alleviate these issues, various interesting case studies have integrated new approaches to cartographic visualization (Cauvin et al. 2010c) such as realistic 3D models (Appleton et al. 2002; Paar 2006; van Lammeren et al. 2010), and have explored the use of historic photography to illustrate land use changes (Kull 2005).

In addition to these realistic 3D examples, technological developments in the mapping industry have enabled the production of new cartographic tools that have yet to be explored in the communication of LUC information. Three areas are in need of further research and implementation. First, the current predominance of digital maps that are viewed through a computer device equipped with speakers, contrasts with the almost complete absence of research into sound mapping applied to LUC analysis. Second, the limited interactive capacity of LUC digital maps makes it difficult to compare them. And third, the possibilities offered by the computerised environment for visualizing animations, perhaps the most efficient tool for communicating changes over time, have yet to be applied in LUC change studies.

9 Conclusions

In this chapter we have reviewed the main cartographic methods for representing and communicating LUC and LUCC information. The maps produced must comply with basic cartographic rules and must therefore have an appropriate cartographic projection, a balanced level of generalization, MMU and attribute details, a suitable set of visual variables and, in the case of quantitative data, a proper method for the classification of the thematic variable.

Even the maps that comply with these rules are often not fully comprehensible for their final users. This may be because the scale used in the final printed maps, the format in which most decision-makers receive the information, is too small or simply because not all the actors involved "speak" the cartographic language.

In order to overcome these issues, new cartographic methods including geovisualization techniques like realistic 3D mapping, are being explored. Other technological advances like sound mapping, fully interactive mapping or animated mapping are still underused in LUC studies. The integration of realistic 3D models with animation and sound will enable the inclusion of moving living creatures (like animals or people), human-made moving objects (like vehicles or windmills), vegetation, topography, buildings, and variations in the atmosphere or the light. Progress of this kind in LUC representation will make LUC maps more realistic and will enhance their communication capabilities, which in turn will help ensure better-informed decision-making processes.

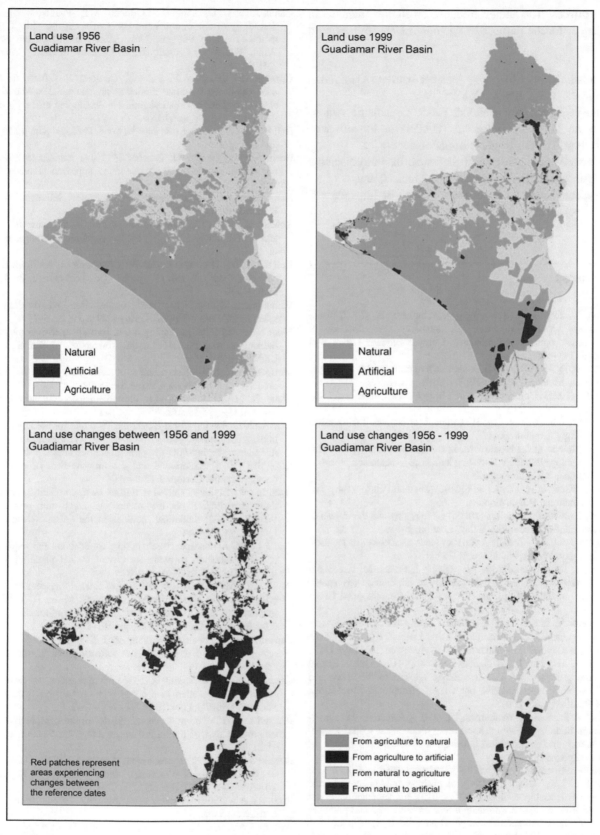

Fig. 10 Cartographic representation of LUC changes

Data Sources The author produced all figures included in this chapter for the purpose of this book. Data sources used are:

- Spanish National Mapping Agency: Instituto Geográfico Nacional (IGN) at www.ign.es;
- Spanish Agency for National Parks: Organismo Autónomo de Parques Nacionales (OAPN) at https://www.miteco.gob.es/es/parques-nacionales-oapn/;
- Spanish National Bureau of Statistics: Instituto Nacional de Estadística (INE) at https://www.ine.es/; and
- European Environment Agency (EEA) at https://www.eea.europa.eu/.

References

Appleton K, Lovett A, Sünnenberg G, Dockerty T (2002) Rural landscape visualisation from GIS databases: a comparison of approaches, options and problems. Comput Environ Urban Syst 26(2–3):141–162

Bertin J (1967) Sémiologie graphique. Les diagrammes, les réseaux et les cartes, Mouton, Gauthier-Villars, Paris, p 431

Brewer C (2021) ColorBrewer Tool. http://www.ColorBrewer.org. Accessed 8 July 2021

Brooks CEP, Carruthers N (1953) Handbook of statistical methods in meteorology. London, p 412

Camacho Olmedo MT, Paegelow M, Mas J-F, Escobar F (eds) (2018) Geomatic approaches for modeling land change scenarios. Springer International Publishing, p 525

Cauvin C, Escobar F, Serradj A (2010a) Thematic cartography and transformations. Wiley, London, p 463

Cauvin C, Escobar F, Serradj A (2010b) Cartography and the impact of the quantitative revolution. Wiley, London, p 408

Cauvin C, Escobar F, Serradj A (2010c) New approaches in thematic cartography. Wiley, London, p 291

Cebrecos A, Domínguez-Berjón MF, Duque I, Franco M, Escobar F (2018) Geographic and statistic stability of deprivation aggregated measures at different spatial units in health research. Appl Geogr 9:9–18

Di Gregorio A, Jansen L (2000) Land Cover Classification System (LCCS): Classification concepts and user manual, FAO, http://ww.fao.org/3/x0596e/X0596e00.htm#P-1_0. Accessed 1st July 2021

Eagleson S, Escobar F, Williamson I (2002) Hierarchical Spatial Reasoning theory and GIS technology applied to the automated delineation of administrative boundaries. Comput Environ Urban Syst 26(2002):185–200

Eagleson S, Escobar F, Williamson I (2003) Automating the administration boundary design process using hierarchical spatial reasoning theory and geographical information systems. Int J Geogr Inf Sci 17(2):99–118

European Environment Agency (2017) Technical specifications for the CORINE Land Cover (CLC) pilot projects implemented in the Eastern Partnership countries (2017–2019), Implementation of the Shared Environmental Information System principles and practices in the Eastern Partnership countries (ENI SEIS II East), p 15

García-Álvarez D (2018) Aproximación al estudio de la incertidumbre en la modelización del Cambio de Usos y Coberturas del Suelo (LUCC). PhD Dissertation. Universidad de Granada, p 445

García-Álvarez D, Camacho Olmedo MT, Paegelow M (2019) Sensitivity of a common Land Use Cover Change (LUCC) model to the Minimum Mapping Unit (MMU) and Minimum Mapping Width (MMW) of input maps. Comput Environ Urban Syst 78:101389

Guzman LA, Escobar F, Peña J, Cardona R (2020) A cellular automata-based land-use model as an integrated spatial decision support system for urban planning in developing cities: the case of the Bogotá region. Land Use Policy 92:1–13

Hall ET (1971) La dimension cachée, Seuil, Paris, 1st edn, p 256, 1966, 2nd edn, p 254

Hewitt R, van Delden H, Escobar F (2014) Participatory land use modelling, pathways to an integrated approach. Environ Model Softw 52:149–165

Huntsberger DV (1961) Elements of statistical inference. Allyn & Bacon, Boston, p 528

Koláčny A (1977) Cartographic information. A fundamental concept and term in modern cartography, Cartographica, monograph no 19, pp 39–45

Kull CA (2005) Historical landscape photography as a tool for land use change research. Norsk Geografisk Tidsskrift-Norwegian J Geogr 59:253–268

Monmonier MS (1982) Flat laxity optimisation and rounding in the selection of class intervals. Cartographica 19(1):16–27

Monmonier MS (1993) Mapping it out. Expository cartography for the humanities and social science. University of Chicago Press, Chicago, p 301

Morrison JL (1976) The science of cartography and its essential processes. Int Yearbook Cartogr 16:85–97

Muller JC (1975) Association in choropleth map comparison. Annals Ass Am Geogr (AAAG) 65(3):403–413

NOAA (2011) Digital Coast GeoZone. Tech talk for the Digital Coast. https://geozoneblog.wordpress.com/2011/10/28/resolution-vs-minimum-mapping-unit-size-does-matter/ Accessed 1 July 2021

Olson JM (1975) Experience and the improvement of cartographic communication. Cartogr J 12:94–108

Openshaw S, Taylor PJ (1979) A million or so correlation coefficients: three experiments on the modifiable areal unit problem. In: Wrigley N (ed) Statistical methods in the spatial sciences. Pion, London, pp 127–144

Paar P (2006) Landscape visualizations: applications and requirements of 3Dvisualization software for environmental planning. Comput Environ Urban Syst 30(6):815–839

Pontius RG, Malizia NR (2004) Effect of category aggregation on map comparison. In: Egenhofer MJ, Freksa C, Miller HJ (eds) Third international conference of the geographic information science (GIScience). Springer, Adelphi, pp 251–268

Rajabifard A, Escobar F, Williamson I (2000) Hierarchical spatial reasoning applied to spatial data infrastructures. Cartography 29 (2):41–50

Raposo P, Brewer C, Sparks K (2016) An impressionistic cartographic solution for base map land cover with coarse pixel data. Cartogr Perspect. https://doi.org/10.14714/CP83.1351

Ratajski L (1978) The main characteristics of cartographic communication as a part of theoretical cartography. Int Yearbook Cartogr 18:21–32

Robinson AH (1952) The look of maps: an examination of cartographic design. University of Wisconsin, Madison, US, p 105. Reprinted in The American Cartographer, 13:3, 280, (1986)

Robinson AH (1953) Elements of cartography, 1st edn. John Wiley & Sons, New York, p 245

Robinson AH, Sale R, Morrison J, Muehrcke PC (1984) Elements of cartography, 5th edn. John Wiley & Sons, New York, p 541

Robinson AH (1969) Elements of cartography, 2nd edn. John Wiley & Sons, New York, p 333

Robinson GM (1998) Methods & techniques in human geography. John Wiley & Sons, England, p 390

Slocum TA, Mcmaster RB, Kessler FC, Howard HH (2005) Thematic cartography and geography visualization. Pearson Prentice Hall, p 518

van Lammeren R, Houtkamp J, Colijn S, Hilferink M, Bouwman A (2010) Affective appraisal of 3D land use visualization. Comput Environ Urban Syst 34:465–475

White T (2017) Symbolization and the visual variables. the geographic information science & technology body of knowledge, 2nd Quarter. In: Wilson JP (ed) https://doi.org/10.22224/gistbok/2017.2.3

White R, Engelen G (1993) Cellular dynamics and GIS: Modelling spatial complexity. Geograph Syst 1:237–253

Wu J, Harbin L (2006) Concepts of Scale and Scaling. In: Wu J, Jones KB, Li H, Loucks OL (eds) Scaling and uncertainty analysis in ecology: methods and applications. Springer, Dordrecht, The Netherlands, pp 3–15

Sample Data for Thematic Accuracy Assessment in QGIS

Miguel Ángel Castillo-Santiago, Edith Mondragón-Vázquez, and Roberto Domínguez-Vera

Abstract

We present an approach that is widely used in the field of remote sensing for the validation of single LUC maps. Unlike other chapters in this book, where maps are validated by comparison with other maps with better resolution and/or quality, this approach requires a ground sample dataset, i.e. a set of sites where LUC can be observed in the field or interpreted from high-resolution imagery. Map error is assessed using techniques based on statistical sampling. In general terms, in this approach, the accuracy of single LUC maps is assessed by comparing the thematic map against the reference data and measuring the agreement between the two. When assessing thematic accuracy, three stages can be identified: the design of the sample, the design of the response, and the estimation and analysis protocols. Sample design refers to the protocols used to define the characteristics of the sampling sites, including sample size and distribution, which can be random or systematic. Response design involves establishing the characteristics of the reference data, such as the size of the spatial assessment units, the sources from which the reference data will be obtained, and the criteria for assigning labels to spatial units. Finally, the estimation and analysis protocols include the procedures applied to the reference data to calculate accuracy indices, such as user's and producer's accuracy, the estimated areas covered by each category and their respective confidence intervals. This chapter has two sections in which we present a couple of exercises relating to sampling and response design; the sample size will be calculated, the distribution of sampling sites will be obtained using a stratified random scheme, and finally, a set of reference data will be obtained by photointerpretation at the sampling sites (spatial units). The accuracy statistics will be calculated later in Sect. 5 in chapter "Metrics Based on a Cross-Tabulation Matrix to Validate Land Use Cover Maps" as part of the cross-tabulation exercises. The exercises in this chapter use fine-scale LUC maps obtained for the municipality of Marqués de Comillas in Chiapas, Mexico.

Keywords

Single map validation • Sample size • Sampling design • Systematic sampling • Random sampling • Reference data

1 Sample Size Estimation and Spatial Distribution of Sampling Sites in a Stratified Randomised Design

When conducting error assessment, it is important to strike a balance between the theoretical requirements and the practical reality of implementation (Congalton 1991). In the map validation process, it is therefore crucial to have the right sample size and to use the right number of spatial units in order to ensure that reliable accuracy indices can be obtained without incurring high costs.

There is no single right way to calculate the ideal sample size; in general, this task could be regarded as a process of successive approximations, in which criteria such as the availability of resources, levels of sampling error, or the desired degree of accuracy all play an important role. The expertise of the user and his/her interest in certain thematic classes are also important factors in the success of the estimation process.

An initial estimation of the most appropriate sample size can be made with the formulae used in statistical sampling. Equations for the validation of thematic maps have often

M. Á. Castillo-Santiago (✉) · E. Mondragón-Vázquez · R. Domínguez-Vera
Departamento de Observación y Estudio de la Tierra, la Atmósfera y el Océano, El Colegio de la Frontera Sur, San Cristóbal de las Casas, Mexico
e-mail: mcastill@ecosur.mx

D. García-Álvarez et al. (eds.), *Land Use Cover Datasets and Validation Tools*,
https://doi.org/10.1007/978-3-030-90998-7_6

been taken from the original work by Cochran (1977) and for a simple, stratified randomised design, Stehman and Foody (2019) propose the following formula:

$$n = \frac{z^2 O(1 - O)}{d^2}$$

where O = accuracy expressed as a proportion (in the case of simple random sampling O is the anticipated overall accuracy, whilst in stratified sampling it is the anticipated user's accuracy); n = number of sampling sites; z = percentile from the standard normal distribution (z = 1.96 for a 95% confidence interval); and d = desired half-width of the confidence interval of O. It can also be expressed as $z * s\left(\widehat{O}\right)$,

In the case of stratified random sampling, Olofsson et al. (2014) recommend the following formula:

$$n = \frac{\left(\sum W_i S_i\right)^2}{\left[S\left(\widehat{O}\right)\right]^2 + \left(\frac{1}{N}\right) \sum W_i S_i^2} \approx \left(\frac{\sum W_i S_i}{S\left(\widehat{O}\right)}\right)^2$$

where S(Ô) = standard error of estimated overall accuracy; Wi = mapped proportion of the area of class I; Si = standard deviation of class i, $S_i = \sqrt{Ui(1 - Ui)}$; and Ui = User's accuracy for class i.

Note that in both cases, it is necessary for the user to define certain parameters in advance, such as the permissible level of error (S(Ô)) or the user's accuracy values. These data should be obtained from prior or approximate knowledge regarding the quality of the map or from previous experience in producing maps with similar characteristics.

Sometimes it may be difficult to estimate user's accuracy, so practical recommendations for sample size calculation may be useful. Hay (1979) proposed allocating 50 sampling sites per thematic class. Congalton (1988, 2016), based on a series of Monte Carlo simulations, also recommended allocating 50 sampling sites per thematic class but only when map extent was under 500,000 ha and there were 12 or fewer thematic classes. In more complex situations, i.e. when the map extent was over 500,000 ha or it had more than 12 classes, he proposed allocating 75–100 samples per thematic class. According to his approach, therefore, total sample size is dependent on the size of the thematic map and the number of classes it contains.

Sampling design is another important factor to consider. Frequently used types include systematic, simple random and stratified random sampling. Traditionally, the cost or

ease of fieldwork was a criterion for preferring some designs over others. With the increased availability of high-resolution imagery, in many cases, it is no longer essential to obtain data directly in the field. Reference data can now be interpreted from the imagery, so reducing costs dramatically.

The systematic sampling is easier to implement in the field, but has the disadvantage that it cannot be used to construct an unbiased variance estimator (Stehman 2009). Randomised designs can be more effective at estimating accuracy parameters (Congalton 1988) and can adapt more easily to changes in sample size without losing their probability sampling character (Stehman 2009). In the case of stratified random sampling, once the sample size has been calculated, rules must be established to allocate the sampling sites to each of the strata or thematic classes. These rules normally apply one of the following criteria: an equal number of sites in each class, a number proportional to the size of the class, or a number that depends on both the size of the class and the expected user's accuracy for this class (optimal allocation). The allocation criterion affects the precision of some of the accuracy parameters. For example, with optimal allocation, the variance of the overall accuracy and the user's accuracy for rare classes decreases. By contrast, with equal distribution, more precise estimates of the user's accuracy for rare classes can be obtained, whilst in large classes, the precision decreases (Stehman 2012).

Regardless of the design chosen, a problem that sometimes arises is the under-representation of small or·rare thematic classes in the sample. In other words, once the sample has been calculated and distributed, it may leave some classes with too few sites (<50). Some authors (Olofsson et al. 2014; Finegold et al. 2016) suggested a two-step solution for the specific case of stratified random sampling. First, calculate and distribute the sample according to the proportional or optimal criteria, and if any class turns out to have a small number of sampling sites (<50), then allocate 50 sites to it and recalculate the total sample size.

Once all the different stages of the accuracy assessment process have been performed, the precision values obtained should be reviewed, e.g. the magnitude of the overall accuracy standard error or the width of the user's accuracy confidence intervals. Even if there are some variations from the expected values, if the values obtained meet the analyst's targets as regards accuracy, then there would be no need to repeat the analysis (Stehman and Foody 2019). If not, it would be necessary to try again, adding new sites to increase the sample size.

QGIS Exercise: To calculate sample size and to distribute sample sites using a random stratified approach

Next, we present a practical way to carry out the sampling design for obtaining reference data. In this exercise, we will estimate the sample size for a stratified random design, for which we will have to specify the expected standard error of the overall accuracy and to provide an a priori estimate of user's accuracy values. Sometimes, these figures may be difficult to provide in which case we can use the default values provided by the tool we will be using.

Available tools

- MapAccurAssess plugin
- Semi-Automatic Classification Plugin (SCP)
- AcATaMa plugin

There are several useful tools in QGIS for statistical sampling design. All of them are external plugins such as Semi-Automatic Classification Plugin (SCP), AcATaMa and MapAccurAssess. The MapAccurAssess plugin is a trial version specifically developed in the context of this book, which is not yet available in the official QGIS repositories.

SCP, which was developed by Congedo (2020), is a toolset for the classification and validation of land-cover and land use maps. With this plugin, the sample size and the allocation to each class must be calculated externally using a spreadsheet or other software. Once the number of sites per class has been defined, the plugin allows for a random distribution per thematic class. The size of the spatial units is indicated in number of pixels. Both the map and the samples must be in raster format.

AcATaMa was developed by the Group from the Forest and Carbon Monitoring System for the validation of single LUC maps (Llano 2019). It consists of a set of tools that guide the user through a series of steps: (a) sampling design (stratified or simple); (b) sample classification; and (c) calculation of the confusion matrix and accuracy statistics. In the sample classification step, the spatial unit is a pixel (or points in the GeoPackage or shapefile format), which is not very convenient for those who prefer to use a different spatial unit, such as group of pixels or polygons. At that stage (classification), a set of tools is enabled to zoom in on each of the samples, and four windows are created to display images of interest. An editable attribute table is also created to classify the samples.

In this exercise, we will be using MapAccurAssess, a plugin developed by the authors of this chapter, which includes several of the suggestions proposed by Olofsson et al. (2013) and Finegold et al. (2016). It is available at https://doi.org/10.5281/zenodo.5419130 with its associated documentation. For more information on the plugin, readers are referred to Chapter "About This Book".

This plugin provides several functions for calculating sample size in a stratified random design, using Neyman's optimal allocation to calculate the number of sampling sites in each thematic class. If, after that, any class has less than 50 sampling sites, it must be assigned between 50 and 100 sites depending on the complexity of the map. The result is a layer of points (shapefile) that are distributed over all the thematic categories of the map according to the stratified random design criteria. The points can be further modified to represent a polygon using QGIS functions.

Materials

Marqués de Comillas Land Use Cover Map 2019

Requisites

To calculate the area of each thematic class the LUC map must be projected in any cartographic projection (not geographic coordinates). The plugin has been tested on map projections with distance units in metres (rather than feet for example).

Execution

Step 1

Install the MapAccurAssess plugin. All the relevant information regarding the installation of the plugin can be found in Chapter "About This Book" and the plugin's manual, which is included in the plugin's download.

Step 2

Go to the Plugins Menu, select the Accuracy Assessment and *Random point* options. Alternatively, you can click on the Random Point icon (Fig. 1).

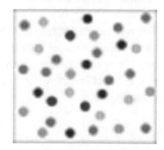

Fig. 1 Exercise 1. Step 2. MapAccurAssess plugin icon

Fig. 2 Exercise 1. Step 3.
MapAccurAssess plugin

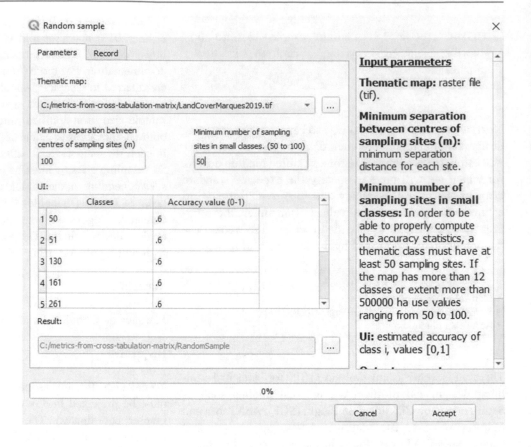

Step 3

In the dialogue box in Fig. 2, fill in the map filename (LUC map of Marqués de Comillas) and modify or accept the suggested values. A minimum distance between the centres of the spatial units must be specified in order to prevent overlapping, e.g. if the spatial units are square polygons of one ha, the minimum distance between their centres must be 100 m.

The Ui values (User's Accuracy for class i) refer to an a priori estimation of accuracy for the thematic class, which could be based on expert judgement or on previous assessments. If there are any doubts about these values, the default values can be retained. Whilst Ui can vary between 0 and 1, a value of 0.5 was allocated to a large number of sites. Values of over 0.5 will generate a smaller sample size. The last stage is to select the folder where the results will be saved.

Results and Comments

The results are displayed in the Record tab, and two types of output are generated and saved in the selected folder: (i) a. csv file with statistics about the thematic classes and (ii) a point shapefile where the points represent the centres of the sampling sites.

The .csv file contains a row for each thematic class and four columns showing id, area, the number of sampling sites estimated using Neyman's optimal criteria and the suggested number of sampling sites, adjusted to ensure that none of the classes have less than 50 sites (Table 1). If the area covered by a particular class is so small that 50 sites cannot be placed on it, the adjusted value will also be less than 50. Classes like this should be merged into other similar classes.

The vector point layer contains the spatial location of the centres of randomly distributed sampling sites (Fig. 3). Each point has two attributes, a unique identifier (id) and the thematic class value recovered from the LUC map.

2 Collection of Reference Data for Assessing the Accuracy of a Thematic Map

One of the major challenges of map evaluation is to obtain a reliable reference dataset with minimal positional errors and with the same date as the LUC maps. The aim is to obtain a data subset that faithfully represents the population from which it was extracted, so as to obtain confident accuracy estimates (Stehman 2009).

Table 1 Results from Exercise 1. Number of sampling sites per thematic classes

	Classes	Area (ha)	samples_neyman	samples_adjusted
0	50	26,009	77	77
1	51	32,875	98	98
2	130	252	0	50
3	161	6943	20	50
4	261	13,504	40	50
5	290	116,429	347	347
6	301	2357	7	50
7	420	2021	6	50

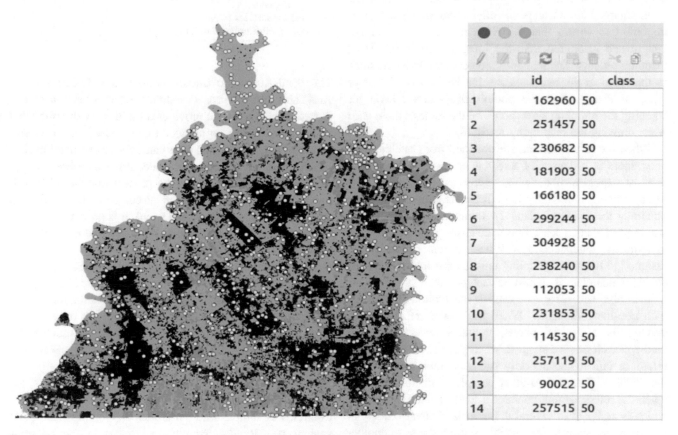

Fig. 3 Result from Exercise 1. Map showing the spatial distribution of sampling sites and the corresponding table of attributes

The collection of reference data requires the prior definition of several aspects relating to the size of the sampling area and the characteristics of the information we want to obtain (Olofsson et al. 2014): (a) characteristics of the spatial assessment units; (b) sources of reference data; (c) labelling protocol; and finally (d) classification agreements. Spatial assessment units refer to the sampling areas where the reference and map values are compared. Traditionally, the chosen spatial unit was a pixel or a polygon, or even a group of pixels, although there is still no consensus regarding the best size (Stehman and Czaplewski 1998; Olofsson et al. 2014; Stephen and Wickham 2011). What is certain is that various factors must be taken into account. For example, when a pixel is used as the spatial unit, it must be decided whether the land-cover label to be assigned will be exclusively what is observed on each individual pixel or whether the surrounding context will be taken into account, so as to reduce possible georeferencing errors. If we use a polygon or group of pixels, it will be necessary to define their size, for example, one hectare or blocks of three by three pixels. The advantage of using an area larger than one single pixel is that the incorrect assignment of labels due to georeferencing errors is minimised. The major drawback is that each spatial unit can contain several different land-cover classes, which means that rules must be drawn up to assign the land-cover to the right class (Stehman and Wickham 2011). The minimum mapping unit of the map must also be taken into account, given that the spatial unit must not be smaller than the minimum mapping unit. In the end, each user will have to opt for the spatial unit size that best suits his or her purposes.

The reference source can be either observed field data or data interpreted from satellite imagery and aerial photographs. Although data collected in the field is always preferable, this method is much more expensive, and the interpretation of aerial photographs and satellite images is often regarded as an acceptable alternative. In this case, it is important to ensure that the reference data has a higher quality and resolution than the images used in the initial mapping process. The labelling protocol should be the same as that used in the mapping, i.e. the land-cover classes or types of change, and the photointerpretation criteria for labelling the sampling sites should be the same as those used when drawing the map being assessed.

When the reference data are obtained from satellite imagery, there is a degree of uncertainty associated with the level of expertise of the photointerpreter. This uncertainty can be reduced if classification criteria are established before obtaining the reference data. To minimize interpreter bias, we suggest that at least two specialists perform the class assignment independently. When different labels are assigned to the same sampling unit, a third interpreter must decide which class it should be assigned to.

It is also necessary to establish the criteria for dealing with non-ideal situations. When the spatial reference unit, defined as a set of pixels, contains several different land-cover classes we suggest, when possible, assigning the reference unit to the majority category, representing more than 50% of the area. Another complex situation could be when the reference unit contains a linear feature or corridor which is assigned to several different land-cover classes. In this case, we suggest moving the sampling site to another place in which there is less uncertainty regarding class allocation. The producer and the person(s) assessing the map should always reach agreement on such decisions and document them, so as to avoid biases in the accuracy assessment.

QGIS Exercise: To collect reference data

This exercise is a guide to collecting reference data. Instead of fieldwork, high-resolution satellite imagery, available on Google Earth, is used together with various QGIS tools. The result of this exercise is a set of comparisons of land-cover observation taken from the high-resolution image (reference data) and the land-covers extracted from the LUC map under evaluation. The output data are formatted to compute the error matrix and accuracy statistics. These calculations are explained in Part III of this book (Sect. 5 in chapter "Metrics Based on a Cross-Tabulation Matrix to Validate Land Use Cover Maps").

Available tools
• SCP plugin
• AcATaMa plugin
• Vectorial menu
Geoprocessing tools
Buffer
• QuickMapServices plugin
• Google Earth Engine Data Catalog plugin

The Semi-Automatic Classification Plugin (SCP) and the AcATaMa plugin have a module for the collection of reference data. AcATaMa provides a multi-view interface that allows spatial units to be added and revised in an orderly manner. However, spatial units can only be one pixel in size. For its part, the process for collecting the reference data using SCP is very similar to the process that would be followed if just QGIS tools were used. Notwithstanding, as SCP uses a unique data format (.scp), it is quite complicated to add other types of data or to use information from other platforms.

Both plugins have valuable tools that assist in the capture of reference data. However, as we intend to use larger spatial units than one pixel and wish to keep the installation of new interfaces and formats to a minimum, we will only use the basic QGIS (*Buffer*) tools, and other data services such as the Google Earth Engine Data catalog and QuickMapServices.

Materials

Centroids of sample sites—Marqués de Comillas (the point vector layer RandomSample.shp created in the previous exercise that contain the centres of the sample sites)

Execution

Step 1

Before data collection can begin, the size and shape of the spatial unit must be established, i.e. the area over which we will be making the comparison between the thematic map values and the reference values. The minimum mappable area of the thematic map to be used in this exercise is 1 ha, and it is generally recommended that the spatial unit should be of a similar size. Accordingly, in this exercise, we will be using square polygons of 1 ha as the spatial unit.

The point layer containing the centroids (Centroids of sample sites—Marqués de Comillas) will be used to create the spatial assessment units. To form square polygons centred on each of the points in the point layer, use first the *Buffer* tool in the Geoprocessing Tools menu. The input will be the point layer, and the distance value depends on the desired size of the square. In this case, 50 m. Change the End cap style to "Square" and leave the rest of the parameters unchanged (Fig. 4).

The newly created layer will have two attributes: the id and the value of the thematic class (inherited from the previous exercise). To avoid bias in the photointerpretation decision-making process, we advise hiding the *class* column (the value of the thematic class taken from the LUC map). To hide a column in the attribute table without deleting it definitively, right-click on the area of the attribute table headers, select the option *Organize Columns*, and then select the columns to hide (Fig. 5).

Step 2

Ideally, to identify the land-cover type of each sampling unit, it would be necessary to overlay them on high spatial resolution images with the same (or similar) date as the images used in the mapping. If such data are available, photointerpretation of the spatial units can proceed directly. However, acquiring high-resolution images to verify extensive areas could be expensive. In this regard, sometimes the resources are limited, which restricts the use of this source of imagery.

In the following steps, we propose a partial solution to this problem based on the combined use of image servers (Google Earth, Bing, ESRI) with high spatial and temporal resolution. However, in these servers it is impossible to identify and select scenes according to their acquisition dates. One way to estimate the dates of these data is to compare them with images with higher temporal resolution, which have a known acquisition time, and for which a longer

Fig. 4 Exercise 2. Step 1. Buffer

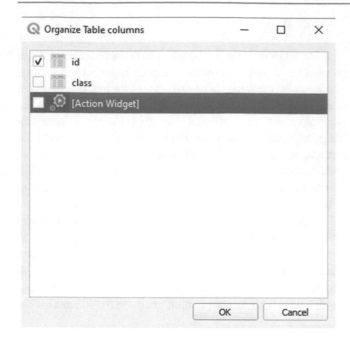

Fig. 5 Exercise 2. Step 1. How to hide columns in the attribute table

historical record is available, such as Sentinel or Landsat images. For this purpose, we will install two plugins with which we can access the high spatial resolution image servers of Google, Bing and ESRI (QuickMapServices) and Sentinel, Landsat, Aster and other images (Google Earth Engine Data Catalog). To see how to install these plugins in QGIS, see Chapter "About This Book".

Step 3

Once QuickMapServices is installed, open the plugin and select *Setting*. Then select the *More Services* tab and click on the *Get contributed* pack. To add images with high spatial resolution to the QGIS Project, in the Web option in the main menu select QuickMapService, then Google and finally

Google Satellite. After selecting these options, the Google Satellite images become available in the Layers menu.

Step 4

Add Sentinel-2 images from the Google Earth Engine Data Catalog plugin. The plugin requires to define the product type, date and cloud cover percentage (Fig. 6). The images can be saved temporarily or in permanent files. The images to be added should be dated as close as possible to the date of the evaluated map.

Step 5

To facilitate the collection of reference data, we suggest creating multiple windows to display images with different dates or resolutions, a good way to work when assessing land-cover change maps.

In the Layers panel, select the Image and vector layer (Centroids of sample sites—Marqués de Comillas) that will be added to the second window, click on "Manage Maps Themes" button (represented as an eye) and select "Add Theme". Name the theme "Image 1". Then go to "View" in the main menu, and select "New Map View". This will create a new display window. Enter the new window and do the following: (a) Select the layers set to display (Image 1) by clicking on the on "Manage Maps Themes"; (b) Synchronise the windows by selecting the "view settings" tool, and then click on the "Synchronise scale" option. The example in Fig. 7 shows a Google server image (2) and a true-colour Sentinel image (1).

Step 6

To capture the reference data, the "Centroids of sample sites—Marqués de Comillas" vector layer must be edited and a new field (integer type) must be added. We suggest naming it "Refer_data".

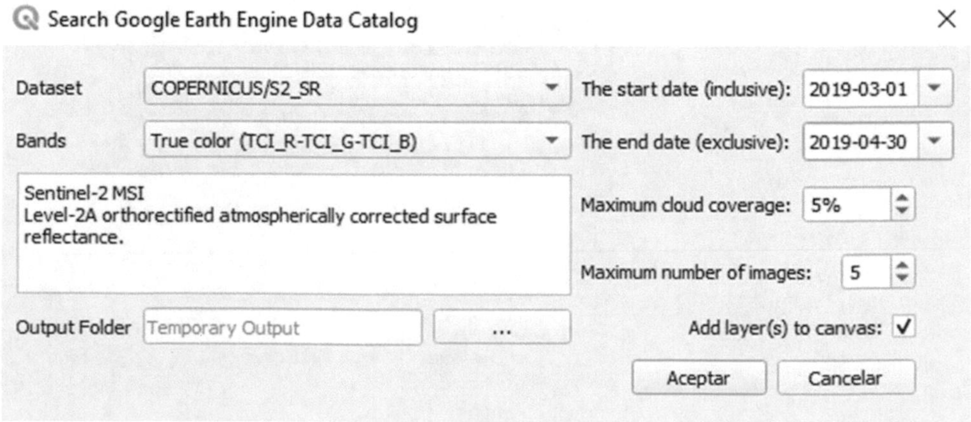

Fig. 6 Exercise 2. Step 4. Google Earth Engine Data Catalog plugin

Fig. 7 Exercise 2. Step 5. Synchronising windows

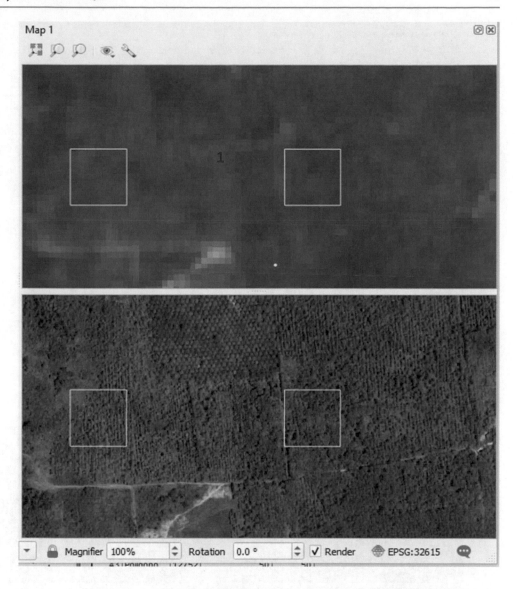

Step 7

To make reference data collection easier, we recommend displaying the *Attribute Table* as a form and anchoring it to the main window, displaying only the selected data. To do this, open the *Attribute Table*, select the "Dock Attribute Table" icon, select the "switch to form view" button and then "show Selected Feature" (Fig. 8).

Step 8

If you have completed Steps 1–8 successfully, you can now start photointerpreting high-resolution satellite images. The exercise involves identifying the predominant land-cover type in each sampling unit and recording the corresponding code in the "Refer_data" column of the attribute table (Fig. 9). The meanings of the codes are described in the auxiliary data distributed with the Marqués de

Comillas LUC map, available at https://doi.org/10.5281/zenodo.5418318.

Photointerpreting all the spatial units of the sample can be a lengthy process, so we suggest that you try to photointerpret at least 10 to 20 spatial units and then compare your results with the "Photo-interpreted reference dataset—Marqués de Comillas 2019", a reference dataset was prepared by the authors. It is available, together with all book's data, at https://doi.org/10.5281/zenodo.5418318. For more information, see Chapter "About This Book".

Results and Comments

The result of this exercise should be a shapefile with an attribute table in which the columns class (map class code) and refer-data (photointerpreted class code) are filled in, as shown in Fig. 10. From the attribute table, you can now calculate the error matrix and the map accuracy statistics, as is done later in Sect. 5 in

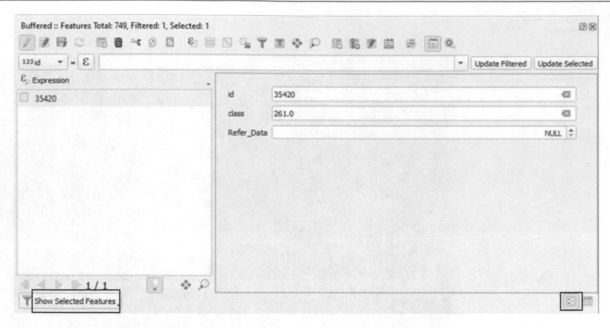

Fig. 8 Exercise 2. Step 7. Displaying selected data

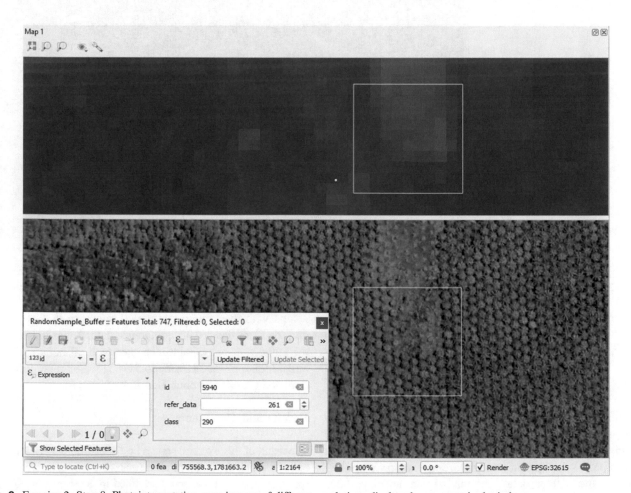

Fig. 9 Exercise 2. Step 8. Photointerpretation over images of different resolutions displayed on syncronised windows

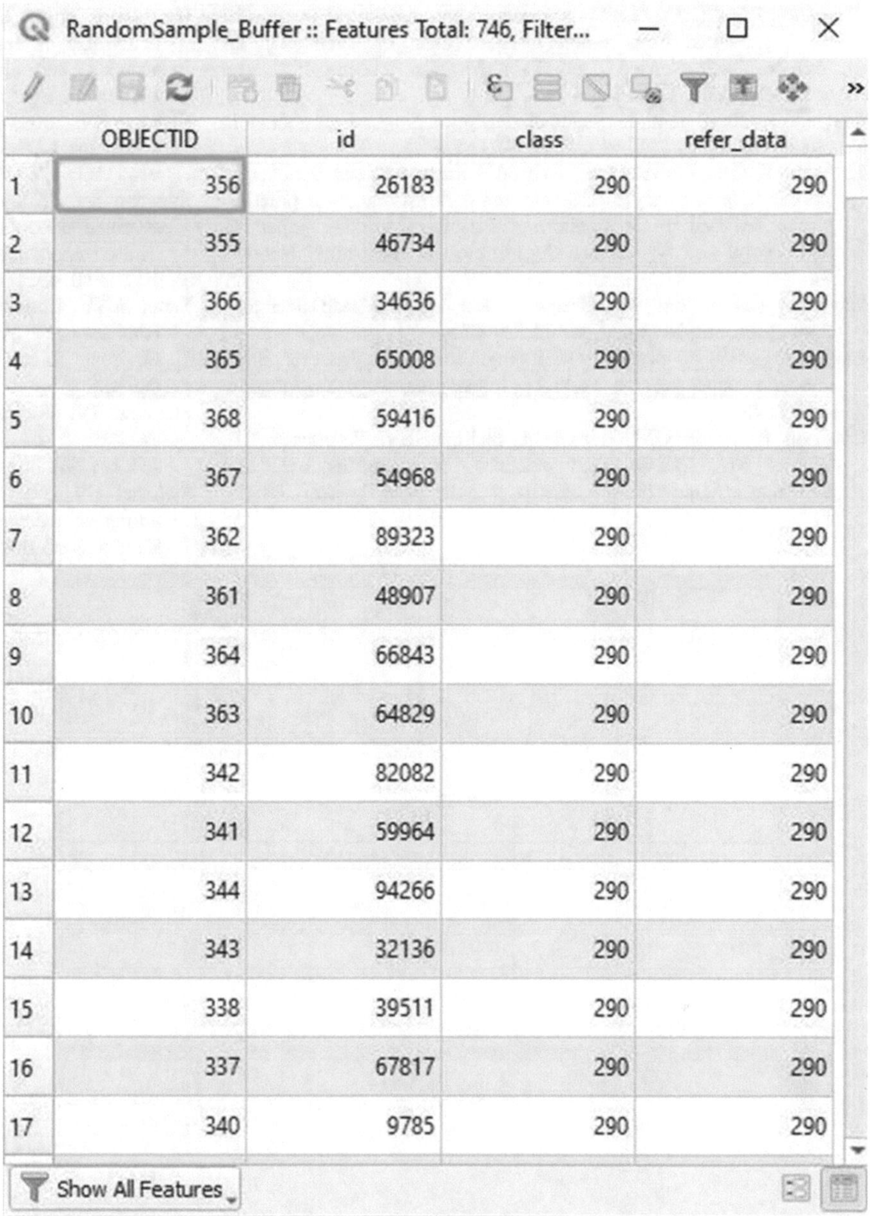

Fig. 10 Results from Exercise 2. Table of attributes with the map codes and the photointerpreted codes

	OBJECTID	id	class	refer_data
1	356	26183	290	290
2	355	46734	290	290
3	366	34636	290	290
4	365	65008	290	290
5	368	59416	290	290
6	367	54968	290	290
7	362	89323	290	290
8	361	48907	290	290
9	364	66843	290	290
10	363	64829	290	290
11	342	82082	290	290
12	341	59964	290	290
13	344	94266	290	290
14	343	32136	290	290
15	338	39511	290	290
16	337	67817	290	290
17	340	9785	290	290

RandomSample_Buffer :: Features Total: 746, Filter...

Show All Features

chapter "Metrics Based on a Cross-Tabulation Matrix to Validate Land Use Cover Maps" of this book.

If the images used in the validation were acquired at a different time than the one for which the LUC map represents the covers on earth, this must be taken into account when assigning labels. This date mismatch may increase the uncertainty of the reference data, a situation that should be avoided.

It is worth remembering that in the absence of high spatial resolution imagery, medium resolution imagery, such as Landsat or Sentinel, can provide sufficient information to validate maps, especially small-scale maps.

Although the spatial assessment unit used in this exercise is widely used and recommended, it may contain several land-cover types. This means that clear rules should be established when deciding the category to which the unit should be allocated in these circumstances.

References

Cochran WG (1977) Sampling techniques. Wiley

Congalton RG (1988) A Comparison of sampling schemes used in generating error matrices for assessing the accuracy of maps generated from remotely sensed data. Photogramm Eng Remote Sens 54:593–600

Congalton RG (1991) A review of assessing the accuracy of classification of remotely sensed data. Remote Sens Environ 37:35–46. https://doi.org/10.1016/0034-4257(91)90048-B

Congalton RG (2016) Assessing positional and thematic accuracies of maps generated from remotely sensed data. In: Thenkabail PS (ed) Remotely sensed data characterisation, classification, and accuracies. Edit. CRC Press, NY

Congedo L (2020) Semi-automatic classification plugin documentation. https://doi.org/10.13140/RG.2.2.25480.65286/1

Finegold Y, Ortmann A, Lindquist E, d'Annunzio R, and Sandker M (2016) Map accuracy assessment and area estimation: a practical guide. National forest monitoring assessment working paper No 46/E. Food and Agriculture Organization of the United Nations. Rome

Hay AM (1979) Sampling design to test land-use map accuracy. Photogram Eng Remote Sens 45:529–533

Llano XC (2019) AcATaMa—QGIS plugin for accuracy assessment of thematic maps, version 19.11.21. https://plugins.qgis.org/plugins/AcATaMa/

Olofsson P, Foody GM, Herold M, Stehman SV, Woodcock CE, Wulder MA (2014) Good practices for estimating area and assessing accuracy of land change. Remote Sens Environ 148:42–57

Olofsson P, Foody GM, Stehman SV, Woodcock CE (2013) Making better use of accuracy data in land change studies: estimating accuracy and area and quantifying uncertainty using stratified estimation. Remote Sens Environ 129:122–131

Stehman SV (2009) Sampling designs for accuracy assessment of land cover. Int J Remote Sens 30:5243–5272. https://doi.org/10.1080/01431160903131000

Stehman SV (2012) Impact of sample size allocation when using stratified random sampling to estimate accuracy and area of land-cover change. Remote Sens Lett 3:111–120. https://doi.org/10.1080/01431161.2010.541950

Stehman SV, Czaplewski SV (1998) Design and analysis for thematic map accuracy assessment: fundamental principles. Remote Sens Environ 64:331–344. https://doi.org/10.1016/S0034-4257(98)00010-8

Stehman SV, Foody GM (2019) Key issues in rigorous accuracy assessment of land cover products. Remote Sens Environ 231:111199. https://doi.org/10.1016/j.rsc.2019.05.018

Stehman SV, Wickham JD (2011) Pixels, blocks of pixels, and polygons: choosing a spatial unit for thematic accuracy assessment. Remote Sens Environ 115:3044–3055

Part III

Tools to Validate Land Use Cover Maps: A Review

Basic and Multiple-Resolution Cross-Tabulation to Validate Land Use Cover Maps

María Teresa Camacho Olmedo and David García-Álvarez

Abstract

In this chapter, we describe the fundamental principles and the normal procedure followed when cross-tabulating two datasets. Cross-Tabulation analysis (Sect. 1) is usually the first step in the validation of Land Use Cover (LUC) data. It compares two datasets to observe their spatial relationship, i.e. their degree of spatial (dis) agreement. Results are usually displayed in the form of maps, tables and other statistical measures. Multiple-Resolution Cross-Tabulation (Sect. 2) compares two raster datasets at multiple spatial resolutions. Basic Cross-Tabulation can compare raster and vector data, while Multiple-Resolution Cross-Tabulation only works with raster data, which is what we use in the exercises provided as examples. In the exercises, raster data were obtained from vector data previously rasterized at different spatial resolutions. As a reference we use LUC maps, although ground points could also be used as reference data for these analyses. Examples of Cross-Tabulation analyses at one and multiple resolutions are presented for four different cases: to validate single LUC maps, to validate the soft maps produced by a model, to validate a simulation exercise and to validate and study land change in a series of LUC maps. In the example exercises, we used CORINE and SIOSE maps from the Asturias Central Area database, as well as maps from the modelling exercises carried out with this database. In the Chapters "Metrics Based on a Cross-Tabulation Matrix to Validate Land Use Cover Maps" and "Pontius Jr. Methods Based on a Cross-Tabulation Matrix to Validate Land Use Cover Maps", we focus on specific analyses that can be carried out on the basis of Cross-Tabulation analyses, such as Land Use Cover Changes (LUCC) Budget or Quantity and Allocation disagreement. These help unleash the full potential of Cross-Tabulation analysis.

Keywords

Cross-Tabulation • Multiple-Resolution • Land Use Cover data • Validation

1 Basic Cross-Tabulation

Description

Cross-Tabulation is a primary analysis that crosses two datasets, either raster or vector, to analyse their spatial relation. This analysis combines the datasets in spatial terms. It produces a map or table that shows how the values of one dataset spatially relate with the values in the other, thereby informing us as to whether the two datasets share the same values at a given location and, if not, with which other values they have established a relation.

Utility

Exercises

1. To validate a map against reference data/map
2. To validate soft maps produced by the model against a reference map
3. To validate a simulation against a reference map
4. To validate a series of maps with two or more time points

Starting with a map and some reference data, we can use Cross-Tabulation to determine to what extent the map we want to validate agrees with the reference data. In this way

M. T. Camacho Olmedo (✉)
Departamento de Análisis Geográfico Regional y Geografía Física, Universidad de Granada, Granada, Spain
e-mail: camacho@ugr.es

D. García-Álvarez
Departamento de Geología, Geografía y Medio Ambiente, Universidad de Alcalá, Alcalá de Henares, Spain

D. García-Álvarez et al. (eds.), *Land Use Cover Datasets and Validation Tools*,
https://doi.org/10.1007/978-3-030-90998-7_7

we can compare the success of a LUC classification exercise or a LUCC modelling exercise against reference data. We can also assess how uncertain a map is with regard to the data used as a reference. Cross-Tabulation can also be used to study the LUC changes between pairs of maps at two or more different points in time, or to validate a chronological series of maps, as it can detect unusual or abnormal changes, which could be due to technical errors.

The Cross-Tabulation matrix provides users with a lot of information from the maps in one single analysis. However, in order to take advantage of the full potential of this analysis, it is important for them to understand what all this information means. This is what we will be explaining in this chapter.

The results of Cross-Tabulation can then be used to make further analyses and to extract other metrics that allow us to take full advantage of this basic analysis. These methods (e.g. LUCC budget, Quantity & Allocation disagreement, the Figure of Merit, Intensity Analysis) (see Sects. 2, 3, 4 and 6 in Chapter "Pontius Jr. Methods Based on a Cross-Tabulation Matrix to Validate Land Use Cover Maps") make it easier for users to interpret the results. However, they also require many further analyses and are therefore more time-consuming. We will now provide an overview of some relevant examples:

- Hagen-Zanker (2009) used a well-known Cross-Tabulation matrix to improve the fuzzy Kappa statistic (see Sect. 3 in Chapter "Metrics Based on a Cross-Tabulation Matrix to Validate Land Use Cover Maps").
- Alcamo et al. (2011) used the Cross-Tabulation function with potential maps from a land use change model.
- Mas et al. (2014) and Paegelow et al. (2014) used Cross-Tabulation in different ways to provide useful information to help them assess land change model robustness.
- Krüger and Lakes (2015) calculated a disagreement index from the Cross-Tabulation matrix used in LUCC modelling exercises.
- Pontius (2018) created an Excel spreadsheet that performs a range of automatic analyses from the Cross-Tabulation matrix.

The maps to be compared or assessed may be in either raster or vector format. For those in raster, we can use both hard and soft maps, such as suitability, transition potential and probabilities maps.

QGIS Exercises

The methods and techniques presented in Chapter "Pontius Jr. Methods Based on a Cross-Tabulation Matrix to Validate Land Use Cover Maps" (e.g. LUCC Budget, Intensity Analysis, Quantity and Allocation disagreement...) are based on this basic Cross-Tabulation analysis. In this chapter, we will therefore only be describing the fundamental principles and the normal procedure followed when performing a Cross-Tabulation between two datasets.

Available tools
• Processing Toolbox
SAGA
Image analysis
Confusion matrix (two grids)
Confusion matrix (polygons/grid)
Raster analysis
Cross-classification and tabulation
• Processing Toolbox
GRASS
Raster
r.cross
• Semi-Automatic Classification Plugin
Tab: Postprocessing
Section: *Cross classification*
Section: *Accuracy*
Section: *Land cover change*

QGIS includes many tools for cross-tabulating spatial data through the associated GRASS and SAGA models. The "Semi-Automatic Classification Plugin" also includes tools to cross-tabulate datasets for different purposes.

Table 1 includes a review of the available Cross-Tabulation tools in QGIS. It provides information of the input and output parameters in each tool. Although the *r.kappa* function also cross-tabulates two raster maps to obtain the Kappa index, we will not be analysing it in this chapter. Those interested in using this tool should refer to the Kappa indices, Sect. 3 in Chapter "Metrics Based on a Cross-Tabulation Matrix to Validate Land Use Cover Maps".

The associated R software can also be used to cross-tabulate pairs of maps. This is done using the *crosstab* function, which is part of the "raster" package.[1] As QGIS already provides many tools for carrying out this analysis, we will not be covering the implementation of this R function in QGIS here.

Of all the tools available in QGIS, the one we will be recommending and using in this book is the "Semi-Automatic

---

[1] https://cran.r-project.org/web/packages/raster/raster.pdf.

Table 1 Review of Cross-Tabulation tools available in QGIS

			Processing toolbox				Plugins		
			SAGA			GRASS	Semi-automatic classification plugin		
			Image Analysis		Raster Analysis	Raster	Postprocessing	Accuracy	Land cover change
			Confusion matrix *(two grids)*	Confusion matrix *(polygons/grid)*	Cross-classification and tabulation *(two grids)*	r.cross	Cross classification	Accuracy	Land cover change
INPUT	Maps	Data	Two raster maps *(two classifications)*	A raster map *(classification)* and a vector map *(polygons)*	Two raster maps *(two grids)*	Multiple raster maps with a maximum of ten	Two maps *(classification and reference map, either raster or vector)*	Two maps *(classification to assess and reference map, either raster or vector)*	Two maps from a chronological series (earlier map used as *reference classification* and later map as *new classification*)
		Options	The user can choose only a few map values to compare (look-up table, minimum, maximum, name…)	The user can choose only a few map values to compare (look-up table, minimum, maximum, name…)	The user can choose a maximum number of classes to compare	The user can choose only no zero data	The user can discard no-data values	No	No
OUTPUT	Maps	Data	Two raster maps (combined classes), one showing all combinations (similarities and differences) and the other showing just the differences	No	One raster map *(cross-classification grid)*	One raster map	One raster map *(output)*	One raster map *(output)*	Two raster maps (combined classes), one showing changes and permanence and the other showing just changes
		Options	No	No	No	Users can choose the extent, spatial resolution, format and metadata format	No	No	No
		Format	.sdat	-	.sdat	Multiple formats	.tiff	.tiff	.tiff

(continued)

Table 1 (continued)

			Processing toolbox			Plugins		
			SAGA		GRASS	Semi-automatic classification plugin		
			Image Analysis	Raster Analysis	Raster	Postprocessing		
OUTPUT	Tables and statistical measures	Data	Confusion matrix (*confusion matrix*); summary per categories and AccProd and AccUsers values (*classes values*); summary table with overall accuracy and kappa values (*summary*)	Cross-classification table	No	Cells and area in square meters for each combination. The area in square meters is also shown in a cross-tabulation matrix	Cells for each combination. Error matrix in cells and square meters (proportional quantities). The standard error (SE), confidence interval (CI), producer's accuracy (PA), user's accuracy (UA), overall accuracy (%) and Kappa hat are also included	Cells for each combination (all possible combinations are shown, not only the real combinations)
		Comb code*	No	No	No	Yes	Yes	Yes
		Units	Cells, percent or area units	Cells, percent or area units	–	Cells and area (m²)	Cells and area (m²)	Cells
		Format	.dbf	.dbf/.shp	–	.csv	.csv	.csv
References			Conrad (2007, 2015)		Shapiro (2003–2020)	Congedo (2016)		

Classification Plugin", which proved to be the most efficient and stable of all those assessed.

Exercise 1. To validate a map against reference data/map

Aim

To validate the CORINE 2011 Land Use map, taking the SIOSE 2011 Land Use map as a reference.

Materials

SIOSE Land Use Map Asturias Central Area 2011
CORINE Land Use Map Asturias Central Area 2011

Requisites

The two maps must have the same extent, spatial resolution, projection and classification legend. If the maps have different classification legends, the user must reclassify the maps in such a way as to unify the two legends.

Execution

Step 1

Open the "Semi-Automatic Classification Plugin" and select the "Postprocessing" tab from the sidebar. Then click on *Accuracy* and select the required parameters: raster to assess (CORINE map) and reference raster (SIOSE map) (Fig. 1).

Results and Comments

Once the function has been executed, QGIS creates an output raster that gives each pixel a code. This code identifies every single possible combination of values between the two input rasters. The meaning of each code is presented in a table in CSV format, which is stored in the same folder as the raster. This information is also displayed in the "output" window of the "Semi-Automatic Classification Plugin" (Fig. 2).

If we analyse the first matrix shown in the "output" window (ErrMatrixCode/Reference/Classified/PixelSum), it will help us understand the meaning of the codes in the raster. The "ErrMatrixCode" is the number that identifies each pixel in the new raster. "Reference" is the code for the category on the reference map (i.e. SIOSE Land Use Map). "Classified" is the code for the category on the compared map (i.e. CORINE Land Use Map), and "PixelSum" refers to the number of pixels for each combination in the new raster.

The ErrMatrixCode 1 identifies 234,164 pixels (PixelSum) in category 0 in SIOSE (Reference) and 0 in CORINE (Classified). The codes for combinations in which the reference and the classified categories are the same (e.g. 0, 0) mean agreement, while those in which the reference and the classified categories are different (e.g. 0, 1) mean disagreement. Code 2 is therefore a disagreement area because the pixel is classified as 0 in SIOSE and as 1 in CORINE.

If we symbolize the obtained raster in such a way that all the codes that refer to combinations of the same classes (1, 15, 29, 43, 57, 71, 85, 99, 113, 127, 141, 155) are labelled as agreement and all the codes that refer to combinations of different classes are labelled as disagreement, we can obtain a

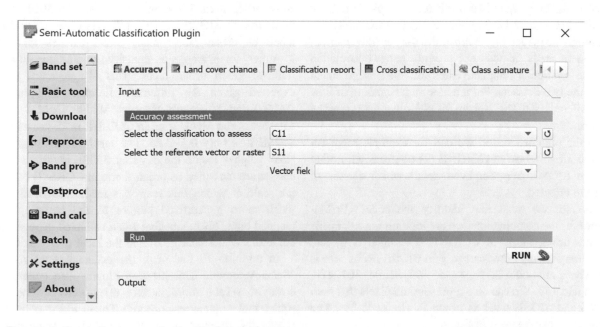

Fig. 1 Exercise 1. Step 1. Semi-automatic classification plugin

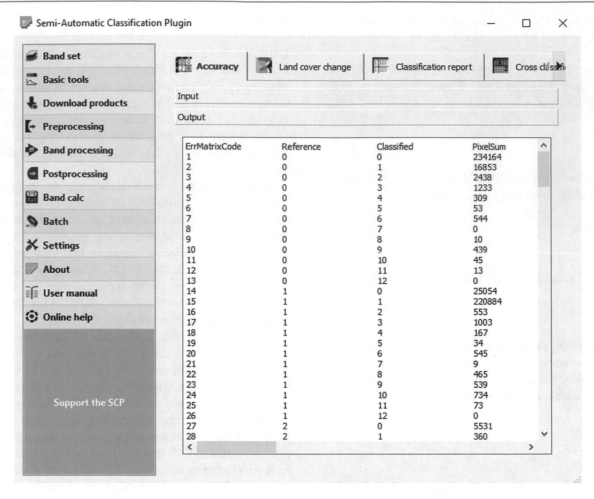

Fig. 2 Results from Exercise 1 displayed in the "output" window of the Semi-Automatic Classification Plugin

map like the one presented in Fig. 3. Code 169 is not represented on this new map because it refers to pixels that are background (category 12) in both SIOSE and CORINE.

Although the map in Fig. 3 illustrates the general pattern of disagreement areas, it does not provide much information about the particular characteristics of the disagreement between the two datasets. For a better understanding of how similar/different CORINE is from SIOSE, other maps must be drawn up.

With the obtained raster, we can for example represent where the urban fabric of CORINE (2) confuses with other classes in SIOSE. We can even detail which classes of SIOSE are affected.

To do so, we must first identify the codes (ErrMatrixCode) for the combinations we are looking for, i.e. pixels which are urban fabric in CORINE (Classified is 2) and which belong to any other category in SIOSE (Reference is not 2). These are codes 3, 16, 42, 55, 68, 81, 94, 107, 120, 133, 146 and 159. We can also represent the pixels that both CORINE and SIOSE label as urban fabric (code 29). The resulting map can be seen in Fig. 4.

This map shows the city of Oviedo and its immediate surrounding area. Most of the city is identified as urban fabric in both SIOSE and CORINE. However, CORINE also labels as urban fabric many small patches that SIOSE identifies, for example, as industrial or commercial areas or artificial green urban areas. This disagreement is to be expected given the different Minimum Mapping Units (MMU) and Minimum Mapping Widths (MMW) of both databases. The MMU used in CORINE is 25 ha, whereas in SIOSE it is only 0.5–2 ha. The result is that many of the small patches inside the city that SIOSE identifies as other classes are classified as urban fabric in CORINE because of the scale at which this map was made. CORINE offers a much more generalized picture of the landscape to be mapped out. This means that when validated against SIOSE, numerous errors emerge due to the level of generalization.

In addition to the map, the accuracy analysis of the "Semi-Automatic Classification Plugin" also generates two error/Cross-Tabulation matrixes, one in cells and the other in square meters (area proportions). The matrix in cells (Fig. 5) shows the number of pixels for each combination. For

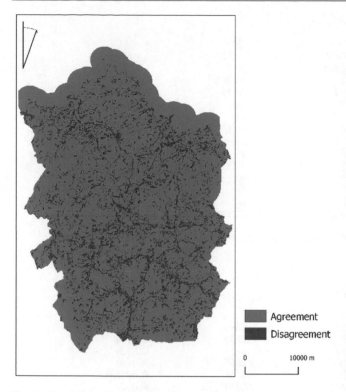

Fig. 3 Result from Exercise 1. Map showing areas of agreement and disagreement between CORINE and SIOSE maps

example, if we look at the combination 0–0, we see that there are 234,164 pixels that have the same value (0) in SIOSE and CORINE. In other words, there are 234,164 pixels classified as agricultural areas on both maps.

The area-based error matrix gives us the same information (the proportion of the total area of the raster represented by each combination), but in different units. Using the example above, the matrix shows that combination 0-0 covers a fraction of 0.2535/1 of the map, i.e. 25.35% of its pixels (Fig. 6).

An analysis of the two tables (area-based error matrix and error matrix pixel count) offers us a detailed picture of how the categories on one map relate with the categories on the other. This highlights the degree of agreement between the reference map and the one we are trying to validate. In both tables, the combinations in which there is agreement can be seen on a diagonal line running across the table. All combinations outside this diagonal mean disagreement (Table 2).

If we look at urban fabric, of a total of 28,110 pixels labelled as urban fabric in CORINE (Total column on the right), 19,455 are also labelled as urban fabric in SIOSE. That is, almost 70% of the pixels identified as urban fabric by CORINE are also considered urban fabric in SIOSE. In the other 30%, CORINE mostly confuses urban fabric with industrial and commercial areas (category 3, 2244 confused pixels), artificial green urban areas (category 9, 1643 confused pixels) and road and rail networks (category 6, 1216 confused pixels).

These results are due to the greater degree of generalization when mapping CORINE, as explained above. On the basis of these results and taking SIOSE as a reference, we can conclude that CORINE maps urban fabric correctly and can be considered a valid map for our exercises.

Users can also carry out more complex analyses with these matrixes using the CSV file generated by the tool. In this way

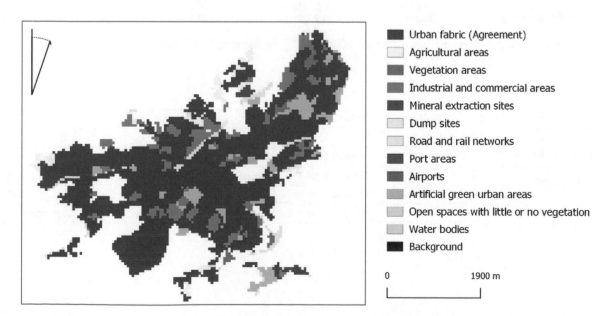

Fig. 4 Result from Exercise 1. Map showing areas of agreement and disagreement between CORINE and SIOSE maps for the CORINE category `urban fabric'. The map specifies with which categories of SIOSE the urban fabric category of CORINE is confussed

```
                    > ERROR MATRIX (pixel count)
                    > Reference
V_Classified        0               1               2               3
0                   234164          25054           5531            1871
1                   16853           220884          360             329
2                   2438            553             19455           2244
3                   1233            1003            1500            13330
4                   309             167             78              75
5                   53              34              23              57
6                   544             545             311             533
7                   0               9               14              117
8                   10              465             0               18
9                   439             539             222             189
10                  45              734             15              24
11                  13              73              5               25
12                  0               0               0               0
Total               256101          250060          27514           18812
```

Fig. 5 Result from Exercise 1 displayed in the "output" window of the Semi-Automatic Classification Plugin: Error matrix in pixels

```
                    > AREA BASED ERROR MATRIX
                    > Reference
V_Classified        0               1               2               3
0                   0.2535          0.0271          0.0060          0.0020
1                   0.0182          0.2392          0.0004          0.0004
2                   0.0026          0.0006          0.0211          0.0024
3                   0.0013          0.0011          0.0016          0.0144
4                   0.0003          0.0002          0.0001          0.0001
5                   0.0001          0.0000          0.0000          0.0001
6                   0.0006          0.0006          0.0003          0.0006
7                   0.0000          0.0000          0.0000          0.0001
8                   0.0000          0.0005          0.0000          0.0000
9                   0.0005          0.0006          0.0002          0.0002
10                  0.0000          0.0008          0.0000          0.0000
11                  0.0000          0.0001          0.0000          0.0000
12                  0.0000          0.0000          0.0000          0.0000
Total               0.2773          0.2708          0.0298          0.0204
Area                640252250       625149756       68784973        47029982
SE                  0.0003          0.0002          0.0001          0.0001
SE area             580258          542766          293597          248955
95% CI area         1137306         1063821         575449          487951
PA [%]              91.4342         88.3324         70.7095         70.8590
UA [%]              85.6793         90.5351         69.2102         64.6742
Kappa hat           0.8018          0.8702          0.6826          0.6394
```

Fig. 6 Result from Exercise 1 displayed in the "output" window of the Semi-Automatic Classification Plugin: Area based error matrix (proportions)

Table 2 Traditional scheme of a Cross-Tabulation matrix, differentiating which cells indicate agreement between the compared maps and which cels indicate disagreement

V_Classified	References			
	0	1	2	3
0	Agreement	Disagreement	Disagreement	Disagreement
1	Disagreement	Agreement	Disagreement	Disagreement
2	Disagreement	Disagreement	Agreement	Disagreement
3	Disagreement	Disagreement	Disagreement	Agreement

the matrixes can be imported in spreadsheet format with software such as Excel or OpenOffice Calc. We can then calculate the agreement and disagreement percentages for the whole raster or for each of the categories under consideration, as we did manually for the urban fabric above.

The error matrixes also provide useful statistical measures (Fig. 6), such as the standard error (SE), confidence interval (CI), producer's accuracy (PA), user's accuracy (UA), overall accuracy (in %) and Kappa (see Sect. 3 in Chapter "Metrics Based on a Cross-Tabulation Matrix to Validate Land Use Cover Maps").

Exercise 2. To validate soft maps produced by the model against a reference map

Aim

To find out whether the urban fabric soft map produced by our model agrees with the urban fabric areas of the reference map for the year of the simulation.

Materials

CORINE Land Use Map Asturias Central Area 2011
Urban fabric suitability map – CORINE model

Requisites

The two maps must have the same extent, spatial resolution and projection. The soft map must be categorical. The Land Use map must only contain information about the category being assessed. For a proper validation, the reference map must refer to the same date on which the landscape was simulated.

Execution

Step 1

Only discrete or categorical maps can be cross-tabulated. As the soft map we want to validate is continuous (continuous values from 0.1 to 1), the first step must be to convert it into a categorical map, using the *Reclassify by table* function (Processing toolbox > Raster analysis > Reclassify by table).

After opening this tool, we select the map we want to reclassify (Urban fabric suitability map) and fill in the "Reclassification table" with the new values that will be replacing the old ones in the raster (Fig. 7). In this case, we are going to reclassify the values on our suitability map

(0–1) into four categories, from low to high suitability. The new categories will be 1 (suitability 0–0.25), 2 (0.25–0.50), 3 (0.50–0.75) and 4 (0.75–1).

Step 2

Given that our objective is to compare the suitability values for urban fabric in the model with the areas classified as urban fabric on the 2011 map, we must ignore all the other categories on the Land Use Cover map. We must therefore obtain a binary map from the initial CORINE map. In this binary map, 1 will mean the category being evaluated (urban fabric) and 0 all the others.

To obtain this binary map, we repeat the same process as in Step 1. In this case, we reclassify the CORINE map, assigning a value of 1 to urban fabric (code 2 in the original map) and a value of 0 to the other categories (codes 0, 1, 3, 4, 5, 6, 7, 8, 9, 10, 11 and 12) (Fig. 8).

Step 3

Once we have obtained the two maps, we then carry out Cross-Tabulation using the "Semi-Automatic Classification Plugin". We click on the "Postprocessing" tab and select the *Cross classification* option.

We then select the required parameters. In "Select the classification" we choose the reference Land Use Cover map obtained after reclassification (Step 2). In "Select the reference vector or raster" we choose the soft map obtained after reclassification (Step 1) (Fig. 9).

Results and Comments

Once the function has been executed, QGIS creates a raster and a CSV file with all the results of the Cross-Tabulation. These are also displayed in the "Output" window (Fig. 10).

The first table provides information about the meaning of each code in the new raster. Pixels with value 2 refer to areas that are urban fabric (Classification is 1) and have a suitability of less than 0.2 (Reference category is 1). This combination occurs in just 2 pixels (PixelSum), which represent an area of 5000 m^2 on the map (Area [metre^2]).

The second table gives an overview of the possible combinations on the two maps and the area, in square meters, covered by these combinations. This shows that the areas that are not urban fabric (Classification is 0.0) and have a suitability of below 0.25 (Reference 1) occupy 2,312,499 m^2.

From all the possible combinations, we can see that most of the pixels that are urban fabric on the reference map fit with the areas with the highest suitability to become urban fabric (26,137 pixels, 65,342,474 m^2). There are relatively few urban fabric pixels with a suitability of between 0.5 and

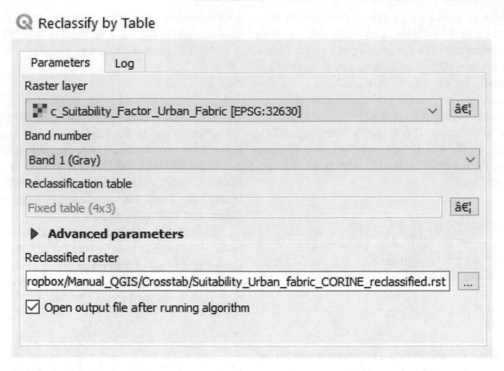

Fig. 7 Exercise 2. Step 1. Reclassify by table

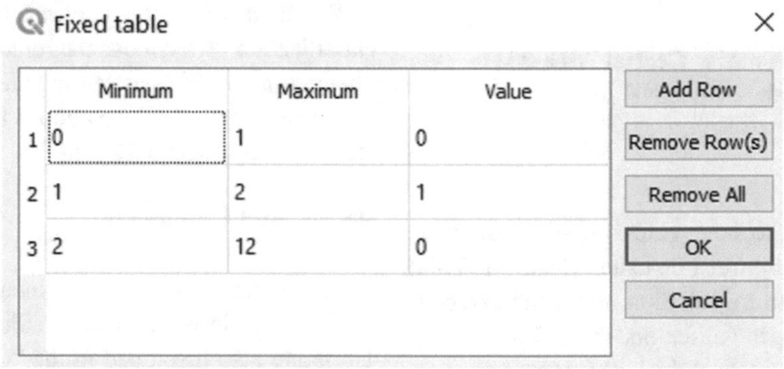

Fig. 8 Exercise 2. Step 2. Table required for the Reclassify by Table tool

0.75 (1971 pixels, 4,927,498 m^2) and an insignificant number with a suitability of less than 0.5.

These results indicate that our suitability map has been validated. In other words, the high suitability values on the soft map correspond with urban fabric areas on the reference map. For their part, the low suitability values correspond to areas where there is no urban fabric on the map. This means that when we use this map in our simulation, it will help us

to correctly identify those areas that can become urban in the future and those that cannot.

Other more sophisticated tools, such as the ROC curve and the Difference in Potential (see Sects. 2 and 3 in Chapter "Validation of Soft Maps Produced by a Land Use Cover Change Model"), can be used to complement this analysis and offer the user a full overview of the validity of their potential maps.

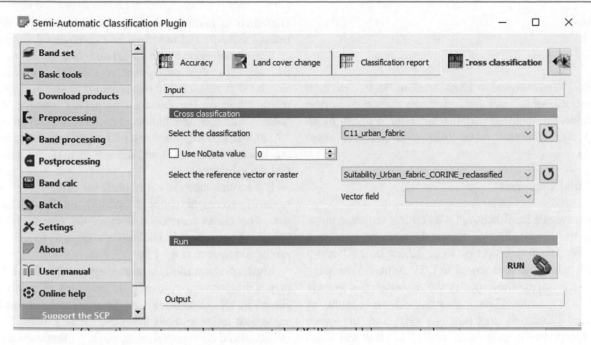

Fig. 9 Exercise 2. Step 3. Semi-Automatic Classification plugin

Output				
CrossClassCode	Reference	Classification	PixelSum	Area [metre^2]
1	1.0	0.0	925	2312499.0978074283
2	1.0	1.0	2	4999.998049313358
3	3.0	0.0	231547	578867274.1621801
4	3.0	1.0	1971	4927498.077598315
5	4.0	0.0	384851	962127124.6381476
6	4.0	1.0	26137	65342474.50745163
	> CROSS MATRIX [metre^2]			
	> Reference			
V_Classification	1.0	3.0	4.0	Total
0.0	2312499	578867274	962127124	1543306897
1.0	4999	4927498	65342474	70274972
Total	2317499	583794772	1027469599	1613581870

Fig. 10 Results from Exercise 2 displayed in the "output" window of the Semi-Automatic Classification Plugin

Exercise 3. To validate a simulation against a reference map

Aim

To validate a simulation for the year 2011, which we obtained through our LUCC modelling exercise with CORINE maps, against a CORINE reference map for the year 2011.

Materials

Simulation CORINE Asturias Central Area 2011
CORINE Land Use Map Asturias Central Area 2011

Requisites

The two maps must have the same extent, spatial resolution, projection and legend. For a proper validation, the reference date must refer to the date on which the landscape was simulated.

Execution

Step 1

Open the "Semi-Automatic Classification Plugin", click on the "Postprocessing" tab and select the section *Accuracy*. Then, select the required parameters: raster to assess (Simulation) and reference raster (CORINE reference map) (Fig. 11).

Results and Comments

When we execute the function, QGIS creates an output raster showing the combination of classes between the two input maps. The function generates three tables in the "output window", which are also stored in CSV format in the same folder as the raster. They specify the meaning of each code in the new raster. They also include a couple of error/Cross-Tabulation matrixes, in cells and in square meters (proportional quantities) (Fig. 12). Statistical measures such as standard error (SE), confidence interval (CI), producer's accuracy (PA), user's accuracy (UA), overall accuracy (%) and Kappa are also provided in the tables.

If we symbolize the raster and focus on the information in the Cross-Tabulation matrix of most interest for assessing our simulation, we can understand the errors we made in our modelling exercise in greater detail.

In our exercise we only actively modelled two categories: urban fabric and industrial areas. In the raster we can identify the simulated areas that show agreement (or disagreement) with the reference map for each of these two categories. To do this, the first step is to identify the code for the combinations involving the two categories being considered: urban fabric (2) and industrial and commercial areas (3).

The combination codes for urban fabric are 3, 16, 29, 42, 81, and 120, while code 29 represents the areas that were simulated as urban fabric (Classified is 2) and also appear as urban fabric on the reference map (Reference is 2). The combination codes for industrial and commercial areas are 4, 17, 30, 43, and 82, while code 43 represents the pixels that are industrial and commercial areas in both the simulation and the reference map.

If we symbolize the raster obtained using these codes in terms of agreement (codes 29 and 43) and disagreement (all the other codes mentioned above), we can visualize the pattern of error in our simulations compared to the map we use as a reference (Fig. 13).

Most of the simulated areas agree with the reference map. Disagreement can only be observed in a few cases. However, this conclusion may be misleading. Most of the agreement refers to areas that were already urban fabric or industrial and commercial areas, i.e. areas that were correctly simulated as permanence.

Simulating permanence for artificial surfaces is very easy. A high rate of success is expected in all cases. If we focus on the areas that actually changed during the simulation period in relation to the reference map and those that were simulated as change, we can detect a higher proportion of errors. However, this cannot be detected on our map. In order to focus on these errors, we should only cross-tabulate the changes in the simulation with respect to the initial map (CORINE 2005) and the changes in the reference map (CORINE 2012) with respect to the initial map (CORINE 2005). Using this method, the new map and the Cross-Tabulation table would only assess those

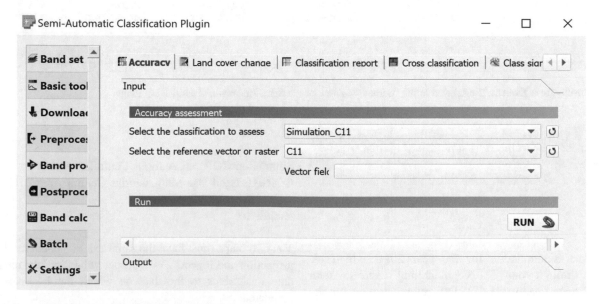

Fig. 11 Exercise 3. Step 1. Semi-Automatic Classification plugin

V_Classified	> ERROR MATRIX (pixel count) > Referencia					
	0	1	2	3	4	5
0	271263	374	543	364	148	11
1	579	243071	61	76	157	109
2	621	60	27402	28	0	0
3	585	33	15	20034	0	0
4	211	327	89	21	2504	0
5	0	0	0	0	0	754
6	44	111	0	0	0	0
7	0	0	0	0	0	0
8	0	0	0	0	0	0
9	0	0	0	0	0	0
10	0	0	0	0	0	0
11	0	0	0	88	0	0
12	0	0	0	0	0	0
Total	273303	243976	28110	20611	2809	874

V_Classified	> AREA BASED ERROR MATRIX > Referencia					
	0	1	2	3	4	5
0	0.2937	0.0004	0.0006	0.0004	0.0002	0.0000
1	0.0006	0.2632	0.0001	0.0001	0.0002	0.0001
2	0.0007	0.0001	0.0297	0.0000	0.0000	0.0000
3	0.0006	0.0000	0.0000	0.0217	0.0000	0.0000
4	0.0002	0.0004	0.0001	0.0000	0.0027	0.0000
5	0.0000	0.0000	0.0000	0.0000	0.0000	0.0008
6	0.0000	0.0001	0.0000	0.0000	0.0000	0.0000
7	0.0000	0.0000	0.0000	0.0000	0.0000	0.0000
8	0.0000	0.0000	0.0000	0.0000	0.0000	0.0000
9	0.0000	0.0000	0.0000	0.0000	0.0000	0.0000
10	0.0000	0.0000	0.0000	0.0000	0.0000	0.0000
11	0.0000	0.0000	0.0000	0.0001	0.0000	0.0000

Fig. 12 Results from Exercise 3 displayed in the "output" window of the Semi-Automatic Classification Plugin

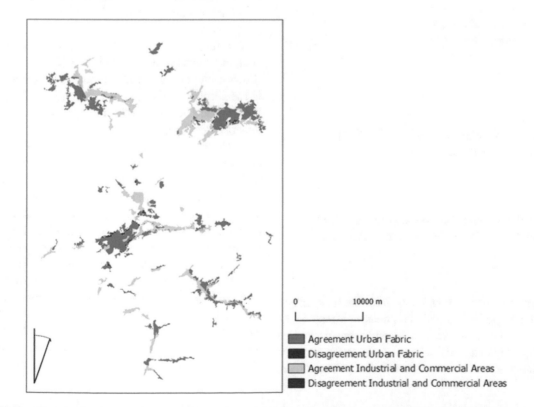

Fig. 13 Result from Exercise 3. Map showing areas of agreement and disagreement between our simulation and the reference map for the two categories actively simulated: urban fabric, industrial and commercial areas

areas that changed between the two dates, so removing unchanged areas from the analysis.

An analysis of the error/Cross-Tabulation matrixes leads to similar conclusions. For urban fabric, out of a total of 28,183 pixels labelled as such on the simulation map (Total column on the right), 27,402 pixels were also classified as urban fabric on the reference map. A total of 621 pixels confuse with agricultural areas, 60 with vegetation areas and 100 with other categories on the reference map. Most of the confusion is therefore with categories where one would expect new urban fabric to develop.

Once again, whereas most of the agreement refers to areas that were already urban fabric in the past and were correctly simulated as persistence, confusion seems to refer above all to areas that were not correctly simulated. That is, agricultural and vegetation areas where new urban fabric was simulated but which, according to the reference map, did not actually undergo any change. We therefore need to repeat the analysis, focusing only on the areas that actually change so as to assess the success of our simulation more effectively.

Other tools, such as the Figure of Merit (see Sect. 4 in Chapter "Pontius Jr. Methods Based on a Cross-Tabulation Matrix to Validate Land Use Cover Maps"), can also be useful to help validate the simulation and overcome some of the limitations we have encountered.

Exercise 4. To validate a series of maps with two or more time points

Aim

To study the land use change between two CORINE maps at two different time points: 2005 and 2011.

Materials

CORINE Land Use Map Asturias Central Area 2005
CORINE Land Use Map Asturias Central Area 2011

Requisites

The two maps must be raster and must have the same extent, spatial resolution, projection and classification legend. If the maps have different classification legends, the user must reclassify the maps in such a way as to unify the two legends. The maps must refer to two different points in time.

Execution

Step 1

The first step is to obtain a raster for the whole study area, showing the areas that changed during the study period and those that remained the same.

To get this map, open the "Semi-Automatic Classification Plugin", click on the "Postprocessing" tab and then select *Land cover change*. Then, complete the required parameters, selecting the older map as the reference classification (CORINE 2005) and the more recent one as the new classification (CORINE 2011). Mark the "Report unchanged pixels" option.

Step 2

To obtain a map that only shows the areas that changed during the study period, we must repeat the same operation, this time leaving the "Report unchanged pixels" option unmarked (Fig. 14).

Results and Comments

After executing Steps 1 and 2, QGIS creates two output rasters, one showing changes and permanent areas (Fig. 15) and the other showing just the changes between the two maps (Fig. 16). Each raster will identify each possible combination between categories or pixel values with a single unique code.

The function also generates a table for each map in the "output" window and stored in CSV format. This table shows each possible combination and the code with which it is represented in the output rasters (Fig. 17). All the combinations are included in the table, even if no pixels actually undergo this change.

Both the rasters and the table can be used to understand the changes in our study area. The table shows those that took place during the study period (Table 2) and includes changes from agricultural areas (category 0 in CORINE 2005), vegetation areas (category 1), urban fabric (2), industrial and commercial areas (3), mineral extraction sites (4), road and rail networks (6) and water bodies (11).

Of the various different transitions of agricultural areas, the one to urban fabric (from category 0 in 2005 to category 2 in 2011) is the most important with a total of 751 pixels. As regards the transitions in vegetation areas (category 1), the most common was the change from vegetation areas to agricultural areas (from category 1 in 2005 to category 0 in 2011), with a total of 588 pixels.

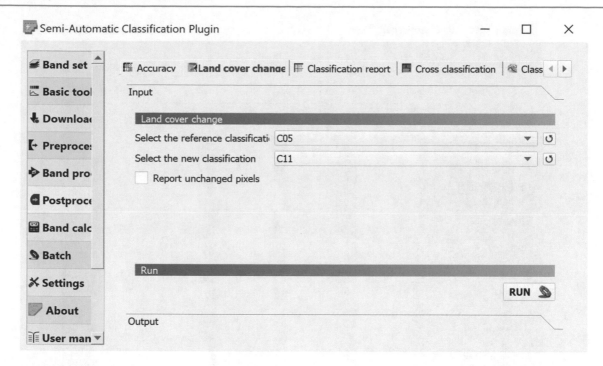

Fig. 14 Exercise 4. Step 2. Semi-Automatic Classification plugin

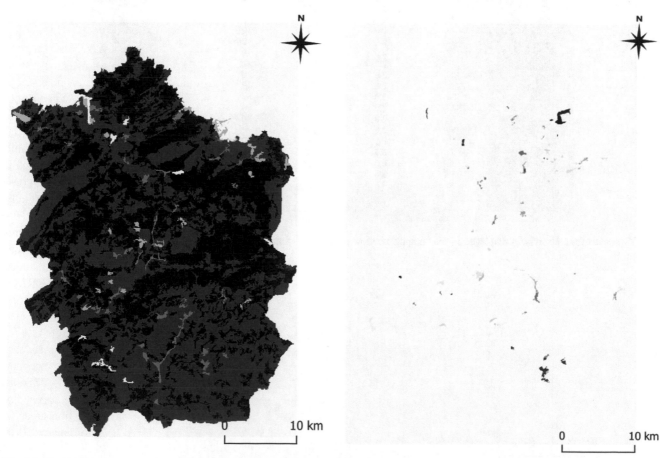

Fig. 15 Result from Exercise 4. Raster displaying the areas that are the same in the two maps compared, that is, the areas of permanence in the time series

Fig. 16 Result from Exercise 4. Raster displaying the areas that are different in the two maps compared, that is, the areas of change in the time series

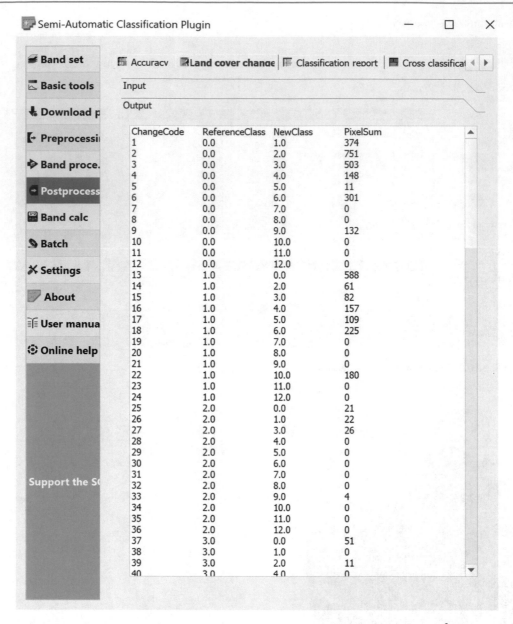

Fig. 17 Results from Exercise 4 displayed in the "output" window of the Semi-Automatic Classification Plugin[2]

---

[2] ReferenceClass and NewClass columns may appear swiched due to the use of a different version of the "Semi-Automatic Classification Plugin".

Table 3 Result from Exercise 4. Table showing the transitions detected between the two maps compared and their size

Code	CORINE 2005	CORINE 2011	Quantity of changes
1	0	1	374
2	0	2	751
3	0	3	503
4	0	4	148
5	0	5	11
6	0	6	301
9	0	9	132
13	1	0	588
14	1	2	61
15	1	3	82
16	1	4	157
17	1	5	109
18	1	6	225
22	1	10	180
25	2	0	21
26	2	1	22
27	2	3	26
33	2	9	4
37	3	0	51
39	3	2	11
49	4	0	211
50	4	1	327
51	4	2	89
52	4	3	21
73	6	0	44
74	6	1	111
136	11	3	88
140	11	7	657

Code column may appear with other values if using a different version of the "Semi-Automatic Classification Plugin"

This change in pixels (Table 3) can be translated into a change in area, by multiplying each pixel by the area it covers. The spatial resolution of our raster is 50 m, so the calculation is easy: a square with a 50 m side covers a surface area of 2500 m^2. This is the area of each pixel. Therefore, the transition from agricultural areas (0) to urban fabric (2) which took place in 751 pixels affected an area of 1,877,500 m^2.

Most of the change in our area was between agricultural and vegetation areas and vice versa and from agricultural and vegetation areas to artificial surfaces. However, there were also various other interesting transitions, such as the conversion of water bodies into port areas (from category 11 to category 7), which affected a total of 657 pixels. This was due to the construction of a dock in Gijón in the north of our study area.

By symbolizing the raster of changes (Fig. 18), we can gain a spatial perspective of what changed. To obtain this map, we must group the changes together according to the new land use. Codes 13, 25, 37, 49 and 73 will show the areas that changed to agricultural areas. Codes 1, 26, 50 and 74 will show changes to vegetation areas. Codes 2, 14, 39 and 51 will show changes to urban fabric. Codes 3, 15, 27, 52 and 136 will show changes to industrial and commercial areas. Codes 4 and 16 will show changes to mineral extraction sites. Codes 5 and 17 will show changes to dump sites. Codes 6 and 18 will show changes to road and rail networks. Code 140 will show changes to port areas. Codes 9 and 33 will show changes to artificial green urban areas. Finally, Code 22 will show changes to open spaces with little or no vegetation.

In the composition of the map in Fig. 18, we also added CORINE 2006 as the base layer, with an opacity of 10%, to enable us to interpret the changes on the map better.

The map shows the changes for the example area of Gijón. In the north, we can observe the new dock built in the port area. Apart from the port, most of the growth in industrial land took place in the south of the city. The same is true for urban fabric, with the construction of a new residential development in Roces. As can be seen on the map, this new residential area is cut off from the existing urban fabric of the city. There is a highway running between the two.

The results of this analysis can also be useful to validate a chronological series of maps. When interpreting the changes, it can help detect unrealistic changes that may be due to errors in the input data. We can also detect changes in the boundaries of the study area which cannot be fully represented on the maps because the study area has been clipped.

Other tools and techniques, such as LUCC budget or Quantity and Allocation disagreement, can also help characterize real changes in the study area and detect areas where no changes have taken place, despite being marked as change areas on the maps. In this way, these techniques can provide useful, complementary information on this question.

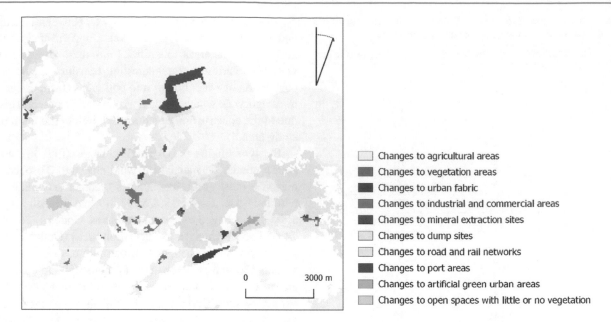

Fig. 18 Result from Exercise 4. Map showing areas of change between the two maps compared, displayed over the map for the oldest year

2 Multiple-Resolution Cross-Tabulation

Description

Multiple-Resolution Cross-Tabulation is based on the same technique as basic Cross-Tabulation (see previous section). It crosses two raster datasets at a minimum of two different spatial resolutions: the original resolution and a coarser one. However, users can compare the dataset at as many different resolutions as they deem fit. These must always be coarser than the original spatial resolution.

The concept of spatial resolution refers to the level of spatial detail available in the spatial data. It applies to data in raster format, where the spatial resolution is defined by the pixel size. This means that, unlike basic Cross-Tabulation, this analysis can only be performed with raster data.

Utility

Exercises
1. To validate a map against reference data/map
2. To validate soft maps produced by the model against a reference map
3. To validate a simulation against a reference map

This technique aims to control the multiscale uncertainty of a validation exercise, which is not considered in basic Cross-Tabulation analyses. It can also be used to evaluate the uncertainty of a LUC classification exercise, a LUC map or a LUCC modelling exercise against reference data.

Maps that show a lot of disagreement at detailed scales can refer to the same information at coarser scales. This technique can therefore be used to discover at which spatial resolution a map is considered least uncertain according to the information provided by a reference map.

This analysis can be used as a complement to fuzzy logic tools (Fritz and See 2005), which evaluate the agreement between maps by considering spatial near-hits. A near-hit occurs when two pixels that share the same value are not in the same spatial position, but close to each other.

Multiple-Resolution Cross-Tabulation can only be carried out with raster data. However, we can make the comparison with either hard- or soft-classified raster maps, such as suitability, transition potential or probabilities maps. In the last case, we must always reclassify the soft-classified raster maps in a set of categories. It is not possible to cross-tabulate rasters with a continuous range of values.

As in the case of basic Cross-Tabulation, if we want to explore the full potential of the results of these analyses, we can use other complementary metrics such as Land Use Cover Change budget (LUCC budget, see Sect. 2), Quantity and Allocation disagreement or the Figure of Merit (see Sects. 3 and 4 in Chapter "Pontius Jr. Methods Based on a Cross-Tabulation Matrix to Validate Land Use Cover Maps").

In addition to the basic Multiple-Resolution Cross-Tabulation presented in this section, some more sophisticated variants have been proposed by other authors. These include:

- Costanza (1989), who proposed a method to determine the goodness of fit between model output and spatial and/or time series data based on the idea that the measurements at one resolution are not sufficient to describe more complex patterns. In his method, an expanding window is used to gradually degrade the resolution of the data, establishing, among the lack of fit, situations of "registration", "resolution" and residual components.

- Kok et al. (2001), who proposed a multiscale land use change modelling procedure, applied at five spatial resolutions, and demonstrated that results improve strongly as spatial resolution decreases.
- Pontius and Cheuk (2006) proposed a method for computing a Cross-Tabulation matrix at multiple scales, focusing on soft-classified pixels. This Multiple-Resolution method resolves difficulties due to traditional Cross-Tabulation approaches and fuzzy methods, proposing a Composite operator.

QGIS Exercises

Available tools

- Processing Toolbox
  GRASS
    Raster
      r.resample
- Processing Toolbox
  GDAL
    Raster projections
      Warp (reproject)
- Processing Toolbox
  SAGA
    Raster tools
      Resampling
- Layer
      Save As...

QGIS does not include a tool to cross-tabulate maps at multiple resolutions. To carry out this analysis, it is therefore necessary to combine raster resampling tools with the basic Cross-Tabulation tools. For detailed information of the tools available in QGIS for performing Cross-Tabulation, please refer to Sect. 1.

Various different tools can be used to resample raster maps in QGIS. The GRASS module provides a tool (*r.resample*) for resampling the raster according to the Nearest Neighbour method. The GDAL module provides a tool to reproject rasters (*Warp (reproject)*) that also enables resampling through different methods, including the Nearest Neighbour. For its part, the SAGA toolbox provides a tool for resampling rasters with similar options. In addition, the

QGIS interface allows the user to resample maps by making a copy of a displayed map via the option "Save raster layer as…" (Layer > *Save as*).

For categorical maps such as Land Use Cover maps, two resampling strategies are usually applied: Nearest Neighbour and Majority Rule. We decided to apply Nearest Neighbour because this is the method that best preserves the landscape composition and configuration or in other words, the proportions of the different categories and their patterns.

The four resampling tools available in QGIS are all equally valid. In this case we decided to use the tool that becomes available when making a copy of an existing raster (*Save as…*) because of its simplicity and efficiency. Nevertheless, users must be aware that the resampled rasters will vary slightly depending on the method chosen, and are therefore not fully comparable. Once a method or tool has been selected, all the resampling procedures must be performed using this same method or tool.

Exercise 1. To validate a map against reference data/map

Aim

To validate the CORINE 2011 Land Use map, taking the SIOSE 2011 Land Use map as the reference and determining the resolution at which the maps show most agreement.

Materials

SIOSE Land Use Vector Map Asturias Central Area 2011
CORINE Land Use Vector Map Asturias Central Area 2011

Requisites

The two maps must have the same extent, projection and classification legend. If the maps have different classification legends, the user must reclassify the maps in such a way as to unify the two legends.

Execution

Step 1

Given that to carry out Cross-Tabulation at multiple resolutions we need to have maps in raster format, the first thing we have to do is rasterize our vector maps. If you would like

Fig. 19 Exercise 1. Step 1. Rasterize (Vector to Raster)

to perform this analysis by resampling original raster maps, please refer to Exercise 2 Step 1.

We are going to convert our original vector file to raster at four different spatial resolutions: 25, 50, 75 and 100 m. Our analysis will be based on the same four spatial resolutions.

To rasterize vector data, we use the *Rasterize (Vector to raster)* tool. Once inside this tool, we begin by indicating the vector layer we want to rasterize (SIOSE 2011 map). Then, we go to "Field to use for burn-in value [optional]" where we indicate the field in the attribute table of the vector layer that will give the raster the pixel values (Metro) (Fig. 19).

We must also set the spatial resolution for the raster we want to create. To do this, we must first define the units for the spatial resolution in the "Output raster size unit" option (Georeferenced Units). Then, we choose the spatial resolution or pixel size through the "Width/Horizontal resolution"

(25) and "Height/Vertical resolution" options (25). We must also specify the extent of the raster that will be created in the option "Output extent (xmin, xmax, ymin, ymax)". We are going to use the extent of the layer we are rasterizing (SIOSE 2011) through the submenu on the right (Use layer extent...).

The final stage is to assign a value to the background, i.e. the pixels that are not covered by any polygon in the vector file. Given that the vector already has values from 0 to 11, we will define the background with code 12. We do this via the option "Pre-initiate the output image with value [optional]", available under the "Advanced parameters" options (Fig. 20).

Our background value (12) will also be the nodata value of our raster. We can assign a nodata value for the raster we are going to create using the option "Assign a specified nodata value to output bands [optional]" (Fig. 19).

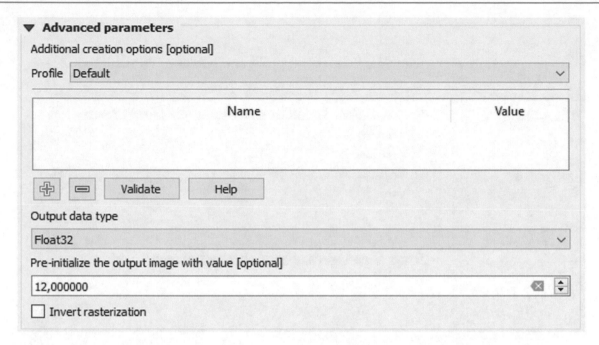

Fig. 20 Exercise 1. Step 1. Advanced parameters of the Rasterize (Vector to Raster) tool

Step 2

Once we have finished the first rasterization, we must repeat the same procedure for the other three spatial resolutions that we need for the SIOSE dataset. Then, we must repeat the whole workflow for the CORINE map. Once all these tasks have been completed, we will have 8 different maps (4 SIOSE and 4 CORINE) at 4 different spatial resolutions (25, 50, 75 and 100 m).

Step 3

Once all the maps have been created, we can start the Cross-Tabulation. To do this, open the "Semi-Automatic Classification Plugin", click on the "Postprocessing" tab and select *Cross Classification*. Then, select the required parameters: raster to assess (CORINE map 25 m) and reference raster (SIOSE map 25 m) (Fig. 21).

Step 4

After the first execution, repeat this process with the other pair of maps (one for CORINE and one for SIOSE) at different spatial resolutions.

Results and Comments

Once we have executed the function four times, QGIS will create an output map for each execution with the combined classes and an error/Cross-Tabulation matrix. These will be stored in the folder we selected earlier when executing the tool. Matrixes are also displayed in the "output" window. For a detailed description of each of these results, please refer to the Sect. 1.

If we compare the results of each of the error matrixes, we can see that there are few differences between them. Error matrixes show the area in square meters covered by each possible combination between classes. The combination that covers most area is always the agreement between agricultural areas: pixels that are 0 (agricultural areas) in both the validated (CORINE) and the reference (SIOSE) maps. At a spatial resolution of 25 m, these areas occupy 585,267,500 m^2; at 50 m, 585,225,000 m^2; at 75 m, 585,815,625 m^2; and at 100 m, 584,660,000 m^2. The differences are therefore very small.

A similar pattern can be observed if we look at the rest of the combinations. This means that at all the spatial resolutions there are very similar levels of agreement and disagreement between the classes on the two maps (CORINE

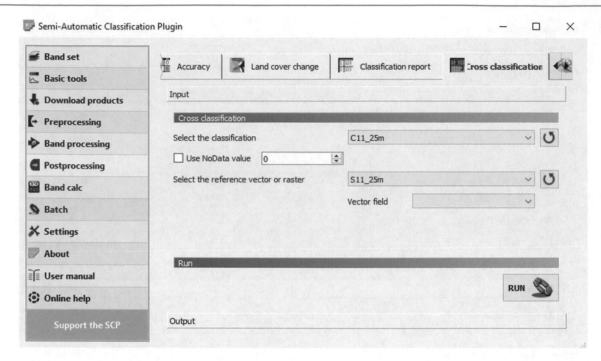

Fig. 21 Exercise 1. Step 2. Semi-Automatic Classification plugin

and SIOSE). We can therefore conclude that the spatial resolution selected to make the analysis has no substantial effect on the results.

That means that the areas classified differently on the two maps are not due to small details drawn on one map that do not appear on the other. Disagreement is not the result of isolated pixels on one map that are not classified in the same category on the other. If this were true, the agreement between the two maps should be higher at coarser resolutions because they are more generalized, so ruling out minor details.

In conclusion, it would seem that the differences between the two maps are structural. In other words, they are not caused by the spatial resolution or level of detail of the maps, and instead result from the fact that each map represents a different reality on the ground. If we generalize both maps and rule out all small details, both maps show a similar level of agreement. Notwithstanding this, we must always remember that most of the areas in both maps agree, as confirmed in the Sect. 1.

When compared with SIOSE, CORINE can be considered a valid map because the agreement between the two is very high. The differences between them are the same regardless of the spatial resolution employed to make the analysis, at least within the resolution range we used (from 25 to 100 m). Thus, although the differences between SIOSE and CORINE are the result of their different scale and Minimum Mapping Unit, they cannot be eliminated simply by generalizing the maps using coarser spatial resolutions. In fact, their agreements and disagreements remain the same,

which suggests that the different scale of production introduces important structural differences in the way the two maps draw the ground land uses and land covers.

Exercise 2. To validate soft maps produced by the model against a reference map

Aim

To evaluate to what extent the urban fabric suitability map of our model agrees with the urban fabric areas of the reference map for the year of the simulation at multiple spatial resolutions, determining the resolution at which there is most agreement.

Materials

CORINE Land Use Map Asturias Central Area 2011
Urban fabric suitability map—CORINE model

Requisites

The two maps must have the same extent, spatial resolution and projection. The soft map must be a categorical map. The Land Use map must only contain information about the category being assessed. For a proper validation, the reference map must refer to the same date as the simulation.

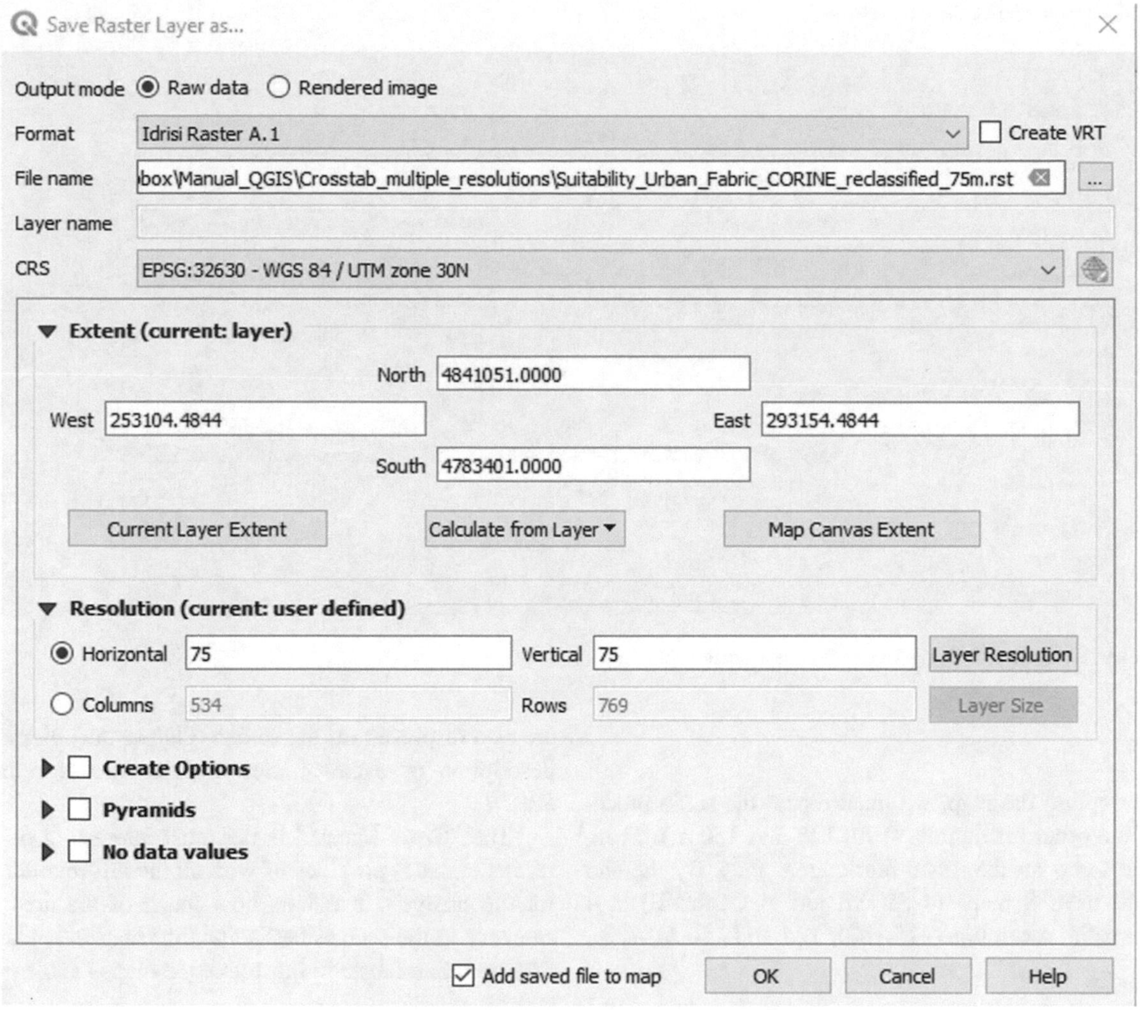

Fig. 22 Exercise 2. Step 3. Save Raster Layer as... tool

Execution

Step 1

We begin by converting our soft map into a categorical one to comply with the requirements of the Cross-Tabulation tool. This is done using the *Reclassify by table* function (Processing toolbox > Raster analysis > Reclassify by table).

There are no standard criteria for the reclassification of soft maps and users can apply whatever thresholds they think best. In this case, we will use the same thresholds we used in Exercise 2 of the Sect. 1. We will therefore reclassify the map into four new categories: 1 (suitability 0–0.25), 2 (0.25–0.50), 3 (0.50–0.75) and 4 (0.75–1).

Step 2

As stated in the requisites, we will cross-tabulate the reclassified soft map with a map that only shows the Land Use Cover category of interest, i.e. urban fabric. To this end, we must extract the urban fabric areas from the LUC map (CORINE) using the same function as in Step 1 (*Reclassify by table*). In the reclassification, we will assign a value of 1 to urban fabric (code 2 in the original map) and a value of 0 to the other categories (codes 0, 1, 3, 4, 5, 6, 7, 8, 9, 10, 11 and 12). For a detailed explanation of how to carry out these first two steps, readers are referred to Exercise 2 of the Sect. 1.

Step 3

Once we have the two maps, we can then resample them at different spatial resolutions to carry out the Multiple-ResolutionCross-Tabulation. In our case, as the original pixel size is 50 m, we will resample our maps at 75, 100, 125 and 150 m using the *Save As...*tool. In this tool, we need to indicate the name of the map we are going to resample (the reclassified suitability map of urban fabric) and the spatial resolution at which we will resample the maps (Fig. 22), in our case, 75 m.

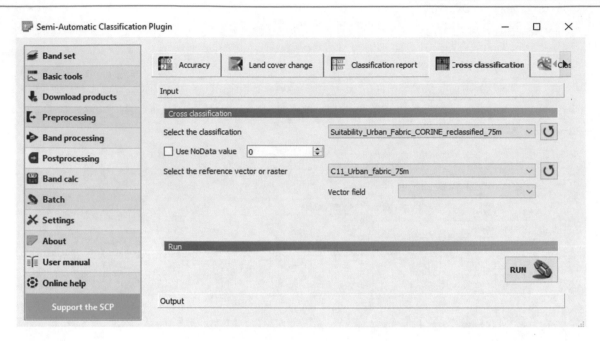

Fig. 23 Exercise 2. Step 5. Semi-Automatic Classification plugin

Step 4

After resampling the map, we must repeat the same procedure for the other resolutions (100, 125 and 150 m). Then, we do the same for the urban fabric areas map. By the end we should have 8 maps (4 SIOSE and 4 CORINE) at 4 different spatial resolutions (75, 100, 125 and 150 m).

Step 5

Once we have obtained all the maps we need, we can then carry out the Cross-Tabulation exercise using the *Cross classification* tool from the "Semi-Automatic Classification Plugin". Once inside the tool, we must indicate the two rasters that we want to cross-tabulate: the soft map (Select the classification) and the land use map for the category of interest (Select the reference vector or raster) (Fig. 23).

Step 6

After we do this for the maps at the original resolution (50 m), we repeat the process at the other 4 spatial resolutions (75, 100, 125 and 150 m).

Results and Comments

After executing the function for each pair of maps at each spatial resolution, the tool produces (for each spatial resolution) an output map with the combination and two matrixes detailing how the values of both maps cross-tabulate. These are stored in the folder we selected and

are also displayed on the screen (Output tab). For a detailed description of each of these results, please refer to the Sect. 1.

"The "Cross Matrix" is the most interesting of all these results in that it provides us with all the information we need for our analysis. It details how much of the area for each category in the reclassified suitability map falls inside areas that are urban fabric in our reference maps (Tables 4, 5, 6, 7 and 8).

For the analysis at a spatial resolution of 50 m, there are 4999 m² of low suitability (suitability below 0.25) that cross-tabulate with areas that are urban fabric in the

Table 4 Result from Exercise 2. Table showing the corresponde between the urban fabric category in CORINE and the different groups of suitability values for urban fabric in the map at 50m of spatial resolution

50 m	0 (Not urban fabric)	1 (Urban fabric)
1 (0–0.25)	2,312,499	**4,999**
3 (0.50–075)	578,867,274	**4,927,498**
4 (0.75–1)	962,127,124	**65,342,474**

Table 5 Result from Exercise 2. Table showing the corresponde between the urban fabric category in CORINE and the different groups of suitability values for urban fabric in the map at 75 m of spatial resolution

75 m	0 (Not urban fabric)	1 (Urban fabric)
1 (0–0.25)	2,277,136	**11,245**
3 (0.50–075)	578,752,547	**4,919,739**
4 (0.75–1)	963,211,926	**65,351,009**

Table 6 Result from Exercise 2. Table showing the corresponde between the urban fabric category in CORINE and the different groups of suitability values for urban fabric in the map at 100 m of spatial resolution

100 m	0 (Not urban fabric)	1 (Urban fabric)
1 (0–0.25)	1,576,662	**738,436**
3 (0.50–075)	578,405,616	**5,208,973**
4 (0.75–1)	961,414,853	**64,373,732**

Table 7 Result from Exercise 2. Table showing the corresponde between the urban fabric category in CORINE and the different groups of suitability values for urban fabric in the map at 125 m of spatial resolution

125 m	0 (Not urban fabric)	1 (Urban fabric)
1 (0–0.25)	2,410,302	**15,651**
3 (0.50–075)	579,223,768	**5,008,419**
4 (0.75–1)	961,507,072	**65,125,110**

Table 8 Result from Exercise 2. Table showing the corresponde between the urban fabric category in CORINE and the different groups of suitability values for urban fabric in the map at 150 m of spatial resolution

150 m	0 (Not urban fabric)	1 (Urban fabric)
1 (0–0.25)	2,296,991	**0**
3 (0.50–075)	580,283,055	**4,841,697**
4 (0.75–1)	960,525,192	**65,216,537**

reference LUC map. If we consider that each pixel represents an area of 2500 m^2 (50 m × 50 m), this means that only 2 pixels of urban fabric cross-tabulate with areas of low suitability on the suitability map. 1971 pixels with medium to high suitability (0.5–0.75) cross-tabulate with areas that are urban fabric. Finally, most of the urban fabric pixels cross-tabulate with areas with the highest suitability (0.75–1): this combination is represented by 26,137 pixels. These data show that there is a positive correlation between suitability and the presence of urban fabric. We can therefore conclude that suitability is a good driver for our model.

Varying the spatial resolution of the analysis did not lead to any major differences in the correlation between the suitability map and the urban fabric areas in the reference maps. At the five spatial resolutions assessed, most of the pixels fell within the highest suitability category (0.75–1).

The dissimilarities between the analyses at different resolutions were very small. At 75 m, just two pixels fell within the areas of lowest suitability (11,245 m^2). At 100 m, there were a lot more: 74 pixels (738,436 m^2). At 125 m there was just 1 pixel (15,651 m^2), and at 150 m, no pixels at all (0 m^2). Similar behaviour can be observed for the other two categories of suitability at all five resolutions.

This indicates that the suitability map for urban fabric in our modelling exercise is correct. It positively correlates with those areas that are urban fabric in our reference map, so helping us to identify the areas in which new urban fabric is most likely to appear. However, no conclusions can be drawn regarding the best spatial resolution at which to carry out the modelling exercise. As the explanatory power of the suitability maps is very similar at all the spatial resolutions assessed, the decision as to which spatial resolution would be best for our modelling exercise should be based on other factors, such as how realistic the pattern looks or what the minimum level of detail might be for the model to be useful for stakeholders and users.

This analysis could be complemented with more sophisticated tools like the ROC curve and the Difference in Potential (see Sects. 2 and 3 in Chapter "Validation of Soft Maps Produced by a Land Use Cover Change Model"). These tools also provide information about how well a model soft map simulates a category of interest, such as urban fabric.

Exercise 3. To validate a simulation against a reference map

Aim

To validate a simulation for the year 2011 against a reference map for the same year at multiple spatial resolutions, determining the resolution at which both maps show the best agreement.

Materials

Simulation CORINE Asturias Central Area 2011
CORINE Land Use Map Asturias Central Area 2011

Requisites

The two maps must have the same extent, spatial resolution, projection and legend. For proper validation, the reference date must refer to the date on which the landscape was simulated.

Execution

Step 1

For Multiple-Resolution Cross-Tabulation, we need first to resample the original rasters (50 m) at other spatial resolutions. In this case, we will resample our simulation at 100,

150 and 200 m, according to the procedure for the *Save As...* tool set out in the previous exercise (Exercise 2, Execution - Step 2). Once inside the tool, we fill in the required parameters: name of the raster to be sampled (Simulation CORINE) and spatial resolution (100 m).

Step 2

Once we have resampled the first map, we then repeat the procedure for the other spatial resolutions (150 and 200 m) and for the reference map. By the end, we should have 8 maps (4 simulations and 4 reference maps) at 4 spatial resolutions (50, 100, 150 and 200 m).

Step 3

With all these resampled maps, we can then carry out the Cross-Tabulation exercise at multiple resolutions. To do this, open the "Semi-Automatic Classification Plugin", click on the "Postprocessing" tab and select *Accuracy*. Fill in the required parameters: raster to assess (Simulation CORINE 11 map at 50 m) and reference raster (CORINE 11 map at 50 m) (Fig. 24).

Step 4

Repeat the same procedure for the other pairs of maps at 100, 150 and 200 m.

Results and Comments

After this function has been executed for each spatial resolution, QGIS will create an output map, a couple of matrixes and some statistical measures. All the tables and statistics can be consulted in the "output window" and all the results will be saved in the folder we selected earlier. For a detailed description of each of these results, please refer to the Sect. 1.

The analysis of the matrixes at the different spatial resolutions shows no important differences between resolutions, and very similar results in all cases. In general, there is a high level of agreement between the simulation and the reference map, as studied above in the Sect. 1 when conducting the analysis at the original resolution of the modelling exercise.

If we take Overall Accuracy as a summary metric describing the similarity between the two maps, we can see that similarity is very high in all cases (Table 9). Only the exercise at 100 m shows a lower agreement rate. This may be due to multiple causes, but it does indicate that coarsening the spatial resolution of the simulation does not ensure higher levels of agreement between the simulated landscape and the reference landscape.

We must also bear in mind the limitations for this exercise mentioned in the Sect. 1. Validating a simulation by cross-tabulating the simulated exercise with a reference map may be misleading. Most of the areas in both maps agree

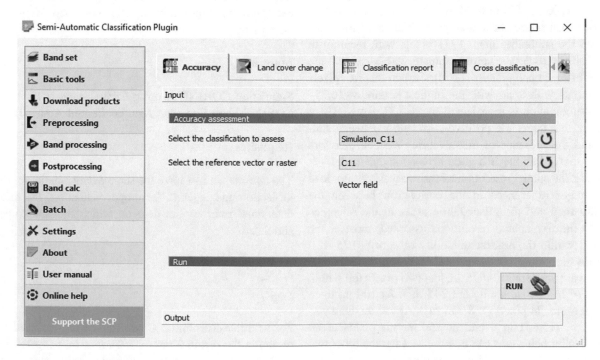

Fig. 24 Exercise 3. Step 3. Semi-Automatic Classification plugin

Table 9 Results from Exercise 3. Overall accuracies of the simulation, when assessed against a reference map, at four spatial resolutions: 50, 100, 150 and 200 m

	50 m	100 m	150 m	200 m
Overall accuracy (%)	99.3	93.6	99.3	99.3

because most of the areas in the simulated landscape remain the same during the modelling period.

The best way to validate the changes modelled in our exercise is to focus exclusively on the simulated changes and on a map of reference showing the changes on the ground. In this case, the Multiple-Resolution exercise could provide very interesting insights, as agreement between simulated and reference changes may be higher at coarser spatial resolutions.

References

Alcamo J, Schaldach R, Koch J et al. (2011) Evaluation of an integrated land use change model including a scenario analysis of land use change for continental Africa. Environ Model Softw 26:1017–1027. https://doi.org/10.1016/j.envsoft.2011.03.002

Congedo L (2016) Semi-automatic classification plugin documentation. https://doi.org/10.13140/RG.2.2.29474.02242/1

Conrad O (c) (2007, 2015) SAGA-GIS module library documentation

Costanza R (1989) Model goodness of fit: a multiple resolution procedure. Ecol Modell 47:199–215. https://doi.org/10.1016/0304-3800(89)90001-X

Fritz S, See L (2005) Comparison of land cover maps using fuzzy agreement. Int J Geogr Inf Sci 19:787–807. https://doi.org/10.1080/13658810500072020

Hagen-Zanker A (2009) An improved fuzzy Kappa statistic that accounts for spatial autocorrelation. Int J Geogr Inf Sci 23:61–73. https://doi.org/10.1080/13658810802570317

Kok K, Farrow A, Veldkamp A, Verburg PH (2001) A method and application of multi-scale validation in spatial land use models. Agric Ecosyst Environ 85:223–238. https://doi.org/10.1016/S0167-8809(01)00186-4

Krüger C, Lakes T (2015) Bayesian belief networks as a versatile method for assessing uncertainty in land-change modeling. Int J Geogr Inf Sci 29(1):111–131. https://doi.org/10.1080/13658816.2014.949265

Mas J-F, Kolb M, Paegelow M et al. (2014) Inductive pattern-based land use/cover change models: a comparison of four software packages. Environ Model Softw 51:94–111. https://doi.org/10.1016/j.envsoft.2013.09.010

Paegelow M, Camacho Olmedo MT, Mas J-F, Houet T (2014) Benchmarking of LUCC modelling tools by various validation techniques and error analysis. Cybergeo. https://doi.org/10.4000/cybergeo.26610

Pontius Jr RG, Cheuk ML (2006) A generalized cross-tabulation matrix to compare soft-classified maps at multiple resolutions. Int J Geogr Inf Sci 20:1–30. https://doi.org/10.1080/13658810500391024

Pontius Jr RG (2018) Pontius matrix. Available at https://www2.clarku.edu/faculty/rpontius/PontiusMatrix42.xlsx

Shapiro M, U.S. Army Construction Engineering Research Laboratory © 2003–2020 GRASS Development Team, GRASS GIS 7.8.3dev Reference Manual

Metrics Based on a Cross-Tabulation Matrix to Validate Land Use Cover Maps

Jean-François Mas, David García-Álvarez, Martin Paegelow, Roberto Domínguez-Vera, and Miguel Ángel Castillo-Santiago

Abstract

The overlaying of two map layers is a standard GIS procedure. As we saw in the previous chapter, it enables us to compute the intersection between two feature classes and cross-tabulate either the area or the pixel count of the intersecting features depending on whether raster or vector data are being used. Cross-tabulation can be used to evaluate different topics depending on the nature of the input data. In this chapter, cross-tabulation is used to assess land cover changes, the spatial agreement between maps and map accuracy. In Sect. 1, Land use/cover changes (LUCC) are quantified by comparing two LUC maps, computing different indices of change and creating a change matrix. In Sect. 2, we used various metrics to evaluate the spatial agreement between two maps. This procedure was applied to compare a LUC map with a reference map, a simulated LUC map with a reference map and a simulated LUCC map with a reference map of changes. Section 3 introduces the Kappa indices, which allow us to assess the agreement between two datasets, given the agreement expected by random coincidence. We used the indices to compare observed or simulated maps with a reference map. In Sect. 4 we evaluate the agreement between maps at a global level (the entire map) by focusing on a specific feature such as a smaller area or a particular category (stratum level). Finally, in Sect. 5, the cross-tabulation between a map and reference sample data is used to assess the thematic accuracy of the map by calculating various different accuracy indices. We present examples of analyses based on cross-tabulation for four different cases: To validate a series of maps with two or more time points, to validate a map against a reference map, to validate a simulation against a reference map and to validate simulated changes against a reference map of changes. In the example exercises, we use CORINE and SIOSE maps from the Asturias Central Area and Ariège Valley datasets and maps of the Marqués de Comillas region of south-eastern Mexico (MarquesLUC dataset). The cross-tabulation techniques proposed by Robert Gilmore Pontius Jr. are applied in Chapter "Pontius Jr. Methods Based on a Cross-Tabulation Matrix to Validate Land Use Cover Maps".

Keywords

Cross-tabulation • Changes • Spatial agreement • Accuracy

J.-F. Mas (✉)
Laboratorio de Análisis Espacial, Centro de Investigaciones en Geografía Ambiental, Universidad Nacional Autónoma de México, Morelia, Mexico
e-mail: jfmas@ciga.unam.mx

D. García-Álvarez
Departamento de Geología, Geografía y Medio Ambiente, Universidad de Alcalá, Alcalá de Henares, Spain

M. Paegelow
Département de Géographie, Aménagement et Environnement, Université Toulouse Jean Jaurès, Toulouse, France

R. Domínguez-Vera · M. Á. Castillo-Santiago
Departamento de Observación y Estudio de la Tierra, la Atmósfera y el Océano, El Colegio de la Frontera Sur, San Cristóbal de las Casas, Mexico

1 Change Statistics

Description

Land use/cover change (LUCC) can be quantified by comparing two maps or two classified images that represent land cover at two different dates.

Absolute change (AC) is the difference in the area covered by a category (category area) between two dates and is usually expressed in hectares or square kilometres.

$$AC = A_2 - A_1$$

D. García-Álvarez et al. (eds.), *Land Use Cover Datasets and Validation Tools*,
https://doi.org/10.1007/978-3-030-90998-7_8

where A_1 and A_2 are the category areas in question at dates 1 and 2, respectively.

AC can be divided by the number of years between the two dates to obtain the average annual change area over the study period.

Relative change (RC) is obtained by normalizing the absolute change value by the category area at date 1.

$$RC = (A_2 - A_1)/A_1$$

This formula expresses the proportion of the category area that changed over the study period.

Other indices of LUCC include rates of change. The most popular rate of change is the annual rate of deforestation proposed by the FAO (1995). This indicator is based on the compound interest law. It expresses the proportion of the category area that changes in one year.

$$t = \left(\frac{A_2}{A_1}\right)^{1/(t_2 - t_1)} - 1$$

An alternative equation, also based on the compound interest law, was proposed by Puyravaud (2003).

$$r = \frac{1}{(t_2 - t_1)} \ln \frac{A_2}{A_1}$$

Both formulae give similar results except when LUCC is very high, in which case r is significantly higher than t (Puyravaud, 2003).

All the change indices presented above indicate net change, which results from the balance after gross losses have been subtracted from gross gains. For instance, a given forest category could show an absolute change of −2 ha, which could be erroneously interpreted as very little change, but in fact is the result of two opposing processes: the deforestation of 202 ha compensated by the reforestation of 200 ha. A more detailed analysis of change dynamics can be obtained by cross-tabulating the two maps at two different dates and drawing up a change matrix. The change matrix is a cross-tabulated table indicating the area covered by each change (or permanence) between a category at date 1 and another category at date 2. Many change indices can be obtained from this matrix (see, for example, Sect. 2).

Utility

Exercises

1. To validate a series of maps with two or more time points

Indices of change are widely used to assess LUCC. Normalized indices, such as rates of change, enable us to compare the rate of change between regions of different sizes.

QGIS Exercise

Available tools

- Processing R provider Plugin
  Change_Statistics.rsx R script

The indices of change proposed in this document are based on the area statistics for the two maps. These could be efficiently computed using a spreadsheet program. However, we suggest using a simple R script using the QGIS Processing R provider plugin. The script generates a table containing the absolute change (AC) area, the relative change (RC) area (both in hectares), the rates of change based on FAO and Puyravaud (2003) and the change matrix.

Exercise 1. To validate a series of maps with two or more time points

Aim

To assess LUCC in the Ariege study area using the CORINE Land Use maps dated 2000 and 2018.

Materials

CORINE Land Cover Map Val d'Ariège 2000
CORINE Land Cover Map Val d'Ariège 2018

Requisites

All maps must be in raster format and have the same resolution, extent and projection.

Execution

If necessary, install the Processing R provider plugin and download the R script Change_Statistics.rsx into the R scripts folder (processing/rscripts). For more information, see Chapter "About This Book".

Step 1

Then, execute the script and fill in the required parameters (names and dates of the two maps and the output table) as shown in Fig. 1.

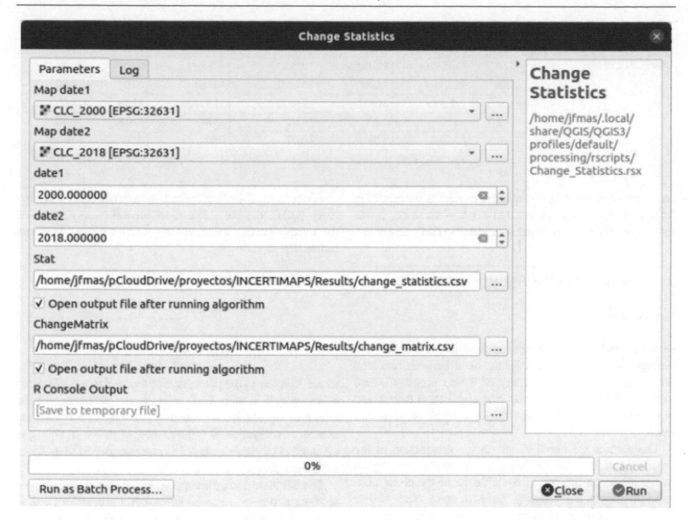

Fig. 1 Exercise 1. Step 1. Change Statistics R script

Table 1 Results from Exercise 1 displayed in the "output" window of the Change Statistics R script. Change indices

	0	1	2	3	4	5	6
Area date 1	74,437	3350	54,558	39,491	12,729	931	76
Area date 2	74,437	5190	52,571	40,344	11,973	943	115
Absolute change (ha)	0	1840	−1987	853	−756	12	39
Relative change (%)	0	54.93	−3.64	2.16	−5.94	1.29	51.32
Annual rate of change t (FAO)	0	2.46	−0.21	0.12	−0.34	0.07	2.33
Annual rate of change r	0	2.43	−0.21	0.12	−0.34	0.07	2.30

The script generates two tables in CSV format: a table showing the change indices (Table 1) and the change matrix (Table 2).

Results and Comments

The two land covers with the most significant absolute change are Categories 1 (built-up) and 2 (agriculture). During the period 2000–2018, the built-up area is increased by 1840 ha, and agriculture lost 1987 ha. The built-up area increased by over 50%. The rates of change resulting from the two equations are very similar. The two categories with the largest rates are built-up (Category 1) and water (Category 6) areas. Over the period 2000–2018, the area covered by these categories increased by around 2.5% a year. Categories 2 and 4 (agriculture and scrubs) present a negative net change rate, indicating that their areas have been shrinking. The change matrix gives us more information about the

Table 2 Results from Exercise 1 displayed in the "output" window of the Change Statistics R script. Change matrix

	1	2	3	4	5	6
1	3301.70	3.41	0.02	8.17	0.00	36.59
2	1853.16	52,059.37	235.25	408.86	0.00	1.57
3	22.74	108.68	39,232.14	127.07	0.00	0.00
4	12.37	399.27	876.72	11,418.38	21.89	0.00
5	0.00	0.00	0.02	10.04	920.81	0.00
6	0.00	0.00	0.00	0.00	0.00	76.40

processes of change. One surprising change is the transition from 1 (built-up) to 6 (water). On closer observation, it was found that pits had been filled with water to create reservoirs.

2 Areal and Spatial Agreement Metrics

Description

Different authors have proposed a series of metrics that evaluate the areal and spatial agreement between two land use/cover maps or between any of their categories. These metrics are obtained from the cross-tabulation matrix and summarize in a single value the agreement between two maps.

The metrics are based either on the comparison of the proportion of total area occupied by a particular category on two maps or on the spatial coincidence of the pixels allocated to any given category on two maps. This review includes some of the most recently developed metrics.

Yang et al. (2017) proposed the overall spatial agreement (A_0) and the individual spatial agreement (A_i) metrics. They are formulated as follows:

$$A_0 = \frac{\sum_1^N XY_{ii}}{M} \times 100$$

$$A_1 = \frac{XY_{ii}}{(X_i + Y_i)/2} \times 100$$

where X_i refers to the number of pixels belonging to category i in map X, Y_i refers to the number of pixels belonging to category i in map Y, XY_{ii} refers to the number of pixels belonging to category i in both maps X and Y, N is the number of categories into which the pixels are classified and M is the number of pixels into which the maps are divided.

The overall spatial agreement (A_0) and the overall spatial inconsistency (OSI) metrics assess the spatial agreement between the categories in two maps. One metric can be obtained from the other. Whereas A_0 shows the spatial agreement (0–100%), the OSI shows the spatial disagreement (0–100%). Added together, they come to 100.

Islam et al. (2019) proposed the overall areal inconsistency (OAI), the individual areal inconsistency (AIC) and the overall spatial inconsistency (OSI) metrics. They are formulated as follows:

$$AIC = |(X_i - Y_i)|/2$$

$$OAI = \sum_i^n AIC$$

$$OSI = \frac{N_{(i \neq j)}}{N} \times 100$$

where X_i refers to the percentage of the total area represented by category i in map X, Y_i refers to the percentage of the total area represented by category i in map Y, n is the total number of categories, N is the number of pixels and $N_{(i \neq j)}$ is the number of pixels assigned to one category in Map X and a different category in Map Y.

Overall areal inconsistency (OAI) shows the agreement between two maps in terms of category proportions and is expressed in values of between 0 and 100. Users can also assess the areal and spatial agreement/disagreement at a category level through the individual areal inconsistency (AIC) and individual spatial agreement (A_i) metrics. The values for the latter range from 0 to 100, and a value of 100 means perfect agreement.

AIC does not have a standard scale of values, as these depend on the proportion of the total area of the map allocated to the category. It is therefore very difficult to compare the values for this metric between classes, so limiting its usefulness.

Utility

Exercises
1. To validate a map against reference data/map
2. To validate a simulation against a reference map
3. To validate simulated changes against a reference map of changes

The areal and spatial agreement metrics assess the similarity between the two maps. They are obtained from the cross-tabulation matrix and therefore do not provide any additional information, in that the values they provide can also be obtained from the matrix. However, they are standard metrics that allow us to measure the agreement between two maps and summarize it in a single figure. In this sense, they are similar to the user's and producer's accuracy metrics and to Kappa indices. They are also complementary to quantity and allocation (dis)agreement metrics, as they can differentiate between spatial and quantity agreements.

These metrics can be used to assess how similar a land use/cover map is to another map used as a reference, i.e. the real situation on the ground. They can also be used to check the similarity between a simulation and the reference map for the same year.

QGIS Exercises

Available tools

• Processing Toolbox
    R
       Areal and spatial agreement metrics
       Individual Areal Inconsistency.rsx
       Individual Spatial Agreement.rsx
       Overall Areal Inconsistency.rsx
       Overall Spatial Agreement.rsx
       Overall Spatial Inconsistency.rsx

QGIS has no specific tool for calculating the metrics proposed by Yang et al. (2017) and Islam et al. (2019). However, these can be easily calculated using the cross-tabulation matrix via the formulae set out above. We have also developed various different tools with R to automatically calculate each metric with QGIS.

When using these R scripts, the categories in LUC rasters must be coded in consecutive numbers, from 1 to the maximum number of categories used in the map. Thus, in a raster with five categories, the categories must be coded as 1, 2, 3, 4 and 5.

Exercise 1. To validate a map against reference data/map

Aim

To validate the CORINE 2011 land use map, take the SIOSE 2011 land use map as a reference. We will be focusing particularly on how the "urban fabric" and "industrial and commercial areas" categories are mapped in CORINE 2011.

Materials

CORINE Land Use Map Asturias Central Area 2011
SIOSE Land Use Map Asturias Central Area 2011

Requisites

All maps must be rasters and have the same resolution, extent, projection and number of categories. LUC categories must be coded consecutively from 1 to the maximum number of categories considered.

Execution

If necessary, install the plugin Processing R provider and download the R scripts indicated above in the "Available Tools" table. Paste the R scripts into the R scripts folder. For more information, see Chapter "About This Book".

Step 1

Our maps do not comply with one of the requisites of the tools we will be using, in that the categories in our LUC maps are coded from 0 (agricultural areas) to 12 (background). The first step is therefore to reclassify the maps to ensure that all the categories are coded consecutively from 1 to 13. This is done using the *Reclassify by table* tool (Figs. 2 and 3).

Step 2

Once the maps comply with the requirements of the tools, the different metrics can then be calculated. To test the overall agreement between the assessed and the reference maps, we will calculate the overall spatial agreement (A_0), the overall areal inconsistency (OAI) and the overall spatial inconsistency (OSI). For their part, individual areal inconsistency (AIC) and individual spatial agreement (A_i) are used to assess agreement specifically for the "urban fabric" and "industrial and commercial areas" categories.

To calculate all these metrics, open the respective tool and select the maps you want to compare (Fig. 4): first the CORINE map and second the SIOSE map, which is used as a reference. In all cases, the background value of the maps (13) must also be indicated. Finally, specify the folder where the result from each tool will be stored.

For class-specific metrics indicate the codes of the classes you want to validate (Fig. 5). In this case, we will be calculating these metrics for two different classes: urban fabric, which is coded 3 after reclassification, and industrial and commercial areas, which is coded 4.

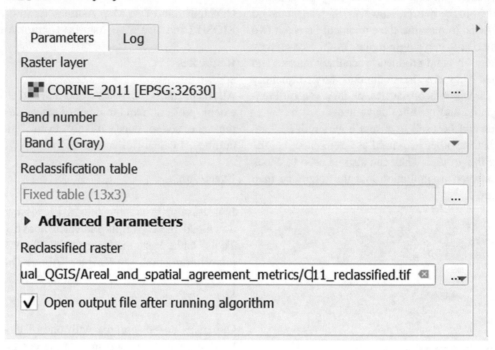

Fig. 2 Exercise 1. Step 1. Reclassify by Table

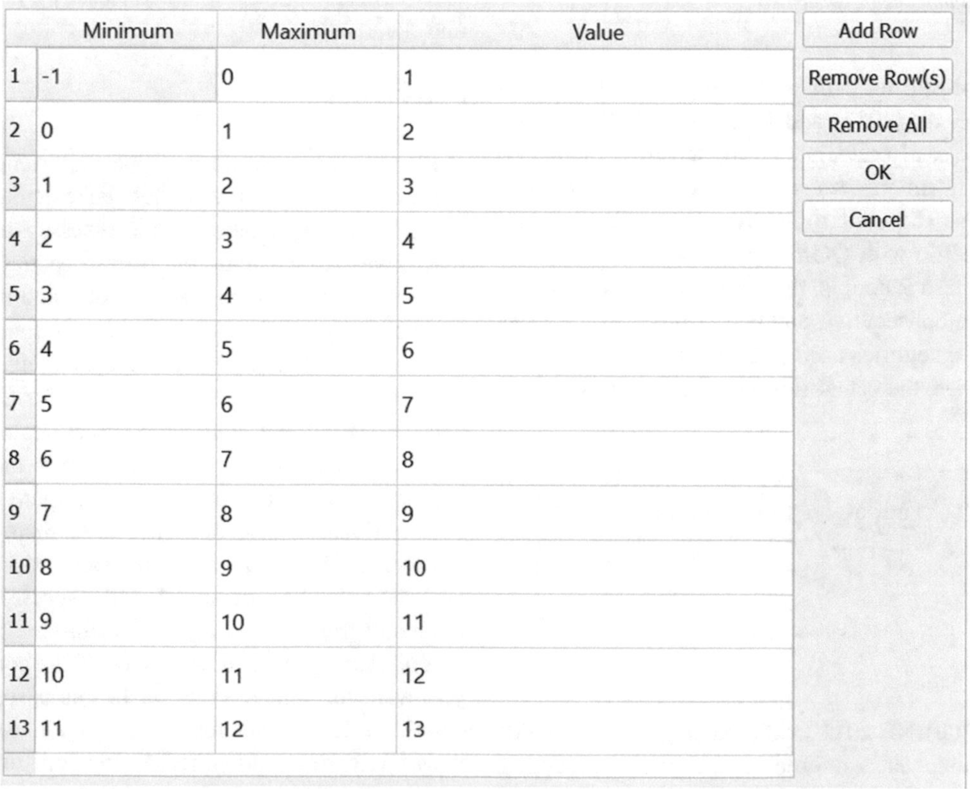

	Minimum	Maximum	Value	
1	-1	0	1	Add Row
2	0	1	2	Remove Row(s)
3	1	2	3	Remove All
4	2	3	4	OK
5	3	4	5	Cancel
6	4	5	6	
7	5	6	7	
8	6	7	8	
9	7	8	9	
10	8	9	10	
11	9	10	11	
12	10	11	12	
13	11	12	13	

Fig. 3 Exercise 1. Step 1. Reclassification table of the Reclassify by Table tool

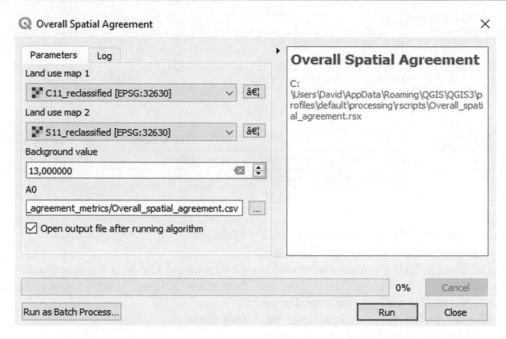

Fig. 4 Exercise 1. Step 2. Overall Spatial Agreement R script

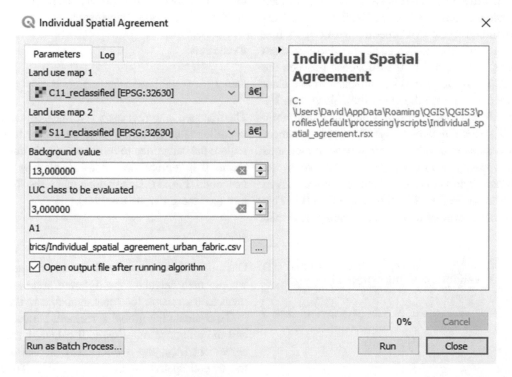

Fig. 5 Exercise 1. Step 2. Individual Spatial Agreement R script

Results and Comments

After calculating all the different metrics, a numerical output is obtained for each one (Tables 3 and 4). This output is also stored in a CSV file in the selected folder.

There is a high overall spatial agreement (close to 90%) between the two maps and low areal inconsistency (around 3%). We can therefore consider the CORINE land cover map for 2011 as validated. The category proportions between CORINE and SIOSE are almost identical and the

Table 3 Results from Exercise 1. Overall agreement indices

Metric	Value
Overall spatial agreement (A_0)	86.85
Overall areal inconsistency (OAI)	3.11
Overall spatial inconsistency (OSI)	13.15

Table 4 Results from Exercise 1. Individual agreement indices

Metric	Urban fabric	Industrial and commercial areas
Individual spatial agreement (A_i)	69.95	67.62
Individual areal inconsistency (AIC)	0.05	0.14

spatial agreement is very high. The disagreements between the two maps are due to their different degree of detail, which draws small features in SIOSE that are not detected at the scale used in CORINE.

At the class level, the picture is slightly different. For the two classes we assessed (urban fabric and industrial and commercial areas) spatial agreement between the two maps to be close to 70%. Although this is a high level of agreement, it is much lower than the overall figure. This could be due to the fact that these two classes are more sensitive than others to the scale difference between SIOSE and CORINE.

In order to interpret the AIC metric, we need to first understand the proportion of total area allocated to each class on the two maps. AIC is half of the difference between the two proportions (i.e. if the proportion allocated to one class is 3% on one map and 4% on the other, the difference is 1% and AIC is 0.5). In our case, the AIC value for urban fabric is less than 0.1, which means a high level of agreement between the two maps regarding the proportion of total area allocated to this category (around 3.9%). The proportion allocated to industrial and commercial areas is around 3% in both maps and the AIC value is slightly more than 0.1. This also indicates a high level of agreement, although less than for urban fabric.

Exercise 2. To validate a simulation against a reference map

Aim

To validate the simulation obtained by our land use/cover change modelling exercise. We will focus on the two categories we have modelled actively: "urban fabric" and "industrial and commercial areas".

Materials

CORINE Land Use Map Asturias Central Area 2011
Simulation CORINE Asturias Central Area 2011

Requisites

All maps must be rasters and have the same resolution, extent, projection and number of categories. LUC categories must be coded consecutively from 1 to the maximum number of categories considered.

Execution

Step 1

The first step is to reclassify our maps to make them comply with the requisites of the tools we will be using. These tools require the categories to be consecutively coded from 1. This means that "agricultural areas" (coded 0) must be given a new code (Fig. 3). This is done using the *Reclassify by table* tool (see the previous exercise).

Step 2

Once the maps comply with the requirements of the tools, we can then calculate the different areal and spatial agreement metrics using the tools available in the R toolbox.

To evaluate the global agreement between the simulation and the reference map, we will calculate the overall spatial agreement (A_0), the overall areal inconsistency (OAI) and the overall spatial inconsistency (OSI). To evaluate agreement for the categories that we actively modelled, we will calculate the individual areal inconsistency (AIC) and the individual spatial agreement (A_i).

To calculate the metrics, open the corresponding tools and indicate the following: the simulation to be evaluated,

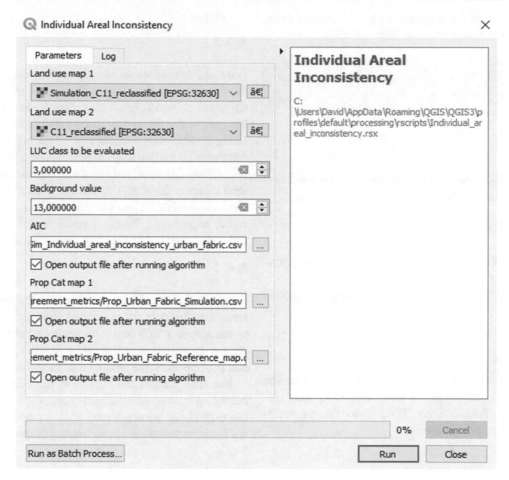

Fig. 6 Exercise 2. Step 2. Individual Areal Inconsistency R script

the reference map (CORINE 2011), the background value of the maps (13) and the folder where the results will be stored. For the class-specific metrics, you must also provide the codes of the classes you want to evaluate: in this case 3 (urban fabric) and 4 (industrial and commercial areas) (Fig. 6).

Results and Comments

Once you have finished the exercise, you will obtain an a CSV file for each metric. The results are summarized in Tables 5 and 6.

The results show almost perfect agreement between our simulation and the reference map. The maps share the same

Table 5 Results from Exercise 2. Overall agreement indices

Metric	Value
Overall spatial agreement (A_0)	99.05
Overall areal inconsistency (OAI)	0.26
Overall spatial inconsistency (OSI)	0.96

Table 6 Results from Exercise 2. Individual agreement indices

Metric	Urban fabric	Industrial and commercial areas
Individual spatial agreement (A_i)	97.35	97.05
Individual areal inconsistency (AIC)	0.006	0.005

LUC in 99% of their area and the areal inconsistency is insignificant (0.26%). A similar pattern is observed in the actively simulated classes.

These results are misleading. There is perfect agreement between our simulation and the reference map in the persistence areas. However, it is not that high for those areas modelled as changes. Because there are relatively few changes in our study area, the disagreement between the two maps in areas where change is predicted has very little impact on the overall high levels of the agreement created by the correct simulation of permanence areas. To correctly validate the changes that we simulated, we should repeat this exercise, focusing exclusively on the areas that changed in the simulation and in the reference map, as compared to the initial map of the simulation (see next exercise).

Exercise 3. To validate simulated changes against a reference map of changes

---

Aim

To validate the changes simulated by our land use/cover change modelling exercise.

Materials

CORINE Land Use Changes Asturias Central Area 2005–2011

Simulated CORINE changes Asturias Central Area 2005–2011

Requisites

All maps must be rasters and have the same resolution, extent, projection and number of categories. LUC categories must be coded consecutively from 1 to the maximum number of categories considered.

Execution

Step 1

Our maps do not comply with the requirements for the tools. In the map of simulated changes, the categories are not consecutively coded from 1. In addition, the reference map of changes has many more categories than the map of simulated changes. Using the *Reclassify by table* tool we can adjust the number of categories on the two maps to the two categories that appear in both (urban fabric and industrial and commercial areas), plus a third category covering non-changing areas and changes that were not simulated. These categories will be assigned codes 1, 2 and 3, respectively. Figures 7 and 8 show the reclassification codes that must be inputted into the *Reclassify by table* tool.

Step 2

After reclassifying the maps, we will calculate the following metrics to validate the simulated changes: individual areal inconsistency (AIC) and individual spatial agreement (A_i). As we are only comparing two categories, the overall metrics provide the same information as the individual ones.

For each metric, we will open the corresponding tool, indicating the map of simulated changes to be validated (Land use map 1), the reference map of changes (Land use map 2), the background value of the maps (0), the category we are

	Minimum	Maximum	Value	
1	-1	0	1	Add Row
2	0	1	1	Remove Row(s)
3	1	2	1	Remove All
4	2	3	2	OK
5	3	4	3	Cancel
6	4	5	1	
7	5	6	1	
8	6	7	1	
9	7	8	1	
10	8	9	1	
11	9	10	1	
12	10	11	1	
13	11	12	1	

Fig. 7 Exercise 3. Step 1. Reclassification table of the Reclassify by Table tool (CORINE changes)

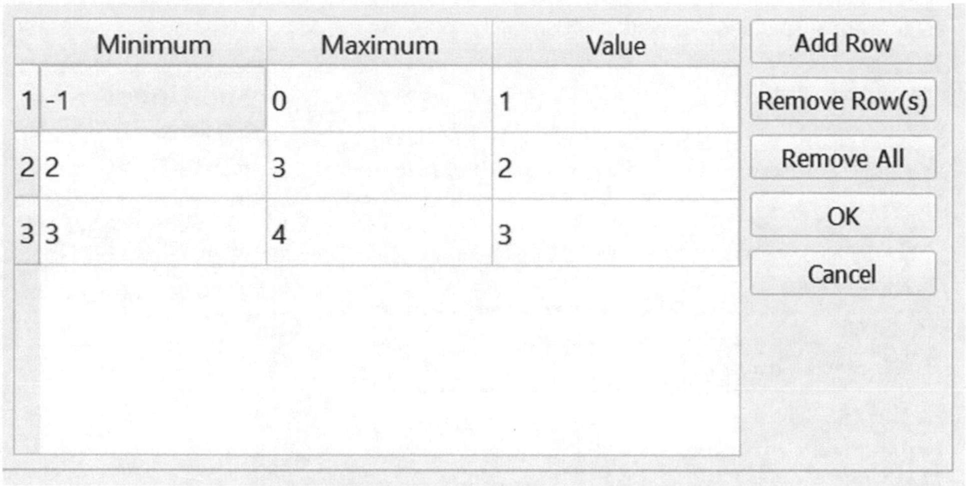

	Minimum	Maximum	Value	Add Row
1	-1	0	1	Remove Row(s)
2	2	3	2	Remove All
3	3	4	3	OK
				Cancel

Fig. 8 Exercise 3. Step 1. Reclassification table of the Reclassify by Table tool (Simulated CORINE changes)

going to evaluate (urban fabric, 2, Fig. 9; industrial and commercial areas, 3, Fig. 10) and the folder where the results of the analysis will be stored. We use 999 as the background value in our maps because no specific value was assigned to the background. 0 means no change, another category that must be considered in this analysis.

Results and Comments

A CSV file will be created for each metric. The results are summarized in Table 7.

The same amount of changes took place in the reference map of changes as in our simulation. There is no disagreement on this point. However, unlike the previous exercise,

the spatial agreement between the simulated and the reference changes was very low. The A_i value for the two categories that were actively simulated was quite similar (less than 25%).

These results mean that only a quarter of the simulated changes were allocated in the same places as the changes observed on the reference map. This result, by itself, is not sufficient to consider the simulation invalid. We need to gain a better picture of the location of the changes that were simulated and their pattern. Even if they were not allocated in exactly the same places as on the reference map, they may be allocated in the same general area and follow a similar pattern, indicating that the model has correctly simulated the processes of change. To assess these aspects, we can perform

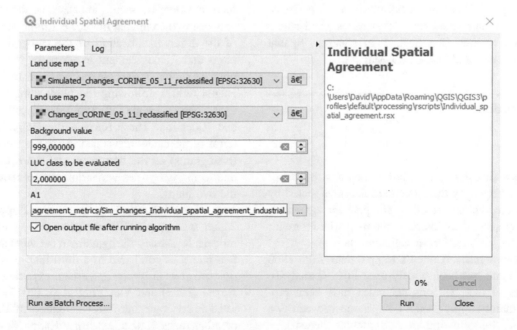

Fig. 9 Exercise 3. Step 2. Individual Spatial Agreemen R script (urban fabric)

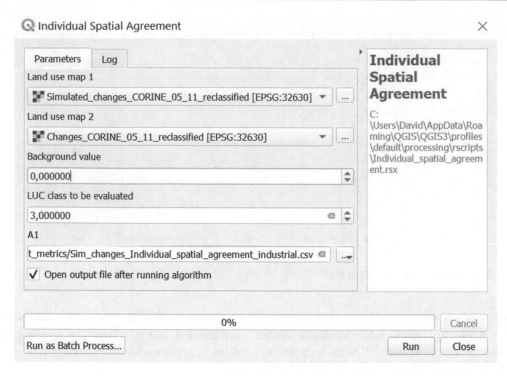

Fig. 10 Exercise 3. Step 2. Individual Spatial Agreement R script (industrial and commercial areas)

Table 7 Results from Exercise 3.Individual agreement indices

Metric	Urban fabric	Industrial and commercial areas
Individual spatial agreement (A_i)	22.37	19.86
Individual areal inconsistency (AIC)	0	0

a visual inspection of the reference and simulated changes on the maps, cross-tabulate them at multiple resolutions (see Sect. 2 in Chapter "Basic and Multiple-Resolution Cross-Tabulation to Validate Land Use Cover Maps") and calculate the spatial metrics (see Sect. 1 in Chapter "Spatial Metrics to Validate Land Use Cover Maps").

3 Kappa Indices

Description

Kappa indices assess the agreement between two sources of spatial data, corrected by the agreement that is expected by chance. They are typically used to compare the agreement between two maps and to compare one map with reference information (e.g. a collection of validation points).

The first Kappa index (Cohen's Kappa) dates from 1960 (Cohen, 1960) and has been widely used in LUC analysis. Many variants of this first original index have been proposed. They mainly apply to the comparison between two maps. Of these, the following are of particular interest:

- Pontius Jr. (2000) split Cohen's Kappa into three indices, called *Kno*, *Kquantity* and *Klocation*. These indices offer more information about the causes of the (dis)agreement between two compared maps, i.e. (dis)agreement in terms of the different allocation of the categories on the two maps and (dis)agreement in terms of the different proportions in which the categories appear on the two maps.
- Hagen (2002), following the work done by Pontius Jr., split Cohen's Kappa into two indices, called Khistogram and Klocation. These refer to the Kappa agreement in terms of the categories appearing in the same proportions (histograms) on the two maps and the Kappa agreement due to the categories appearing in the same location on the two maps.
- Van Vliet et al. (2011) proposed the Kappa simulation, which was specifically designed for validating LUCC models. It assesses the agreement between the changes on two maps, as compared to a third map used as an initial point, corrected by the agreement expected by chance.
- Hagen (2003) and Van Vliet et al. (2013) also incorporated fuzzy logic into the calculation of Kappa indices, creating fuzzy Kappa and fuzzy Kappa simulation. They

took the degree of spatial and thematic mismatch into account when calculating Kappa. In other words, two maps may be said to show partial agreement if the validation pixel or point is close to the compared pixel. The same would apply if the pixels were allocated to different classes, but with similar meanings.

Utility

Kappa indices enable us to test the similarity between two sources of spatial information. If we have one map and reference data, we can determine to what extent the map we want to validate agrees with the reference data.

The main advantage of Kappa indices is that they provide a standard measure. Kappa agreement always ranges between −1 and 1, where 1 means total agreement, −1 total disagreement and 0 random agreement. These are universal measures, which means that the performance of a LUC classification exercise or a LUCC modelling exercise can be compared with the performance typically achieved in these exercises.

There are many critics of the widespread use of Kappa metrics, especially in LUCC modelling validation. There is now a general consensus that these indices should not be the only validation measures used when evaluating modelling exercises and maps. More information about the limitations of Kappa indices and the criticisms levelled against them can be found in Pontius Jr. and Millones (2011) and Van Vliet et al. (2011).

QGIS Exercises

QGIS does not include many tools for calculating Kappa indices. The Cohen's Kappa index can be obtained through the associated GRASS module. The Semi-Automatic Classification plugin also calculates the Kappa index, globally and at the category level, when doing the cross-tabulation (see Chapter "Basic and Multiple-Resolution Cross-Tabulation to Validate Land Use Cover Maps"). The other variants of Kappa are not available through QGIS or any of its pattern software, like R. Those who would like to calculate these indices are referred to the Map Comparison Kit, which is also available for free.[1]

Exercise 1. To validate a map against reference data/map

Aim

To test the validity of the CORINE 2011 land use map, take the SIOSE land use map as a reference. In this way, we can answer the following question: assuming that the SIOSE map shows the true situation, how true is the CORINE map?

Materials

SIOSE Land Use Map Asturias Central Area 2011
CORINE Land Use Map Asturias Central Area 2011

Requisites

The two maps must be rasters and have the same extent, spatial resolution, projection and legend. If they do not have the same legend, the user must reclassify the maps in such a way that they comply with this requirement.

Execution

Step 1

Open the *r.kappa* function and fill in the required parameters: raster to be validated (CORINE map) and reference raster (SIOSE map) (Fig. 11).

Results and Comments

Once the function has been executed, QGIS creates a new text file (.txt) in the specified folder. Users must manually access this folder to open the text file and see the results of

---

[1] http://mck.riks.nl/downloads.

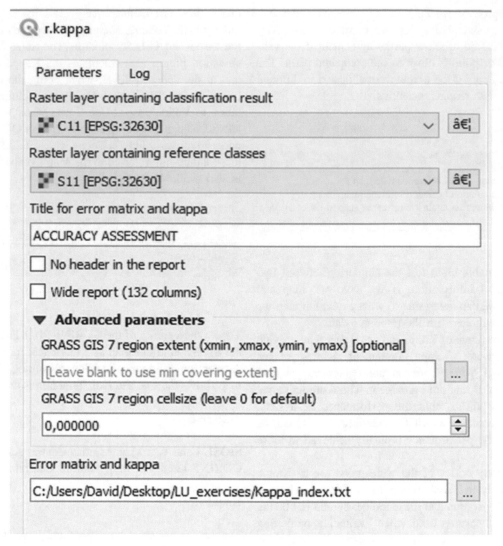

Fig. 11 Exercise 1. Step 1. R.kappa

the analysis. These include a cross-tabulation matrix of the maps, together with the Kappa value. For the two maps assessed, we obtained the following Kappa:

$$Kappa = 0.88$$

where 1 means total agreement, −1 total disagreement and 0 random agreement. A Kappa index value of 0.88 means that the two maps are very similar and therefore that our map has been validated. As a general rule, Kappas above 0.7–0.8 are considered good enough for validation. Kappas above 0.9 indicate very high agreement.

In our case, it is always important to bear in mind that SIOSE is made at a more detailed scale than CORINE. The two maps have different minimum mapping units and minimum mapping widths, which means that perfect agreement is impossible. The SIOSE map will always draw features that are not detected in CORINE because of its coarser scale.

Kappa scores of almost 0.9, like this one, show almost perfect agreement between the two sources.

Users can also assess the agreement between CORINE and SIOSE at the category level so as to obtain more information about the similarities and dissimilarities between the two maps. To compute these metrics, they should refer to Exercise 3, using the Semi-Automatic Classification Plugin instead of *r.kappa*.

Exercise 2. To validate a simulation against a reference map

Aim

To validate the simulation obtained by our land use/cover change modelling exercise.

Materials

Simulation CORINE Asturias Central Area 2011
CORINE Land Use Map Asturias Central Area 2011

Requisites

The two maps being compared must be rasters and have identical resolution, extent, projection and legend. For proper validation, the reference map must refer to the same date for which the landscape was simulated.

Execution

Step 1

Open the *r.kappa* function and fill in the required parameters: raster to be validated (Simulation) and reference raster (CORINE 2011) (Fig. 12).

Results and Comments

QGIS will create a text file in the specified folder. This file contains the Kappa value for our simulation:

$$\text{Kappa} = 0.99$$

where 1 means total agreement. The Kappa value indicates that the two maps are almost the same. However, this does not mean that the changes we simulated are the same as the changes that took place on the reference map (CORINE 2011) as compared to the map used as the starting point for our modelling exercise (CORINE 2006).

In our simulation, most of the landscape remains unchanged. The high Kappa value indicates that we have correctly modelled the persistence of these unchanged areas. However, it is difficult to draw any meaningful conclusions about how closely the changes we simulated fit the changes observed between the CORINE 2011 and 2006 maps. These changes

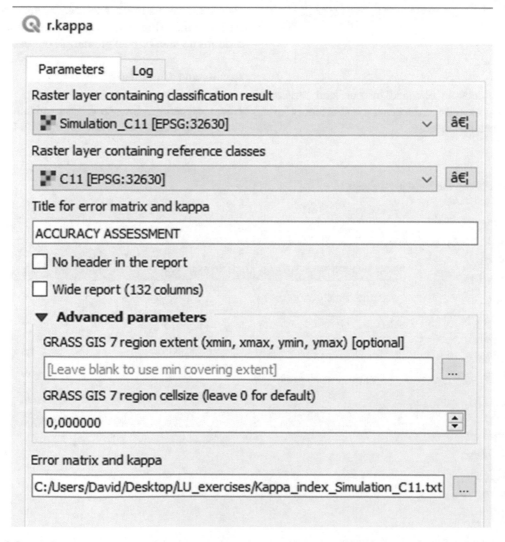

Fig. 12 Exercise 2 Step 1. R.kappa

only affect very small parts of the maps and, therefore, do not have a meaningful impact on the Kappa index when evaluating the agreement between the entire area of the maps.

In order to gain a better picture as to how well the simulated changes fit the changes in the reference maps, other complementary metrics also described in this book can be used, such as the quantity and allocation disagreement or the figure of merit (see Sects. 3 and 4 in Chapter "Pontius Jr. Methods Based on a Cross-Tabulation Matrix to Validate Land Use Cover Maps"). The agreement between simulated and reference changes can also be assessed using Kappa simulation, although this metric is not currently implemented in any tool in QGIS or in its associated software, such as R.

Users can also evaluate the kappa agreement between the simulation and the reference map at the category level, for which purposes they should refer to the next exercise, Exercise 3.

Exercise 3. To validate a simulation against a reference map at the category level

Aim

To validate a simulation obtained by our land use/cover change modelling exercise at the general and category level, focusing on a specific category.

Materials

CORINE Land Cover Map Val d'Ariège 2018
Simulation LCM Val d'Ariège 2018

Requisites

The two maps to be compared must be rasters and have identical resolution, extent, projection and legend. For proper validation, the reference map must refer to the same date for which the landscape was simulated.

Execution

Step 1

The Kappa index can be calculated at the category level for all the categories in our map using the Semi-Automatic Classification Plugin. To this end, open the plugin and select the *Accuracy* (Postprocessing) option from the menu. Then choose the rasters to be assessed, i.e. the simulation and the reference map (Fig. 13). It is also important to indicate the code for no data or background. In our case, the code is 10.

Results and Comments

After executing the tool, we obtain a raster that cross-tabulates the compared maps and a CSV file with the

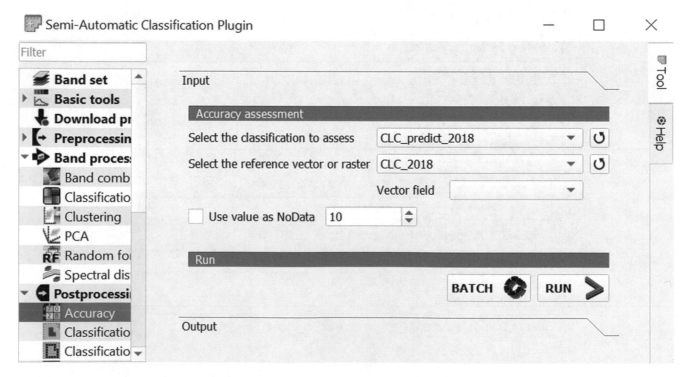

Fig. 13 Exercise 3 Step 1. Semi-Automatic Classification Plugin

Table 8 Results from Exercise 3. Kappa indices: overall and per category

Overall	Built-up areas (1)	Agricultural areas (2)	Forests (3)	Shrub and/or herbaceous vegetation (4)	Open spaces with little or no vegetation (5)	Water surfaces (6)
0.9849	0.9092	0.9699	0.9993	0.9644	1.0000	1.0000

cross-tabulation matrix, the overall, user's and producer's accuracy values and the Kappa indices of agreement, overall and per category. This information will also be displayed in the output window. For detailed information about how to interpret the matrices and the user's and producer's accuracy values, please refer to the Sect. 4 in Chapter "Pontius Jr. Methods Based on a Cross-Tabulation Matrix to Validate Land Use Cover Maps".

The Kappa values for the two maps show high levels of agreement at both a general level and for all categories (Table 8). The Kappa values for the "Open spaces with little or no vegetation" and "Water surfaces" categories are 1, which means perfect agreement. In other words, there are no differences between the two maps for these classes. This makes sense because they were not simulated in our modelling exercise.

The class with the lowest Kappa value is "Built-up areas". This indicates that many of the changes in this category have not been correctly simulated, which is to be expected given the dynamism of this category when compared with others such as forest or water surfaces. It is normally easier to simulate static land categories than changing ones. This explains why "Built-up" areas obtained a very low Kappa score compared to the overall score (Table 8).

Although these results offer some clues as to how well the changes in some categories were simulated, to obtain a more detailed understanding other methods and metrics should be used, such as the quantity and allocation disagreement and the figure of merit (see Sects. 3 and 4 in Chapter "Pontius Jr. Methods Based on a Cross-Tabulation Matrix to Validate Land Use Cover Maps") or the Kappa simulation metrics. Whereas the Kappa metrics calculated here assess the agreement between persistent and changing areas in the compared and the reference maps, the other tools and methods focus on the specific areas that change between the initial and the final year of the simulation. This is a key element for understanding the success of our simulation, as it is easier to model persistence than change.

4 Agreement Between Maps at Overall and Stratum Level

Description

The aim is to assess the agreement between map pairs such as a reference map and a simulation map, at different levels:

overall agreement for the whole map, agreement for a given stratum, a smaller area, formed by a particular territory, LUC category or transition or by sample areas according to a gradient such as distance to a road. The purpose of this validation method is encapsulated in the following question: Does a particular item or area of interest show the same prediction score as the whole map?

Utility

Exercises

1. To validate simulated changes against a reference map of changes

A given map (LUC map, simulation) can be evaluated more precisely at spatial level (specific territory), category level (*Is the simulation closer to the real situation for built-up areas or for forests?*) or specific transitions (*Does the model work better for the transition from forest to agriculture or from forest to pasture?*). In this context, the entire area of interest can be used as a guide for interpreting particular simulation scores.

QGIS Exercise

Available tools

- Raster
  - *Raster calculator*
- Processing Toolbox
  - GRASS
    - Raster (r.*)
      - *r.kappa*
    - Raster analysis
      - *Reclassify by table*
      - *Raster layer unique values report*

Agreement between maps at the overall and stratum levels is more a validation approach than a specific method. Accordingly, there are no specific tools available in QGIS to carry out this analysis, as the used tool will depend on what type of analysis will be carried out at the overall and stratum levels.

For general operations, we will make use of the QGIS *Raster Calculator*. a generic tool performing all kinds of raster calculations. To calculate Kappa indices at the global and stratum levels, we will make use of *r.kappa*. For more information about this tool, please refer to the previous section.

Exercise 1. To validate simulated changes against a reference map of changes

Aim

To find out if the agreement between an observed (reference map) and a simulated transition varies for several distance-based categories resulting from a driver (e.g. distance to roads).

Materials

CORINE Land Cover Map Val d'Ariège 2012
CORINE Land Cover Map Val d'Ariège 2018
Simulation LCM Val d'Ariège 2018
Distance to roads

Requisites

All maps must be in raster format with the same resolution, extent and spatial reference system (SRS).

Execution

Step 1

First, we have to obtain the observed and simulated transitions from agriculture and pasture land to built-up areas over the period 2012–2018. Using the raster calculator, we extract the observed ("CLC_2012@1" = 2 AND "CLC_2018@1" = 1) and the simulated ("CLC_2012@1" = 2 AND "CLC_predict_2018@1" = 1) transition from agriculture and pasture land (Category 2) to built-up areas (Category 1). The result is shown in Fig. 14 (observed change appears in cyan, simulated change in red).

Step 2

The *Reclassify by table* raster analysis tool is used to transform the map showing the continuous distance from roads into

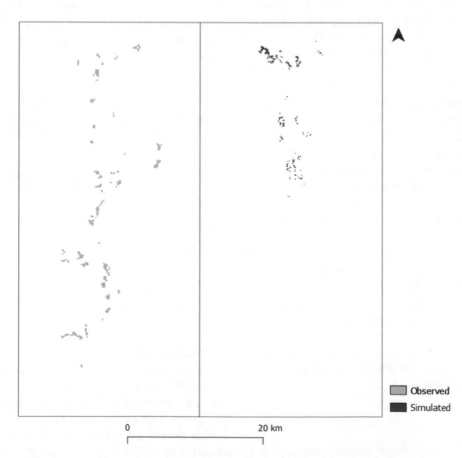

Fig. 14 Exercise 1. Step 1. Intermediate maps showing observed and simulated transitions from agriculture and pasture to built-up areas

Fig. 15 Exercise 1. Step 2. Reclassify by Table

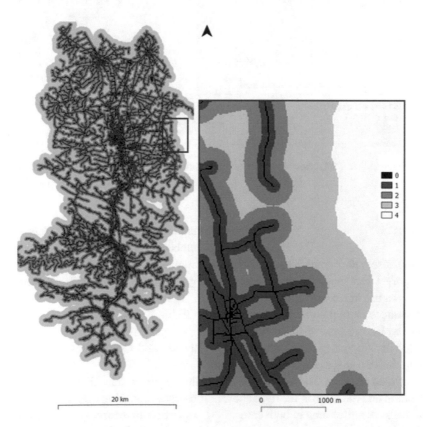

Fig. 16 Exercise 1. Step 2. Intermediate map showing the distance from roads reclassified by intervals

various different classes. Given the dense road network, we intentionally apply a progressive interval as shown in Fig. 15.

Figure 16 shows the general result and the result for a detailed area with the following classes: the road network itself (distance is zero), distance class 1 (less than 100 m), class 2 (100–300 m), class 3 (300–1000 m) and class 4 (more than 1000 m).

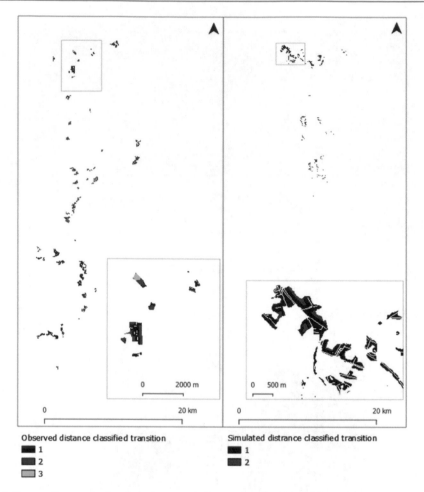

Fig. 17 Exercise 1. Step 3. Intermediate maps showing observed (left) and simulated (right) transition from 2 to 1 as a function of the road distance classes

Step 3

The next step is to compute observed and predicted transitions from Category 2 to Category 1 as a function of the road distance classes. To this end, we use the *Raster calculator* again to calculate: i) the road distance class map multiplied by the observed transition map and ii), the road distance class map multiplied by the simulated transition map. The results can be seen in Fig. 17, in which the two maps show the transition from 2 to 1 as a function of road distance. The map on the left shows the observed transition and the map on the right shows the simulated transition, with a detailed area in both cases.

Step 4

Finally, we compare observed and simulated transitions as a function of distances classes (strata). We use the *Raster layer*

unique values report raster analysis tool to calculate the number of pixels for each road distance category (observed and simulation) for the transition from category 2 to 1 as shown in Fig. 18 (left for observed, right for simulated transition).

The results are then converted into percentage as shown in Table 9.

Results and Comments

The result is that there are almost three times as many observed transitions as predicted transitions. However, the proportion of near-to-road transitions is approximately the same. In conclusion, the model underestimates the quantity of agriculture and pasture land that is transformed into built-up areas, although in the areas close to roads, it accurately predicted what happened in the Ariège Valley between 2012 and 2018.

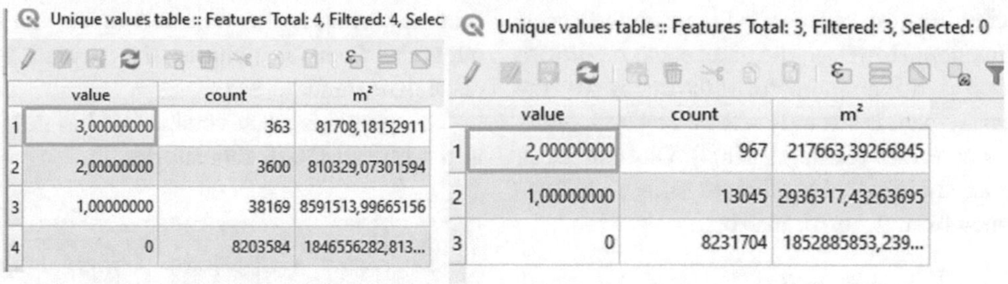

Fig. 18 Exercise 1. Step 4 presented in the "output" window. Number of cells and areas of observed (left) and simulated (right) transition from 2 to 1 as a function of the road distance classes

Table 9 Exercise 1 Step 4. Number and proportion of cells of observed and simulated transition from 2 to 1 as a function of the road distance classes

	Observed		Predicted	
	Pixels	%	Pixels	%
<100 m	38,169	90.59	13,045	93.10
100–300 m	3600	8.54	967	6.9
300–1000 m	363	0.86	–	
Sum	42,132		14,012	

5 Accuracy Assessment Statistics

Description

The thematic accuracy assessment statistics are a set of parameters that measure the degree of agreement between the LUC map and the reference data (for more details about reference data, see Chapter "Sample Data for Thematic Accuracy Assessment in QGIS"). Overall accuracy, user's accuracy and producer's accuracy are reported in many studies. Some additional accuracy measures such as the standard error of overall accuracy and the confidence intervals for the adjusted areas are also helpful.

All these parameters are mainly derived from the error or confusion matrix (see Chapter "Basic and Multiple-Resolution Cross-Tabulation to Validate Land Use Cover Maps"). This matrix is obtained from a cross-tabulation between the reference data and the thematic map. In the resulting table, the reference data are generally shown in the columns and the map data in the rows (Table 10).

In Table 10, n_{ij} refers to the sample count of spatial units in cell (i, j), n_{i+}, n_{+j} denote the sum of n_{ij} in each row and column, and n is the sample size; n_{+j} is the number of spatial assessment units belonging to class j, according to the reference data, and n_{i+} is the number of spatial units belonging to class i according to the thematic map.

Expressing the error matrix in terms of area proportions instead of sample counts enables the calculation of unbiased area estimators. The area proportions $(\widehat{p}_{ij})$ are defined as follows:

$$\widehat{p}_{ij} = W_i \frac{n_{ij}}{n_{i+}}$$

where W_i = (Map area of class i)/(Total area of the map).

Based on these area proportions, the overall estimated accuracy $(\widehat{O})$, user's accuracy $(\widehat{U}_i)$ and producer's accuracy $(\widehat{P}_j)$ are calculated with the following equations:

$$\widehat{O} = \sum_{j=1}^{q} \widehat{p}_{jj}$$

$$\widehat{U}_i = \frac{\widehat{p}_{ii}}{\widehat{p}_{i+}}$$

$$\widehat{P}_j = \frac{\widehat{p}_{jj}}{\widehat{p}_{+j}}$$

Errors of commission and omission are complementary concepts of the user's and producer's accuracy metrics, respectively (i.e. error = 1 − accuracy). An error of commission occurs when a feature is included in a thematic class to which it does not belong. In contrast, an error of omission occurs when a feature is excluded from the thematic class to which it belongs (Finegold et al. 2016).

Errors in the classification process can increase the uncertainty in area estimation. However, the pixel count multiplied by pixel size is often used as an estimator of the true area on the ground. This measurement is strongly affected by both omission and commission errors (Gallego, 2004). Olofsson et al. (2013) proposed an unbiased area estimate using an adjustment factor obtained from the error matrix:

$$\widehat{A}_j = A_{total} \times \widehat{p}_{+j}$$

$\widehat{A}_j$ is the unbiased area estimator or adjusted area. In this case, the area estimator obtained directly from the map (A_{total}) is then adjusted by a factor obtained from the reference data. If there are more samples labelled as class j in the reference sample than in the map, then $\widehat{A}_j$ will be larger than the area obtained directly by pixel counting.

Utility

Exercises
1. To validate a map against reference data/map

The statistics obtained from the thematic accuracy assessment are not only descriptors of the map quality but also represent a fundamental input for calculating unbiased area estimators. Additionally, they provide the necessary elements to decide whether to increase the number of sampling sites in the reference data, if the precision obtained does not meet the initial mapping objectives.

QGIS Exercise

Available tools
• MapAccurAssess Plugin

In QGIS, several plugins, such as Semi-Automatic Classification, AcATaMa and MapAccurAssess, can be used to calculate the map accuracy statistics. All three plugins provide the overall accuracy, producer's accuracy, user's accuracy and the error matrix, although AcATaMa and MapAccurAssess also report some additional statistics about the adjusted areas and their levels of accuracy.

In this exercise, we use the MapAccurAssess plugin because it can use a shapefile directly with the reference data. The results provided by this plugin, based on Olofsson et al. (2013), include the error matrix and a table with the following statistics: the class area, the producer's and user's

accuracy values, the adjusted areas and their confidence intervals. It also includes the overall accuracy and its respective standard error.

This plugin is a test version and has not yet been accepted in the official QGIS repositories.

Exercise 1. To validate a map against reference data/map

Aim

To validate a LUC map for the Marqués de Comillas study area by computing accuracy assessment statistics and the error matrix via cross-tabulation of the reference data and the thematic map.

Materials

Marqués de Comillas Land Use Cover Map 2019
Photointerpreted reference dataset—Marqués de Comillas 2019 (reference dataset resulting from the exercise in Sect. 2 in Chapter "Sample Sata for Thematic Accuracy Assessment in QGIS")

Requisites

In order to compute the areas, the land cover map must be in raster format (GeoTiff) in any cartographic projection. The reference data must be contained in a shapefile with the same type of projection as the map. The shapefile attribute table must contain at least two columns, showing the value for the thematic class obtained from the land cover map and the value according to field ground-truthing or photointerpretation. Both columns must have the same data type (integer or text) to be comparable. Each row of the table corresponds to one reference site.

Execution

Step 1

Install the MapAccurAssess plugin. Should you need help, please see Chapter "About This Book" and the plugin's documentation.

Step 2

If the plugin has been successfully installed, an icon should appear in the main graphics panel. To start the exercise, click on this icon. Alternatively, go to the Complements menu, select *Accuracy Assessment* and then *Accuracy Assessment* again.

Table 10 Confusion matrix

	Class	Reference data				Total
		1	2	…	q	
Mapdata	1	n_{11}	n_{12}	…	n_{1q}	n_{1+}
	2	n_{21}	n_{22}	…	n_{2q}	n_{2+}
	…	…	…	…	…	…
	Q	n_{q1}	n_{q2}	…	n_{qq}	n_{q+}
	Total	n_{+1}	n_{+2}	…	n_{+q}	n

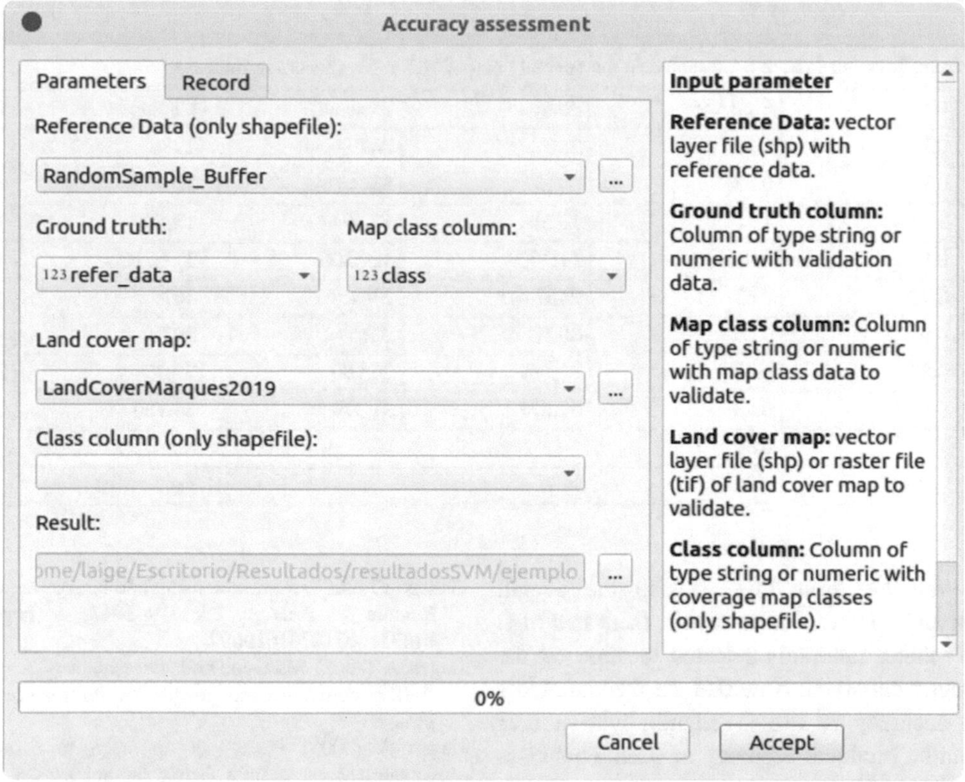

Fig. 19 Exercise 1 Step 3. MapAccurAssess plugin

Step 3

Select the shapefile with the reference samples (Photo-interpreted reference dataset—Marqués de Comillas 2019)[2] and indicate the column with the reference data and the column with the values for the thematic classes used in the map. After that, select the land cover map you want to assess (Marqués de Comillas Land Use Land Cover Map 2019). If the map is in vector format, indicate the column containing the thematic class values. Finally, select a folder where the results will be saved and click "Accept" (Fig. 19).

Results and Comments

The output of this plugin consists of two CSV tables. The first contains the error matrix (Table 11), and the second contains the map accuracy assessment statistics (Table 12). These statistics are as follows: user's accuracy, producer's accuracy, thematic class area (as retrieved from the map), the area adjusted by the error level (Area_adj), the confidence intervals for the adjusted area (CI_sup and CI_inf) and the overall accuracy (O).

---

[2] The photointerpreted reference dataset for Marqués de Comillas (RandomSample_Buffer.shp) was obtained from the exercise in Sect. 2 in Chapter "Sample Data for Thematic Accuracy Assessment in QGIS". This layer has two columns, "class" and "refer_data". The first contains the values for the thematic classes used in the map and the second contains the reference data, which were obtained from the photointerpretation of satellite images.

Table 11 Result from Exercise 1. Error matrix

Classes	130	161	261	290	301	420	50	51
130	35	0	0	14	0	0	1	0
161	0	38	0	6	0	0	0	6
261	0	0	50	0	0	0	0	0
290	2	9	0	328	3	1	1	3
301	0	1	0	13	36	0	0	0
420	0	0	0	1	0	49	0	0
50	0	0	0	5	0	0	64	8
51	0	0	0	3	0	0	15	80

Table 12 Results from Exercise 1, Step 3 presented in the second "output" CSV file (accuracy indices)

Classes	UsAcc	ProdAcc	Area	Area_adj	CI_sup	CI_inf
130	0.7	0.21	252	847	1777	−82
161	0.76	0.63	6943	8344	10,465	6222
261	1	1	13,504	13,504	13,504	13,504
290	0.95	0.96	116,429	114,306	117,716	110,896
301	0.72	0.63	2357	2704	3877	1530
420	0.98	0.86	2021	2316	2979	1654
50	0.83	0.80	26,009	26,990	30,274	23,707
51	0.82	0.86	32,875	31,379	34,739	28,019
Overall Accuracy (O):	0.91					
Std(O):	0.0113					

According to the data from this exercise, the overall accuracy of the map is 0.91. In other words, there is a high probability (91%) that a randomly selected location on the map will be correctly classified. Note that the thematic class with the lowest accuracy is 130 (Wetland), with a user accuracy of 0.7 and a producer accuracy of 0.21. This class covers a small area (252 ha according to the map). We decided to keep this class to show that illogical situations can occur when there is only a small number of sampling sites, e.g. negative areas. However, we recommend merging class 130 with another class of similar characteristics and recomputing.

References

Cohen J (1960) A coefficient of agreement for nominal scales. Educ Psychol Meas 20:37–46

FAO (1995) Forest resources assessment 1990. In: Global synthesis. FAO, Rome

Finegold Y, Ortmann A, Lindquist E, d'Annunzio R and Sandker M (2016) Map accuracy assessment and area estimation: a practical guide. National forest monitoring assessment working paper no. 46/E. Food and Agriculture Organization of the United Nations. Rome

Gallego FJ (2004) Remote sensing and land cover area estimation. Int J Remote Sens 25:3019–3047. https://doi.org/10.1080/01431160310001619607

Hagen A (2002) Multi-method assessment of map similarity. In: 5th AGILE conference on geographic information science. Mallorca, pp 1–8

Hagen A (2003) Fuzzy set approach to assessing similarity of categorical maps. Int J Geogr Inf Sci 17:235–249. https://doi.org/10.1080/13658810210157822

Islam S, Zhang M, Yang H, Ma M (2019) Assessing inconsistency in global land cover products and synthesis of studies on land use and land cover dynamics during 2001 to 2017 in the southeastern region of Bangladesh. J Appl Remote Sens 13:1. https://doi.org/10.1117/1.JRS.13.048501

Olofsson P, Foody GM, Stehman SV, Woodcock CE (2013) Making better use of accuracy data in land change studies: estimating accuracy and area and quantifying uncertainty using stratified estimation. Remote Sens Environ 129:122–131. https://doi.org/10.1016/j.rse.2012.10.031

Pontius RG Jr (2000) Quantification error versus location error in comparison of categorical maps. Photogramm Eng Remote Sensing 66:1011

Pontius RG Jr, Millones M (2011) Death to Kappa: birth of quantity disagreement and allocation disagreement for accuracy assessment. Int J Remote Sens 32:4407–4429. https://doi.org/10.1080/01431161.2011.552923

Puyravaud J-P (2003) Standardizing the calculation of the annual rate of deforestation. For Ecol Manage 177(1–3):593–596

Van Vliet J, Bregt AK, Hagen-Zanker A (2011) Revisiting Kappa to account for change in the accuracy assessment of land-use change models. Ecol Modell 222:1367–1375. https://doi.org/10.1016/j.ecolmodel.2011.01.017

Van Vliet J, Hagen-Zanker A, Hurkens J, Van Delden H (2013) A fuzzy set approach to assess the predictive accuracy of land use simulations. Ecol Modell 261–262:32–42. https://doi.org/10.1016/j.ecolmodel.2013.03.019

Yang Y, Xiao P, Feng X, Li H (2017) Accuracy assessment of seven global land cover datasets over China. ISPRS J Photogramm Remote Sens 125:156–173. https://doi.org/10.1016/j.isprsjprs.2017.01.016

Pontius Jr. Methods Based on a Cross-Tabulation Matrix to Validate Land Use Cover Maps

Martin Paegelow, Jean-François Mas, Marta Gallardo,
María Teresa Camacho Olmedo, and David García-Álvarez

Abstract

Several validation techniques based on the cross-tabulation matrix can be applied to validate Land Use Cover (LUC) maps. The exercises in this chapter focus, in particular, on the cross-tabulation techniques proposed by Robert Gilmore Pontius Jr., who has developed many indices and techniques in this field. Given his major contribution to this family of validation techniques, we have associated his name here with cross-tabulation techniques without this in any way implying that his scientific activity is limited to this field. The null model (Sect. 1) is especially useful for validating simulations, comparing the modelled map to a reference map with full persistence. LUCC budget (Sect. 2) only focusses on changes, which it splits into different components. This method can be used to compare the changes we want to validate with a reference set of changes, so providing interesting information as to how well our maps capture the dynamics of the landscape. Quantity and allocation disagreement (Sect. 3) analyse the differences between the reference map and the map being validated using two indices: disagreement in quantity and disagreement in allocation. The Figure of Merit (FoM) (Sect. 4) technique is used to validate a set of LUC changes by comparing them with a reference, distinguishing between different components of agreement: correctly simulated change, wrongly simulated or missing change. Incidents and States (Sect. 5) allows us to identify illogical transitions in a time series of maps by providing the number of states and transitions that a cell undergoes over the course of the series. Intensity analysis (Sect. 6) and Flow matrix (Sect. 7) also enable us to validate the logic of LUC changes in a time series of maps. Intensity analysis provides information on the speed of changes, identifying those transitions or changes that do not follow a logical trend, while the flow matrix enables us to spot unstable changes in a series of maps. In this chapter, we present examples of how these techniques can be used in different cases: to validate single LUC maps, to validate a series of maps with two or more time points, to validate simulated changes against a reference map of changes and to validate changes simulated by various models. All these techniques are illustrated by exercises using datasets from the Asturias Central Area and the Ariège Valley.

M. Paegelow (✉)
Département de Géographie, Aménagement et Environnement, Université de Toulouse Jean Jaurès, Toulouse, France
e-mail: martin.paegelow@univ-tlse2.fr

J.-F. Mas
Laboratorio de Análisis Espacial, Centro de Investigaciones en Geografía Ambiental, Universidad Nacional Autónoma de México, Morelia, Mexico

M. Gallardo
Departamento de Geografía, Universidad Nacional de Educación a Distancia, Madrid, Spain

M. T. Camacho Olmedo
Departamento de Análisis Geográfico Regional y Geografía Física, Universidad de Granada, Granada, Spain

D. García-Álvarez
Departamento de Geología, Geografía y Medio Ambiente, Universidad de Alcalá, Alcalá de Henares, Spain

Keywords

LUCC budget • Change matrices • Cross-tabulation • Error Analysis • Figure of Merit • Intensity Analysis • Flow matrix

1 Null Model

Description

The null model is a method specifically developed by Pontius and Malanson (2005) to validate LUCC modelling simulations. It assumes that the land use/land cover at the simulation start time (t_1) is exactly the same at the end time (t_2) and that no changes take place. The aim is to evaluate

D. García-Álvarez et al. (eds.), *Land Use Cover Datasets and Validation Tools*,
https://doi.org/10.1007/978-3-030-90998-7_9

whether a landscape with no changes more closely resembles the reference landscape for the year of the simulation (t_2) than the simulated landscape. In other words, we change the date of the initial LUC map while leaving the content unchanged. It then becomes a reference map (no change) with which we can measure the predictive power of the model.

If the agreement between the observed LUC at t_2 and the simulation map at t_2 is higher than that between observed LUC at t_2 and the so-called *null model*, the simulation has greater predictive power than the hypothesis of complete persistence (no change). The agreement between the null model, the simulation and the reference map is usually assessed using common cross-tabulation techniques and Kappa indices (see Sect. 1 in Chapter "Basic and Multiple-Resolution Cross-Tabulation to Validate Land Use Cover Maps" and Sect. 3 in Chapter "Metrics Based on a Cross-Tabulation Matrix to Validate Land Use Cover Maps").

Utility

Exercises

1. To validate simulated changes against a reference map of changes

The null model helps to measure the relative success of a simulation compared to persistence in time. The usefulness of this method depends on the spatiotemporal dynamics of the study area.

The method is based on the hypothesis that a simulation is successful if it gets better validation scores than a landscape in which no changes occur. When simulating change in a study area in which little change is taking place, it may be difficult to correctly simulate these changes in the same positions as on the reference map of changes. As a result, the null model may provide better validation scores than the simulation, in that the null model avoids possible errors when allocating changes and always simulates persistence correctly. This is why the null model is especially useful for validating whether an LUCC model simulates persistence correctly.

QGIS Exercise

Available tools

- Processing Toolbox
  GRASS
    Raster (r.*)
    r.kappa
- Semi-Automatic Classification Plugin
  Tab: Postprocessing
    Section: *Cross-classification*

To calculate the null model, we must use the same techniques as cross-tabulation and Kappa. Please see Sect. 1 in Chapter "Basic and Multiple-Resolution Cross-Tabulation to Validate Land Use Cover Maps" and Sect. 3 in Chapter "Metrics Based on a Cross-Tabulation Matrix to Validate Land Use Cover Maps" for details about how to compute cross-matrices and kappa indices between two raster layers.

Exercise 1. To validate simulated changes against a reference map of changes

Aim

To find out if the prediction score obtained by the simulation map for 2018 is higher than that obtained by the null model.

Materials

CORINE Land Cover Map Val d'Ariège 2012
CORINE Land Cover Map Val d'Ariège 2018
Simulation LCM Val d'Ariège 2018

Requisites

All maps must be rasters and must have the same resolution, extent and projection.

Execution

Step 1

The first step is to calculate the Kappa indices measuring the agreement between the simulation, the null model and the reference map showing observed LUC in 2018. We use the GRASS *r.kappa* raster tool to calculate the kappa values for agreement: (i) between observed LUC in 2012 duplicated in 2018 (null model) and observed LUC in 2018 and (ii) between observed LUC in 2018 and simulated LUC in 2018.

Step 2

We then generate the cross-matrices between the simulation, null model and reference map (CLC_2012 against CLC_2018 and CLC_predict_2018 against CLC_2018) using the *Cross-classification* tool (see Exercise 2 of Sect. 1 in Chapter "Basic and Multiple-Resolution Cross-Tabulation to Validate Land Use Cover Maps"). This method complements the kappa agreement indices and provides additional information about the similarity between the different maps.

Step 3

Once the cross-tabulations are obtained, on a spreadsheet we calculate the sum of cells on the diagonal (pixel-to-pixel correspondence).

Results and Comments

The resulting Kappa values are 0.9849 for the simulation (CLC_predict_2018 related to CLC_2018) and 0.9875 for the null model (CLC_2012 related to CLC_2018). The quantity and allocation correspondence (the proportion of diagonal pixels in the cross-matrices) are 98.22% for the simulation and 98.53% for the null model. Therefore, with both techniques, the null model obtains a slightly higher score than the simulation.

Interpretation of these results is difficult and has to be done carefully due to the limitations of this technique and the criticisms often levelled against it. The results show that persistence is the dominant process (98.5% of the study area did not change between 2012 and 2018; null model). Taking into account that most models simulate persistence better than change, it would be difficult to obtain a higher prediction score for a study area in which so little land use change is taking place. The low proportion of changes makes it difficult to simulate the changes between land use categories correctly. The slightest error diminishes the performance of the simulation compared to the null model.

Other methods, such as the Figure of Merit (see Sect. 4), can provide a better picture on how the model correctly simulated the change.

2 LUCC Budget

Description

LUCC budget is a technique for analysing land use/cover change (LUCC) using the cross-tabulation matrix obtained by overlaying two maps of the same area at two different dates. For each category, the changes are characterized in four components: gross gains, gross losses, net change and swap (Pontius et al. 2004).

Gross gains are the areas gained by each category, and gross losses are the areas lost. Net change is the difference between gains and losses. In categories in which gains and losses are occurring in different places, swap is a measure of the real changes taking place which are not revealed by the net change indicator. It measures the total area in which an equivalent amount of gains and losses have taken place, i.e. if in one category there are gains of 5 ha in one place and losses of 3 ha in another, the 3 ha that it losses in one place

and recoups in another are the swap (swap = 3 + 3 = 6 ha), while the remaining 2 ha (5–3) are the net change.

Utility

Exercises
1. To validate a series of maps with two or more time points

When monitoring landscape changes, the LUCC budget technique helps to identify the most critical land use transitions and should ultimately facilitate linking patterns to process (Pontius et al. 2004). It also allows LUCC simulation models to compare observed LUCC with simulated LUCC in both the calibration and validation steps (Paegelow 2018). In short, LUCC budget enables a more detailed analysis of land use change in a particular area.

QGIS Exercise

Available tools
• Processing R provider plugin   *LUCCBudget.rsx* R script

The components of change computed by the LUCC budget are derived from the cross-tabulation matrix. This matrix can be obtained by overlaying the two maps in QGIS and then calculating the LUCC budget values using a spreadsheet programme. However, we suggest using the *LUCCBudget. rsx* R script with the QGIS Processing R provider plugin. This script will carry out the entire LUCC budget calculation and will generate a table containing the values for the four components of change.

See Chapter "About this Book" for more detailed information about how to integrate R into QGIS and how to use R scripts such as the one applied in this exercise.

Exercise 1. To validate a series of maps with two or more time points

Aim

To carry out LUCC budget analysis in the Ariege study area using the CORINE Land Use maps dated 2000 and 2018.

Materials

CORINE Land Cover Map Val d'Ariège 2000
CORINE Land Cover Map Val d'Ariège 2018

Requisites

All maps must be in raster format and have the same resolution, extent and projection.

Execution

If necessary, install the Processing R provider plugin, and download the *LUCCBudget.rsx* R script into the R scripts folder (processing/rscripts). For more details, see Chapter "About this Book".

Step 1

Then, run the script and fill in the required parameters (names of the two maps and the output table) as shown in Fig. 1.

Results and Comments

The script will generate the cross-tabulation or change matrix as shown in Table 1. This matrix is saved as an intermediate product. The script will also generate a table in CSV format that indicates, for each category, the value of the four components assessed by the LUCC budget technique (Table 2).

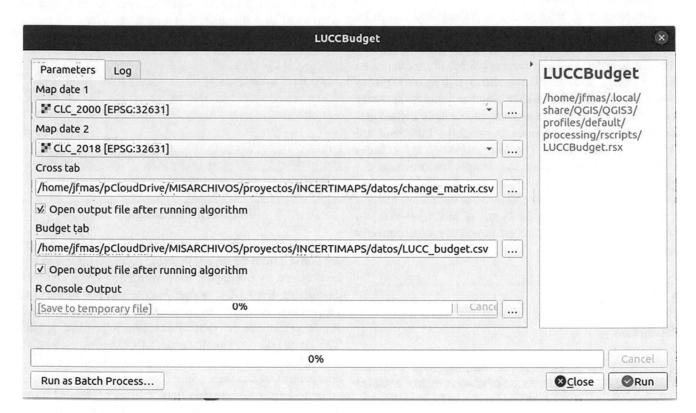

Fig. 1 Exercise 1. Step 1. LUCCBudget R script

Table 1 Result from Exercise 1. Cross-tabulation or change matrix

	0	1	2	3	4	5	6
0	**74,437**	0	0	0	0	0	0
1	0	**3,302**	3	0	8	0	37
2	0	1,853	**52,059**	235	409	0	2
3	0	23	109	**39,232**	127	0	0
4	0	12	399	877	**11,418**	22	0
5	0	0	0	0	10	**921**	0
6	0	0	0	0	0	0	**76**

Table 2 Results from Exercise 1. LUCC budget components

	Gains	Losses	Swap	Net
0	0	0	0	0
1	1,888	48	96	1,840
2	511	2,499	1,023	1,987
3	1,112	258	517	854
4	554	1,310	1,108	756
5	22	10	20	12
6	38	0	0	38

As can be seen in Table 2, the only class in which there are no losses, and consequently no swap is Category 6 (water). Therefore, for this category, the gross change is equal to the net change. Similar behaviour could be expected for Category 1 (built-up) because it is a "definitive" class (with no return), in the sense that it is very unlikely that a built-up area will be converted into another land cover. However, the change matrix (Table 1) shows small areas of transition from Category 1 (built-up) to Categories 2 (agriculture), 4 (scrublands) and 6 (water). These transitions are probably erroneous changes, resulting from misclassifications in the maps. The other categories appear to be more dynamic with both gross losses and gains and significant swap values.

3 Quantity and Allocation Disagreement

Description

Pontius Jr. and Millones (2011) proposed a set of metrics, obtained from the cross-tabulation matrix, which classify the overall change detected between a pair of maps into various components, namely, differences in the quantity of each category and differences in their location.

When analysing a time series (or single maps evaluated against a reference map), this method can differentiate between the changes that are due to differences in the relative importance of certain categories (some increase and others decrease) and those derived from changes in the location of the elements that make up these categories. It also identifies the categories that undergo net changes and swaps. As regards differences in location, this method distinguishes between exchanges between classes and changes in the location of two or more classes.

Utility

Exercises

1. To validate a series of maps with two or more time points

Quantity and allocation disagreement assess how similar a simulation or simulation is to a reference map, differentiating between (dis)agreement that is due to the quantities of different classes and (dis)agreement caused by the allocation of these classes in different places. By providing the same information, this method can also be used to validate an LUC map against a reference map or to assess the LUC changes in a time series of maps and understand whether or not these changes follow a logical trend.

QGIS Exercise

Available tools

- Processing Toolbox
     GRASS
          Raster
               r.cross
               r.kappa
     SAGA
          Confusion matrix
- Pontius matrix (Excel sheet)
          http://www2.clarku.edu/~rpontius/PontiusMatrix41.xlsx
- Semi-Automatic Classification plugin (SCP)
     Tab: Postprocessing
          Section: *Cross-Classification*

For more information about the use of *r.cross*, *r.kappa*, *SAGA Confusion matrix* and *SCP*, please refer to Chapters

"Basic and Multiple-Resolution Cross-Tabulation to Validate Land Use Cover Maps" and "Metrics Based on a Cross-Tabulation Matrix to Validate Land Use Cover Maps". QGIS *Raster Calculator* is a generic tool performing all kinds of raster calculations. It is intended for detailed analysis of the differences in quantity and allocation, rather than global studies.

Exercise 1. To validate a series of maps with two or more time points

Aim

To detect quantity and allocation changes between CORINE LUC maps of the Ariège Valley (southern France) between 2012 and 2018.

Materials

CORINE Land Cover Map Val d'Ariège 2012
CORINE Land Cover Map Val d'Ariège 2018

Requisites

All maps must be in raster format with the same resolution, extent and spatial reference system (SRS).

Execution

Step 1

In order to be able to make this analysis, the CORINE LUC map for 2018 must be polygonized. To this end, use the tool *Polygonize*.

Step 2

After polygonizing the CORINE raster, the next stage is to cross-tabulate the two maps we are going to compare. To this end, open the SAGA *confusion matrix* tool and select the CORINE LUC map for 2012 as Classification 1 layer and the CORINE LUC map for 2018 as Classification 2 layer. Then, fill in the parameters for the following lines—*Value*, *Value (Maximum)* and *Name*—into the function. Do not change any default options (the "Report unchanged classes" box must be ticked; output as "cells" and open the results generated) (Fig. 2). Rather than saving these results in a file, they can be handled as temporary layers.

Step 3

Import the SAGA-generated confusion matrix obtained in the previous stage into a spreadsheet software such as Excel. Then translate the obtained matrix into percentages (Table 3). This is done by dividing each pixel score in the original table by the total number of pixels multiplied by 100.

Step 4

Finally, use the SAGA-generated confusion matrix obtained in Step 2 to calculate the quantity and allocation disagreements in a spreadsheet software such as Excel. For a pixel resolution of 15 × 15 m, 1 ha corresponds to 44.44 pixels. Quantity disagreement is calculated by subtracting column total from row total (quantity disagreement = row total – column total) (Table 4). Allocation disagreement corresponds to all not-diagonal cell values.

Results and Comments

Table 3 shows the SAGA-generated confusion matrix reformatted in Excel and converted into a per cent of the study area. The sum of the diagonal corresponds to the overall persistence between 2012 and 2018. This value is 98.52%, which means that the change rate is 1.48%.

Although the net balance values (2018–2012) provided in Table 4 mask the changes that have taken place in certain classes, we can see from Table 3 that built-up gains (1.01%) result almost exclusively from the conversion of agricultural and pasture land (1.00), whose losses are partially compensated by the conversion of scrubland into agriculture and pasture (0.08). Scrubland is the only category with net losses and no net gains.

Table 4 expresses the amount of change (2018–2012) in ha (for a pixel resolution of 15 × 15 m; 1 ha corresponds to 44.44 pixels). As can be seen, no significant changes took place in mineral and water areas, while losses in scrubland were matched by gains in forest (about 400 ha) and losses in agriculture and pasture were matched by gains in built-up areas (about 1,000 ha).

Allocation disagreement corresponds to all not-diagonal cell values. These may be expressed as gains (2018—intersection 2012 against 2018) and losses (2012—intersection 2012 against 2018). While in some classes there are net changes (e.g. scrubland is the only category with net losses and no net gains), the changes in agriculture and pasture land are almost all losses (1.05), with just a few small gains (0.08%) from scrubland. This means that quantity disagreement shows a negative net balance for agriculture and

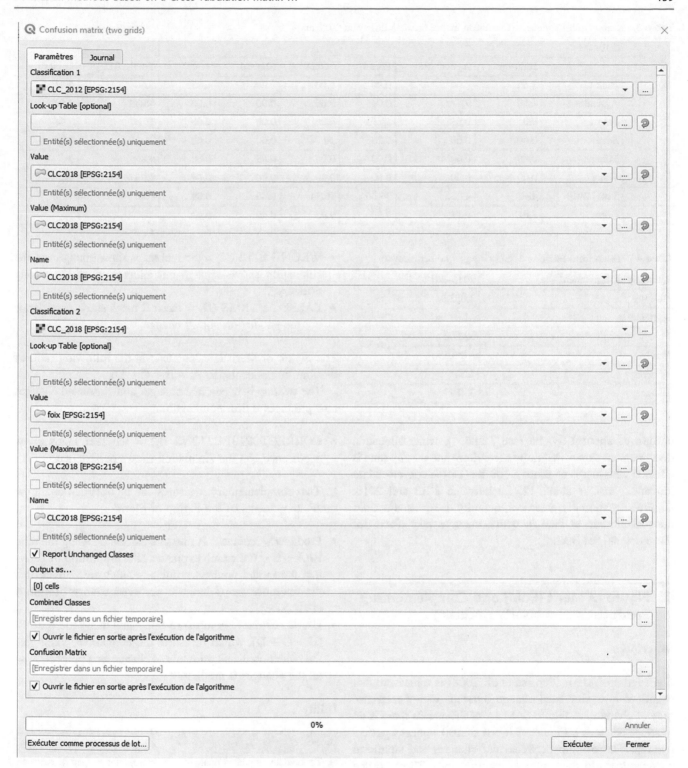

Fig. 2   Exercise 1. Step 2. Confusion matrix (two grids)

Table 3 Result from Exercise 1. Confusion matrix between 2018 and 2012 maps

%	2018	Built-up	Agriculture	Forest	Scrubs	Mineral	Water	Total 2012	Losses
2012	Built-up	*3.63*	0.00	0.00	0.00	0.00	0.03	**3.66**	*0.03*
	Agriculture	1.00	*47.21*	0.04	0.00	0.00	0.00	**48.25**	*1.04*
	Forest	0.01	0.03	*36.03*	0.00	0.00	0.00	**36.07**	*0.04*
	Scrubs	0.00	0.08	0.28	*10.74*	0.00	0.00	**11.11**	*0.37*
	Mineral	0.00	0.00	0.00	0.00	*0.85*	0.00	**0.85**	*0*
	Water	0.00	0.00	0.00	0.00	0.00	*0.06*	**0.06**	*0*
	Total 2018	**4.63**	**47.32**	**36.36**	**10.74**	**0.86**	**0.09**	100	
	Gains	*1.01*	*0.11*	*0.33*	*0.00*	*0.00*	*0.03*		

Table 4 Result from Exercise 1. Net change (ha) per category

Quantity disagreement (ha)	2018–2012
Built-up	1,083.04
Agriculture	−1,037.43
Forest	322.58
Scrubs	−406.55
Mineral	4.30
Water	33.59

pasture of about 1,037 ha (see Table 4), while allocation disagreement shows that more agriculture and pasture land is affected with losses of about 1,160 ha (1.04% converted into ha) and gains of about 123 ha between 2012 and 2018. Unlike allocation disagreement, quantity disagreement hides the real amount of land in which changes take place (for more details, see Sect. 2).

4 Figure of Merit (FoM) and Complementary Producer's and User's Accuracy

Description

The Figure of Merit (Pontius et al. 2008) is a measure that examines how simulated change overlaps with a reference map of changes. A Figure of Merit of 0% means there is no overlap, whereas a Figure of Merit of 100% means perfect overlap. The overlap between real changes and simulated changes leads to four possible combinations. These are the four components of the Figure of Merit:

- MISSES (A) = the real maps show change but the simulation shows persistence.
- HITS (B) = the real maps show change and the simulation shows change.

- WRONG HITS (C) = the real maps show change and the simulation shows change but allocates it to the wrong category.
- FALSE ALARMS (D) = the real maps show persistence but the simulation shows change.

The Figure of Merit is calculated via the following ratio of the four components: B/(A + B + C + D).

The overlap between real changes and simulated changes also produces a fifth combination:

- CORRECT REJECTIONS (E) = the real maps show persistence and the simulation shows persistence.

Two complementary measures can be obtained using the same components of the Figure of Merit:

- Producer's accuracy: A measure calculated using the ratio B/(A + B + C), which expresses "the proportion of pixels that the model predicts accurately as change, given that the reference maps indicate observed change" (Pontius et al. 2008).
- User's accuracy: A measure calculated using the ratio B/(B + C + D), which measures the number of pixels that the model predicts accurately as change as a proportion of all the changes it predicts.

Utility

Exercises
1. To validate simulated changes against a reference map of changes
2. To validate simulated changes against a reference map of changes in a binary format
3. To validate the changes simulated by various models

The Figure of Merit and the complementary Producer's and User's accuracies are very useful measures for validating the change simulated by a model. The different components of the Figure of Merit can give users a better picture of how accurate the simulation is, e.g. if the model estimated more or less changes than those appearing on the reference map. They can also differentiate between quantity and allocation errors (Pontius et al. 2018).

These measures are also highly recommended for comparing several simulations using a standard measure. They can be applied, for example, to assess the congruence of model outputs. This is a form of validation that evaluates the agreement between simulations obtained through different models or between simulations obtained using the same model but parametrized in different ways. The agreement between the simulation maps is measured and the degree of congruence is considered an indicator of the stability of the model and the plausibility of the simulations. The congruence of model outputs provides useful information about model robustness (Paegelow et al. 2014; Camacho Olmedo et al. 2015).

Complementary analyses to the Figure of Merit and the Producer's and User's accuracies include spatial metrics, Kappa indices, the Land Use and Cover budget (LUCC budget) technique and Quantity and Allocation disagreement. These indices are described in Sects. 2 and 3 of this chapter.

QGIS Exercises

Available tools

- Processing Toolbox
  SAGA
    Image analysis
      Confusion matrix (two grids)
      Confusion matrix (polygons/grid)
    Raster analysis
      Cross-classification and tabulation
- Processing Toolbox
  GRASS
    Raster
      r.cross
- Semi-Automatic Classification Plugin
    Tab: Postprocessing
      Section: *Cross-classification*
      Section: *Accuracy*
      Section: *Land cover change*

The Figure of Merit and the complementary Producer's and User's accuracy indices are not calculated directly in QGIS. Producer's and User's accuracy per category can be calculated using the SAGA *Confusion matrix (two grids)* and *Confusion matrix (polygons/grid)* tools and in the "Semi-Automatic Classification Plugin" (*Accuracy*).

Users can calculate the Figure of Merit from the cross-tabulation matrices. As commented in Sect. 1 in

Chapter "Basic and Multiple-Resolution Cross-Tabulation to Validate Land Use Cover Maps", QGIS includes many tools for cross-tabulating spatial data in the GRASS and SAGA toolboxes. The "Semi-Automatic Classification Plugin" also includes cross-tabulation tools.

Of all the tools available in QGIS, in this book, we recommend the "Semi-Automatic Classification Plugin", which is the most efficient, most stable tool of all those assessed.

Exercise 1. To validate simulated changes against a reference map of changes

Aim

To validate the change simulated by a model against a reference map of changes for the same simulation period. The initial map is the CORINE map for 2005 in both cases. The changes from 2005 to 2011 are calculated for the simulation and for the CORINE data as reference.

Materials

CORINE Land Use Map Asturias Central Area 2005
CORINE Land Use Map Asturias Central Area 2011
Simulation LCM Val d'Ariège 2018

Requisites

The maps must have the same extent, spatial resolution, projection and legend. If they do not have the same legend, the maps must be reclassified to meet this requirement. For a proper validation, the latest reference map must refer to the same date as the simulation.

Execution

Step 1

We begin by obtaining two rasters showing the areas that changed in the study area during the period analysed and those that remained the same. This procedure must be done twice: once for the reference map (CORINE 2005–CORINE 2011) and once for the simulated map (CORINE 2005– Simulation 2011).

To obtain these maps, open the "Semi-Automatic Classification Plugin" and the "Postprocessing" tab. Then select *Land cover change* and fill in the required parameters: the earlier map in the reference classification (CORINE 2005) and the more recent map in the new classification (CORINE 2011; Simulation 2011) (Fig. 3). Leave the "Report unchanged pixels" option unmarked so as to obtain a map

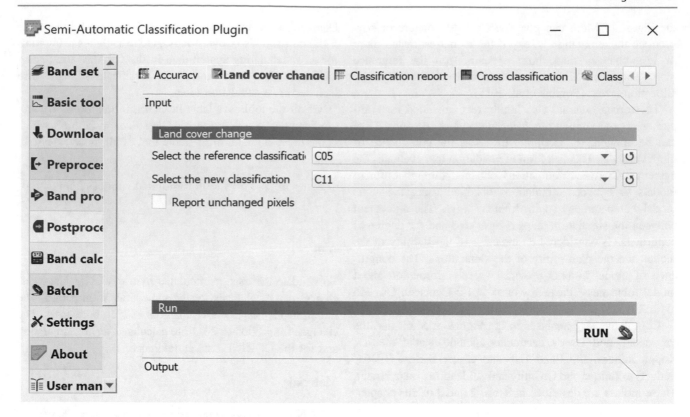

Fig. 3 Exercise 1. Step 1. Semi-Automatic Classification Plugin

that only shows the areas that changed during the study period. If this option is marked, a map showing both change and persistence areas will be obtained.

Run the tool to obtain two output maps showing the changes on the reference map (CORINE) and the changes simulated by the model. Both will refer to the same period (2005–2011).

Step 2

The next stage involves cross-tabulating the two maps of changes. To obtain these maps, open the Semi-Automatic Classification Plugin and in the "Postprocessing" tab, select *Accuracy*. Select the required parameters: classification to assess (simulated changes) and reference raster (CORINE 05–11 changes) (Fig. 4).

Results and Comments

Step 1 produces two maps of changes, which are stored in the folder specified by the user. The function also generates a matrix for each pair of cross-tabulated maps. These matrices appear in the "output" window, stored in CSV format. They show each possible combination between the two cross-tabulated maps and the code under which each combination is represented in the output raster.

Only four transitions (new codes *3, 4, 16* and *17*) are simulated by the model, as expressed in Table 5. Twenty-eight transitions occur between the CORINE maps (Table 2).

Most of the changes predicted in the simulation refer to the transition from agricultural areas (Category 0) to urban fabric (Category 2) and to the transition from agricultural areas to industrial and commercial areas (Category 3). Together, they represent 1,546 of the 1,632 pixels simulated. That is, almost 95% of the simulated pixels. In the reference map, these transitions represent 751 and 503 pixels, respectively, a less significant proportion of total change (in italics in Table 6).

After completing *Step 2*, we now have a cross-tabulation raster and a table showing every possible combination between the two cross-tabulated maps (Table 7).

Following the definitions provided by Pontius et al. (2008), in our case, HITS were only obtained in new codes 12 (old code 3 in the CORINE map of changes and old code 3 in the simulated map of changes), 18 (old codes 4 and 4) and 55 (old codes 17 and 17). HITS are obtained when both the reference map and the simulation show the same change or transition, which is why they both have the same codes.

The WRONG HITS correspond to combinations where both the reference map and the simulation show change, but to different gaining categories. For example, new code 13 (old codes 3 and 4) refers to areas that were agricultural

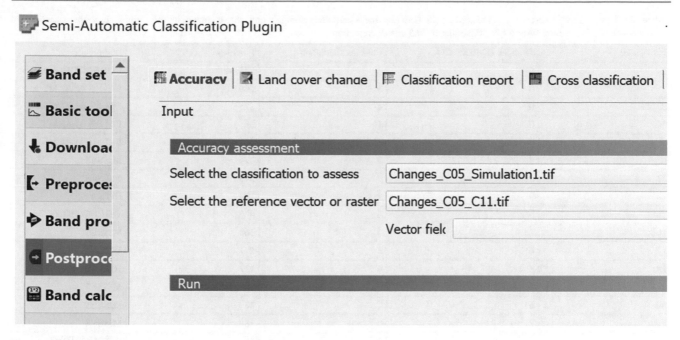

Fig. 4 Exercise 1. Step 2. Semi-Automatic Classification Plugin

Table 5 Result from Exercise 1. Variety and size of the simulated transitions

New codes	CORINE 05 category	Simulation category	Pixel sum
3	0	2	874
4	0	3	672
16	1	2	38
17	1	3	48

Table 6 Result from Exercise 1. Size of transitions between CORINE 2005 and CORINE 2011 maps

New codes	CORINE 05 category	CORINE 11 category	Pixel sum
2	0	1	374
3	0	2	751
4	0	3	503
5	0	4	148
6	0	5	11
7	0	6	301
10	0	9	132
14	1	0	588
16	1	2	61
17	1	3	82
18	1	4	157
19	1	5	109
20	1	6	225
24	1	10	180
27	2	0	21

(continued)

Table 6 (continued)

New codes	CORINE 05 category	CORINE 11 category	Pixel sum
28	2	1	22
30	2	3	26
36	2	9	4
40	3	0	51
42	3	2	11
53	4	0	211
54	4	1	327
55	4	2	89
56	4	3	21
79	6	0	44
80	6	1	111
147	11	3	88
151	11	7	657

areas that changed to urban fabric in the simulation and to industrial and commercial areas in the reference map (Tables 5 and 6).

FALSE ALARMS refer to areas that are marked as persistence in the reference map and as change in the simulation. Examples include new code 2 (old codes 0 and 3). Areas with that code refer to pixels that were simulated as urban fabric in the simulation, but do not show change in the reference map. Code 0 does not appear among the codes in Table 6 summarizing all the possible transitions between the original (CORINE 2005) and the reference map (CORINE 2011). It must therefore refer to persistence.

Table 7 Result from Exercise 1. (Dis)agreement between the simulated changes and the changes in the reference maps classified in five categories: misses, hits, wrong hits, false alarms and correct rejections

New codes	Changes CORINE 05–11	Changes simulation	Pixel sum	Interpretation
1	0	0	577,949[1]	CORRECT REJECTION
2	0	3	600	FALSE ALARMS
3	0	4	525	FALSE ALARMS
4	0	16	38	FALSE ALARMS
5	0	17	33	FALSE ALARMS
6	2	0	374	MISSES
11	3	0	543	MISSES
12	3	3	204	HITS
13	3	4	4	WRONG HITS
16	4	0	364	MISSES
17	4	3	2	WRONG HITS
18	4	4	137	HITS
21	5	0	148	MISSES
26	6	0	11	MISSES
31	7	0	280	MISSES
32	7	3	15	WRONG HITS
33	7	4	6	WRONG HITS
36	10	0	79	MISSES
37	10	3	53	WRONG HITS
41	14	0	579	MISSES
45	14	17	9	WRONG HITS
46	16	0	61	MISSES
51	17	0	76	MISSES
55	17	17	6	HITS
56	18	0	157	MISSES
61	19	0	109	MISSES
66	20	0	225	MISSES
71	24	0	180	MISSES
76	27	0	21	MISSES
81	28	0	22	MISSES
86	30	0	26	MISSES
91	36	0	4	MISSES
96	40	0	51	MISSES
101	42	0	11	MISSES
106	53	0	211	MISSES
111	54	0	327	MISSES
116	55	0	89	MISSES
121	56	0	21	MISSES
126	79	0	44	MISSES
131	80	0	111	MISSES
136	147	0	88	MISSES
141	151	0	657	MISSES

[1] The result of 577,949 pixels classified as CORRECT REJECTIONS was calculated by subtracting the 339,103 pixels of no data from the 917,052 pixels coded as 1.

MISSES refer to the areas where the reference map shows change but the simulation shows persistence. Examples include code 16 (old code 4 and 0). Finally, CORRECT REJECTION refers to the pixels marked as persistence in the reference map that were correctly simulated as persistence (new code 1, old codes 0 and 0).

In total, HITS account for 347 pixels, WRONG HITS for 89 pixels, FALSE ALARMS for 1,196 pixels and MISSES for 4,869 pixels (Table 7). Therefore, the simulation produced a lot more FALSE ALARMS than HITS and the vast majority of the predictions were MISSES. This makes sense because most of the landscape remained unchanged over the simulation period.

With all the above information, we can finally calculate the Figure of Merit (B/(A + B + C + D)) for the model. It is 5.340%. This is a very low Figure of Merit, far below the 100% that would mean perfect overlap. However, perfect overlap is almost impossible. In most cases, low Figures of Merit are the norm.

We must also consider that the Figure of Merit compares the simulated changes with all the changes in the reference map. In our simulation, we only modelled two categories actively (urban fabric and industrial and commercial areas). This means that the changes in all the other categories were not even simulated and no agreement can therefore be expected. This limitation must be borne in mind when evaluating the Figure of Merit.

The best way to obtain a Figure of Merit that offers objective information about the validity of our modelling exercise is to repeat the same exercise, focusing exclusively on the actively modelled transitions (from agricultural and vegetation areas to urban fabric and industrial and commercial areas).

Producer's accuracy (B/(A + B + C)) is 6.54% and expresses the number of pixels that the model accurately predicts as change as a proportion of total observed change. For its part, User's accuracy (B/(B + C + D)) measures the number of pixels that the model predicts accurately as change as a proportion of total predicted change, in this case 21.26%.

As regards the four simulated changes, shown in Table 5, the Producer's and User's accuracy values for Categories 3 and 4 are higher than for Category 17, and are zero in Category 16 (Table 8).

Exercise 2. To validate simulated changes against a reference map of changes in a binary format

Aim

To validate the change simulated by a model against a reference map of changes for the same simulation period. To do this, we overlay two maps that show change versus non-change over the same period. The initial map in both cases is the CORINE dataset for 2005. The changes from 2005 to 2011 are calculated for the simulation and for the CORINE dataset as reference. In this exercise we do not evaluate the WRONG HITS.

Materials

CORINE Land Use Map Asturias Central Area 2005
CORINE Land Use Map Asturias Central Area 2011
Simulation CORINE Asturias Central Area 2011

Requisites

The maps must have the same extent, spatial resolution, projection and legend. If they do not have the same legend, the maps must be reclassified so as to meet this requirement. For a proper validation, the latest reference map must refer to the same date as the simulation.

Execution

Step 1

The first step is to obtain two rasters showing the areas that changed and those that remained the same over the period being analysed: one for the reference map (CORINE 2005–CORINE 2011) and one for the simulation (CORINE 2005–Simulation 2011). To obtain these maps, follow the instructions in Exercise 1 *Step* 1 above.

Step 2

Once the two maps have been obtained, they must be reclassified into binary format, i.e. into a map with two possible values: 0 (persistence) and 1 (changes). This is done using the *Reclassify by table* tool.

Table 8 Results from Exercise 1. Producer's and User's accuracy values

Categories in changes simulation	3	4	17	16
Producer's accuracy %	27.1638	27.2366	7.3171	0.000
User's accuracy %	23.3410	20.3869	12.5000	-

Fig. 5 Exercise 2. Step 2. Intermediate map showing the areas of change in the reference maps

Fig. 6 Exercise 2. Step 2. Intermediate map showing the areas of change in the simulation

Figures 5 and 6 show the change areas (value 1) in black and the persistence areas (value 0) in white, for both the reference map (Fig. 5) and the simulation (Fig. 6).

Step 3

Finally, the two binary maps must be cross-tabulated. To do so, open the "Semi-Automatic Classification Plugin" and, in the "Postprocessing" tab, select the *Cross-classification* option. Fill in the required parameters: classification (binary changes from the simulation) and reference raster (binary changes from CORINE) (Fig. 7).

Results and Comments

Once we have completed *Step 3*, the QGIS creates an output raster that shows all possible combinations between the two binary change maps. The function also generates a table showing all possible combinations between the two input maps. This table appears in the "output" window, stored in CSV format. This table also lists the codes with which each combination is represented in the output raster.

Table 9 presents the four possible combinations obtained from the two binary maps crossed in *Step 3*. As 0 was used to represent persistent areas and 1 areas that changed, new code 1 (0/0) refers to pixels that the model correctly simulated as persistence (CORRECT REJECTIONS). New code 4 (1/1) refers to pixels that the model correctly simulated as change (HITS), while codes 2 and 3 refer to pixels in which the model does not agree with the reference map. Code 2 (0/1) corresponds to FALSE ALARMS: the model simulated change but the reference map shows persistence. Code 3 (1/0) stands for MISSES: the model simulated persistence but the reference map shows change.

The sum of MISSES plus HITS (5,305 pixels) represents the change in the reference map (CORINE) for the period 2005–2011. These pixels cover just 0.9077% of the total study area. Very little change therefore took place in the reference map for our study area.

HITS plus FALSE ALARMS (1,632 pixels) gives all the pixels in which the simulation predicted change. These pixels cover 0.2792% of the total study area. This means that fewer changes were simulated than actually took place on

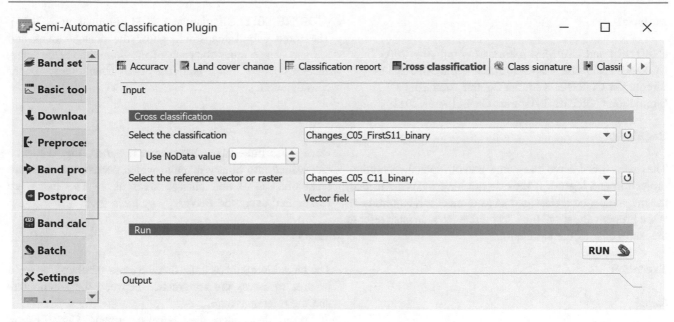

Fig. 7 Exercise 2. Step 3. Semi-Automatic Classification Plugin

Table 9 Result from Exercise 1. (Dis)agreement between the simulated changes and the changes in the reference maps classified in five categories: misses, hits, wrong hits, false alarms and correct rejections

New codes	Binary CORINE changes	Binary simulated changes	Pixel sum	Interpretation
1	0	0	577,949[2]	CORRECT REJECTIONS
2	0	1	1,196	FALSE ALARMS
3	1	0	4,869	MISSES
4	1	1	436	HITS

the reference map. This makes sense given that in our simulation we only simulated the transitions from agricultural and vegetation areas to urban fabric and industrial and commercial areas, while the reference map also considered many other changes between all the other categories represented on the map, which were not simulated in our modelling exercise.

The Figure of Merit (B/(A + B + C + D)) for our simulation is very low at 6.7%. This indicates that the simulation did not simulate most of the changes that took place in the reference map correctly. This is partly due to the fact that we only actively modelled two categories, while the reference map showed the changes that took place between all categories. As a result, overlap between the two maps is impossible in many areas. Even so, the general level of overlap between the simulated changes and those observed on the reference maps is still quite low. Other metrics and tools must therefore be used in order to interpret the simulation and the performance of the modelling exercise better.

The Figure of Merit in this exercise is a bit better than in the previous one because we did not take WRONG HITS into account. In this case, we only compared changes,

without taking into account the type of change that happened in the simulation period.

Exercise 3. To validate the changes simulated by various models

Aim

To compare and validate the change simulated by two models. For this purpose, we overlay three maps that show change versus non-change over the same interval. The initial map in all cases is the CORINE dataset for 2005. The changes from 2005 to 2011 are calculated for the simulation from model 1, for the simulation from model 2 and for the CORINE dataset as reference. WRONG HITS are not evaluated in this exercise.

---

[2] There are 339,103 pixels of no data. If we subtract them from the 917,052 pixels coded as 1, the result is 577,949 pixels in which there were CORRECT REJECTIONS.

Materials

CORINE Land Use Map Asturias Central Area 2005
CORINE Land Use Map Asturias Central Area 2011
Simulation CORINE Asturias Central Area 2011
Simulation CORINE 2 Asturias Central Area 2011

Requisites

The maps must have the same extent, spatial resolution, projection and legend. If they do not have the same legend, the maps must be reclassified so as to meet this requirement. For a proper validation, the latest reference map must refer to the same date as the simulation.

Execution

Step 1

The first step is to obtain three rasters for the study area showing the areas that changed and those that remained the same over the period being analysed. In this way, we obtain: (i) the map of changes for the reference map (CORINE 2005–CORINE 2011), (ii) the map of changes for the first simulation (CORINE 2005–Simulation 1 2011) and (iii) the map of changes for the second simulation (CORINE 2005–Simulation 2 2011).

To obtain these maps, open the "Semi-Automatic Classification Plugin" and, in the "Postprocessing" tab, select *Land cover change*. Then, fill in the required parameters: the earliest map in the reference classification (CORINE 2005) and the more recent maps in the new classifications

(CORINE 2011, Simulation 1 2011, Simulation 2 2011). The three output maps will show the change areas and the persistence areas for each of the three maps (the reference CORINE map and the two simulations) under consideration.

Step 2

Once these three maps have been obtained, they must be reclassified into binary maps in which persistence areas are reclassified as 0 and change areas as 1. The maps are reclassified using the *Reclassify by table* tool.

Step 3

The three binary maps must then be cross-tabulated, so as to be able to assess the congruence between the simulations and the reference map.

To do this, open the "Semi-Automatic Classification Plugin" and the "Postprocessing" tab, and then select *Cross-classification*. Start by cross-tabulating the two simulations you want to compare. To this end, fill in the following parameters: classification (binary map of changes from simulation 1) and reference raster (binary map of changes from simulation 2) (Fig. 8).

Step 4

The procedure is repeated again, this time cross-tabulating the raster obtained in the previous step with the reference map. In this case, open the tool and fill in the parameters as follows: classification (raster obtained after running the tool

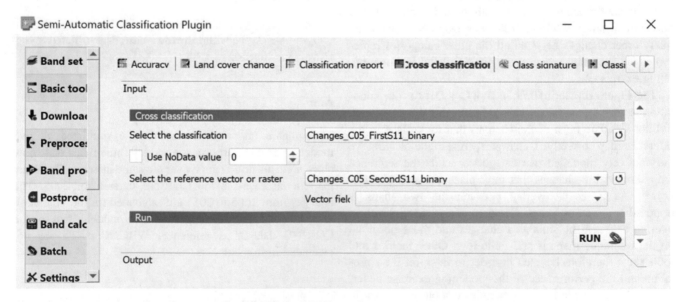

Fig. 8 Exercise 3. Step 3. Semi-Automatic Classification Plugin

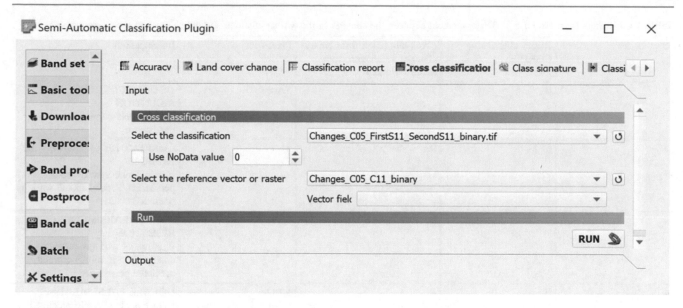

Fig. 9 Exercise 3. Step 4. Semi-Automatic Classification Plugin

as explained in the previous step) and reference raster (CORINE 05–11 binary map of changes) (Fig. 9).

Results and Comments

After carrying out *Steps 3* and *4*, QGIS creates two output rasters. The function also generates a table for each raster, which appears in the "output" window in CSV format. This table shows every possible combination between the values of the cross-tabulated maps. It also lists the codes under which each combination is represented in the output raster.

The raster obtained in *Step 3* measures the agreement between the two simulations (Table 10). In the binary maps, 0 was used to refer to persistent areas whereas 1 referred to areas that changed. New code 1 (previous codes 0/0) therefore refers to the pixels in which both models predicted persistence, while new code 4 (1/1) refers to the pixels where both models predicted change. Finally, new codes 2 and 3

represent areas in which the simulations do not agree: one shows persistence, whereas the other shows change.

The raster obtained in *Step 4* was produced by cross-tabulating a reference change map with the raster obtained after cross-tabulating the change maps produced by the two simulations. This cross-tabulation therefore produces eight possible combinations (Table 11).

In order to interpret the results of this second cross-tabulation correctly, we need to understand the values of the two rasters that were cross-tabulated. In the reference change map, 0 refers to persistent areas and 1 to areas that changed during the period under consideration. The meanings of the new codes in the raster obtained in *Step 3* are detailed in Table 9.

This enables a better interpretation of the results of the last raster generated. New code 1 (previous codes 0/1) refers to areas in which persistence was observed on the reference map of changes (code 0) and was also simulated by the two

Table 10 Results from Exercise 3. (Dis)agreement between the changes in the two simulations that have been compared

New codes	Binary changes from simulation 1	Binary changes from simulation 2	Pixel sum	Interpretation
1	0	0	581,158[3]	Both models predicted persistence
2	0	1	64	First model predicted persistence/Second model predicted change
3	1	0	1,660	First model predicted change/Second model predicted persistence
4	1	1	1,568	Both models predicted change

[3] There are 339,103 pixels of no data. If we subtract them from the 920,261 pixels coded as 1, the result is 581,158 pixels.

Table 11 Results from Exercise 3. (Dis)agreement between the changes in the two simulations and the changes in the reference maps

New codes	Binary changes CORINE	Cross-tabulation from binary changes simulation from models 1 and 2	Pixel sum	Interpretation
1	0	1	576,588[4]	DOUBLE CORRECT REJECTION Both models correctly predicted persistence
2	0	2	54	CORRECT REJECTION/FALSE ALARMS First model correctly predicted persistence/Second model wrongly predicted change
3	0	3	1,361	FALSE ALARMS/CORRECT REJECTION First model wrongly predicted change/Second model correctly predicted persistence
4	0	4	1,142	DOUBLE FALSE ALARMS Both models wrongly predicted change
5	1	1	4,570	DOUBLE MISSES Both models wrongly predicted persistence
6	1	2	10	MISSES/HITS First model wrongly predicted persistence/Second model correctly predicted change
7	1	3	299	HITS/MISSES First model correctly predicted change/Second model wrongly predicted persistence
8	1	4	426	DOUBLE HITS Both models correctly predicted change

models (code 1) (see Table 9 to understand the meaning of this code). Those cases in which the two models and the reference map all simulated persistence are referred to as DOUBLE REJECTIONS (Camacho Olmedo et al. 2015).

New code 4 (previous codes 0/4) refers to areas where the two models simulated change (code 4) and the reference change map showed persistence. These are known as DOUBLE FALSE ALARMS.

New code 5 (1/1) corresponds to areas where both models simulated persistence and the reference map showed change (DOUBLE MISSES). New code 8 (1/4) refers to areas where the two models and the reference map also showed change (DOUBLE HITS). Finally, the other four combinations refer to areas where each simulation shows a different agreement with the reference map (Table 11).

These eight possible combinations are expressed as two maps. The first map (a zoomed area is shown in Fig. 10, on the left) shows the four possible combinations for the areas on the CORINE map in which persistence was observed. Pixels simulated as persistence are therefore

CORRECT REJECTIONS, while those simulated as change areas are FALSE ALARMS. The areas that changed are masked in white. The second map (a zoomed area is shown in Fig. 11, on the right) shows the four possible combinations for the areas on the CORINE map in which change was observed. Pixels simulated as change are HITS, while those simulated as persistence are MISSES. The persistence areas are masked in white.

According to all the above results, it seems that the two simulations are very similar in terms of predictive accuracy. The vast majority of the pixels on the map are DOUBLE CORRECT REJECTIONS, which means that both models are very accurate when predicting persistence. This makes sense in that persistence is very easy to simulate in a highly stable area like the one we simulated. The most challenging task is to correctly simulate change. The best

---

[4] There are 339,103 pixels of no data. If we subtract them from the 915,691 pixels coded as 1, the result is 576,588 pixels classified as DOUBLE CORRECT REJECTIONS.

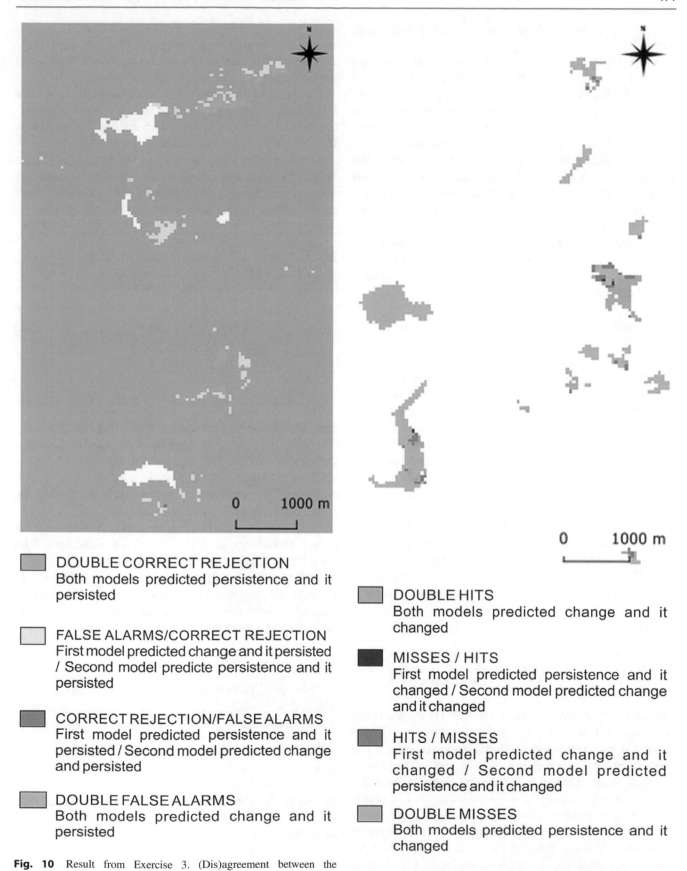

DOUBLE CORRECT REJECTION
Both models predicted persistence and it persisted

FALSE ALARMS/CORRECT REJECTION
First model predicted change and it persisted / Second model predicte persistence and it persisted

CORRECT REJECTION/FALSE ALARMS
First model predicted persistence and it persisted / Second model predicted change and persisted

DOUBLE FALSE ALARMS
Both models predicted change and it persisted

DOUBLE HITS
Both models predicted change and it changed

MISSES / HITS
First model predicted persistence and it changed / Second model predicted change and it changed

HITS / MISSES
First model predicted change and it changed / Second model predicted persistence and it changed

DOUBLE MISSES
Both models predicted persistence and it changed

Fig. 10 Result from Exercise 3. (Dis)agreement between the simulations and the reference maps for the areas where persistence was observed

Fig. 11 Result from Exercise 3. (Dis)agreement between the simulations and the reference maps for the areas where change was observed

models are therefore those that simulate change most accurately.

If we focus exclusively on the areas that changed, the accuracy is very low. 86.1451% of the pixels were DOUBLE MISSES, while in the remaining pixels there were HITS in one or both models. This means that in the vast majority of cases, our models incorrectly simulated change. These simulations cannot therefore be validated, although other validation tools can be used to check whether the simulated pattern is valid. In this regard, even if a hard comparison does not show a high level of agreement between a simulation and the reference map, the pattern of the simulated changing areas may be logical or correct. The models can therefore be considered valid in a qualitative sense.

5 Incidents and States

Description

Incidents and states are terms proposed by Pontius Jr. et al. (2017) to characterize land use cover changes in a series of three or more maps. States refer to the number of land uses or land covers a pixel is assigned in the series of maps. There can be as many states as there are maps in the series. Hao and Gen-Suo (2014) used the term "land use classification variety" for this metric when applying it to validate Land Use Cover maps (MODIS Land Cover product).

Incidents refer to the number of times a pixel changes category over the course of a time series. There can be as many incidents as there are stages in the time series. In a series of 5 maps, there are 4 time-stages. The series may therefore have between 0 and 4 incidents, i.e. the pixel may change category between 0 and 4 times. The number of incidents can also be referred to as "Transition frequency".

Utility

Exercises
1. To validate a series of maps with two or more time points

The number of incidents and states assigned to the pixels in a time series of Land Use Cover maps can help us identify the changes that take place for technical reasons, i.e. erroneous or spurious changes which do not really happen on the ground.

When obtained from satellite imagery classification, Land Use Cover maps usually have important sources of uncertainty. Various different Land Use and Cover categories can have very similar levels of reflectance. If the imagery is obtained at different times of the year, or under different atmospheric conditions, the reflectance of a pixel can vary to a similar extent to the difference in reflectance between two

Land Use Cover categories. The same pixel could therefore be classified under different categories over the course of the time series. The number of incidents and states of the pixel can potentially help us to identify these errors.

For example, in a time series of six maps, if a pixel has five incidents, but only two states, it means that it alternates between these two categories at each stage in the time series. If we discover which categories are involved in the transitions we can determine to what extent these changes are logical. Incidents and states can also be used to validate a series of simulations, when working with modelling exercises to obtain scenarios for more than two time points.

QGIS Exercise

Available tools

• Processing Toolbox
   GRASS
     Raster
       r.series

The GRASS toolbox associated with QGIS has a tool for calculating the number of states in a time series of Land Use Cover maps. QGIS does not provide any specific tool to calculate the number of incidents in the time series, so this metric must be calculated manually. This is done using the raster calculator and a raster reclassification tool.

QGIS offers several raster calculators and reclassification tools. Although they are all valid, in this exercise we will be using the ones from the core QGIS toolbox.

Pontius et al. (2017) also developed a tool in Excel to automatically calculate the incidents and states of a series of Land Use Cover raster maps in .rst format. It is available online free of charge.[5]

Exercise 1. To validate a series of maps with two or more time points

Aim

To find out if technical changes may have taken place in the last series of CORINE Land Cover maps produced for the Asturias Central Area.

---

[5] The tool is available on R. G. Pontius Jr's personal website: http://www2.clarku.edu/~rpontius/.

Materials

CORINE Land Use Map Asturias Central Area 2005
CORINE Land Use Map Asturias Central Area 2011
CORINE Land Use Map Asturias Central Area 2018

Requisites

All maps must be rasters and have the same resolution, extent and projection.

Execution

Step 1

In order to calculate the number of states per pixel, we must open the *r.series* tool and select all the maps that form part of the series of Land Use Cover maps we are analysing ("Input raster layer(s)"). In this case, we select the three maps in our series: CORINE Land Cover 2005, 2011 and 2018.

In the "Aggregate operation [optional]" option, select "Diversity". This will count the number of different categories to which a pixel is assigned over the course of the time series.

In "Advanced parameters", indicate the range of values of the Land Use Cover maps introduced as input, i.e. the minimum and maximum values. In our case, the minimum value for a category is 0 and the maximum value is 12 (Fig. 12).

The final stage is to indicate where the new map will be saved.

Step 2

There is no specific tool for calculating the number of incidents in a pixel over the course of a time series. This operation must therefore be carried out manually. The first

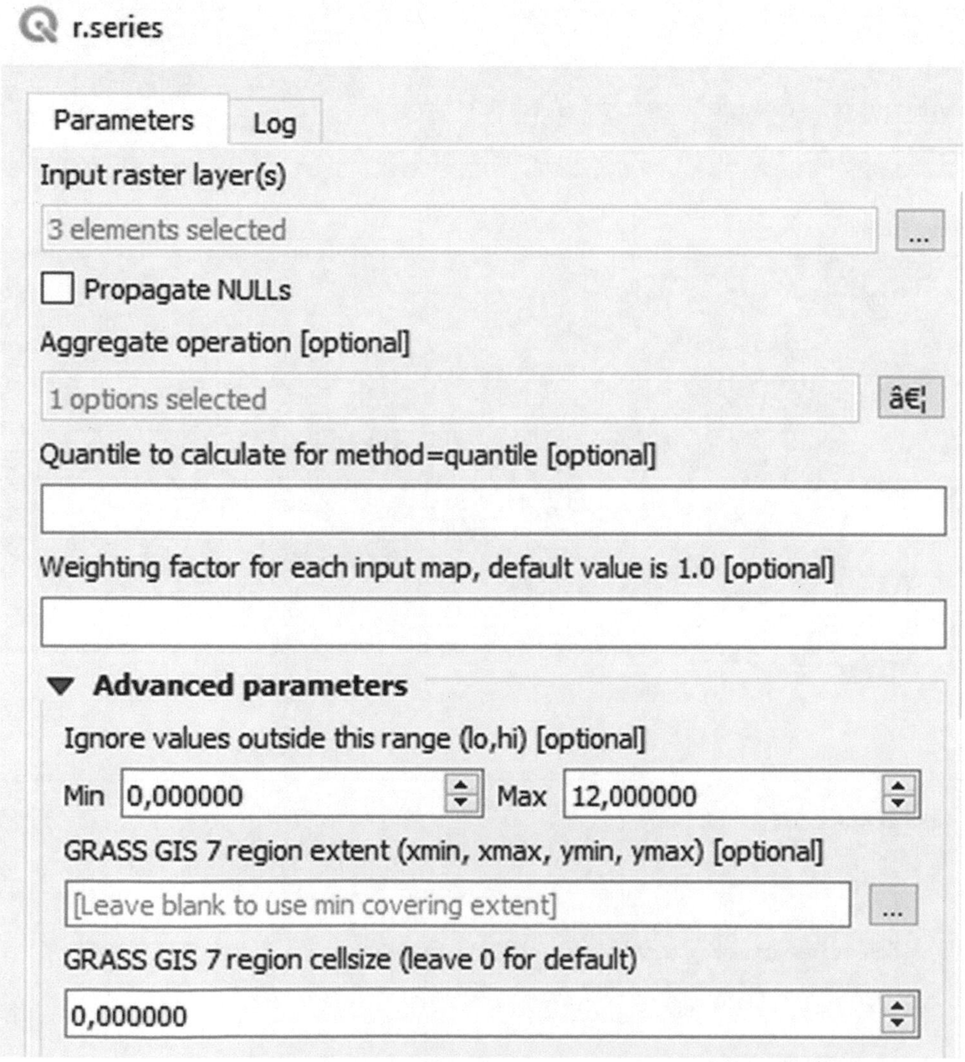

Fig. 12   Exercise 1. Step 1. R.series

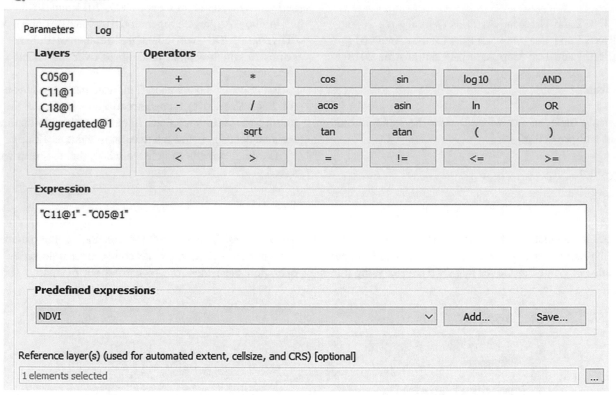

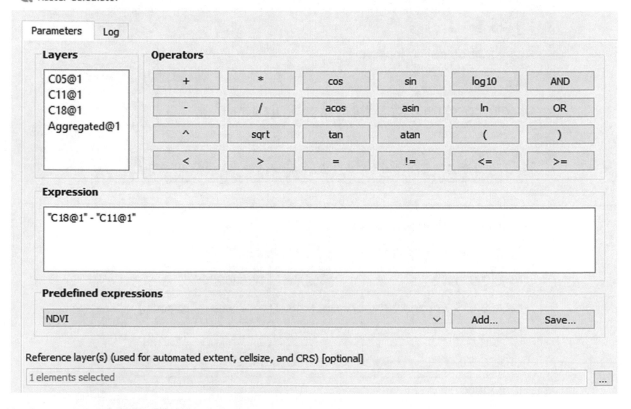

Fig. 13 Exercise 1. Step 2. Raster Calculator

step is to identify where the changes happened. For each pixel, we must then calculate the number of times it underwent change (or not). To carry out these operations, we have to work with pairs of maps: first 2005 and 2011 and then 2011 and 2018.

To identify where the changes happened, for each pair of rasters we must subtract one raster from the other. If a pixel does not change, the result of the subtraction will be a value of 0 for that pixel. If the pixel changes, the result of the subtraction will be a value other than 0.

The subtraction operation is carried out using the *Raster calculator*, in which we must write the following subtraction expression for each pair of maps:

$$t2\,map - t1\,map$$

We also need to indicate which raster is the reference map that will be used to define the characteristics (extent, spatial resolution and projection) of the new raster obtained after the calculation. In this case, we will be using the first map in our series (CORINE 2005). This must be indicated in the "Reference layer(s) (used for automated extent, cell size and CRS) [optional]" option (Fig. 13).

Step 3

Once the previous step has been completed, the maps obtained must be reclassified to enable us to identify the pixels where an incident took place (values other than 0) and the pixels that were incident-free at each stage (0 values).

To identify all pixels in which incidents took place with a value of 1, we reclassify all values other than 0 as 1 using the *Reclassify by table* tool (Fig. 15). The first stage in the reclassification process is to indicate the two rasters that must be reclassified. Then, detail the reclassification criteria

using the "Reclassification table" option. In the window that opens for selecting the reclassification criteria, add two rows using the "Add row" button. Then, introduce the following values (Fig. 14):

That means that all values between −999 and −1 will be reclassified with the value 1. The same will be true for all values between 1 and 999. If as a result of the raster subtraction we get bigger negative values than −999 or bigger positive values than 999 we will need to adjust the values in the reclassification table accordingly.

Step 4

The last step is to count the number of incidents for each pixel over the course of the time series. This is done using the *Raster calculator*, which adds together the rasters we reclassified in the previous step using the following expression:

$$Incidents\_C05\_C11 + Incidents\_C11\_C18\,(Fig.\,5)$$

The CORINE 2005 map will be used as a reference to define the characteristics of the output raster (Fig. 16).

Results and Comments

After completing all the operations described above, two different maps will be obtained: one with the number of states per pixel and another with the number of incidents per pixel.

The above maps (Fig. 17) show the number of incidents and states for a specific part of the Asturias Central Area. Most of the areas that change over the period 2005–2018 underwent just one LUCC transition (one incident and one state). However, we discovered a couple of cases in which there were two incidents and two states. This means that, for

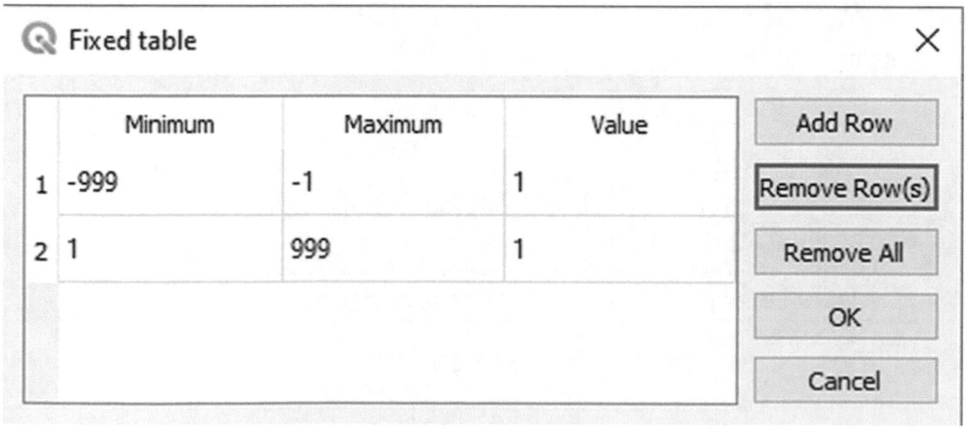

	Minimum	Maximum	Value	
				Add Row
1	-999	-1	1	Remove Row(s)
2	1	999	1	Remove All
				OK
				Cancel

Fig. 14 Exercise 1. Step 3. Reclassification table of the Reclssify by Table tool

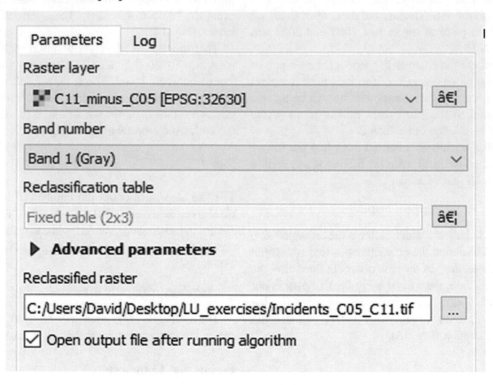

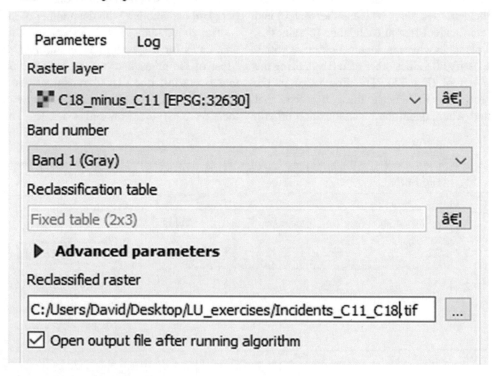

Fig. 15 Exercise 1. Step 3. Reclassify by Table

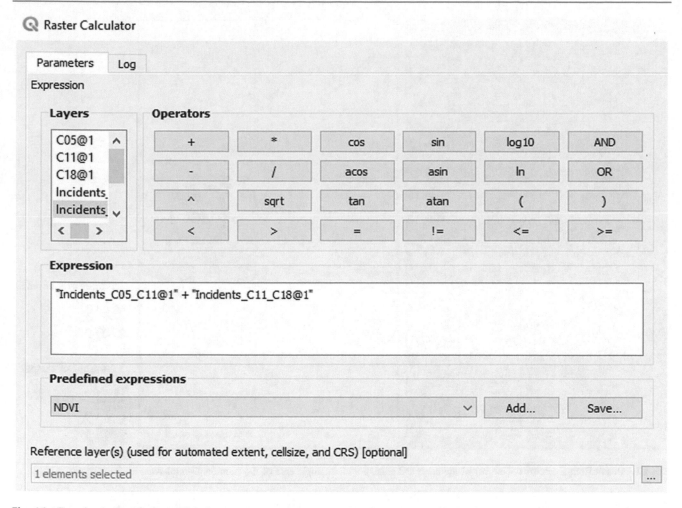

Fig. 16   Exercise 1. Step 4. Raster Calculator

the 3 years analysed (2005, 2011 and 2018), there were two changes or transitions, but these only involved two land uses or covers. In other words, the area changed from its original land use or cover in 2005 to a different one in 2011 and then reverted to the original in 2018.

If we refer back to the original maps, we can identify the transitions that took place. The changing area on the right (1) (Fig. 17) underwent a transition from "Agricultural areas" in 2005 to "Urban fabric" in 2011 and then changed back to "Agricultural areas" in 2018. It is highly unlikely that an agricultural cover could change to an artificial cover

and then revert to its original state a few years later. It must therefore have been an error (technical or spurious change).

The changing area on the left (2) (Fig. 17) underwent a transition from "Agricultural areas" in 2005 to "Vegetation areas" in 2011, before changing back to "Agricultural areas" in 2018. This transition, although unlikely, seems more logical. So, before labelling it as an error or technical change, we should confirm whether these changes really took place in the area in question during the timeframe analysed. This can be done by photointerpretation of aerial imagery.

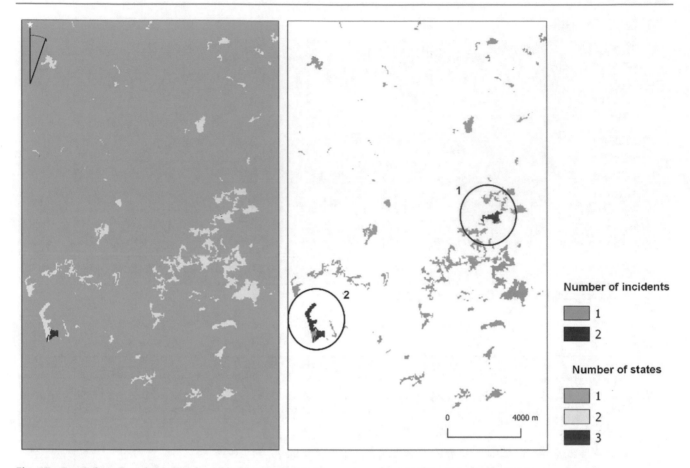

Fig. 17 Result from Exercise 1. Number of incidents and states for an example area of the Asturias Central Area

6 Intensity Analysis

Description

Intensity analysis, proposed by Aldwaik and Pontius (2012), enables us to assess the rate or intensity at which change takes place during each time interval in a time series of LUC maps. It also helps identify apparently random or uniform processes. It is a three-stage analysis process, which identifies: (i) periods of relatively slow/fast change; (ii) relatively dormant/active land use categories and (iii) the transitions that are actively avoided/targeted by a given land use category. A series of maps with three or more time points are needed for this analysis.

During the first stage of this process, the overall rate of land use change over each time interval is analysed to assess whether change was relatively fast or slow. To this end, the average annual rate of change for each time interval is compared with the average annual rate of change for the whole period.

The second stage analyses the intensity of change at category level within each time interval relative to the

overall change rate for the interval calculated in stage one. It measures the gross losses and gross gains in area for each category so as to analyse whether the category shows a similar, stable pattern across the various time intervals in terms of the intensity of gains and losses. These observed intensities for each category are compared with an average annual rate of gains/losses that would exist if the changes within each interval were distributed uniformly over the entire time interval. This shows which categories are relatively dormant or active.

The final stage is at transition level. It examines the intensity of a particular transition over a given time interval, taking into account the different sizes of the categories and relative to the results of the category-level analysis. The gains made by a specific category may vary in size and intensity among the different categories from which it makes these gains. By comparing the observed rate of gains from each category with a uniform rate of gains that would exist if the gains were made uniformly from among all the available categories, we can identify those categories that are intensively avoided or targeted. Losses can be analysed in a similar way.

Intensity analysis also allows us to determine whether a particular transition occurs at a stable rate or occurs more intensely over a particular time interval within the series. If the same category is targeted (or avoided) over all the different time intervals, then this transition is said to be stationary.

Utility

Exercises
1. To validate a series of maps with two or more time points

Intensity analysis analyses the size and intensity of land changes. It also checks for stationarity and takes the relative size of the categories into account, rather than just the absolute gains or losses they may undergo.

At the interval level, users can identify how quickly or slowly LUC change is taking place during each time interval as compared to the average annual rate of change over the whole time series. At the category level, intensity analysis allows users to identify which categories are dormant versus active in terms of gains or losses in the size of each category. At the transition level, when a given category makes gains or losses, users can identify which other categories are most intensively targeted or avoided.

QGIS Exercise

Available tools
• Aldwaik and Pontius matrix (Excel sheet) https://sites.google.com/site/intensityanalysis/ • R Package *Intensity.analysis* • Processing R provider Plugin *Intensity_analysis.rsx* R script

There is not any specific tool available in QGIS to make intensity analysis, although this has been implemented in an R package (intensity.analysis) (Pontius and Khallaghi, 2019). Based on this package, we have developed an R script that allows to integrate this analysis in QGIS. This package will carry out the entire analysis and will generate three tables containing the results at each level of analysis (overall, category and transition) and a plot showing the results at the interval level.

See Chapter "About this Book" for more detailed information about how to integrate R into QGIS and how to use R scripts such as the one applied in this exercise.

Exercise 1. To validate a series of maps with two or more time points

Aim

To study land change in the Ariège study area using the CORINE Land Use maps dated 2000, 2012 and 2018. The results of this exercise can also be used to validate land change.

Materials

CORINE Land Cover Map Val d'Ariège 2000
CORINE Land Cover Map Val d'Ariège 2012
CORINE Land Cover Map Val d'Ariège 2018

Requisites

All maps must be in raster format and have the same resolution, extent and projection.

Execution

If necessary, install the Processing R provider plugin and download the *Intensity analysis.rsx* R script into the R scripts folder (processing/rscripts). See Chapter "About this Book" of this book for further information about how to use the QGIS R script.

Step 1

The land use maps need to be stacked into a multilayer file in chronological order. The first map is the oldest map. The second map is the next oldest and so on. This can be done with the *Merge* tool in the Raster tab.

Step 2

Run the script and fill in the required parameters (path and name of the time-series stack, null value, the path to the folder where the results will be saved, the path and name of the output plot) as shown in Fig. 18.

Results and Comments

The script will generate three files in the results folder: IntervalLevel.csv, CategoryLevel.csv and TransitionLevel.

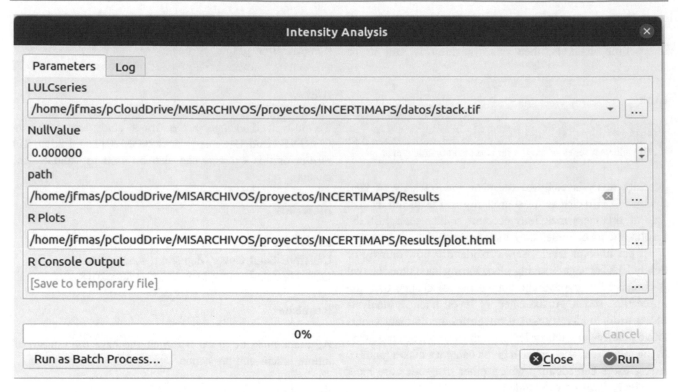

Fig. 18 Exercise 1. Step 2. Intensity Analysis R script

▲	A	B	C	D	E	F	G	H
1	,"Change Size","Annual change","Uniform Change across Intervals","Interval Behavior"							
2	1_2,112424,0.0227104877429948,0.0187202341993798,Fast							
3	2_3,72918,0.0147299806557647,0.0187202341993798,Slow							
4								

Fig. 19 Result. from Exercise 1. Average annual rate of change for each time interval and for the entire period

csv. A plot of the interval level is also produced. Plots of both the category and transition level have to be created from the Excel data sheet.

The first Excel file, called IntervalLevel.csv (Fig. 19), shows the average annual rate of change for each time interval (in this case there are two) and the average annual rate of change for the entire period. When the average rate for each interval is compared with the overall average rate, we can assess whether the interval in question was one of slow or fast change.

The automatically generated plot is shown in Fig. 20. The results show that land use change was more intense in the first time period than in the second. The average change rate over the entire period was 1.8, which means that change was relatively fast over the first period and relatively slow over the second.

The CategoryLevel.csv document (Fig. 21) contains information regarding gross losses and gross gains and the amount of loss intensities and gain intensities for each land use category (in this case there are six categories). If these gains or losses are compared with the average annual rate that would exist if the change within each interval were distributed uniformly over the entire time series, we can see which land categories are relatively dormant/active.

This table may be used to calculate the plots at the category level for each time interval. Figure 22 shows the result for the first time interval.

This figure shows the intensity of change in the different categories, regardless of their relative size within the study area. The categories with short bars to the left of the blue line representing average, uniform intensity are relatively inactive or dormant, whereas those that extend to the right are relatively active. For example, Category 1 showed the highest intensity in terms of land use gains, while Category 4 underwent more intense gains and losses than the average. At the other end of the scale, Category 3 was relatively

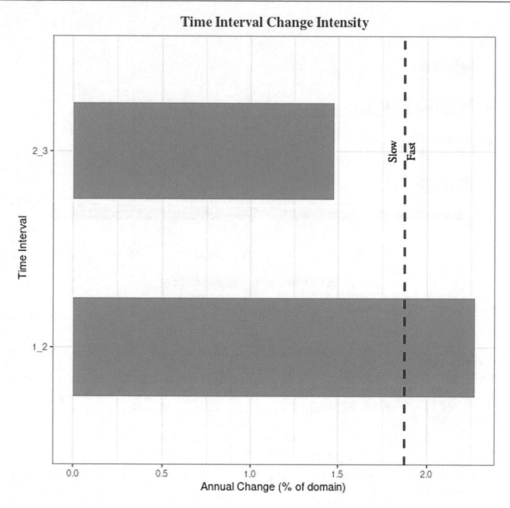

Fig. 20 Result from Exercise 1. Time interval change intensity plot

	A	B	C	D	E	F	G	H	I	J
	Category level Intensity Analysis for interval: 1 - 2									
	,"Gross Loss","Gross Gain","Loss Intensity","Gain Intensity","Uniform Category Intensity","Loss Behavior","Gain Behavior"									
	1,"516","33840","0.00345807420115805","0.185384025419086","0.0227104877429948","Dormant","Active"									
	2,"60134","18721","0.0247443942018108","0.00783700896770504","0.0227104877429948","Active","Dormant"									
	3,"10483","33408","0.00595948030921306","0.0187477798460244","0.0227104877429948","Dormant","Dormant"									
	4,"40844","25480","0.0720383226097754","0.0461918885013379","0.0227104877429948","Active","Active"									
	5,"447","975","0.010780436040903","0.0232187083253953","0.0227104877429948","Dormant","Active"									
	6,"0","0","0","0","0.0227104877429948","Dormant","Dormant"									

Fig. 21 Result from Exercise 1. Gross gains and losses and amount of loss and gain intensities for each category

dormant compared to the other land use categories, as both gain and loss intensity are located to the left of the blue line.

Finally, the TransitionLevel.csv (Fig. 23) shows which transitions are intensively avoided or targeted taking into account the relative size of all the individual categories in the landscape. It compares the observed rate of gains from each category with a uniform rate of gains that would exist if the gains were made uniformly from among all the available

categories, so allowing us to identify those categories that are intensively avoided or targeted. This information may be used to calculate different plots showing the intensity for each transition and time interval.

Figure 24 shows the first level of information in Fig. 23, that is, the annual transition size for gains in Category 1 in the first interval or period of time. The vertical blue line shows the uniform transition intensity. Categories on the left

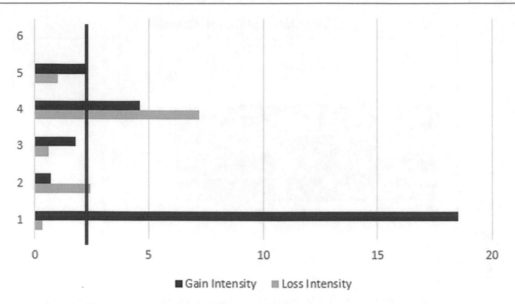

Fig. 22 Result from Exercise 1. Plot of gain and loss intensities per category

⧨	A	B	C	D	E	F	G	H	I	J	K
1	Transition level Intensity Analysis for interval: 1 - 2										
2	,"Annual Transition Size for Gain of 1","Transition Intensity for Gain of 1","Uniform Transition Intensity","Transition Behavior for Gain of 1"										
3	2,"33352","0.0137239338048158","0.00704839061747568","Target"										
4	3,"488","0.000277423103204805","0.00704839061747568","Avoid"										
5	4,"0","0","0.00704839061747568","Avoid"										
6	5,"0","0","0.00704839061747568","Avoid"										
7	6,"0","0","0.00704839061747568","Avoid"										
8											
9	,"Annual Transition Size for Gain of 2","Transition Intensity for Gain of 2","Uniform Transition Intensity","Transition Behavior for Gain of 2"										
10	1,"152","0.0010186575166202","0.00742865872652132","Avoid"										
11	3,"4405","0.00250419829839583","0.00742865872652132","Avoid"										
12	4,"14164","0.0249816570719043","0.00742865872652132","Target"										
13	5,"0","0","0.00742865872652132","Avoid"										
14	6,"0","0","0.00742865872652132","Avoid"										
15											
16	,"Annual Transition Size for Gain of 3","Transition Intensity for Gain of 3","Uniform Transition Intensity","Transition Behavior for Gain of 3"										
17	1,"0","0","0.0104685726605053","Avoid"										
18	2,"7703","0.00316968883720605","0.0104685726605053","Avoid"										
19	4,"25705","0.0453370160288972","0.0104685726605053","Target"										
20	5,"0","0","0.0104685726605053","Avoid"										
21	6,"0","0","0.0104685726605053","Avoid"										
22											
23	,"Annual Transition Size for Gain of 4","Transition Intensity for Gain of 4","Uniform Transition Intensity","Transition Behavior for Gain of 4"										
24	1,"364","0.00243941668453785","0.00581292422027424","Avoid"										
25	2,"19079","0.00785077155978894","0.00581292422027424","Target"										
26	3,"5590","0.00317785890761242","0.00581292422027424","Avoid"										
27	5,"447","0.010780436040903","0.00581292422027424","Target"										
28	6,"0","0","0.00581292422027424","Avoid"										
29											
30	,"Annual Transition Size for Gain of 5","Transition Intensity for Gain of 5","Uniform Transition Intensity","Transition Behavior for Gain of 5"										
31	1,"0","0","0.000198620939169434","Avoid"										
32	2,"0","0","0.000198620939169434","Avoid"										
33	3,"0","0","0.000198620939169434","Avoid"										
34	4,"975","0.00171964950897392","0.000198620939169434","Target"										
35	6,"0","0","0.000198620939169434","Avoid"										
36											

Fig. 23 Results from Exercise 1. Comparison of the observed rate of gains with an uniform rate of gains, differentiating between transitions that are intensively avoided and transitions that are intensivily targeted

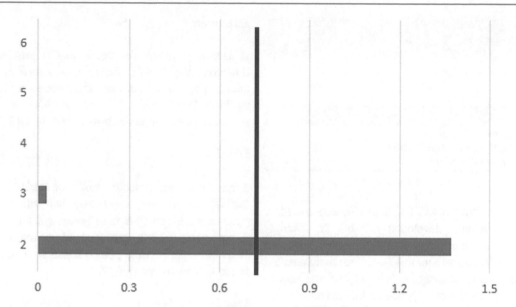

Fig. 24 Result from Exercise 1. Graph with the annual transition size for gains in category 1 in the first period of time

of this line tend to avoid this transition (for example, the change from Category 3 to Category 1) while the categories that extend to the right of the blue line tend to target this transition (for example, the transition from Category 2 to Category 1).

These analyses can also be used to validate land change in a series of maps with two or more time points. If there are large differences at the interval, category and/or transition level between the different time intervals, this means it would be difficult to validate the time series for simulating future trend scenarios, as the intensity of change over the time series has not been sufficiently stable or uniform to provide a base for future predictions. These differences may also be due to errors in the maps, which must be verified.

7 Flow Matrix

Description

The Flow Matrix was developed by Runfola and Pontius (2013) to quantitatively measure the instability of annual land use change between time intervals. The aim was to identify anomalies relative to the total amount of change over the time series. Flow Matrix exercises require a series of maps with at least three time points.

The Flow Matrix is a cross-tabulation matrix that shows the proportion of the study area that transitions from one category to another, excluding persistence. It assumes linear change over time during each time interval. It allows us to calculate: (a) the annual proportion of the study area that

changes during each time interval and (b) the uniform annual proportion of the study area that changes over the entire time series, and the proportion of change that would have to be reallocated to different time intervals in order for change to be perfectly stable (R). When change is perfectly stable, R is zero. This value increases as change becomes more unstable.

A vertical bar chart is produced showing the amount of annual land use change during each time interval as compared to the uniform annual change.

Utility

Exercises
1. To validate a series of maps with two or more time points

The Flow Matrix provides an analysis of the temporal extent at which phenomena are stable. It can be used to find out whether land use change takes place at a uniform rate over the course of the entire study period or if more change takes place during certain intervals. It can also be used to detect errors. If one particular interval is very different from the others in terms of its annual change rate, this may be due to errors in the mapping or the methodology.

The Flow Matrix can also be used in the selection of particular calibration intervals when developing future historical trend simulations, as the data should show the greatest possible uniformity in past land use change. It can also be used to assess whether the results of a trend scenario are consistent, i.e. whether the model simulates much more or much less change than actually happened in the historical series.

QGIS Exercise

Available tools

- Processing R provider Plugin
  Stable_change_flow_matrix.rsx R script
  Flow_matrix_graf.rsx R script

No specific tool is available in QGIS to calculate the Flow Matrix. We have developed two R scripts (*Stable_change_flow_matrix.rsx* and *Flow_matrix_graf.rsx*) to this end. See Chapter "About this Book" for more detailed information about how to integrate R into QGIS and how to use R scripts such as the one applied in this exercise.

The first script will generate two tables in CSV format with the stable and unstable data that would exist for the whole study period, respectively. The second script will generate two tables, in CSV format, presenting the annual change for each interval and the uniform rate, respectively. It also produces a plot showing this annual change and the uniform rate for the entire time series.

Exercise 1. To validate a series of maps with two or more time points

Aim

To study and validate land change in the Ariège Valley study area using CORINE Land Use maps dated 2000, 2012 and 2018.

Materials

CORINE Land Cover Map Val d'Ariège 2000
CORINE Land Cover Map Val d'Ariège 2012
CORINE Land Cover Map Val d'Ariège 2018

Requisites

All maps must be in raster format and have the same resolution, extent and projection.

Execution

If necessary, install the Processing R provider plugin and download the *Stable_change_flow_matrix.rsx* and *Flow_-matrix_graf.rsx* R scripts into the R scripts folder (processing/ rscripts). See Chapter "About this Book" of this book for further information about how to use the QGIS R script.

Step 1

Then, run the stable and unstable change script (*stable_change_flow_matrix.rsx*) and fill in the required parameters: number of time points (in this case, 3), background value (in this case, 0), land use maps and number of years between the time points. Make sure you save the files in the correct folder (Fig. 25).

Step 2

Now, run the Annual Change Rates script (*Flow_matrix_-graf.rsx*). Fill in the parameters as in the previous section (Fig. 25) to generate the plot.

Results and Comments

Step 1

generates two CSV files containing the data regarding unstable change (Fig. 26) and stable change (Fig. 27). The first file shows the proportion of change that would have to be reallocated to different time intervals in order for change to be perfectly stable (R). If change is perfectly stable, then R is zero. The R value increases as change becomes more unstable. In our case, R is 0.06, which means that 6% of change is unstable.

Stable change is the percentage of change that is stable in our study area between the first and the second intervals. This data is used to calculate the R value (R = 1 − stable change). In this case R = 1 − 0.94; R = 0.06.

Step 2

produces a chart showing the annual amount of land use change (expressed as a proportion of the study area) during each time interval and the uniform rate that would exist if the annual changes were distributed uniformly across the entire time period. This is shown as a horizontal line in Fig. 28. It also generates a CSV file showing the uniform change

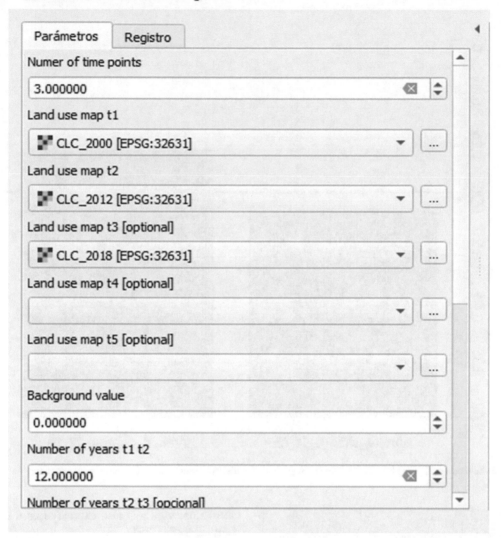

Fig. 25   Exercise 1. Step 1. Stable and unstable change R script

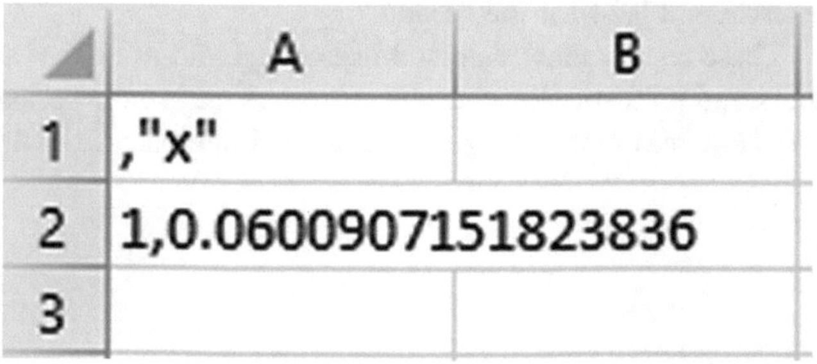

	A	B
1	,"x"	
2	1,0.0600907151823836	
3		

Fig. 26   Result from Exercise 1. Rate of unstable changes

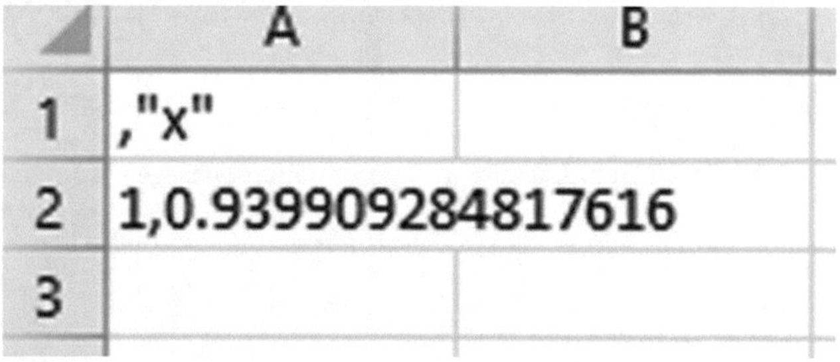

	A	B
1	,"x"	
2	1,0.939909284817616	
3		

Fig. 27   Result from Exercise 1. Rate of stable changes

Annual Change Rate

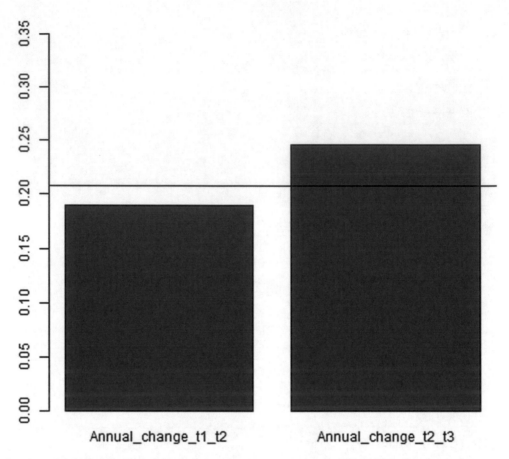

Fig. 28 Result from Exercise 1. Graph with the annual change rate for the two time periods that have been analysed

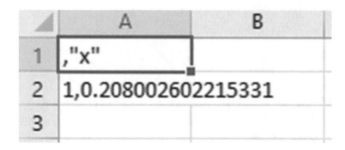

Fig. 29 Result from Exercise 1. Rate of uniform change

calculation, which is also expressed as a proportion of the study area (Fig. 29).

The tool also provides us with data about the annual land use change during each interval, as a percentage of the study area (Fig. 30). In our example, this is 0.19 for the first time interval and 0.24 for the second.

These results show that land use change did not occur at the same uniform rate over the course of the study period and there was more change in the second interval. It should

	A	B	C	D
1	,"Annual_change_t1_t2","Annual_change_t2_t3"			
2	1,0.189254064524957,0.2454996775960079			
3				

Fig. 30 Result from Exercise 1. Annual land change rates for each time period

be noted than if one time interval is very different from the others in terms of the amount of annual change (this did not happen in our case), this may be due to potential mapping errors.

The maps validated here could be used for simulating future trend scenarios, as there is not much difference between the intervals in terms of the annual rate of land use change.

References

Aldwaik SZ, Pontius RG (2012) Intensity analysis to unify measurements of size and stationarity of land changes by interval, category and transition. Landsc Urban Plan 106:103–114

Camacho Olmedo MT, Pontius RG Jr, Paegelow M, Mas JF (2015) Comparison of simulation models in terms of quantity and allocation of land change. Environ Model Softw 69(2015):214–221. https://doi.org/10.1016/j.envsoft.2015.03.003

Hao G, Gen-Suo J (2014) Assessing MODIS land cover products over china with probability of interannual change. Atmos Ocean Sci Lett 7:564–570. https://doi.org/10.1080/16742834.2014.11447225

Paegelow M (2018) LUCC Budget. In: Camacho OM, Paegelow M, Mas JF, Escobar F (eds) Geomatic approaches for modeling land change scenarios. Springer, Lecture notes in geoinformation and cartography, pp 437–440

Paegelow M, Camacho Olmedo MT, Mas J-F, Houet T (2014) Benchmarking of LUCC modelling tools by various validation techniques and error analysis. Cybergeo. https://doi.org/10.4000/cybergeo.26610

Pontius RG Jr, Shusas E, McEachern M (2004) Detecting important categorical land changes while accounting for persistence. Agr Ecosyst Environ 101:251–326

Pontius RG Jr, Malanson J (2005) Comparison of the structure and accuracy of two land change models. Int J Geogr Inf Sci 19:243–265. https://doi.org/10.1080/13658810410001713434

Pontius Jr RG, Boersma W, Castella J-C, Clarke K, de Nijs T, Dietzel C, Duan Z, Fotsing E, Goldstein N, Kok K, Koomen E, Lippitt CD, McConnell W, MohdSood A, Pijanowski B, Pithadia S, Sweeney S, Trung TN, Veldkamp AT, Verburg PH (2008) Comparing input, output, and validation maps for several models of land change. Ann Reg Sci 42(1):11e47

Pontius RG Jr, Millones M (2011) Death to Kappa: birth of quantity disagreement and allocation disagreement for accuracy assessment. Int J Remote Sens 32:4407–4429. https://doi.org/10.1080/01431161.2011.552923

Pontius RG Jr, Krithivasan R, Sauls L et al (2017) Methods to summarize change among land categories across time intervals. J Land Use Sci 12:218–230. https://doi.org/10.1080/1747423X.2017.1338768

Pontius Jr RG et al (2018) Lessons and challenges in land change modeling derived from synthesis of cross-case comparisons. In: Behnisch M, Meinel G (eds) Trends in spatial analysis and modelling. Geotechnologies and the environment, vol 19. Springer, Cham

Pontius Jr RG, Khallaghi S (2019) Intensity of change for comparing categorical maps from sequential intervals, R package intensity.analysis version 1.0.6. https://cran.r-project.org/web/packages/intensity.analysis/intensity.analysis.pdf

Runfola DSM, Pontius RG Jr (2013) Measuring the temporal instability of land change using the flow matrix. Int J Geogr Inf Sci 27:1696–1716. https://doi.org/10.1080/13658816.2013.792344

Validation of Soft Maps Produced by a Land Use Cover Change Model

María Teresa Camacho Olmedo, Jean-François Mas, and Martin Paegelow

Abstract

In Land Use Cover Change (LUCC) modelling, soft maps are often produced to express the propensity of an area to land use change. These maps are generally prepared in raster format, and have values of between 0 and 1, indicating the propensity of each pixel to change. In the literature, they are referred to as suitability, change potential or change probability maps. These maps are sometimes considered as the final product of a model (e.g. map of deforestation risk), but they can also serve as intermediate products that simulate the changes from which a hard-simulated land use/cover map can later be prepared using, for example, a cellular automaton. In both cases, it is essential to evaluate the soft map's ability to identify the areas that are most susceptible to change. One way of assessing this ability is to compare the spatial coincidence between the real changes observed on the ground and the values estimated by the soft map. One would expect real change areas to coincide with high change potential values (near 1) and real no-change areas with low change potential values (near 0). This comparison can be made using various statistical approaches including Correlation Coefficient (Sect. 1), the Receiver Operating Characteristic (ROC) (Sect. 2) and the Difference in Potential (DiP) (Sect. 3). Other measures, such as total uncertainty, quantity uncertainty and allocation uncertainty (Sect. 4), are used exclusively in the analysis of soft maps. In this chapter, we describe the fundamental steps involved in these four statistical approaches to validating the soft maps produced by a model. The four sections are illustrated with specific cases: to validate soft maps produced by the model, to validate soft maps produced by the model against a reference map and to validate soft maps produced by various models against a reference map. We use the Ariège database to validate the different soft maps (change potential and suitability maps) produced by the model by comparing them with real land use maps of the Ariège Valley for two dates (CORINE 2012 and 2018). All these validation techniques are carried out using raster data. As commented earlier, the soft maps produced by the model are continuous, ranked variables. We designed exercises using this original format. In other chapters of this book, the soft maps produced by the model are validated after reclassification of the original maps.

M. T. Camacho Olmedo (✉)
Departamento de Análisis Geográfico Regional y Geografía Física, Universidad de Granada, Granada, Spain
e-mail: camacho@ugr.es

J.-F. Mas
Laboratorio de Análisis Espacial, Centro de Investigaciones en Geografía Ambiental, Universidad Nacional Autónoma de México, Morelia, Mexico

M. Paegelow
Département de Géographie, Aménagement, Environnement, Université Toulouse Jean Jaurès, Toulouse, France

Keywords

Soft maps • Correlation • Receiver Operating Characteristic • Difference in Potential • Uncertainty • Validation

Preliminary QGIS Exercise

Available tools

• Semi-Automatic Classification Plugin
  Tab: Postprocessing
    Section: *Land cover change*

Before beginning the exercises in this chapter, we need to obtain a map of the real transitions that took place between two land use categories (Category 2 to Category 1 and Category 3 to Category 1) between 2012 and 2018. Of all the various tools offered by QGIS (see Sect. 1 in Chapter "Basic and Multiple-Resolution Cross-Tabulation to Validate Land Use Cover Maps", about basic Cross-Tabulation), in this exercise we will be using *Land cover change* from the "Semi-Automatic Classification Plugin".

Exercise 1. To create binary change maps for two transitions

Aim

To create binary change maps for two transitions (2 to 1 and 3 to 1) using CORINE Land Use maps for the years 2012 and 2018. For each transition, each pixel on the map is allocated a value of 1 or 0 depending on whether or not the transition occurred.

Materials

CORINE Land Cover Map Val d'Ariège 2012
CORINE Land Cover Map Val d'Ariège 2018

Requisites

All maps must be raster and have the same resolution, extent and projection.

Execution

Step 1

To create the map of real change, open the "Semi-Automatic Classification Plugin" and, in the tab "Postprocessing", select the option *Land cover change*. Then, fill in the required parameters: the earlier map in the reference classification (CORINE 2012) and the more recent map in the new classification (CORINE 2018). Check the option "Report unchanged pixels".

QGIS then creates an output raster and a table, stored in CSV format, showing all the different combinations observed between the two input maps and the code with which these combinations are represented in the output raster. These combinations (those with 1 or more pixels) and the number of pixels affected by them are presented in Fig. 1.

Step 2

The raster obtained in *Step 1* is reclassified twice: (i) to represent the areas in which a change was observed from 2 to 1 and those in which there was no change and (ii) to do the same for the transition from 3 to 1.

To reclassify the raster, open the *Reclassify by table* tool and allocate a new code 1 to ChangeCode 16 (transition from 2 to 1) and a new code 0 to ChangeCodes 17, 18, 19

ChangeCode	ReferenceClass	NewClass	PixelSum
1	0	0	3,315,666
9	1	1	180,908
10	1	2	1
11	1	3	1
14	1	6	1,630
16	2	1	49,869
17	2	2	2,336,938
18	2	3	1,914
19	2	4	3
21	2	6	70
23	3	1	402
24	3	2	566
25	3	3	1,780,932
26	3	4	71
31	4	2	4,173
32	4	3	14,217
33	4	4	533,222
39	5	3	1
41	5	5	41,991
49	6	6	3,403

Fig. 1 Results from Exercise 1. Step 1. Combinations observed between two input maps, code in the output raster and number of pixel affected

and 21 (pixels that belonged to Category 2 in 2012 but did not change to Category 1 in 2018). All these ChangeCodes are highlighted in the Fig. 1 in green. All the other ChangeCodes (i.e. those with a reference class other than 2 which cannot possibly undergo the transition from 2 to 1 and are therefore considered as No Data) must be allocated a new code −99. Save the output raster as TrueChange2to1.

As regards the transition from 3 to 1, allocate a new code 1 to ChangeCode 23 (pixels that were Category 3 in 2012 and changed to Category 1 in 2018) (in bold type) and a new code 0 to ChangeCodes 24, 25, and 26 (pixels that were Category

3 in 2012 but did not change to Category 1 in 2018). All the candidate ChangeCodes are highlighted in orange in the Fig. 1. A new code −99 must be allocated to the remaining ChangeCodes (i.e. those which cannot undergo this transition). Save the output raster as TrueChange3to1.

In Fig. 2, the areas that changed from 2 to 1 are shown in white, the Category 2 areas that did not change to 1 are shown in grey and the non-candidate areas (i.e. those with a reference class other than 2) appear in black. In Fig. 3, the areas that changed from 3 to 1 are shown in white, the Category 3 areas that did not change to 1 in grey and the non-candidate areas in black.

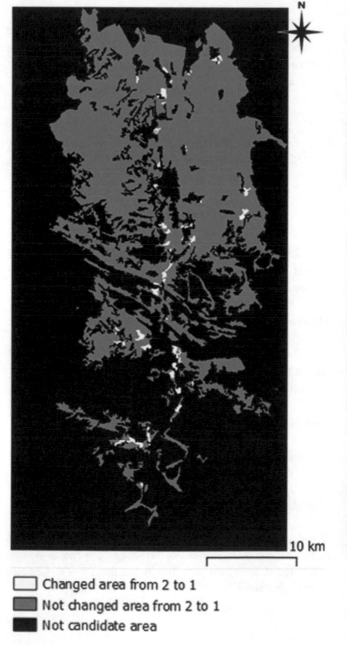

☐ Changed area from 2 to 1

▨ Not changed area from 2 to 1

■ Not candidate area

Fig. 2 Results from Exercise 1. Binary change map for transition from 2 to 1

☐ Changed area from 3 to 1

▨ Not changed area from 3 to 1

■ Not candidate area

Fig. 3 Results from Exercise 1. Binary change map for transition from 3 to 1

1   Correlation

Description

Correlation is a statistical measure that evaluates the extent to which two variables are related. This means that when one variable changes in value, the other variable also tends to change. Correlation coefficients are quantitative metrics that measure both the strength and the direction of this tendency of two variables to vary together.

The correlation coefficients range from 1 to −1. A coefficient of 1 shows a perfect positive correlation, while a coefficient close to zero indicates that there is no relationship between the variables. A coefficient of minus 1 indicates a perfect negative correlation, that is, as one variable increases, the other decreases.

The Pearson correlation measures the linear correlation between two variables. Spearman's correlation is the non-parametric version of the Pearson correlation and is based on the rank order of the variables rather than on their values. Spearman's correlation is often used to evaluate non-linear relationships or relationships involving ordinal variables.

Utility

Exercises
1. To validate soft maps produced by the model against a reference map of changes

Correlation analysis is useful for making a rapid assessment of a soft map expressing the propensity of an area to change. Assuming that the change map is coded 0 for no change and 1 for change, we would expect a positive value close to 1 if the soft map is successfully attributing high change potential values to change areas and low change potential values to no-change areas. A correlation coefficient of 0 indicates a completely random model. A negative coefficient indicates that the model is making incorrect predictions in that it produces soft maps in which low change potential values are awarded to areas in which changes are in fact taking place (Bonham-Carter 1994; Camacho Olmedo et al. 2013).

We used Pearson and Spearman correlations to assess the correlation between the map showing real changes and its respective change potential map. The correlation between a binary variable and a continuous variable is known as a Point-Biserial Correlation and measures the strength of association between the two.

QGIS Exercise

Available tools
• Processing R provider plugin *Correlation.rsx* R script

We have created an R script to calculate in QGIS the Pearson and Spearman correlation coefficients. This script performs a sampling of the images and calculates both the Pearson and Spearman correlations.

Exercise 1. To validate soft maps produced by the model against a reference map of changes

Aim

To calculate the correlation between the map showing real change from 2 to 1 and the corresponding map of potential change.

Materials

TrueChange2to1 (calculated in the preliminary QGIS exercise in this chapter)
Transition potential map from agricultural to artificial areas

Requisites

All maps must be in raster format and have the same resolution, extent and projection.

Execution

If necessary, install the Processing R provider plugin and download the *Correlation.rsx* R script into the R scripts folder (processing/rscripts). For more details, see Chapter "About this Book".

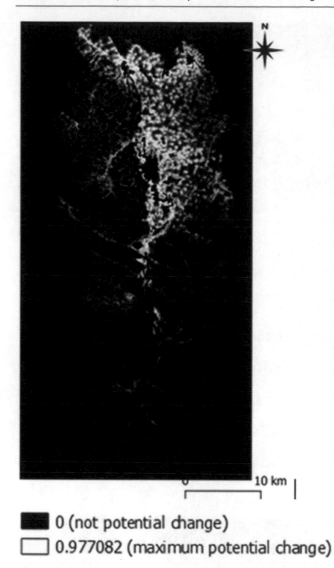

■ 0 (not potential change)

☐ 0.977082 (maximum potential change)

Fig. 4 Exercise 1. Initial map: Change potential map from 2 to 1

Initial maps

The initial maps for comparison are the TrueChange2to1 map (see the preliminary QGIS exercise in this chapter) and the map showing the change potential from 2 to 1 (Fig. 4). Values range from 0 (black) to 0.977082 (white), the value for the areas with maximum change potential. The areas with a value of 0 (black) are those in which there is no potential for change.

Step 1

Run the script and fill in the required parameters (names of the two maps, proportion of pixels to be sampled, Null value) as shown in Fig. 5. The Null value enables us to exclude part of the image from the calculations, for instance, the pixels with no potential for change.

The script samples the images in order to speed up the computing of the correlation coefficients. It then displays both the Pearson and Spearman correlation coefficients and a scatterplot in the log files (Figs. 6 and 7).

Results and Comments

As can be seen in Fig. 6, both maps show a low positive correlation (Pearson = 0.13, Spearman = 0.12), which means that the real changes tend to occur more frequently in the areas with higher change potential values. However, as can be seen in Fig. 7 and by the low value of the coefficients, the difference between the potential values for change and no-change areas is quite small.

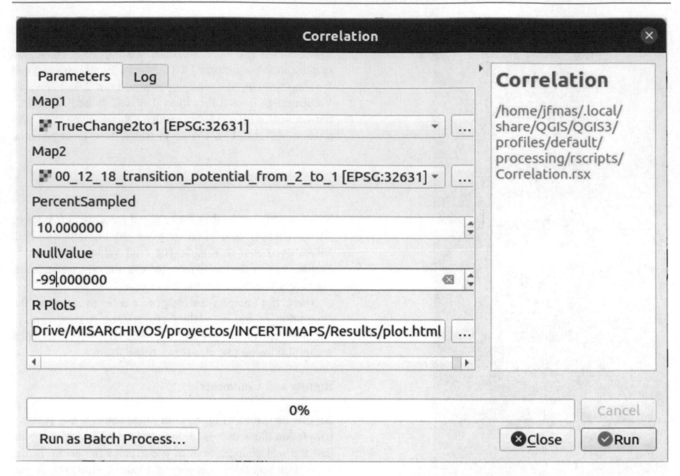

Fig. 5 Exercise 1. Step 1. Correlation R script

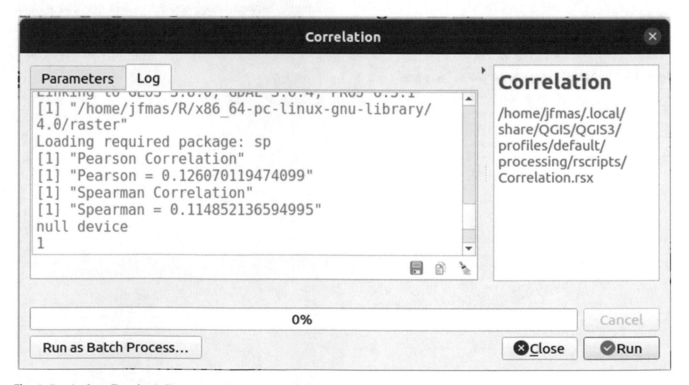

Fig. 6 Results from Exercise 1. Pearson and Spearman correlation coefficients

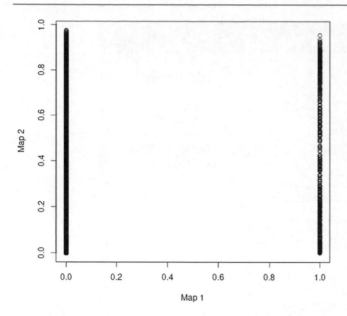

Fig. 7 Result from Exercise 1. Scatterplot

2 Receiver Operating Characteristic (ROC)

Description

Receiver Operating Characteristic (ROC) analysis enables users to evaluate binary classifications with continuous output or rank-order values. In spatial modelling, ROC analysis is used to assess soft maps such as probability or suitability maps, which present the sequence in which the model selects cells to determine the occurrence of binary events, such as change versus no change (Camacho et al. 2013). The probability map can be compared with the observed binary event map so as to assess the spatial coincidence between the event and the probability values. An accurate predictive model would produce a probability map in which the highest ranked probabilities coincide with the actual event.

ROC applies thresholds to the probability map to produce a sequence of binary predicted event maps and assess the coincidence between predicted and real events. A curve is obtained in which the horizontal axis represents the false positive rate (proportion of no-event cells modelled as an event) and the vertical axis the true positive rate (proportion of true event cells modelled as an event).

A standard metric based on the ROC curve is the area under the curve (AUC). If the actual events coincide perfectly with the highest ranked probabilities, then the AUC is equal to one. A random probability map produces a curve in which the true positive rate equals the false positive rate at all threshold points, and AUC is therefore 0.5. Probability

maps that produce a ROC curve below the diagonal (AUC < 0.5) have less predictive accuracy than a random map (Mas et al. 2013; Pontius and Parmentier 2014).

Utility

Exercises
1. To validate soft maps produced by the model against a reference map of changes

The main application of ROC analysis in spatial modelling is in the assessment of maps that predict events such as land use/cover change, species distribution, disease and disaster risks.

QGIS Exercise

Available tools
• Processing R provider plugin • ROCR package   *ROCAnalysis.rsx* R script

QGIS does not provide any tool for ROC analysis, although R provides several packages to this end. We implemented the *ROCAnalysys.rsx* R script in QGIS using the QGIS Processing R provider plugin and the ROCR package to plot the ROC curve and calculate the AUC (Sing et al. 2005). This script resamples the images to reduce the number of observations and carry out the standard ROC analysis.

Exercise 1. To validate soft maps produced by the model against a reference map of changes

Aim

To assess the accuracy of a change potential map using ROC analysis.

Materials

TrueChange2to1 (calculated in the preliminary QGIS exercise in this chapter)

Transition potential map from agricultural to artificial areas

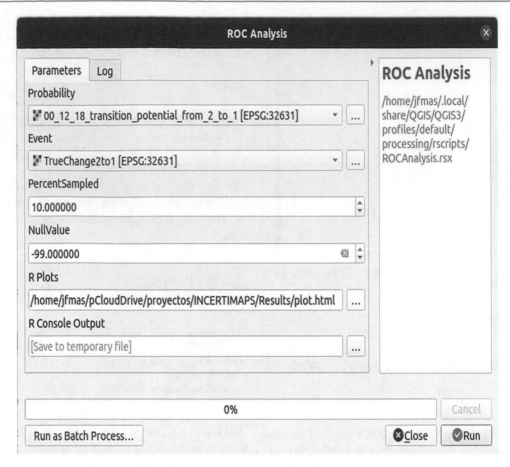

Fig. 8 Exercise 1. Step 1. ROC Analysis R script

Requisites

All maps must be in raster format and have the same resolution, extent and projection.

Execution

If necessary, install the Processing R provider plugin, and download the *ROCAnalysis.rsx* R script into the R scripts folder (processing/rscripts). For more details, see Chapter "About this Book".

Step 1

Then run the script and fill in the required parameters (Fig. 8): Probability map is a soft prediction map for the event; Event is a binary map that indicates the occurrence, or not, of the event. This map can have NullValue cells for the areas that are not affected by the prediction.

The maps have large numbers of both "event" and "non-event" cells, although there are normally more "event" cells than "non-event" cells. The PercentSampled parameter uses random sampling to reduce the number of non-event cells observed.

Results and Comments

The script carries out a sampling of the cells, plots the ROC curve and calculates the AUC. The ROC curve (Fig. 9) is saved, and the AUC value is displayed in the R console.

We assessed the change potential map for the transition from Category 2 (agriculture) to Category 1 (built-up) using ROC analysis. An AUC of 0.74 was obtained. We can therefore conclude that this predictive map was reasonably successful at identifying the agricultural areas that were most likely to be converted to built-up over the period 2012–18.

3 Difference in Potential (DiP)

Description

DiP is based on the Peirce Skill Score (PSS):

$$PSS = H - F$$

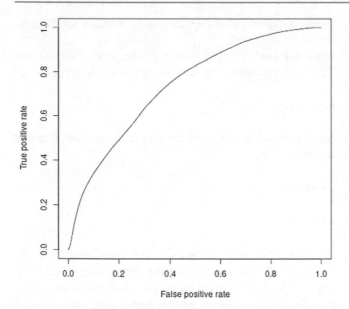

Fig. 9 Results from Exercise 1. The ROC curve

where H = HITS, i.e. pixels in which both the real maps and the simulation show change and F = FALSE ALARMS, i.e. pixels in which the real maps show persistence but the simulation shows change.

In DiP, proposed by Eastman et al. (2005), the simulation maps are soft (change potential, suitability maps…) rather than hard maps. DiP therefore compares the relative weight of the potential (in a generic sense) that is allocated to areas that changed, i.e. HITS, and the relative weight of the potential allocated to areas that did not change, i.e. FALSE ALARMS. Results are normally between 1.0 (perfect predictor) and 0 (prediction no better than random). Negative values are also possible (prediction systematically incorrect). In other words, DiP is defined as the difference between the mean potential in the change areas and the mean potential in the no-change areas (Pérez-Vega et al. 2012).

Utility

Exercises
1. To validate soft maps produced by the model against a reference map of changes
2. To validate soft maps produced by various models against a reference map of changes

DiP is used as a tool for validating soft maps in a modelling exercise, by assessing their predictive accuracy. Users can validate and compare several soft maps simulated by the same model or several soft maps simulated by different models. Pérez-Vega et al. (2012) validated a map of overall change potential created by superimposing all the potential maps produced by a model.

As these soft maps are rank-order indices, but real land use typically includes a categorical legend, we would expect each real category or transition to be allocated where the values are highest in soft-classified maps, whereas other categories or transitions would be allocated where the values are lower. The validation methods therefore have to compare a rank image with a Boolean image in which the real category or transition is located.

Compared to other assessment techniques such as ROC analysis (see previous section), which is based on a relative threshold, DiP analysis is a measure of absolute threshold. As Eastman et al. (2005) suggested, PSS, DiP and similar procedures could be used in models based on absolute performance, while ROC could be used in models based on relative performance. DiP and ROC present a different picture, in that in DiP the results show greater variability between the potential maps and models.

QGIS Exercises

Available tools

• Processing Toolbox
   Raster Analysis
      Raster layer zonal statistics
• LecoSPlugin
   Landscape statistics
      Zonal statistics

The Difference in Potential is a simple subtraction between average values from two maps. The required functions are related to zonal statistics which is why in these exercises we will be using the *Raster layer zonal statistics* tool.

Exercise 1. To validate soft maps produced by the model against a reference map of changes

Aim

To validate and compare two change potential maps (soft maps), obtained from the same model, against a reference map-CORINE Land Use map of real changes (from 2012 to 2018).

Materials

TrueChange2to1 (calculated in the preliminary QGIS exercise in this chapter)

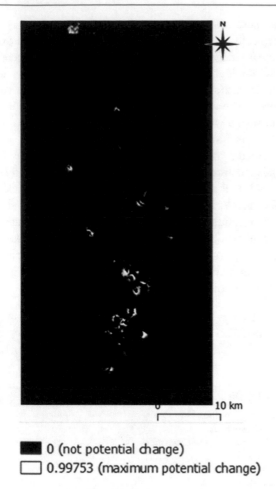

Fig. 10 Exercise 1. Initial map: change potential map from 3 to 1

TrueChange3to1 (calculated in the preliminary QGIS exercise in this chapter)

Transition potential map from agricultural to artificial areas

Transition potential map from forests to artificial areas

Requisites

All maps must be raster and have the same resolution, extent and projection.

Execution

Initial maps

In this exercise, we will be using the TrueChange2to1 and the TrueChange3to1 maps (see the preliminary QGIS exercise in this chapter), the change potential map from 2 to 1 (see Sect. 1) and the change potential map from 3 to 1 (Fig. 10). Values are from 0 (black) to 0.99753 (white), the latter corresponding to the areas with the maximum potential for change. The areas in which this change is not predicted are allocated a value of 0.

Step 1

We open *Raster layer zonal statistics* (located in the Processing Toolbox) to extract the mean values from the change potential map from 2 to 1 (Input layer) using the TrueChange2to1 map as the Zones layer (Fig. 11).

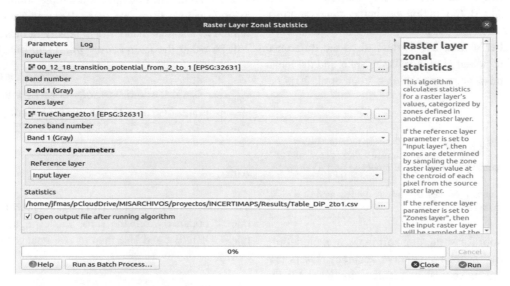

Fig. 11 Exercise 1. Step 1. Raster layer zonal statistics

Step 2

We then repeat the exercise with the change potential map from 3 to 1 (Input layer) using the TrueChange3to1 map as the Zones layer.

Results and Comments

The mean value for change potential from 2 to 1 in the candidate areas that actually change to Category 1 is 0.43, while in the candidate areas that did not change, the mean value is 0.20. Therefore, the Difference in Potential is 0.23. In spite of the fact that the change potential is twice as high in the areas that changed to Category 1 than in those that did not change, the absolute potential (about 0.43) is quite low.

As regards the change from 3 to 1, the mean value for change potential in the candidate areas that change to Category 1 is 0.31, while in the candidate areas that did not change, the mean value is 0.02. Therefore, the Difference in Potential is 0.29. In spite of the fact that the absolute difference is quite low, it is important to highlight that the change potential value in the candidate areas that did not change is almost zero. From this point of view DiP throws up interesting results.

The fact that these soft maps have similar DiP values means that they have similar predictive capacity. This is slightly higher in the map charting potential change from 3 to 1, although we should also bear in mind that the change from 3 to 1 affects just one small, contiguous area.

> **Exercise 2. To validate soft maps produced by various models against a reference map of changes**

Aim

To validate and compare two soft maps obtained from two different models against a reference map-CORINE Land Use map of real changes (from 2012 to 2018).

Materials

TrueChange2to1 (calculated in the preliminary QGIS exercise of this chapter)
Transition potential map from agricultural to artificial areas
Markovian probability map for artificial areas Ariège Valley

Requisites

All maps must be raster and have the same resolution, extent and projection.

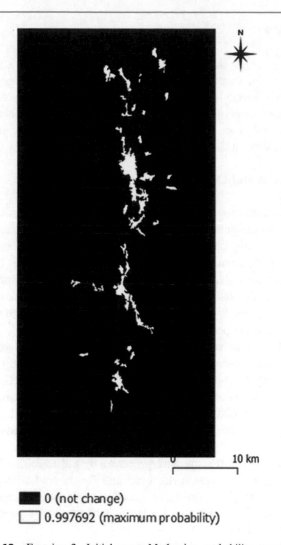

0 (not change)
0.997692 (maximum probability)

Fig. 12 Exercise 2. Initial map: Markovian probability map for Category 1

Execution

Initial maps

In this exercise, we will be using the TrueChange2to1 (see the preliminary QGIS exercise in this chapter), the change potential map from 2 to 1 (see Sect. 10.1) and the Markovian probability map for Category 1 (Fig. 12), with values from 0 (black) to 0.997692 (white), the latter corresponding to the areas with the highest probability to be Category 1.

Step 1

In order to obtain the mean values from the change potential map for the transition from 2 to 1, follow the process set out in Exercise 1 of this section.

Step 2

We now use the *Raster layer zonal statistics* tool to extract the mean values from the probability map for Category 1 (Input layer) using the TrueChange2to1 map as zones (Zones layer). In other words, in both soft maps (change potential map and probability map) we extract the mean values using the same map as zones.

Results and Comments

As commented in Exercise 1 of this section, the mean value for change potential from 2 to 1 in the candidate areas that did actually change to Category 1 is 0.43; while in the candidate areas that did not change, the mean value is 0.20. This means that the Difference in Potential is 0.23. In spite of the fact that the change potential is twice as high in the areas that changed to Category 1 than in those that did not change, the absolute potential (about 0.43) is quite low.

The mean value for the probability of Category 1 in the candidate areas that did actually change from Category 2 to 1 is 0.013, while in the candidate areas that did not change, the mean value is 0.0098. The Difference in Potential is therefore 0.0032. This very small difference means that the only Markovian-generated probability map has no predictive value.

The two soft maps, each generated by a different model to predict the changes in land use and cover, produce highly varying results: some areas considered to have high change potential by one model are attributed low change potential by the other.

In this case, it is important to remember that we are comparing two quite different change potential maps. Firstly, a change potential map in which only one specific transition is evaluated (in this case from 2 to 1) and therefore only one source category (Category 2) is considered for its potential for change to the target category (Category 1). Secondly, a suitability map, which generates the probability of any part of the study area belonging to a particular target category (in this case Category 1) at the end of the period regardless of its original source category. However, when comparing the outputs of these models, we evaluated the same transition in both soft maps and validated them against the same real change.

The second main difference is that the change potential map is based not only on two LUC maps but also on selected drivers, while the Markov Probability map is computed without additional knowledge (drivers). The conclusion is that when comparing different maps, it is important to bear in mind that the data may have been obtained in different ways.

4 Total Uncertainty, Quantity Uncertainty, Allocation Uncertainty

Description

In an exhaustive state of the art on the accuracy of model outputs, Krüger and Lakes (2016) proposed an uncertainty measurement for probability maps such as soft predictions, which could be considered equivalent to the disagreement indices for hard prediction maps.

These authors proposed a measurement of the probability of predictions being misses (PM) (also called omissions) or false alarms (PF) (also called commissions) for soft prediction maps. They also introduced three uncertainty measures:

$$QU \; \text{Quantity uncertainty} = 2 \times (PM - PF)$$
$$AU \; \text{Allocation uncertainty} = 4 \times (PF)$$
$$TU \; \text{Total Uncertainty} = QU + AU$$

where PM = the average for the values less than 0.5 (pixel values equal to or higher than 0.5 are previously set to zero); PF = average of soft prediction map where values less than 0.5 are set to zero while values equal to or higher than 0.5 are converted into their complement to 1 (0.8 becomes 0.2; 0.51 becomes 0.49).

Utility

Exercises
1. To validate soft maps produced by the model

The uncertainty indices proposed by Krüger and Lakes (2016) for probability maps such as soft predictions are equivalent to disagreement indices for hard classified maps such as hard predictions. Theses indices allow us to evaluate the uncertainty of soft predictions by comparing the level of uncertainty in soft prediction outputs with the level of disagreement in hard prediction outputs.

QGIS Exercise

Available tools
• Raster
Raster Calculator
Raster Layer Statistics

There is not any specific tool implemented in QGIS or R that allows to directly calculate the uncertainty indices proposed

Fig. 13   Exercise 1. Initial map: Soft prediction map

by Krüger and Lakes (2016). However, these can be easily obtained through common spatial analysis tools, such as *Raster Calculator* and *Raster Layer Statistics*.

Exercise 1. To validate soft maps produced by the model

Aim

To validate the soft map produced by the LCM model for the Ariège Valley case study.

Materials

Soft prediction LCM Val d'Ariège 2018

Requisites

The map must be raster.

Execution

Initial maps

Figure 13 shows the generated soft prediction map (2018) generated by Land Change Modeler (LCM) for Ariège Valley, based on CLC 2000 and CLC 2012 training dates. The map values range from 0, which means minimal probability to change, to 1, which means maximal probability to change.

Step 1

To calculate the PM map (probability of being a miss), we use the Raster Calculator twice. First, we generate an intermediate map in which all pixel values less than 0.5 are coded as 1: calculator expression = "CLC_predict_2018_-soft_UTM@1" < 0.5. Then, we multiply this mask (intermediate map named "TMP_1") by the soft prediction map: calculator expression = "TMP_1@1" * "CLC_predict_2018_soft_UTM@1". As a result, we obtain the PM map (Fig. 14).

Step 2

To calculate the PF (probability of being a false alarm) map, we need to use the *Raster Calculator* again. With the calculator, we can first compute an intermediate map in which all pixel values equal to or greater than 0.5 are coded as 1: calculator expression = "CLC_predict_2018_soft_UTM@1" > = 0.5. Then, we subtract the values of the soft prediction map from 1 before multiplying it by the mask (intermediate map, here named "TMP_2"): calculator expression = (1-"CLC_predict_2018_soft_UTM@1") * "TMP_2@1". As a result, we obtain the PF map (Fig. 15).

Step 3

Finally, we use the *Raster Layer statistics* tool to calculate the average PM and PF values from the corresponding maps.

$$2018\_PM_{average} = 0.00963$$
$$2018\_PF_{average} = 0.00577$$

Step 4

Once we have obtained the PM and PF values, we can calculate the Quantity Uncertainty (QU), Allocation Uncertainty (AU) and Total Uncertainty (TU) following the formulas provided by Krüger and Lakes (2016):

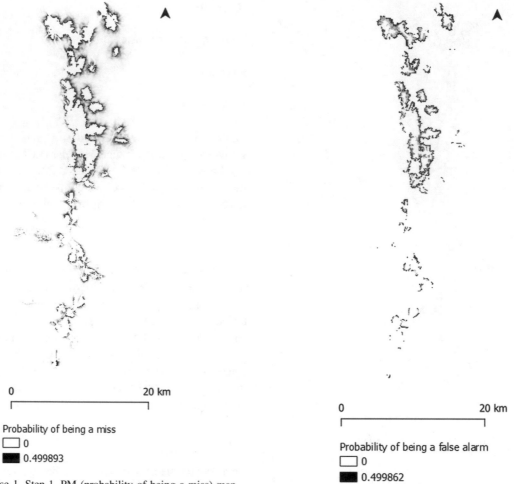

Fig. 14 Exercise 1. Step 1. PM (probability of being a miss) map

Fig. 15 Results from Exercise 1. Step 2. PF (probability of being a false alarm) map

$$QU = 2 \times (0.00963 - 0.00577) = 0.00772$$
$$AU = 4 \times (0.00577) = 0.02308$$
$$TU = 0.00772 + 0.02308 = 0.0308$$

Results and Comments

All three uncertainty indices are very low because only a small proportion of the pixels change category. The soft prediction map (Fig. 13) indicates that persistence is the dominant trend and there are very few high-probability soft-predicted changes.

For this dataset, quantity uncertainty is about three times lower than allocation uncertainty. It is important to bear in mind that areas with low rates of change also have lower uncertainty rates, so limiting the significance of these indices.

References

Bonham-Carter GF (1994) Tools for map analysis: map Pairs. In: Bonham-Carter GF (ed) Geographic information systems for geoscientists. Pergamon, pp 221–266. ISBN 9780080418674. https://doi.org/10.1016/B978-0-08-041867-4.50013-8

Camacho Olmedo MT, Paegelow M, Mas J-F (2013) Interest in intermediate soft-classified maps in land change model validation: suitability versus transition potential. Int J Geogr Inf Sci 27 (12):2343–2361. https://doi.org/10.1080/13658816.2013.831867

Clarklabs (2021) https://clarklabs.org/terrset/land-change-modeler/

Eastman JR, Van Fossen ME, Solarzano LA (2005) Transition potential modelling for land cover change. In: Maguire D, Goodchild M, Batty M (eds) GIS, spatial analysis and modeling. ESRI Press, Redlands, California

Krüger C, Lakes T (2016) Revealing uncertainties in land change modeling using probabilities. Trans GIS 20:526–546. https://doi.org/10.1111/tgis.12161

Mas J-F, Soares-Filho BS, Pontius RG Jr, Gutiérrez MF, Rodrigues H (2013) A suite of tools for ROC analysis of spatial models. ISPRS Int J Geo Inf 2(3):869–887. https://doi.org/10.3390/ijgi2030869

Pérez-Vega A, Mas JF, Ligmann-Zielinska A (2012) Comparing two approaches to land use/cover change modeling and their implications for the assessment of biodiversity loss in a deciduous tropical forest. Environ Model Softw 29(1):11–23. https://doi.org/10.1016/j.envsoft.2011.09.011

Pontius RG Jr, Parmentier B (2014) Recommendations for using the relative operating characteristic (ROC). Landscape Ecol 29(3):367–382. https://doi.org/10.1007/s10980-013-9984-8

Sing T, Sander O, Beerenwinkel N, Lengauer T (2005) ROCR: visualizing classifier performance in R. Bioinformatics 21 (20):7881. https://doi.org/10.1093/bioinformatics/bti623

Spatial Metrics to Validate Land Use Cover Maps

David García-Álvarez and Martin Paegelow

Abstract

When validating Land Use Cover (LUC) products, pattern analysis can be used to assess the agreement between the patterns of two maps. It therefore complements other methods and techniques that focus exclusively on the quantity (proportions) and allocation agreement between the categories. Spatial metrics are the first step for any analysis of the patterns of categorical maps. With the wide range of spatial metrics available, it is possible to fully characterize the pattern of any map. It can also be characterized in greater detail using other more complex techniques, as explained in the next chapter of this book (Chap. "Advanced Pattern Analysis to Validate Land Use Cover Maps"). This chapter provides an introduction to the essentials of pattern analysis by explaining the theory behind the calculations of spatial metrics. To this end, we offer examples of how to use spatial metrics to validate LUC maps (either single maps or series) and Land Use Cover Change (LUCC) simulations from modelling exercises. We also include two example exercises illustrating how spatial metrics can be used for general purposes of pattern characterization without validation. Despite all the spatial metrics currently available, in this chapter we will be focusing exclusively on the most common and most suitable metrics for carrying out the type of analyses being performed here. Most of the spatial metrics proposed in the literature are closely related. This means that users must select the metrics that provide most information for their specific cases, so as to avoid reiteration and make sure that clear conclusions are reached. The example exercises were drawn up with maps (CORINE, SIOSE) and modelling exercises from the Asturias Central Area and Ariège Valley databases.

Keywords

Pattern analysis • Landscape Ecology • Landscape Metrics • Spatial Metrics

1 Spatial Metrics

Description

Spatial metrics are a set of indices or metrics that were first developed within the field of landscape ecology (Forman 1995), which is why they are often referred to as landscape metrics. Landscape metrics were designed to quantitatively characterize the pattern of a landscape, and its relationship with landscape processes. Nowadays, they are also widely used to characterize the pattern of categorical maps. When used for this purpose, they are generally referred to as spatial metrics (Herold et al. 2005).

Spatial metrics were initially developed for raster data, although some of them have also been adapted for calculation with vector data, for which the polygon is the unit of measurement. For raster data, the reference concept for calculating the metrics is the patch.

A patch is defined as a contiguous area of pixels belonging to the same category. The number and shape of the patches in a raster will depend on the neighbourhood rule applied (Fig. 1). Under the 4-cell neighbourhood rule, two pixels with the same value are considered to belong to the same patch if one is immediately above, below or adjacent to another pixel. An 8-cell neighbourhood will also consider pixels that are diagonal to each other as part of the same patch.

Spatial metrics can be calculated at three different levels: per patch, per category or for the whole map (landscape

D. García-Álvarez (✉)
Departamento de Geología, Geografía y Medio Ambiente, Universidad de Alcalá, Alcalá de Henares, Spain
e-mail: David.garcia@uah.es

M. Paegelow
Département de Géographie, Aménagement, Environnement, Université Toulouse Jean Jaurès, Toulouse, France

D. García-Álvarez et al. (eds.), *Land Use Cover Datasets and Validation Tools*,
https://doi.org/10.1007/978-3-030-90998-7_11

Fig. 1 Examples of patch
delineation according to 4-cell
and 8-cell neighbourhood rules
for an example landscape

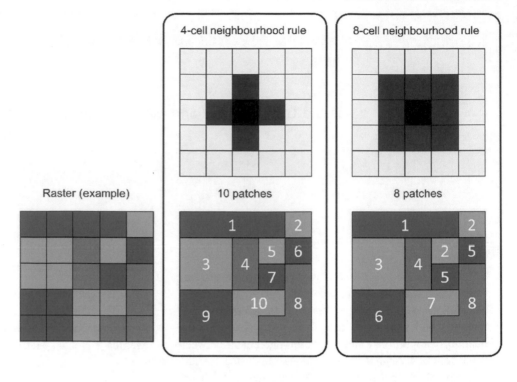

level). In the first case, each metric is calculated for every single patch. In the second case, the metrics are calculated for all the patches belonging to every single category on the map. In the last case, the metrics are calculated for the map as a whole. Not all metrics can be calculated for the three levels of analysis, but some of them are only available for certain levels of analysis.

There is a wide variety of metrics available, and new ones are regularly being proposed (Botequilha Leitao et al. 2006; Jaeger 2000; Mcgarigal 2018). Most of them are closely correlated. This means that despite the wide number of metrics available, many of them provide the same or very similar information.

Spatial metrics are usually classified into groups according to the information they provide: area, density and edge metrics; shape metrics; connectivity metrics and diversity metrics. The first group (area, density and edge) provides information about the area and perimeter of the patches. Shape metrics assess the complexity of the shape of the patches, based on their area and perimeter, while connectivity metrics quantify the degree to which patches relate to each other (how connected they are) and are usually calculated at the category level. Finally, diversity metrics quantify the heterogeneity of the map and can only be computed at a landscape level.

For an overview of the range of metrics available and a description, please see Botequilha Leitao et al. 2006, Jaeger 2000; Mcgarigal 2018.

Utility

Exercises
1. To validate a map against reference data / map
2. To validate a simulation against a reference map
3. To validate simulated changes against a reference map of changes
4. To validate a series of maps with two or more time points
5. To validate a series of maps with two or more time points (vector)
6. To validate a series of maps with two or more time points (raster)

Spatial metrics are some of the most popular tools for analysing the pattern of categorical maps. Using the wide diversity of spatial metrics currently available, we can obtain numerous quantitative measurements of the fragmentation, shape complexity and heterogeneity of the landscape.

Spatial metrics can be calculated for the whole map or for certain specific features. In the case of Land Use Cover Change analyses, including LUCC modelling, spatial metrics can be specifically used to characterize the pattern of the elements that change.

Spatial metrics are usually highly dependent on the scale of analysis (Šímová and Gdulová 2012). Scale refers not only to the cartographic scale at which the map was drawn but also to its spatial and thematic resolutions. This makes them useful for evaluating the impact of changes in the scale on the way a landscape is represented on a map. They can also be used to assess the impact of resampling categorical maps. However, this also makes them very uncertain tools when comparing two or more maps that have different resolutions or were obtained at different scales. In these cases, the results must be treated with caution.

For maps at the same scale, spatial metrics can be used to assess to what extent their patterns differ. In other words, they assess the relative complexity of their shapes and perimeters, the degree to which they are fragmented, or how close patches belonging to the same categories are to each other.

QGIS Exercises

As mentioned earlier, there are a lot of spatial metrics available, many of which are highly correlated. Despite this, there is a wide range of different metrics that characterize map patterns in different ways.

It would be impossible to present example exercises for all the available spatial metrics in the literature, as there would be enough material to fill an entire book. This is why, in the exercises proposed here, we focus on the metrics most commonly used for validating maps or analysing their uncertainties. These metrics are also suitable for many other exercises that users may typically wish to perform. However, they should be aware that other metrics are available which may be more suitable or useful in certain specific cases.

Available tools

- Raster
  - Landscape ecology
    - *Landscape statistics*
    - *Landscape vector overlay*
- Processing toolbox
  - LecoS
    - Landscape modifications
      - *Neighbourhood Analysis (Moving Window)*
    - Landscape statistics
      - *Count raster cells*
      - *Landscape-wide statistics*
      - *Patch statistics*
      - *Zonal statistics*
    - Landscape vector overlay
      - *Overlay raster metrics (Polygon)*
- Processing toolbox
  - SAGA
    - Raster analysis
      - *Pattern analysis*
- Processing toolbox
  - SAGA
    - Vector polygon tools
      - *Polygon shape indices*

Despite the widespread use of spatial metrics, QGIS offers few tools for calculating them. For vector maps, we have the *Polygon shape indices* tool, which characterizes the area, perimeter and shape compactness of polygons. Metrics are calculated for each polygon, i.e. at patch level.

Of the tools available for raster maps, we highlight two: the *LecoS* plugin (Jung 2016) and the SAGA "Pattern analysis" tool. GRASS also provides a suite of tools for calculating spatial metrics: *r.li* tools. However, there are certain problems with their integration in the QGIS environment that prevent their normal use. This is why we have not considered them as an option for calculating spatial metrics in this book.

The SAGA tool only allows the user to calculate a few metrics (relative richness, diversity, dominance, fragmentation, number of different classes, centre versus neighbours), although these are not amongst the most frequently used when comparing map patterns. These metrics can only be calculated for the entire landscape or study area and are not available at patch or class level. In addition, although the user may select the window at which the spatial metrics are calculated (3×3, 5×5 or 7×7), the 8-cell neighbourhood rule is applied by default and cannot be changed.

The "LecoS" plugin offers a wider set of metrics and two levels of analysis: per class and for the entire map. It also provides a few extra tools with which to manipulate the maps and extract specific elements that may be of interest to users. The plugin also allows us to calculate the metrics for specific areas of the map that overlay a vector layer defined by the user. Nonetheless, these spatial metrics cannot be calculated per patch and the 8-cell neighbourhood used by default for the calculation cannot be changed. For full information about the plugin and the various possibilities it offers, readers should consult the *Lecos* website and the paper by Jung (2016).

The R package "landscapemetrics"[1] provides almost all the options currently available for calculating spatial metrics (Hesselbarth et al. 2019). It supplies many more metrics than the "LecoS" plugin, allows the user to select the neighbourhood rule and includes the three levels of analysis (patch, category or whole landscape). R offers the full workflow available through FRAGSTATS (McGarigal 2015),[2] a free, very user-friendly software, which is widely regarded as the software of reference for calculating spatial metrics.

Although the R package offers us all the options currently available for calculating spatial metrics, in this chapter we will be focusing exclusively on the *LecoS* plugin. This is because it provides enough tools for the exercises we

---

[1] https://r-spatialecology.github.io/landscapemetrics/
[2] https://www.umass.edu/landeco/research/fragstats/fragstats.html.

propose, and is a tested, efficient software which allows us to perform these analyses easily and quickly.

Exercise 1. To validate a map against reference data/map

Aim

To assess to what extent the pattern of the CORINE map is similar to the pattern of the reference SIOSE map, which charts the real situation on the ground.

Materials

SIOSE Land Use Map Asturias Central Area 2011
CORINE Land Use Map Asturias Central Area 2011

Requisites

The two maps must be raster. The background class must be 0 or no data.

Execution

Step 1

One of the requisites of the "LecoS" plugin is that no category, apart from the background, is coded with the number 0. In our maps, the category "agricultural areas" is coded 0. The first step is therefore to reclassify the maps, so the background is coded 0 (currently it is coded 12) and all other categories have different codes other than 0.

The maps are reclassified using the *Reclassify by table* tool (Processing toolbox > Raster analysis > Reclassify by table). After opening the tool, indicate the map you want to reclassify (CORINE map) and fill in the "Reclassification table" with the values that will replace the existing values in the raster (Fig. 2). When filling in this table, bear in mind that the tool will search for values that are less than or equal to the maximum and greater than the minimum. In other words, if you reclassify as 2 (new value) the values with a maximum value of 1 and a minimum value of 0, all the pixels with value 1 will be reclassified as 2. 1 is the only value greater than 0 that is also less than or equal to 1.

Bearing these criteria in mind, fill in the reclassification table and run the tool (Fig. 3).

Step 2

After running the tool, you will obtain a reclassified map that meets the requirements of the *LecoS* plugin. You are now in a position to calculate the spatial metrics. This is done by accessing the *Landscape statistics* option of the "LecoS" plugin via the following route: Raster > Landscape ecology > Landscape statistics.

Once there, in the "Landcover grid" box indicate the raster for which you want to calculate the spatial metrics (CORINE reclassified), the "No-data" value (0, which is the background) and the spatial resolution of the raster (50 m, which you can check in the layer properties). You must also select the particular metrics you want to obtain (Fig. 4).

Several spatial metrics can be selected at the same time, using the "Select multiple metrics" tab. In this case, we selected the following: Land cover; Landscape proportion; Number of patches; Greatest patch area; Smallest patch area; Mean patch area; Median patch area; Fractal dimension index; Like adjacencies; Patch cohesion index. Once you have done this, run the function.

If your computer is unable to calculate all the metrics at the same time, split the task into two (e.g. two groups of five metrics). In this case, after running the tool for the second time, the results must be gathered together in a single file, as the plugin creates one file for each time you run the tool.

Step 3

The last step is to repeat the whole workflow for the reference raster, i.e. for the SIOSE map. In this case, you will probably need to split the spatial metrics calculation into different steps as the plugin may be not able to handle all the information at once. As the SIOSE map is made up of a larger number of patches, the plugin will need more time to make all the calculations.

Results and Comments

Once the spatial metrics for each of the maps have been calculated, the results of the analysis will be stored in CSV files in the folder of your choice.

You will have one file for each time you have run the tool. The first step will therefore be to gather all the information together in one file to make it easier to compare the spatial metrics for the two maps (Table 1). This is done using a spreadsheet program such as OpenOffice Calc or Microsoft Excel. Once the results have been correctly organized, you can now compare the pattern of the two maps (Tables 1, 2, 3 and 4).

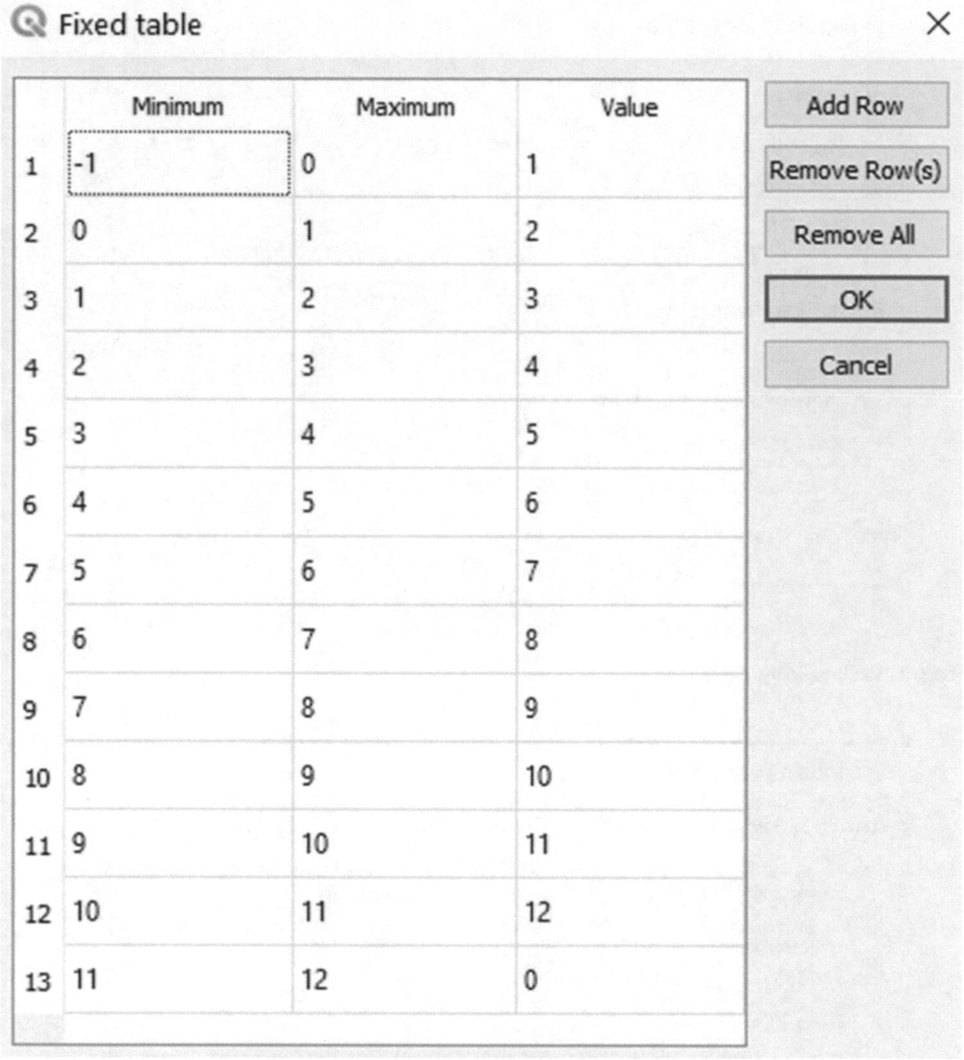

	Minimum	Maximum	Value
1	-1	0	1
2	0	1	2
3	1	2	3
4	2	3	4
5	3	4	5
6	4	5	6
7	5	6	7
8	6	7	8
9	7	8	9
10	8	9	10
11	9	10	11
12	10	11	12
13	11	12	0

Fig. 2 Exercise 1. Step 1. Table required for the "Reclassify by Table" tool

The "Land cover" and "Landscape proportion" metrics (Table 1) offer information about the space occupied by each category on the map. This gives us an insight into the composition of the landscape, i.e. the proportions or areas occupied by each category on the map, regardless of exactly where these categories are allocated.

The "Land cover" metric indicates the surface area in square metres occupied by each category. The "Landscape proportion" gives the proportion of the entire map (out of 1) occupied by each category. If the two maps have the same extent, both metrics will provide the same information, albeit in different units (square metres and percentage). Comparing maps with different extents is not recommended and could lead to important issues in the interpretation of the analysis.

In our case, the landscape composition of the two maps is very similar. All the categories are represented in similar proportions. Nonetheless, some differences were observed in the case of mineral extraction sites (Category 5 after reclassification), dump sites (Category 6) or road and rail networks (Category 7), among others.

The "Number of patches" (Table 1) indicates how many patches (contiguous areas with the same pixel value) make up each category. This metric is easy to understand and provides us very useful information about how fragmented a particular category is, so giving us an insight into the configuration of the landscape, i.e. about the way each category is allocated in the map.

Unlike landscape composition, important differences can be observed between the two maps in terms of landscape configuration. The SIOSE map is much more fragmented than the CORINE one. This difference is very significant for example in the road and rail networks category (Category 7

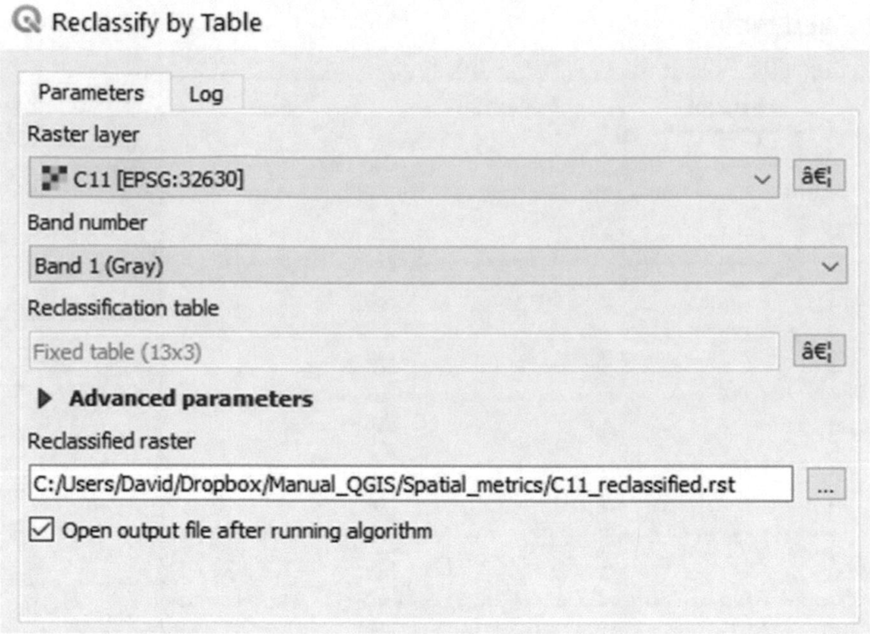

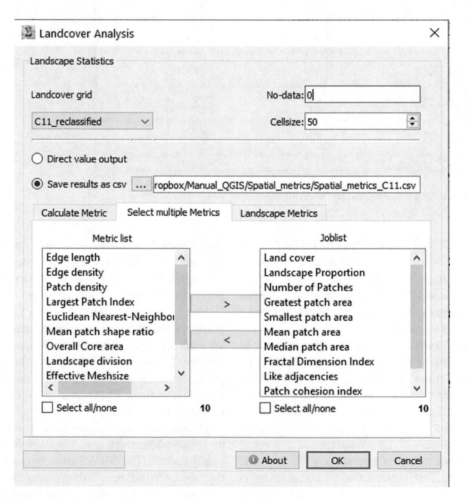

Fig. 3 Exercise 1. Step 1. Reclassify by Table

Fig. 4 Exercise 1. Step 2. LecoS plugin

after reclassification). Whereas in CORINE this class is made up of just 28 patches, in SIOSE it is much more fragmented with 2,464 patches (Table 1).

These differences are to be expected given that the CORINE and SIOSE maps use different Minimum Mapping Units (MMU) and Minimum Mapping Widths (MMW). SIOSE represents homogenous areas covering over 0.5-2 ha with a minimum width of 15 m, whereas CORINE only shows areas of over 25 ha with a minimum width of 100 m. Many small patches that appear in SIOSE do not therefore appear on the CORINE map.

Those land use categories that usually appear on the ground in small areas, such as small dump sites, or with linear features such as most of the road network, are not represented on the CORINE map, although they do appear in SIOSE. This explains the differences between the two maps in terms of the areas or proportions of certain classes referred to above.

It would be wrong therefore to conclude that CORINE does not map these areas of disagreement between the two maps well. They do not appear in CORINE simply because it has different MMU and MMW rules.

The "Greatest patch area" and "Smallest patch area" metrics (Table 2) help us to characterize the degree of fragmentation referred to earlier. The first metric measures the area (in square metres) of the largest patch on the map. The second metric does the same for the smallest patch on the map.

These two metrics highlight CORINE's simpler pattern and higher level of generalization. With a few exceptions, the largest patch in CORINE is usually larger than its counterpart in SIOSE. For the smallest patch, there are small differences between the maps. In most cases, the smallest patch occupies 2,500 m^2 in both maps. In other words, the smallest patch covers a single pixel with a 50 m edge ($50 \times 50 = 2,500 \ m^2$). It does not comply with the MMU and MMW rules of CORINE. This may be due to the presence of isolated pixels on the edge of the map after clipping it or due to the rasterization process.

The "Mean patch area" and "Median patch area" metrics (Table 3) also help us characterize the fragmentation of the map. These metrics measure the mean area and the median area of all the patches belonging to a particular category. As one might imagine, mean and median patch area are always smaller for SIOSE than for CORINE because of SIOSE's higher fragmentation. This is because the SIOSE map, due to its smaller MMU and MMW, draws more small polygons than CORINE, which tends to group them together in larger polygons.

The "Fractal dimension index" (Table 4) measures the mean shape complexity of the patches that make up each category. Values closer to 1 indicate simple geometries, more closely resembling a square, whereas values closer to 2 indicate more complex geometries, which are less like the simple shape of a square.

Contrary to what might be expected, and with the exception of the port areas (Category 9 after reclassification), patch shapes were more complex in CORINE than SIOSE. This seems illogical given that SIOSE is made at a finer scale (1:25,000) than CORINE (1:100,000) and delimits land use areas more accurately.

In our case, CORINE has more complex patch shapes than SIOSE because of the rasterization of the CORINE and

Table 1 Results from Exercise 1. Table showing the spatial metrics (Land Cover, Landscape proportion; Number of patches) for each category of the two maps that have been analysed (CORINE and SIOSE)

	Land cover (m^2)		Landscape proportion		Number of patches	
	CORINE	SIOSE	CORINE	SIOSE	CORINE	SIOSE
1	683,257,500	640,252,500	0.42	0.40	245	971
2	609,940,000	625,150,000	0.38	0.39	255	1,768
3	70,275,000	68,785,000	0.04	0.04	92	896
4	51,527,500	47,030,000	0.03	0.03	61	610
5	7,022,500	8,950,000	0.00	0.01	15	94
6	2,185,000	4,740,000	0.00	0.00	5	116
7	11,807,500	31,102,500	0.01	0.02	28	2,462
8	4,955,000	5,127,500	0.00	0.00	7	16
9	1,892,500	737,500	0.00	0.00	1	4
10	9,670,000	13,332,500	0.01	0.01	22	274
11	8,592,500	13,030,000	0.01	0.01	18	260
12	152,457,500	155,345,000	0.09	0.10	8	329

Table 2 Results from Exercise 1. Table showing the spatial metrics (Greatest patch area; Smallest patch area) for each category of the two maps that have been analysed (CORINE and SIOSE)

	Greatest patch area (m²)		Smallest patch area (m²)	
	CORINE	SIOSE	CORINE	SIOSE
1	468,677,500	217,527,500	2,500	2,500
2	160,860,000	137,540,000	2,500	2,500
3	14,812,500	13,975,000	2,500	2,500
4	11,737,500	6,610,000	2,500	2,500
5	1,145,000	985,000	5,000	2,500
6	930,000	1,062,500	285,000	2,500
7	1,602,500	2,027,500	2,500	2,500
8	3,480,000	3,340,000	2,500	2,500
9	1,892,500	660,000	1,892,500	10,000
10	910,000	710,000	2,500	2,500
11	1,965,000	1,767,500	12,500	2,500
12	150,047,500	149,600,000	2,500	2,500

Table 3 Results from Exercise 1. Table showing the spatial metrics (Mean patch area; Median patch area) for each category of the two maps that have been analysed (CORINE and SIOSE)

	Mean patch area (m²)		Median patch area (m²)	
	CORINE	SIOSE	CORINE	SIOSE
1	2,788,806.12	659,374.36	252,500	37,500
2	2,391,921.57	353,591.63	192,500	10,000
3	763,858.70	76,768.97	271,250	15,000
4	844,713.11	77,098.36	385,000	20,000
5	468,166.67	95,212.77	375,000	45,000
6	437,000.00	40,862.07	322,500	20,000
7	421,696.43	12,633.02	332,500	2500
8	707,857.14	320,468.75	252,500	48,750
9	1,892,500.00	184,375.00	1,892,500	33,750
10	439,545.45	48,658.76	371,250	20,000
11	477,361.11	50,115.38	390,000	10,000
12	19,057,187.50	472,173.25	450,000	5000

SIOSE vector databases, which reduced the complexity of the SIOSE polygons, resulting in more regular shapes.

Finally, "Like adjacencies" and the "Patch cohesion index" (Table 4) provide information about the compactness of the categories in a map. Values closer to 0 mean than the patches belonging to a particular category are very scattered. Values closer to 1 (Like adjacencies) or to 10 (Patch cohesion index) indicate that they are tightly clustered.

The "Like adjacencies" metric is based on the number of adjacencies between pixels, whereas the "Patch cohesion index" is obtained by calculating the ratio between the area and the perimeter of the patches. This means that although they provide information on a similar subject (compactness), they complement each other.

These metrics show that land uses are represented in a more compact (more clustered) manner in the CORINE database. This makes sense because of the lower degree of fragmentation and the greater generalization of CORINE compared to SIOSE.

All in all, even if important differences between the two maps could be identified in terms of landscape configuration, most of these are due to the different criteria used in the drawing of each map. This also applies to the small differences in terms of landscape composition. Our CORINE map must therefore be considered validated after comparison with SIOSE.

However, in order to be able to validate CORINE with certainty and to interpret the results of the spatial metrics

Table 4 Results from Exercise 1. Table showing the spatial metrics (Fractal dimension index; Like adjacencies; Patch cohesion index) for each category of the two maps that have been analysed (CORINE and SIOSE)

	Fractal dimension index		Like adjacencies		Patch cohesion index	
	CORINE	SIOSE	CORINE	SIOSE	CORINE	SIOSE
1	1.10	1.09	0.85	0.75	9.96	9.93
2	1.10	1.07	0.84	0.75	9.93	9.91
3	1.11	1.07	0.78	0.58	9.77	9.71
4	1.10	1.06	0.79	0.63	9.72	9.53
5	1.11	1.08	0.73	0.62	9.37	9.03
6	1.11	1.06	0.73	0.52	9.32	8.98
7	1.13	1.03	0.66	0.28	9.43	9.20
8	1.11	1.09	0.78	0.74	9.66	9.64
9	1.09	1.12	0.87	0.52	9.63	9.34
10	1.11	1.06	0.72	0.55	9.33	8.80
11	1.13	1.07	0.71	0.54	9.43	9.23
12	1.08	1.06	0.96	0.93	9.95	9.95

more effectively, we should always compare the maps via visual inspection. In this case, visual inspection reveals that the differences identified by the spatial metrics are mostly due to the different criteria used in the drawing of each map, and not because they interpret land use in different ways. Complementary tools must therefore be used to contextualize the results of our validation or uncertainty analysis. If this is not done, there is a high chance that we will make incorrect assumptions due to not having all the relevant information.

Exercise 2. To validate a simulation against a reference map

Aim

To assess to what extent the pattern of our simulation is similar to the pattern of a reference map for the same year, which accurately reflects the real situation on the ground.

Materials

Simulation CORINE Asturias Central Area 2011
CORINE Land Use Map Asturias Central Area 2011

Requisites

The two maps must be raster. The background class must be 0 or no data.
For a proper validation, the reference map and the simulation must refer to the same year.

Execution

Step 1

In order to comply with the requirements of the "LecoS" plugin, which assumes that pixels with the value 0 are No Data or background, we must first reclassify the two maps we are going to compare. The background, which is coded as 12, must be reclassified as 0. Agricultural areas, which were coded as 0, must be reclassified as 1. All the remaining classes must be reclassified following the same criteria (new code = original code + 1).

The *Reclassify by table* (Processing toolbox > Raster analysis > Reclassify by table) tool will be used to reclassify the maps (Fig. 5). First, indicate the map you want to reclassify and, then, fill in the "Reclassification table" with the new category codes that will replace the existing ones in the raster.

Step 2

Once the two maps have been reclassified, the next stage is to calculate the spatial metrics for each map: first for the simulation and then for the reference map. This is done using the *Landscape statistics* option in the "LecoS" plugin (Raster > Landscape ecology > Landscape statistics) (Fig. 6).

In "Landcover grid" select the raster for which you want to obtain the spatial metrics. You must also indicate the value of the background (No-data) and its spatial resolution (Cellsize). Finally, select the spatial metrics you are going to calculate.

Several spatial metrics can be selected at the same time using the "Select multiple metrics" tab. In this case, we selected the following metrics: Land cover; Landscape

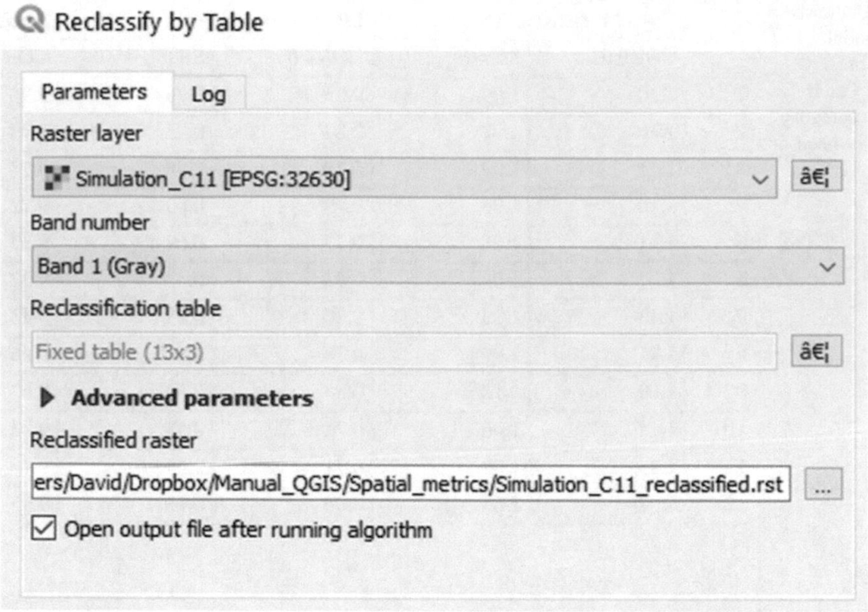

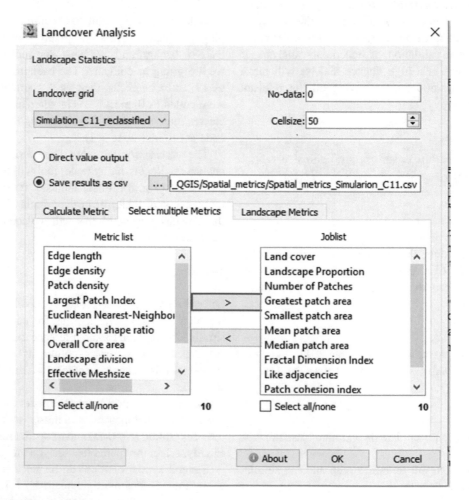

Fig. 5 Exercise 2. Step 1. Reclassify by Table

Fig. 6 Exercise 2. Step 2. LecoS plugin

proportion; Number of patches; Greatest patch area; Smallest patch area; Mean patch area; Median patch area; Fractal dimension index; Like adjacencies; Patch cohesion index.

Results and Comments

Once the spatial metrics for each of the maps have been calculated, the results of the analysis will be stored in CSV files in the folder of your choice. To make it easier to interpret and compare the spatial metrics, the two files must be merged into one. This can be done using a spreadsheet program such as OpenOffice Calc or Microsoft Excel. This will display the results in a table similar to Table 5.

At first sight, the differences between the patterns of the two maps do not seem very significant. This makes sense in that we are calculating the metrics for the whole area of the maps. However, land use changes only affect a small portion of maps, usually less than 10% or even 5% of the studied areas. The changes we simulated or that actually happened on the ground according to the reference map will not therefore have a dramatic impact on the spatial metrics for the whole map.

Even so, some differences can be observed. Agricultural areas (Category 1 after reclassification) and vegetation areas (Category 2) are made up of a larger number of patches in the simulation than in the reference map (Table 5). By contrast, urban fabric (Category 3) and industrial and commercial areas (Category 4) are made up of a slightly smaller number of patches.

These trends may indicate that the changes simulated as transitions to urban fabric and to industrial and commercial areas have made these classes more compact (patches that were not previously connected have now become connected with the simulated changes). That is, these classes did not grow in an isolated way, but via the expansion of previously existing patches. The slight differences between the reference map and the simulation in the "Like adjacencies" and "Patch cohesion index" metrics for industrial and commercial areas (Category 4 after reclassification) also point in this direction.

In the process of expansion of urban fabric and industrial areas, some patches of agricultural and vegetation areas could become isolated, so increasing the fragmentation of the category. This would explain why there are more patches in these categories in the simulation than in the reference map.

The difference in pattern between the simulation and the reference map can best be calculated using spreadsheet software, as described in the example for Table 6. In this table, we have subtracted the spatial metric for each category in the simulation from the value for the same metric in the reference map. Thus, for instance, the reference map has 602,500 m^2 more agricultural areas (Category 1) than the simulation. By contrast, the simulation has 1,205,000 m^2 more vegetation areas (Category 2) than the reference map. In our simulation, more space is also allocated to urban fabric (Category 3) and industrial and commercial areas (Category 4) than in the reference map.

Table 5 Results from Exercise 2. Table showing the spatial metrics (Number of patches; Like adjacencies; Patch cohesion index) for each category of the simulation and the reference map

	Number of patches		Like adjacencies		Patch cohesion index	
	Reference	Simulation	Reference	Simulation	Reference	Simulation
1	245	288	0.85	0.85	9.96	9.96
2	255	259	0.84	0.84	9.93	9.93
3	92	89	0.78	0.78	9.77	9.77
4	61	58	0.79	0.80	9.72	9.73
5	15	24	0.73	0.73	9.37	9.33
6	5	5	0.73	0.71	9.32	9.23
7	28	27	0.66	0.65	9.43	9.43
8	7	7	0.78	0.73	9.66	9.52
9	1	1	0.87	0.87	9.63	9.63
10	22	22	0.72	0.72	9.33	9.31
11	18	17	0.71	0.71	9.43	9.43
12	8	8	0.96	0.96	9.95	9.95

Table 6 Results from Exercise 2. Difference in the value of the spatial metrics (Land Cover; Greatest patch area; Mean patch area) calculated for the simulation and the reference map. The results on the table indicate how far or close are the values of the spatial metrics in the two maps

	Land cover (m^2)	Greatest patch area (m^2)	Mean patch area
	Dif Simulation − Ref map	Dif Simulation − Ref map	Dif Simulation − Ref map
1	−602,500	347,500	−418,476.26
2	1,205,000	−750,000	−32,288.37
3	182,500	177,500	27,798.60
4	155,000	−45,000	46,364.48
5	857,500	−270,000	−139,833.34
6	−300,000	−242,500	−60,000.00
7	−927,500	32,500	−18,733.47
8	−1,642,500	−1,642,500	−234,642.85
9	0	0	0.00
10	−340,000	0	−15,454.54
11	−450,000	0	1609.48
12	1,862,500	1,857,500	232,812.50

These differences in the total area allocated to each category help us understand how the model calculates the changes it simulates. If the model had simulated the same amount of change that actually occurred on the maps, no differences would be noticed.

In our simulation, we did not actively model the vacant classes. Thus, whereas according to the reference map there were many vegetation areas that changed to agricultural areas, in our simulation this did not happen. As a consequence, our simulation has more vegetation areas, but less agricultural areas than the reference map.

The "Greatest patch area" metric shows that we did not model one of the biggest industrial developments in the study area correctly. The largest patch in our simulation is 450,000 m^2 smaller than the one in the reference map. The opposite was true in the case of urban fabric. According to the model, many pixels were considered to have changed as a result of the expansion of large pre-existing patches, when this trend was in fact not that strong according to the reference map.

If we focus on the "Mean patch area" metric for the two categories we modelled actively (3 and 4) we can see how in both cases the mean area of patches is always bigger in the simulation than in the reference map. This may be due to the same process as in urban fabric, i.e. most of the changes are simulated as expansions of pre-existing large patches.

In all other categories apart from the first 4 (1, 2, 3, 4), there are important differences between the two maps. However, as changes in these categories were not modelled in the simulation (they remained invariant), the differences between the maps are due to changes that took place in reference map but were not simulated.

To sum up, it is difficult with the information available to us to understand whether the pattern of the changes we simulated is valid or not. We have various clues about the pattern of the changes (more compact and connected than in the reference map), but these trends are best confirmed by visual inspection. Calculating the spatial metrics solely for the areas that changed is also highly recommended and can provide additional insight.

Exercise 3. To validate simulated changes against a reference map of changes

Aim

To assess to what extent the pattern of the changes we simulated is similar to the pattern of a reference map of changes for the same year, which accurately reflects the real situation on the ground.

Materials

CORINE Land Use Changes Asturias Central Area 2005–2011
Simulated CORINE changes Asturias Central Area 2005–2011

Requisites

The two maps must be raster. The background class must be 0 or no data.

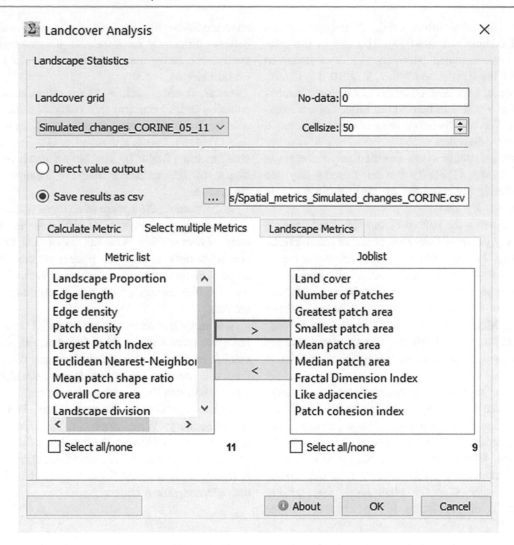

Fig. 7 Exercise 3. Step 1. LecoS plugin

For a proper validation, the changes in the reference map must refer to the same time period as the simulation period.

Execution

Step 1

Given that the background is already coded 0 in the two maps charting changes, we do not need to take any preliminary steps prior to calculating the spatial metrics. This can be done directly using the *Landscape statistics* option in the "LecoS" plugin (Raster > Landscape ecology > Landscape statistics).

In the tool, we must indicate the raster for which we want to calculate the spatial metrics (Landcover grid), the value of the background in our maps (No-Data) and their spatial resolution (Fig. 7). Several spatial metrics can be selected at the same time, using the "Select multiple metrics" tab.

In this analysis, we will be calculating the following metrics: Land cover; Number of patches; Greatest patch area; Smallest patch area; Mean patch area; Median patch area; Fractal dimension index; Like adjacencies; Patch cohesion index.

Step 2

We repeat this process for the second map.

Results and Comments

Once we have run the tool twice, once for each map, we will have two CSV files with the metrics for each of the change maps. These will be saved in the specified folder.

The reference map of changes includes land use changes for many categories (1, 2, 5, 6, 7, 8, 10 and 11) that are not drawn on the simulated map of changes. This is because we only actively simulated urban fabric (Category 3) and industrial and commercial areas (4). The rest of the categories were only simulated passively (1, 2) or remained invariant during the simulation (5, 6, 7, 8, 9, 10, 11, 12). As a result, the map of simulated changes only includes patches from Categories 3 and 4 (urban fabric, industrial and commercial areas). We will therefore only compare the spatial metrics for these categories.

The changes we simulated are quantitatively the same as the reference changes (Table 7). We can therefore say that our model correctly predicted the quantity of changes that happened in our study area.

On the other hand, the pattern of the simulated changes seems to be very different from the pattern of the reference map of changes. In the reference map, the changes took place in just a few patches and most of the pixels that changed are allocated close to each other. In the simulation, the changes are fragmented in many different patches (Table 7). The "Mean patch area" and "Median patch area" metrics confirm this trend (Table 8). The simulated changes take place in very small patches, made up of just a few pixels.

When working with Cellular Automata models, change usually takes place organically as an expansion of existing patches. In the real world, however, changes in urban and industrial areas tend to happen at the same time over entire cadastral parcels. Often, these parcels are quite big, comprising a large number of pixels. However, as CA models usually simulate change at the pixel level, they are not normally capable of simulating big patches of change covering large numbers of pixels. Our model therefore behaves differently from the real processes taking place on the ground, hence the disagreements in the pattern of simulated changes.

Other metrics, such as "Like adjacencies" and "Patch cohesion index" confirm this behaviour. The pixels in the reference map are better grouped than those in the simulated map (Table 9). This is also manifested by the "Greatest patch area" metric (Table 7). The largest patch is always much bigger in the reference map of changes than in the simulation.

In conclusion, the pattern of changes we simulated is very different to the pattern of changes in the reference map. However, this does not mean that the changes we simulated have altered the pattern of the simulated landscape. On the contrary, as we discovered in the previous exercise, the pattern of the whole landscape remains very similar.

It is important to remember here that we are only calculating the pattern of the areas that changed, without viewing them in any larger context. By contrast, when we calculate the spatial metrics for the whole map, we also consider the context and can therefore assess whether the changes have altered the pattern of the map. Thus, both analyses are complementary. We recommend users to carry out both analyses when validating the pattern of their simulations.

Finally, a qualitative validation through visual inspection is highly recommended for contextualizing the results and understanding them better.

Table 7 Results from Exercise 3. Table showing the spatial metrics (Land Cover; Number of patches; Greatest patch area) for each actively simulated category in the simulation and the reference map

	Land cover (m^2)		LNumber of patches (m^2)		LGreatest patch area (m^2)	
	Reference	Simulation	Reference	Simulation	Reference	Simulation
3	2,280,000	2,280,000	20	121	572,500	190,000
4	1,800,000	1,800,000	24	172	295,000	182,500

Table 8 Results from Exercise 3. Table showing the spatial metrics (Mean patch area; Median patch area) for each actively simulated category in the simulation and the reference map

	Mean patch area (m^2)		Median patch area (m^2)	
	Reference	Simulation	Reference	Simulation
3	114,000	18,842.97	40,000	5,000
4	75,000	10,465.12	55,000	2,500

Table 9 Results from Exercise 3. Table showing the spatial metrics (Fractal dimension index; Like adjacencies; Patch cohesion index) for each actively simulated category in the simulation and the reference map

	Fractal dimension index		Like adjacencies		Patch cohesion index	
	Reference	Simulation	Reference	Simulation	Reference	Simulation
3	1.07	1.04	0.63	0.37	9.11	8.26
4	1.09	1.03	0.53	0.30	8.67	8.24

Aim

To test the consistency of the pattern of land uses in a series of LUC maps made up of two different time points.

Materials

CORINE Land Use Map Asturias Central Area 2005 v.0
CORINE Land Use Map Asturias Central Area 2011

Requisites

The two maps must be raster. The background class must be 0 or no data.

Execution

Step 1

In order to comply with the requirements of the "LecoS" plugin, the maps must be reclassified to ensure that the background code is 0 and all other categories have a positive code different from 0. This is done using the *Reclassify by table* tool (Processing toolbox > Raster analysis > Reclassify by table) (Fig. 8).

After opening the tool, we indicate the map we want to reclassify and then fill in the "Reclassification table" with the new category codes that will replace the existing ones in the raster (Fig. 9).

Step 2

Once the categories have been reclassified, the spatial metrics for each map can be calculated using the *Landscape statistics* option in the "LecoS" plugin (Raster > Landscape ecology > Landscape statistics) (Fig. 10).

After opening the tool, we select the raster for which we wish to obtain the metrics (Landcover grid), the background value of the raster (No-Data) and its spatial resolution (cellsize). We then select the different metrics we want to calculate in the "Select multiple metrics" tab. In this case we selected the following: Land cover; Landscape proportion; Number of patches; Greatest patch area; Smallest patch area; Mean patch area; Median patch area; Fractal dimension index; Like adjacencies; Patch cohesion index.

Results and Comments

After running the tool, the metrics are displayed in two CSV files which are saved in the specified folder.

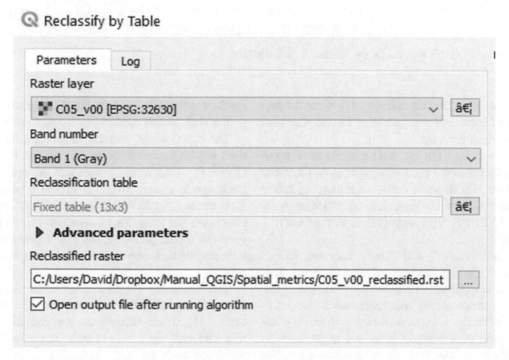

Fig. 8 Exercise 4. Step 1. Reclassify by Table

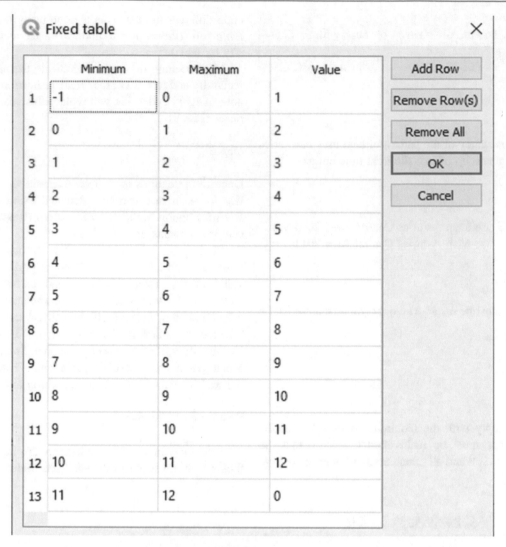

Fig. 9 Exercise 4. Step 1. Table required for the "Reclassify by Table" tool

The metrics reveal important differences between the two maps in terms of landscape configuration, i.e. the way land uses are allocated on each map.

The categories in the CORINE 2011 map are made up of many more patches than the same categories in the CORINE 2005 map (Table 10). In some cases, such as urban fabric (Category 3 after reclassification), there are twice as many patches in the CORINE 2011 map (92) as in the CORINE 2005 map (44).

The "Like adjacencies" and "Patch cohesion index" metrics also show slight differences between the maps. This is unusual when comparing a time series of land use maps, as these metrics are not usually sensitive to small changes in the landscape. With the exception of highly dynamic environments, in most of the study areas we might wish to assess, change affects less than 5% of the landscape. We should not therefore expect meaningful differences in the spatial metrics that characterize the landscape over a short period such as that used in our example (2005–2011).

The "Land cover" metrics show big differences between the maps in terms of the areas covered by each category (Table 11). One would not expect the composition of the landscape to change so much in just 6 years. Agricultural areas occupy 28,710,000 m^2 more in the CORINE 2005 map than in the 2011 one. That means that 11,484 pixels changed over the 6-year period. However, a process of change of such magnitude was not observed on the ground in the study area.

The "Greatest patch area" and "Mean patch area" metrics also differ greatly for the two maps in the time series (Table 11). These differences are also much bigger than might be expected due to changes in the landscape over the timeframe analysed.

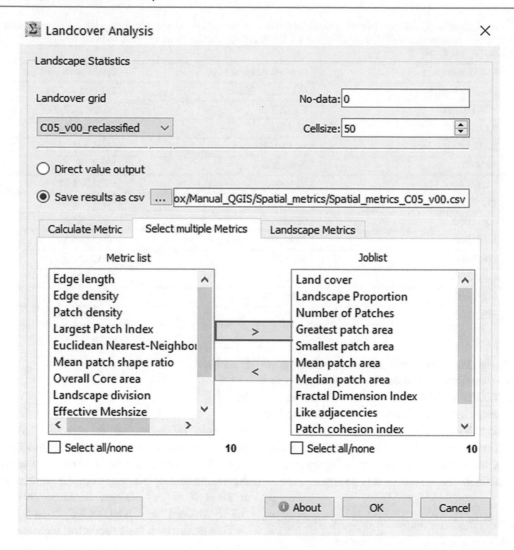

Fig. 10 Exercise 4. Step 2. LecoS plugin

Table 10 Results from Exercise 4. Table showing the spatial metrics (Number of patches; Like adjacencies; Patch cohesion index) for each category of the two maps that have been analysed (CORINE 2005 and CORINE 2011)

	Number of patches		Like adjacencies		Patch cohesion index	
	C05	C11	C05	C11	C05	C11
1	126	245	0.87	0.85	9.96	9.96
2	173	255	0.86	0.84	9.91	9.93
3	44	92	0.85	0.78	9.81	9.77
4	33	61	0.84	0.79	9.73	9.72
5	14	15	0.78	0.73	9.39	9.37
6	2	5	0.79	0.73	9.45	9.32
7	15	28	0.67	0.66	9.35	9.43
8	3	7	0.80	0.78	9.60	9.66
9	1	1	0.88	0.87	9.58	9.63
10	12	22	0.77	0.72	9.26	9.33
11	12	18	0.71	0.71	9.27	9.43
12	5	8	0.96	0.96	9.95	9.95

Table 11 Results from Exercise 4. Difference in the value of the spatial metrics (Land Cover; Greatest patch area; Mean patch area) calculated for the two maps that have been analysed (CORINE 2005 and CORINE 2011). The results on the table indicate how far or close are the values of the spatial metrics in the two maps

	Land cover (m^2)	Greatest patch area (m^2)	Mean patch area (m^2)
	Dif C11 – C05	Dif C11 – C05	Dif C11 – C05
1	−28,710,000	7,802,500	−2,861,729.59
2	9,125,000	12,570,000	−1,080,997.50
3	3,430,000	−4,810,000	−755,345.84
4	580,000	2,482,500	−699,150.53
5	−1,272,500	62,500	−124,333.33
6	722,500	−120,000	−294,250.00
7	4,760,000	667,500	−48,136.90
8	1,710,000	1,182,500	−373,809.53
9	370,000	370,000	370,000.00
10	4,882,500	−7500	40,587.12
11	4,695,000	1,027,500	152,569.44
12	−292,500	−427,500	−11,492,812.50

These results indicate that there are many differences between the two maps that are not due to real changes in the landscape. These differences may be due to technical issues within the time series in that different methods were used to produce CORINE 2005 and 2011.

These conclusions were confirmed by a visual inspection of the two maps, an additional check that is highly recommended to complement the results of this analysis.

Exercise 5. To validate a series of maps with two or more time points (vector)

Aim

To study the pattern of a specific transition (from scrubland to forest) in our study area (Ariège Valley) for a given period (2000–2018).

Materials

CORINE Land Cover Map Val d'Ariège 2000
CORINE Land Cover Map Val d'Ariège 2018

Requisites

All raster maps must have the same resolution, extent and projection.

Execution

Step 1

We begin by extracting the changes we want to study (transition from scrub to forest) with the *Raster Calculator* (Fig. 11). In the raster calculator expression box, we write an expression to obtain a map with the features that were scrub in 2000 (Category 4) and forest in 2018 (Category 3): "CLC_2000@1" = 4 AND "CLC_2018@1" = 3.

This produces a raster showing the areas that underwent this transition (Fig. 12).

Step 2

Once the raster for this transition has been obtained, it must be converted into vector format (polygons) using the *Polygonize* GDAL tool. When making this conversion, the "Use 8-connectedness" option must be selected (Fig. 13). In this way, the tool considers all pixels diagonal to other pixels as part of the same polygon. If this option is not selected, pixels situated diagonal to other pixels are considered as separate polygons.

Step 3

Once the polygons that undergo this transition have been obtained in vector format, we can then calculate their spatial metrics using the SAGA *Polygon Shape Indices* tool (Fig. 14).

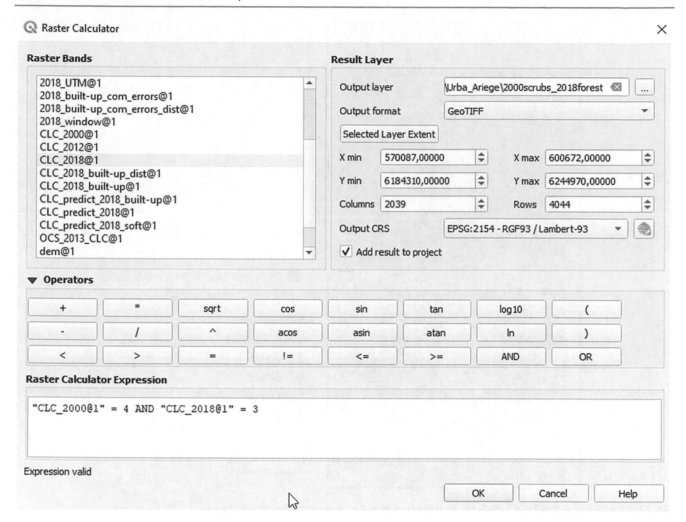

Fig. 11 Exercise 5. Step 1. Raster calculator

After calculating the spatial metrics, we obtain a vector file. The values for these metrics are calculated for each polygon and are stored in the attribute table of the vector (Fig. 15). The metrics used in this case were: area; perimeter; ratio perimeter / area; ratio perimeter / square root area; maximum distance; maximum distance / area; maximum distance / square root area; and shape index.

Step 4

In order to better interpret the general pattern of all the polygons that undergo this transition, the results of the metrics can be exported to a spreadsheet where statistics such as the mean, standard deviation, minimum and maximum can be calculated (Table 12).

Results and Comments

The pattern of the areas that transition from scrubland (2000) to forest (2018) is very diverse, with patches of varying size, capacity and shape complexity. The smallest polygon covers only 224 m^2, while the largest occupies 2,462,094.65 m^2. Perimeter lengths also vary enormously: from almost 60 m to 13,155.37 m. These results indicate that the areas that transition from scrubland to forests have very different sizes and shapes.

The perimeter / area (P/A) ratio is a measure of the compactness of the patches. Lower P/A values mean more compact polygons, whereas higher P/A values mean elongated or less compact polygons. The maximum distance metric indicates the longest segment of a polygon. The maximum distance / area (D/A) ratio is a measure of how

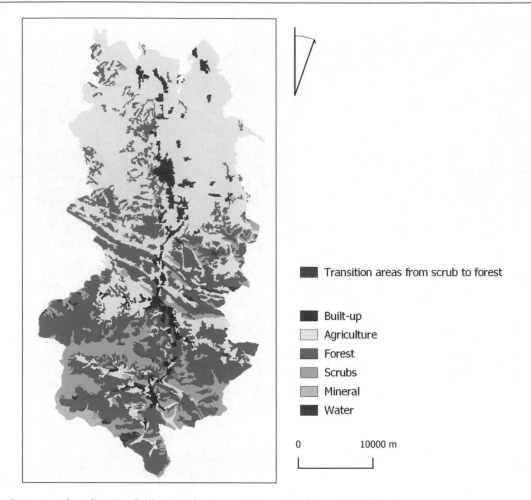

Transition areas from scrub to forest

Built-up
Agriculture
Forest
Scrubs
Mineral
Water

0 10000 m

Fig. 12 Intermediate output from Exercise 5. Map showing areas that transition from scrub to forest

elongated the polygon is. Lower values indicate more compact, less elongated polygons, whereas higher values mean the opposite. Finally, the shape index measures the shape complexity of a patch, using the following formula: Perimeter/(2 * Square Root(PI * Area).

The metrics calculated in this exercise can be compared with the metrics obtained and analysed in Exercise 6 below, which carries out the same analysis with raster data. The comparison will offer an insight into how data format (vector or raster) can affect the results of a pattern analysis.

Exercise 6. To validate a series of maps with two or more time points (raster)

Aim

To study the pattern of a specific transition (scrub into forest) in our study area (Ariège Valley) for a given period (2000–2018).

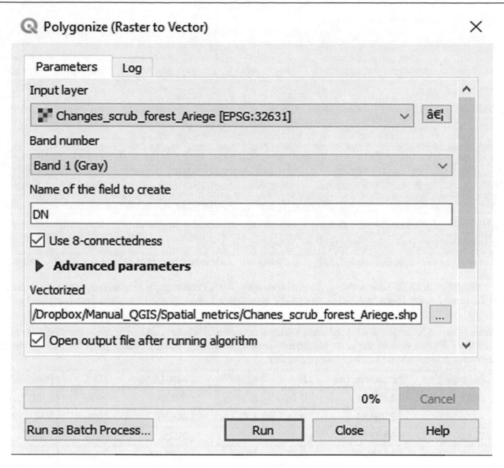

Fig. 13 Exercise 5. Step 2. Polygonize (Raster to Vector)

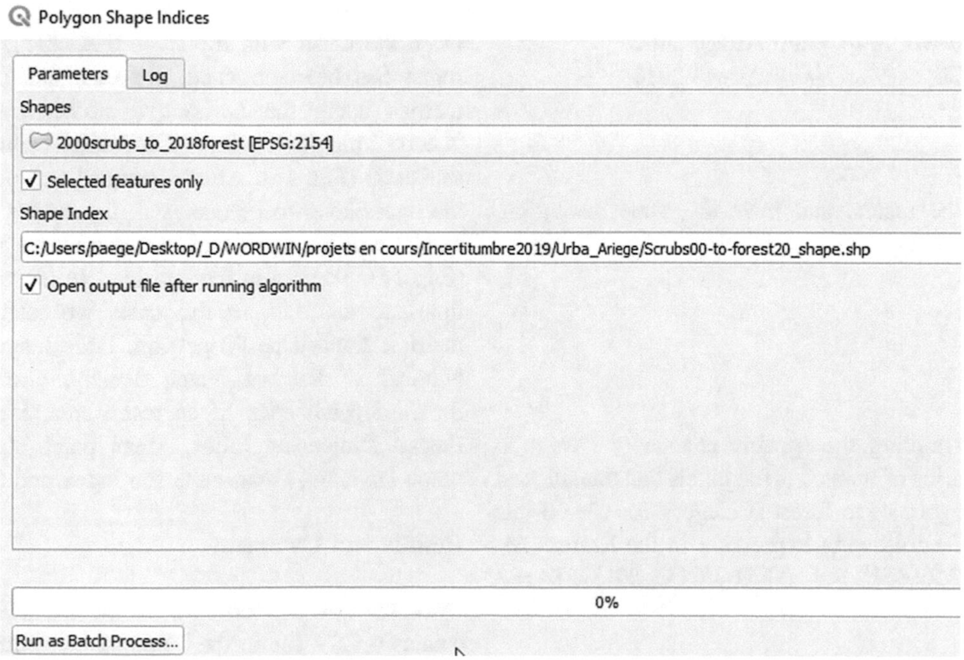

Fig. 14 Exercise 5. Step 3. Polygon shape indices

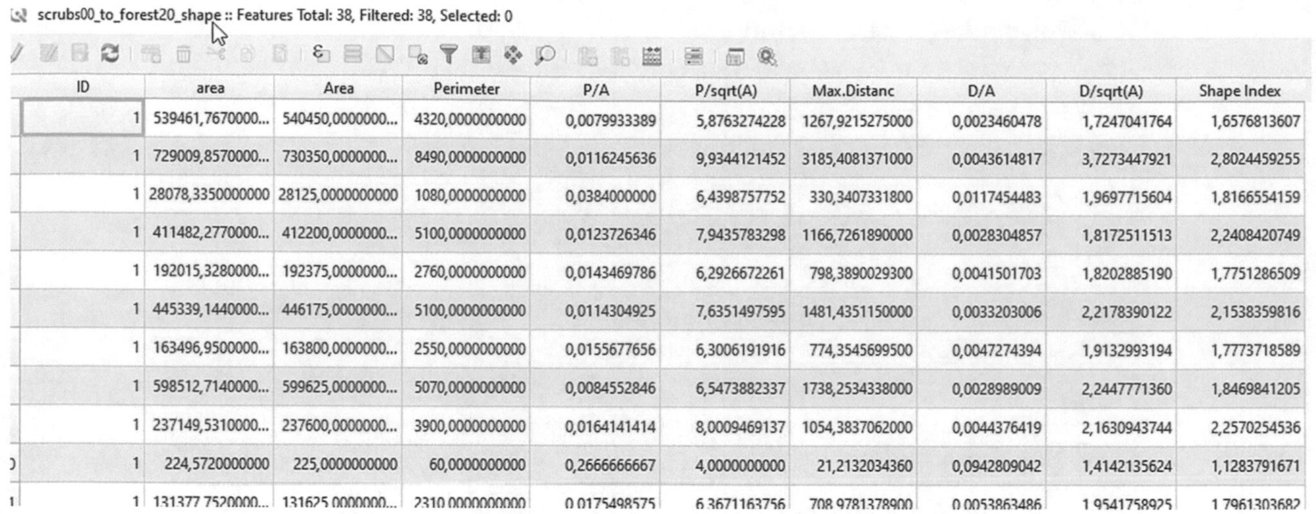

scrubs00_to_forest20_shape :: Features Total: 38, Filtered: 38, Selected: 0

ID	area	Area	Perimeter	P/A	P/sqrt(A)	Max.Distanc	D/A	D/sqrt(A)	Shape Index
1	539461,7670000...	540450,0000000...	4320,0000000000	0,0079933389	5,8763274228	1267,9215275000	0,0023460478	1,7247041764	1,6576813607
1	729009,8570000...	730350,0000000...	8490,0000000000	0,0116245636	9,9344121452	3185,4081371000	0,0043614817	3,7273447921	2,8024459255
1	28078,3350000000	28125,0000000000	1080,0000000000	0,0384000000	6,4398757752	330,3407331800	0,0117454483	1,9697715604	1,8166554159
1	411482,2770000...	412200,0000000...	5100,0000000000	0,0123726346	7,9435783298	1166,7261890000	0,0028304857	1,8172511513	2,2408420749
1	192015,3280000...	192375,0000000...	2760,0000000000	0,0143469786	6,2926672261	798,3890029300	0,0041501703	1,8202885190	1,7751286509
1	445339,1440000...	446175,0000000...	5100,0000000000	0,0114304925	7,6351497595	1481,4351150000	0,0033203006	2,2178390122	2,1538359816
1	163496,9500000...	163800,0000000...	2550,0000000000	0,0155677656	6,3006191916	774,3545699500	0,0047274394	1,9132993194	1,7773718589
1	598512,7140000...	599625,0000000...	5070,0000000000	0,0084552846	6,5473882337	1738,2534338000	0,0028989009	2,2447771360	1,8469841205
1	237149,5310000...	237600,0000000...	3900,0000000000	0,0164141414	8,0009469137	1054,3837062000	0,0044376419	2,1630943744	2,2570254536
1	224,5720000000	225,0000000000	60,0000000000	0,2666666667	4,0000000000	21,2132034360	0,0942809042	1,4142135624	1,1283791671
1	131377,7520000...	131625,0000000...	2310,0000000000	0,0175498575	6,3671163756	708,9781378900	0,0053863486	1,9541758925	1,7961303682

Fig. 15 Results from Exercise 5. QGIS table showing the spatial metrics (Area, Perimeter, Perimeter/Area; Perimeter / Square root of the area; Maximum distance; Distance / Area; Distance / Square root of the area; Shape index) for each transition area (polygon) of the analysed map

Table 12 Results from Exercise 5. Table showing the mean, standard deviation, minimum and maximum of the spatial metrics (Area, Perimeter, Perimeter/Area; Perimeter / Square root of the area; Maximum distance; Distance / Area; Distance / Square root of the area; Shape index) calculated for the areas that underwent the scrub to forest transition

	Area (m^2)	Perimeter (m)	P/A	P/sqrt(A)	Max.Distanc	D/A	D/sqrt(A)	Shape Index
Mean	236,950.95	2,625.73	0.10	6.29	759.01	0.03	1.96	1.77
Standard Dev	427,332.78	2,769.54	0.11	1.74	758.96	0.04	0.56	0.49
Min	224.50	59.93	0.01	4.00	21.19	0.00	1.41	1.13
Max	2,462,094.65	13,155.37	0.27	10.00	3,200.44	0.09	3.75	2.82

Materials

CORINE Land Cover Map Val d'Ariège 2000
CORINE Land Cover Map Val d'Ariège 2018

Requisites

All maps must be rasters and have the same resolution, extent and projection.

Execution

Step 1

We begin by extracting the specific changes we want to study from our series of maps, i.e. the pixels that transitioned from scrub (Category 4) to forest (Category 3). We do this by introducing the following expression in the Raster Calculator: "CLC_2000@1" = 4 AND "CLC_2018@1" = 3 (Fig. 11).

Step 2

Once the raster with the areas that changed from scrub to forest has been obtained, we then calculate their spatial metrics using the *Landscape statistics* option from the "LecoS" plugin (Raster > Landscape ecology > Landscape statistics) (Fig. 16). After opening the tool, we must select the raster layer to be analysed, the output folder where the results will be saved and the metrics we want to calculate (Fig. 17). To choose the metrics, we must select the "Multiple metrics" tab. In this case, we selected 14 different metrics: Landscape Proportion, Edge length, Edge density, Number of Patches, Patch density, Greatest patch area, Smallest patch area, Mean patch area, Median patch area, Fractal Dimension Index, Mean patch shape ratio, Landscape Division, Patch cohesion index and Splitting index.

Results and Comments

Once the spatial metrics have been calculated, the plugin creates a CSV file in the output folder with the results.

Fig. 16 Exercise 6. Step 2. Landscape statistics option of the LecoS plugin

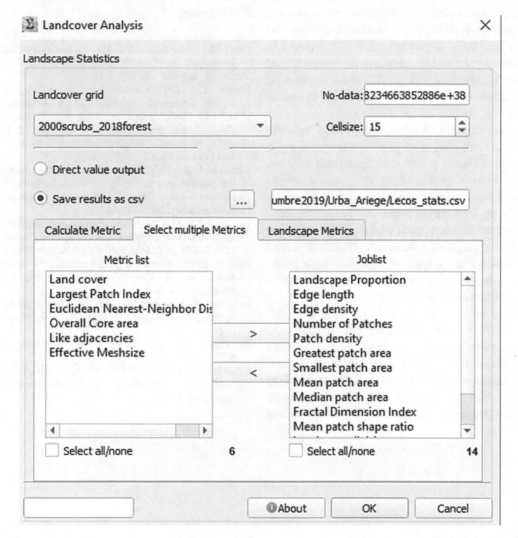

Fig. 17 Exercise 6. Step 2. LecoS plugin

Table 13 Results from Exercise 6. Table showing the spatial metrics (Landscape proportion; Edge length; Edge density; Number of patches; Patch density; Greatest patch area; Smallest patch area; Mean patch area; Median patch area; Fractal dimension index; Mean patch shape ratio; Landscape division; Patch cohesion index; Splitting index) for the areas that underwent the scrub to forest transition

Landscape Proportion	Edge length	Edge density	Number of Patches	Patch density	Greatest patch area	Smallest patch area
1	97,260	0.011069	37	0	2,467,575	225

Mean patch area	Median patch area	Fractal Dimension Index	Mean patch shape ratio	Landscape division	Patch cohesion index	Splitting Index
237,478.378	111,600	1.06899650	1.42935958	0.88744375	9.82890312	225

The results show that 37 different patches underwent the transition from scrub to forest, as shown in the "Number of patches" metric in Table 13. These patches have different sizes, varying from 225m^2 to 2,467,575m^2. There are a few small patches, but most patches are big, as revealed by the mean (237,539m^2) and median (111,600m^2) metrics.

The "Landscape proportion" metric indicates the percentage of the studied landscape occupied by the category in question. As we are only considering one category in our analysis (the areas that transition from scrubland to forests), this category occupies 100% of the studied landscape and therefore has a landscape proportion value of 1 (Table 13). The fractal dimension index informs about the complexity of the patches in the specified category. Values closer to 2 mean more complex shapes, whereas values closer to 1 mean simpler shapes.

The landscape division, patch cohesion and splitting indices assess the compactness or fragmentation of the patches that make up a class, i.e. how well aggregated they are. A "Landscape division" value close to 1 means a very fragmented landscape, whereas values close to 0 indicate a landscape made up of a single patch. A "Patch cohesion" value of 0 means one isolated patch, whereas values closer to 100 mean more aggregated patches. A "Splitting index" value of 1 indicates a landscape made up of a single patch, while splitting index values of more than 1 indicate a progressively more fragmented landscape.

If we compare these results to those obtained in vector format (Exercise 5), we can see that the same values were obtained for comparable measures (e.g. mean area, greatest / smallest area), while other measures use different formulas. These include the shape and compacity indices (standardized or not, area-weighted or not, completed by a constant or not). The LecoS plugin also offers complementary indices which are not calculated in vector format, such as the fractal dimension or the splitting index. In addition, whereas the spatial metrics in vector can be calculated individually for each patch or polygon (Exercise 5), this is not possible in raster format when using the LecoS plugin. The plugin usually calculates the mean values of all the patches for each metric.

References

Botequilha Leitao A, Miller J, Ahern J, McGarigal K (2006) Measuring landscapes: a planner's handbook. Island Press, Washington, Covelo, London

Forman RTT (1995) Land mosaics: the ecology of landscapes and regions. Cambridge University Press, Cambridge, United Kingdom

Hesselbarth MHK, Sciaini M, With KA, Wiegand K, Nowosad J (2019) Landscapemetrics: an open-source R tool to calculate landscape metrics. Ecography 42(10):1648–1657. https://doi.org/10.1111/ecog.04617

Herold M, Couclelis H, Clarke KC (2005) The role of spatial metrics in the analysis and modeling of urban land use change. Comput Environ Urban Syst 29:369–399. https://doi.org/10.1016/j.compenvurbsys.2003.12.001

Jaeger JAG (2000) Landscape division, splitting index, and effective mesh size: New measures of landscape fragmentation. Landsc Ecol 15:115–130. https://doi.org/10.1023/A:1008129329289

Jung M (2016) LecoS—a python plugin for automated landscape ecology analysis. Ecol Inform 31:18–21. https://doi.org/10.1016/j.ecoinf.2015.11.006

McGarigal K, Cushman SA, Neel MC, Ene E (2015) FRAGSTATS: spatial pattern analysis program for categorical and continuous maps

Mcgarigal K (2018) Landscape metrics for Categorical Map Patterns, p 77

Šímová P, Gdulová K (2012) Landscape indices behavior: a review of scale effects. Appl Geogr 34:385–394. https://doi.org/10.1016/j.apgeog.2012.01.003

Advanced Pattern Analysis to Validate Land Use Cover Maps

Martin Paegelow and David García-Álvarez

Abstract

In this chapter we explore pattern analysis for categorical LUC maps as a means of validating land use cover maps, land change and land change simulations. In addition to those described in Chap. "Spatial Metrics to Validate Land Use Cover Maps", we present three complementary methods and techniques: a Goodness of Fit metric to measure the agreement between two maps in terms of pattern (Map Curves), the focus on changes on pattern borders as a method for validating on-border processes and a technique quantifying the magnitude of distance error. Map Curves (Sect. 1) offers a universal pattern-based index, called Goodness of Fit (GOF), which measures the spatial concordance between categorical rasters or vector layers. Complementary to this pattern validation metric, the following Sect. 2 focuses specifically on the changes that take place on pattern borders. This enables changes to be divided into those that take place on the borders of existing features and those that form new, disconnected features. Bringing this chapter on landscape patterns to a close, Sect. 3 presents a technique for quantifying allocation errors in simulation maps and more precisely on the minimum distance between the allocation errors in simulation maps and the nearest patch belonging to the same category on the reference map. The comparison between a raster-based and a vector-based approach brings us back to the differences in measurement inherent in the representation of entities in raster and vector mode. These techniques are applied to two datasets. Section 1 uses the Asturias Central Area database, where CORINE maps are compared to SIOSE maps and simulation outputs. For their part, the techniques described in Sects. 2 and 3 are applied to the Ariège Valley database. CORINE maps for 2000 and 2018 are used as reference maps in comparisons with simulated land covers.

Keywords

Allocation distance error • Change on pattern borders • Map Curves • Pattern shape and size indices

1 Map Curves

Description

This is a quantitative method proposed by Hargrove et al. (2006) to evaluate the spatial concordance between different categorical raster or vector datasets. It calculates the Goodness of Fit (GOF) (Fig. 1), a standard metric that evaluates the spatial concordance between the patches of two or more rasters or the polygons of two or more vectors. Unlike other methods, it does not evaluate spatial agreement at cell level, and instead focuses on agreement at patch level in rasters or at polygon level in vectors. Consequently, this method is independent of spatial resolution.

M. Paegelow (✉)
Département de Géographie, Aménagement et Environnement,
Université Toulouse Jean Jaurès, Toulouse, France
e-mail: martin.paegelow@univ-tlse2.fr

D. García-Álvarez
Departamento de Geología, Geografía y Medio Ambiente,
Universidad de Alcalá, Alcalá de Henares, Spain

© The Author(s) 2022
D. García-Álvarez et al. (eds.), *Land Use Cover Datasets and Validation Tools*,
https://doi.org/10.1007/978-3-030-90998-7_12

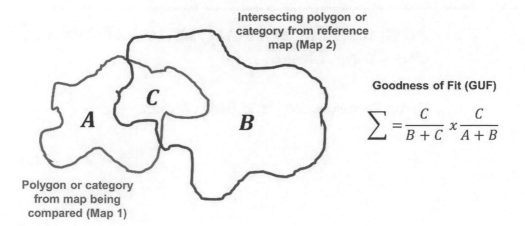

Fig. 1 Goodness of Fit (GOF) algorithm, where $\sum$ refers to all the polygons or patches in Map 2 intersecting each polygon or patch in Map 1; A refers to the area of each polygon or patch in Map 1 that is not intersected with polygons or patches in Map 2; B refers to the area of each polygon or patch in Map 2 that is not intersected with polygons or patches in Map 1; and C refers to the area of intersection between polygons or patches from Maps 1 and 2

GOF values range from 0 to 1. Maximum GOF (1) is obtained when there is full overlap between two polygons or patches. If there is no overlap, GOF is 0. If overlap affects half the area of the polygons or patches, GOF will be 0.5.

When comparing pairs of maps, the GOF value may vary depending on whether the assessed map is evaluated against the reference map or the reference map is evaluated against the assessed map. Map Curves calculates the GOF values for both these operations. It then uses the highest of these two GOF values in the comparison.

GOF values may be obtained either for the whole dataset or for the set of patches or polygons that make up each category on the map. Although it is technically possible to calculate a GOF for each individual polygon or patch, it is computationally very demanding and is not normally done.

Based on the GOF metrics at the category level, the results of the map comparison may be expressed in a graph, which shows the percentage of the categories in the map that have a specific GOF value. For example, if there are 10 categories and 2 of these have a GOF value of ≥ 0.8, the graph will show that 20% of the categories have GOF values of ≥ 0.8.

Utility

Exercises
1. To validate a map against reference data/map
2. To validate a simulation against a reference map
3. To validate simulated changes against a reference map of changes
4. To validate a series of maps with two or more time points

Map Curves provides a simple metric for assessing the extent to which two datasets share the same spatial structure, i.e. the same number and shape of polygons or patches. Unlike many other metrics, GOF evaluates the spatial agreement between maps at a polygon or patch level. In most cases, this type of analysis is based on raster data and comparisons are made at cell level. However, polygons or patches reflect the real structure of a landscape better than cells. GOF therefore provides a better, more realistic method for validating the similarity between maps than cell-based metrics.

GOF provides a standard and, therefore, comparable metric. The GOF value in one validation exercise may be compared with the GOF value obtained in another. Consequently, when using this metric to assess validity, we can establish a general minimum acceptable GOF threshold above which the map can be considered valid.

Map Curves gives an overview of the pattern agreement for the whole landscape and at category level. However, it does not provide information about the agreement per polygon. This means that a few polygons that do not show good overlap when comparing the maps could be hidden in the general analysis. Thus, as currently implemented, this technique only provides information on spatial agreement at a category level and does not shed light on disagreements occurring at more detailed scales of analysis.

The fact that GOF is unaffected by the spatial resolution used in the analysis should be considered an important strength, as spatial resolution is one of the main sources of uncertainty associated with any validation exercise. Nonetheless, at very coarse spatial resolutions, the area and shape of some polygons and patches can become very distorted, and this could affect the results of the analysis. Therefore, when used with rasters, GOF can be considered independent of spatial resolution below a certain threshold.

We do not recommend validating the spatial structure of a map by comparing it with another map obtained at a different resolution. Changes in spatial resolution or scale will always result in changes in the spatial structure of the maps.

The results of the analysis will highlight not only the differences between the original maps in the way they represent LUC in the landscape, but also the differences produced by changes in the spatial resolution.

Although Map Curves could be a useful tool for comparing the agreement of the spatial pattern between different maps, its results must be treated with caution when validating the pattern of the maps. This is because Map Curves only assesses the degree of overlap between the patches or polygons belonging to each category in the two maps compared. If the overlap is low, the GOF score obtained by Map Curves analysis will also be low. However, this only means that their classes do not overlap well and does not imply that the two maps being compared have completely different patterns.

Spatial metrics (see Chap. "Spatial Metrics to Validate Land Use Cover Maps") are more suitable for validating the pattern of the map. Even if there is no spatial overlap, they provide objective information about the fragmentation of the landscape or the complexity of the polygons/patches, which can be used when comparing two maps. Spatial metrics therefore allow us to compare pattern agreement between maps, even if they do not locate land uses in the same positions.

QGIS Exercises

Available tools

- Processing Toolbox
  R
     Pattern evaluation
        Map Curves raster R script
        Map Curves vector R script

There is no default tool in QGIS for carrying out Map Curves analysis. It is however implemented in R. We have developed two R tools for QGIS to perform the Map Curves analysis for either raster or vector data. To learn how to configure QGIS to work with R scripts, see Chap. "About This Book" of this book. This also explains how to install the different R scripts required to do some of the exercises presented in the book.

The *Map Curves raster* script is based on the code developed by Professor Emiel van Loon from the University of Amsterdam.[1] The script provides full Map Curves results. These consist of: (i) the GOF value of the analysis, with details of the map used as a reference; (ii) the table for the

GOF between categories; and (iii) the Map Curves graph. The R code of the *Map Curves raster* script also allows us to compare raster and vector maps. However, the vector option is unstable and does not always produce correct results. Its use is therefore not recommended.

The *Map Curves vector* script, which can only be employed to compare vector maps, is based on the "Sabre" R package.[2] Unlike the previous script, it only provides information on the overall GOF between the two maps and the map used as a reference when obtaining it.

The *Map Curves raster* script provides more information than the *Map Curves vector* script. It is also much faster and more efficient. We therefore recommend that this analysis be carried out with raster data.

Exercise 1. To validate a map against reference data/map

Aim

To check the agreement between the SIOSE and CORINE maps, considering SIOSE as a valid reference. We will assess to what extent the spatial structure of the CORINE map (number of polygons, shape) is similar to the SIOSE map.

Materials

SIOSE Land Use Map Asturias Central Area 2011
CORINE Land Use Map Asturias Central Area 2011

Requisites

The two maps must be raster and have the same projection. Although the tool does work with raster maps at different extents and with different thematic resolutions, we recommend comparing rasters with the same or very similar extents and thematic resolutions, so as to avoid results that may not be particularly meaningful.

Execution

If necessary, install the Processing R provider plugin, and download the *MapCurves_raster.rsx* R script into the R scripts folder (processing/rscripts). For more details, see Chap. "About This Book" of this book.

---

[1] The code is available on the Professor's personal website: https://www.uva.nl/en/profile/l/o/e.e.vanloon/e.e.vanloon.html.

[2] Full details of this R package and the functions it includes can be found at: https://cran.r-project.org/web/packages/sabre/index.html.

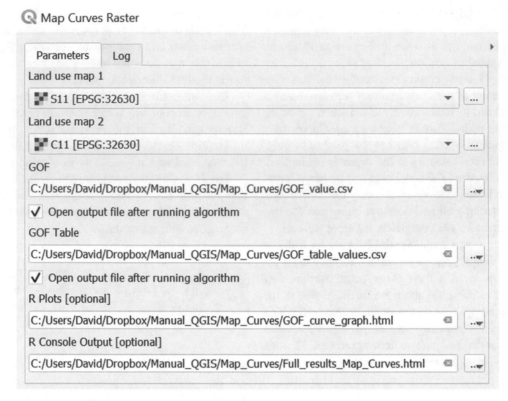

Fig. 2 Exercise 1. Step 1. Map Curves Raster R script

Step 1

Open the *Map Curves Raster* function and fill in the required parameters. These are basically the two LUC maps to be compared: "Land Use map 1" (SIOSE) and "Land Use map 2" (CORINE) (Fig. 2).

Results and Comments

After running the function, we obtain two tables and one graph. All the information, with the exception of the graph, will also be displayed in the "Log" window (Fig. 3).

The GOF value is a measure of the general agreement between the two maps being compared. This value ranges from 0 to 1, with 0 meaning no agreement and 1 total agreement. The GOF value for our comparison (0.54) indicates that the agreement between the two maps is significant, although not very high. The patches of the same categories partially overlap.

The reference map ($Refmap) value informs us as to which map was used as the reference when obtaining the GOF value. If value "A" is obtained, it means that "Land use map 1" was used as the reference map in the comparison. If value "B" appears, it means that "Land use map 2" was used. Therefore, in our case, a GOF of 0.54 was obtained when comparing SIOSE and CORINE and taking CORINE as the

reference. If SIOSE had been taken as the reference, agreement (GOF value) would have been lower.

The GOF table details the GOF value for agreement per category, so providing a measure of how similar the pattern for a particular category is in the two maps. It therefore answers the following question: to what extent do the patches that make up a particular category overlap in the two maps being compared?

In our case, the category that shows the greatest pattern agreement between the two maps is water bodies (Category 11), with a GOF value of 0.968. Agricultural areas (Category 0; GOF 0.783) and vegetation areas (Category 1; GOF 0.800) also show high levels of agreement. By contrast, agreement between the two maps is very low for road and rail networks (Category 6; GOF 0.112).

If we observe the two maps, most of the agreement and disagreement is due to the fact that they follow different Minimum Mapping Unit (MMU) and Minimum Mapping Width (MMW) criteria. Thus, if a patch is larger than the MMU and MMW of both maps, it will be similarly mapped in both cases. However, if a patch is drawn in SIOSE, but is too small for the MMU and MMW of CORINE, this will lead to disagreement between the two maps.

This explains the results for Category 6 (road and rail networks). Whereas many patches representing road and rail networks are mapped in SIOSE, most of them are not

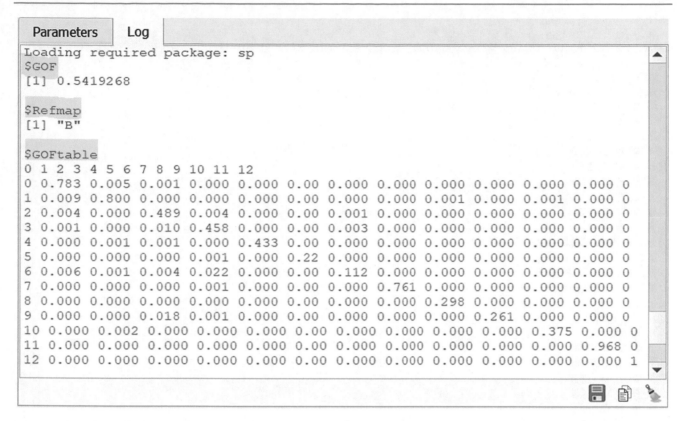

Fig. 3 Results from Exercise 1 displayed in the Log window of the Map Curves Raster script. General GOF value and GOF table

mapped in CORINE because they are less than 100 m wide and therefore do not comply with its MMW criterion (Fig. 3). As a result, the agreement for this category in terms of overlapping patches is very low. Although in the few patches for this category in which the two maps overlap the agreement is high, in most cases the SIOSE road and rail networks patches do not overlap with patches in CORINE, and the agreement is null. Overall, the agreement for this category in the two maps is very low, with a GOF of just 0.112.

In this exercise, the GOF values for the different categories did not indicate a high degree of similarity between the category patterns on the two maps. On the contrary, they indicated different patterns of fragmentation for each category because of the different MMU and MMW rules applied in each map.

In addition to the overall GOF and the GOF table detailing the GOF agreement per category, the Map Curves function also produces two extra tables: the $BMC_A2B and the $BMC_B2A (Fig. 4).

Unlike the other two tables, these tables are only displayed in the "Log" window and are not stored in any folder. For each category, they indicate the category with which it shows most agreement (GOF) on the other map. Whereas, the information in the first table ($BMC_A2B) was obtained using map A (Land use map 1) as the reference, the information in the second table ($BMC_B2A) was obtained using map B (Land use map 2) as the reference.

When Land use map 1 (SIOSE) was used as the reference map, the agricultural areas (category 0) in SIOSE showed the best agreement with the agricultural areas (category 0) in CORINE. The GOF value was 0.783, which indicates a very high overlap between the patches of this category on the two maps.

For Land use map 2 (CORINE), the agricultural areas (category 0) showed the best agreement with the agricultural areas (category 0) of SIOSE. The GOF value was the same as that obtained when SIOSE was used as the reference. In this category it therefore makes no difference which map is used as the reference map.

All the categories showed their best agreement with the same category on the other map. In other words, agricultural areas in Map 1 showed their best agreement with agricultural areas in Map 2, and vegetation areas in Map 1 showed their best agreement with vegetation areas in Map 2 etc. This indicates that the two maps are thematically consistent, i.e. the categories are distributed in a similar way in both maps.

Finally, the last result provided by the Map Curves function is the Map Curves graph (Fig. 5), which is stored in .png format in the folder specified when running the tool (R plots). The graph presents the same information provided in the GOF table. It represents the percentage of categories that

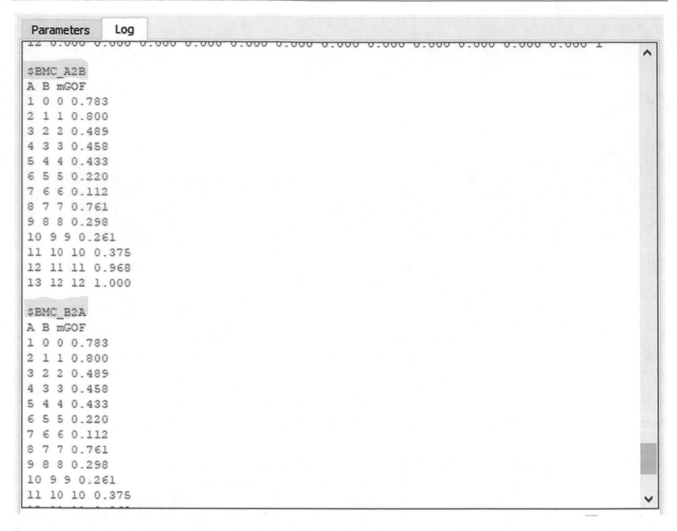

```
Parameters    Log
12 0.000 0.000 0.000 0.000 0.000 0.000 0.000 0.000 0.000 0.000 0.000 0.000 1

$BMC_A2B
A  B  mGOF
1  0  0  0.783
2  1  1  0.800
3  2  2  0.489
4  3  3  0.458
5  4  4  0.433
6  5  5  0.220
7  6  6  0.112
8  7  7  0.761
9  8  8  0.298
10 9  9  0.261
11 10 10 0.375
12 11 11 0.968
13 12 12 1.000

$BMC_B2A
A  B  mGOF
1  0  0  0.783
2  1  1  0.800
3  2  2  0.489
4  3  3  0.458
5  4  4  0.433
6  5  5  0.220
7  6  6  0.112
8  7  7  0.761
9  8  8  0.298
10 9  9  0.261
11 10 10 0.375
```

Fig. 4 Results from Exercise 1 displayed in the Log window of the Map Curves Raster R script. Tables indicating the categories with wich each category in the reference map show the highest agreement

reach or exceed a specific GOF threshold. Thus, all the categories (100%) always have a GOF score higher than 0. However, only around 40% of the categories in this map have a GOF score of over 0.5 and none of the categories show perfect agreement (0% of the categories have a GOF score of 1) (Fig. 5).

The graph provides the GOF scores using either Land use map 1 (A) or Land use map 2 (B) as a reference. It is therefore a good summary of the pattern agreement between the two maps.

In summary, in this exercise we have noted that although the GOF value is not very high, CORINE has a very similar pattern to SIOSE. The lower GOF is the result of different pattern fragmentation in the two maps: SIOSE maps have many small patches that do not appear in CORINE. However, if we look at the maps, the polygons from the same category usually overlap very well and have a similar pattern structure. In addition, thematic agreement, as we noted in the $BMC_A2B and $BMC_B2A tables, seems to be very high.

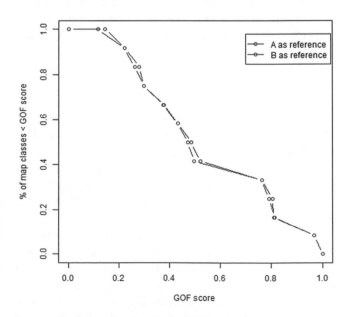

Fig. 5 Result from Exercise 1. Map Curves graph

Exercise 2. To validate a simulation against a reference map

Aim

To assess the similarity between the spatial structure of a simulation and the spatial structure of a map used as a reference.

Materials

Simulation CORINE Asturias Central Area 2011
CORINE Land Use Map Asturias Central Area 2011

Requisites

The two maps must be raster and have the same projection. Although the tool works with raster maps at different extents and with different thematic resolutions, we recommend that raster maps with the same or very similar extents and thematic resolutions be compared so as to avoid results that may be not fully informative. For a proper validation, the reference map must be for the same year as the simulation.

Execution

If necessary, install the Processing R provider plugin and download the *MapCurves_raster.rsx* R script into the R scripts folder (processing/rscripts). For more details, see Chap. "About This Book".

Step 1

Open the *Map Curves Raster* function and fill in the required parameters: "Land Use map 1" (CORINE simulation) and "Land Use map 2" (CORINE reference map) (Fig. 6).

Results and Comments

After running the tool, a GOF value was obtained for the whole maps compared and broken down per pair of classes (GOF table). The GOF values are stored in different tables and displayed in the "Log" window ($GOF, $GOFtable). The GOF values per pair of classes are also represented in the Map Curves graph, which is stored in the specified folder (R Plots).

The GOF value for our comparison is very high (0.92). This is logical given that most of the simulated landscape did not change over the simulation period and, therefore, remained the same. Permanence is one of the easiest processes to simulate in LUC modelling. This means that the reference and the simulated maps look very similar. The patterns of the two maps are very similar because most of the pattern remains unchanged over the simulation period and was correctly simulated as such.

The agreement (GOF) per category was always very high. The minimum scores were for port areas (0.669) and mineral extraction sites (0.708). In the modelling exercise, these categories were treated as features (categories that remained

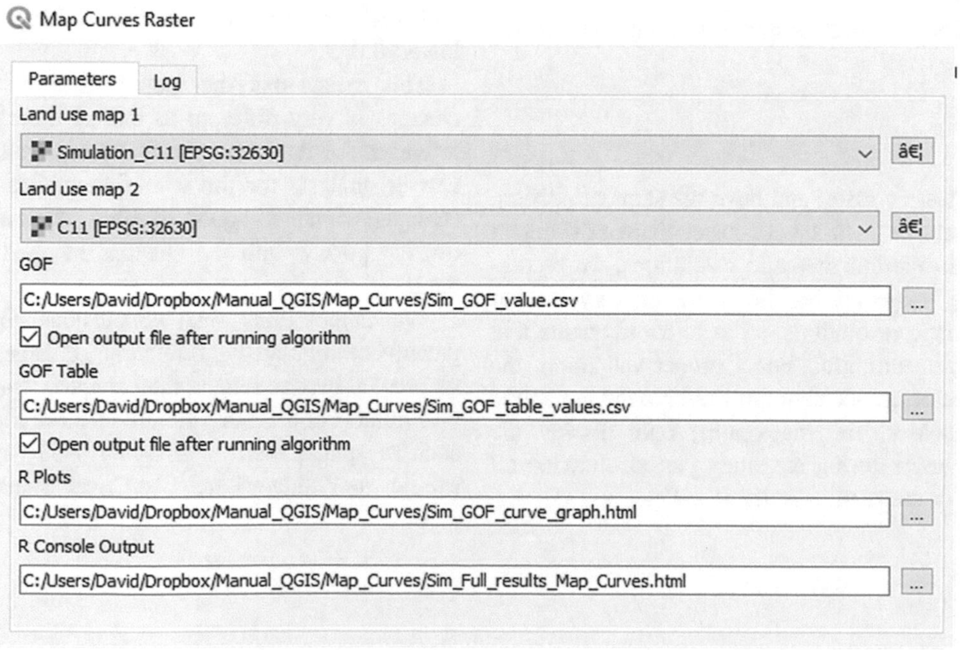

Fig. 6 Exercise 2. Step 1. Map Curves Raster R script

invariant during the simulation) and were therefore not simulated. However, a few changes did in fact occur in these categories in the reference map. As a result, the Map Curves analysis produced a relatively poor fit for these categories when comparing the simulation with the reference map. Whereas no change occurred in these categories in the simulation, a few changes did take place in the reference map. Given that these categories consist of a very small number of patches, even a small number of changes can reduce the GOF values substantially.

All in all, this analysis is not particularly meaningful. It confirms that the two compared maps have very similar patterns because most of the landscape was correctly simulated as permanence. However, more meaningful results could be obtained by focusing exclusively on the areas that were simulated as change. Hence, for a proper validation of the simulation, the simulated changes must be compared with the changes observed on the reference maps.

Exercise 3. To validate simulated changes against a reference map of changes

Aim

To evaluate how similar the changes we simulated in our modelling exercise are to those observed on the reference map.

Materials

CORINE Land Use Changes Asturias Central Area 2005–2011
Simulated CORINE changes Asturias Central Area 2005–2011

Requisites

The two maps must be raster and have the same projection. Although the tool does work with raster maps at different extents and with different thematic resolutions, we recommend comparing rasters with the same or very similar extents and thematic resolutions, so as to avoid results that may not be very meaningful. For a proper validation, the simulation and the reference map must refer to the same time period. In both cases, the maps must only display the changes that occurred during the study period, showing all other areas as 0 or some other suitable code.

Execution

If necessary, install the Processing R provider plugin and download the *MapCurves_raster.rsx* R script into the R scripts folder (processing/rscripts). For more details, see Chap. "About This Book".

Step 1

Open the *Map Curves Raster* function and fill in the required parameters: "Land Use map 1" (Simulated CORINE changes) and "Land Use map 2" (CORINE changes) (Fig. 7).

Results and Comments

After running the function, we get the overall GOF ($GOF) value, the GOF value per category ($GOFtable) and the Map Curves graph (R Plots). In this case, the only results that might be useful for interpreting the validity of the simulated changes are the results per category.

The general GOF value is 0.3, but this is artificially high due to the almost perfect overlap of class 0 (areas with no change) which has a GOF value of 0.993 (Table 1). A high level of agreement between areas of permanence is always expected, as explained in detail in the previous exercise (Exercise 2). In this case, however, we want to assess the agreement between simulated changes and reference map changes for the two classes that were modelled actively: urban fabric and industrial and commercial areas.

The spatial overlap between these two categories in the two maps is very low. The GOF value for urban fabric (Category 3 in the maps) is only 0.05. In the case of industrial and commercial maps (Category 4) it is even lower: 0.039.

This means that the spatial structure of the simulated changes is very different to that of the changes used as a reference for the same period. Thus, even though the Map Curves analysis for the whole simulation (persistence and changes) obtained good results, the simulated changes overlap poorly with the changes mapped in the reference data.

We cannot draw final conclusions about the different patterns of simulated and reference changes. Even if there is no overlap between them, their shape or fragmentation could be similar. For a clearer picture of these aspects, other tools, such as spatial metrics, must be used (see Chap. "Spatial Metrics to Validate Land Use Cover Maps").

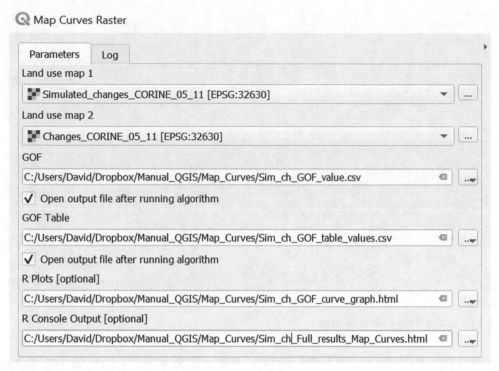

Fig. 7 Exercise 3. Step 1. Map Curves Raster R script

Table 1 Result from Exercise 3 showing the class GOF values between observed and simulation land use

	0	1	2	3	4	5	6	7	8	10	11
0	**0.993**	0.001	0.001	0.001	0	0	0	0.001	0.001	0	0
3	0	0	0	**0.05**	0	0	0	0	0	0.023	0
4	0	0	0	0	**0.039**	0	0	0	0	0	0

Exercise 4. To validate a series of maps with two or more time points

Aim

To test the consistency of the pattern of land uses in a series of LUC maps made up of two different time points.

Materials

CORINE Land Use Map Asturias Central Area 2005 v.0
CORINE Land Use Map Asturias Central Area 2011

Requisites

The two maps must be raster and have the same projection. It is also recommended that they have similar extents and thematic resolutions.

Execution

If necessary, install the Processing R provider plugin and download the *MapCurves_raster.rsx* R script into the R scripts folder (processing/rscripts). For more details, see Chap. "About This Book".

Step 1

Open the *Map Curves Raster* function and fill in the required parameters: "Land Use map 1" (CORINE 2005) and "Land Use map 2" (CORINE 2011) (Fig. 8).

Results and Comments

The results show the level of overall agreement between the pair of maps compared ($GOF), the agreement per category ($GOFtable), the best matches between categories ($BMC_A2B, $BMC_B2A) and the Map Curves graph (R plots). All results are displayed in the "Log" window and stored in the preselected folders.

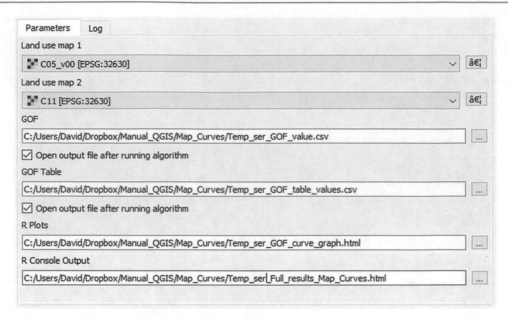

Fig. 8 Exercise 4. Step 1. Map Curves Raster R script

The overall agreement between our maps is 0.5, which is not high. This means that there is only partial overlap between the categories in the two maps. In a series of two or more Land Use maps, persistence is the norm and one would expect almost perfect overlap between the maps for most of the landscape. Landscapes must be very dynamic to experience changes affecting more than 10% of the study area. The Asturias Central Area is not a dynamic landscape of this kind. The low GOF score therefore suggests that a lot of the differences between the two maps are due to technical changes or errors.

When agreement was assessed at the category level, the only very high values were for water bodies (Category 11),

with a GOF of 0.961 (Fig. 9), and background (Category 12), with a GOF of 1. The background is therefore identical in the two maps, whereas the water bodies have an almost perfect overlap. The small difference between the two maps for the water bodies category (0.039) could be due to spurious or erroneous changes, although real changes in the areas covered by water may also have taken place.

The agricultural areas (0.709), vegetation areas (0.704) and airports (0.778) show a high level of agreement between the two maps. However, there are still important differences between them that cannot be explained solely by the normal land use dynamism of the study area, in which only small changes usually take place.

```
$GOFtable
   0   1   2   3   4   5   6   7   8   9  10  11  12
0  0.709 0.020 0.005 0.001 0.000 0.000 0.001 0.000 0.000 0.001 0.000 0.000 0
1  0.016 0.704 0.000 0.000 0.000 0.000 0.000 0.000 0.000 0.000 0.004 0.000 0
2  0.001 0.000 0.460 0.016 0.000 0.000 0.001 0.000 0.000 0.010 0.000 0.000 0
3  0.000 0.000 0.005 0.467 0.001 0.000 0.013 0.000 0.000 0.000 0.000 0.000 0
4  0.000 0.001 0.000 0.000 0.300 0.003 0.000 0.000 0.000 0.000 0.002 0.000 0
5  0.000 0.000 0.000 0.000 0.000 0.183 0.000 0.000 0.000 0.000 0.000 0.000 0
6  0.001 0.000 0.001 0.001 0.000 0.004 0.113 0.000 0.000 0.000 0.000 0.000 0
7  0.000 0.000 0.000 0.001 0.000 0.000 0.000 0.407 0.000 0.000 0.000 0.000 0
8  0.000 0.000 0.000 0.000 0.000 0.000 0.000 0.000 0.778 0.000 0.000 0.000 0
9  0.000 0.000 0.000 0.000 0.000 0.000 0.000 0.001 0.000 0.245 0.000 0.000 0
10 0.000 0.000 0.000 0.000 0.000 0.000 0.000 0.000 0.000 0.000 0.155 0.000 0
11 0.000 0.000 0.000 0.000 0.000 0.000 0.000 0.004 0.000 0.000 0.000 0.961 0
12 0.000 0.000 0.000 0.000 0.000 0.000 0.000 0.000 0.000 0.000 0.000 0.000 1
```

Fig. 9 Result from Exercise 4. GOF matrix

```
$BMC_A2B              $BMC_B2A
A B mGOF              A B mGOF
1  0  0  0.709       1  0  0  0.709
2  1  1  0.704       2  1  1  0.704
3  2  2  0.460       3  2  2  0.460
4  3  3  0.467       4  3  3  0.467
5  4  4  0.300       5  4  4  0.300
6  5  5  0.183       6  5  5  0.183
7  6  6  0.113       7  6  6  0.113
8  7  7  0.407       8  7  7  0.407
9  8  8  0.778       9  8  8  0.778
10 9  9  0.245       10 9  9  0.245
11 10 10 0.155       11 10 10 0.155
12 11 11 0.961       12 11 11 0.961
13 12 12 1.000       13 12 12 1.000
```

Fig. 10 Results from Exercise 4. Tables indicating for each category in the reference map the category in the compared map with wich it shows the highest agreement. On the right, agreements when using map A as the reference. On the left, agrements when using map B as the reference

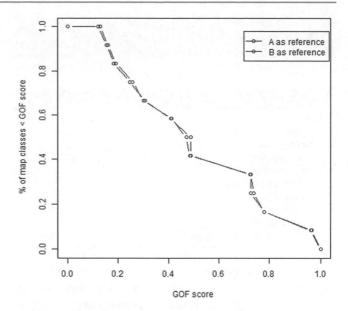

Fig. 11 Result from Exercise 4. Map Curves graph

For all the other categories, agreement is low or very low. Nonetheless, there is no evidence of systematic confusion between one category on the first map and a different category on the second. This is confirmed by the tables showing the best matches between categories (Fig. 10) in which the best match for each category (i.e. the largest overlap or agreement) was always with the same category on the other map.

The low agreement or overlap between the categories in the two maps is also summarized in the Map Curves graph (Fig. 11), which shows that only around 40% of the classes on the maps obtained a GOF score of over 0.5. This means that more than half the categories show poor overlaps, i.e. most of the categories are mapped very differently on each map.

All in all, we can conclude that the time series we assessed has many errors and uncertainties and is therefore affected by many erroneous or spurious changes. These are changes that did not really happen on the ground and arose due to technical reasons, such as different production methods. In a coherent time series of LUC maps, high GOF scores of 0.9 or over would be expected.

The low agreement in our exercise is due to the change in the methodology used to produce the Spanish CORINE Land Cover maps between 2006 and 2011. The CORINE 2005 map (v.00) used in this exercise was obtained using photointerpretation of satellite imagery. However, from 2011 onwards the CORINE maps were obtained by generalizing more detailed Land Use maps (SIOSE). This change

in the production method resulted in LUC maps with important differences from their predecessors. In order to solve this problem, the Copernicus service produced another CORINE map for 2005 in Spain according to the new methodology, which was consistent and comparable with the CORINE 2011 map. This more recent version of the CORINE 2005 map is the one normally used in the different exercises of this book.

2 Change on Pattern Borders

Description

In pairs of maps or time series, this technique is used to identify the changes taking place on the edges of patches. The allocation of changes (on the edge of an existing patch or a new disconnected one) provides useful information about the nature of change dynamics: the expanding or shrinking of existing boundaries or the appearance of new land use patches.

Utility

Exercises
1. To validate a series of maps with two or more time points

By detecting the changes taking place on the edges of the patches, we can assess both the type of landscape dynamics taking place and the data errors resulting from different data sources, classifiers or spectral responses.

QGIS Exercise

Available tools

- Raster
    - *Raster Calculator*
  Conversion
    - *Polygonize*
- Vector Overlay
    - *Extract by location*
- Vector Table
    - *Field Calculator*
- Vector Analysis
    - *Basic statistics for fields*

For the sake of simplicity, we will only be presenting the tools used in this exercise, although we are aware that there are many other tools that could be used to carry out this analysis.

Exercise 1. To validate a series of maps with two or more time points

Aim

To focus on gains taking place on the edges of patches for a specific land use/cover category. We can then assess the proportion of change taking place on the edges of existing patches compared to the change that appears in new, disconnected areas.

Materials

CORINE Land Cover Map Val d'Ariège 2000
CORINE Land Cover Map Val d'Ariège 2018

Requisites

All maps must be rasters and have the same resolution, extent and projection.

Execution

Step 1

First, we extract forests in 2000 (Fig. 12) and then new forested locations in 2018 (non-forest in 2000 AND forest in 2018) using the *Raster Calculator* (Fig. 13).

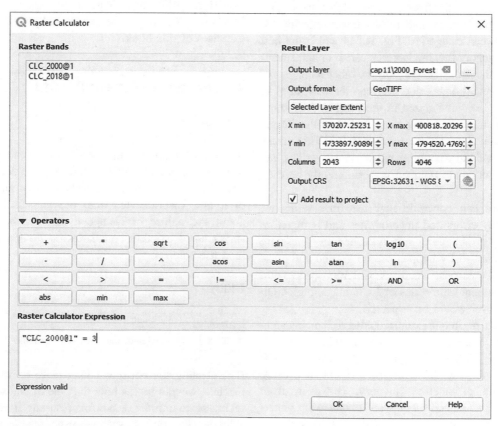

Fig. 12 Exercise 1. Step 1. Raster Calculator

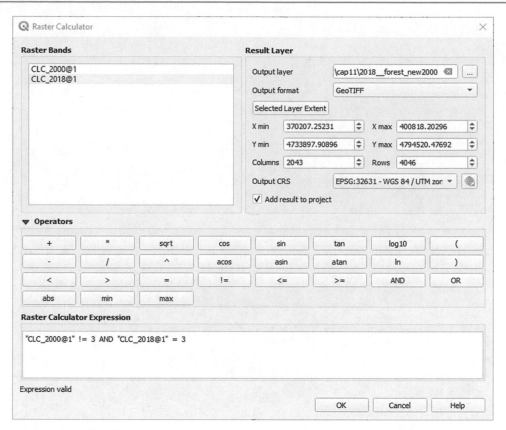

Fig. 13 Exercise 1. Step 1. Raster Calculator

Figure 14 shows the result as an overlay of the two maps obtained: forest in 2000 in light green and forest gains between 2000 and 2018 in dark green.

Step 2

We then vectorize the binary raster maps computed in Step 1 using the *Polygonize* Raster Conversion function with no specific parameters.

Step 3

We now isolate the forest gains on the edge of the pattern. The aim is to distinguish between new areas of forest in 2018 (i.e. that did not exist in 2000) which are contiguous with forests that existed in 2000 and others that are not. For this purpose, we use the *Extract by location* Vector Selection tool with the 'touch' operator (Fig. 15).

Figure 16 shows a detail from the resulting layer: the forests that existed in 2000 are shown in light green, while the new forests that appeared in 2018 separately from existing forests are in dark green. The new forests that appeared in connection with forests that already existed in 2000 are overlaid in brown.

Step 4

In this step we will isolate the new forests that are not connected to forests that existed in 2000. This step is optional insofar as new forest patches not connected to forests that existed in 2000 can be obtained simply by subtracting new connected forests from the total area for new forests.

To get an independent layer of new forest in 2018 that is not connected to forests that existed in 2000, we use the same *Extract by location* tool, opting this time for the 'disjoint' operator (Figs. 17 and 18).

Forest 2000

Forest gains 2000-2018

0   20 km

Fig. 14 Exercise 1. Step 1. Intermediate map displaying the overlay of forest areas in 2000 in light green and the overly of forest gains between 2000 and 2018 in dark green

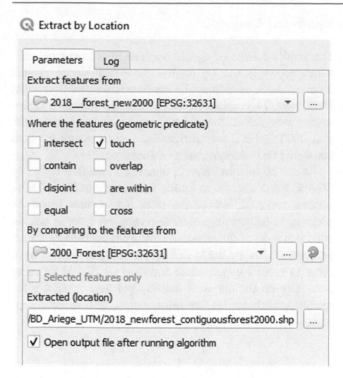

Fig. 15 Exercise 1. Step 3. Extract by Location tool

Step 5

The next step is to calculate the area covered by new connected/unconnected forests. We use the Vector table *Field Calculator* tool to create a new attribute called area_ha (decimal number), selecting the $area operator, divided by 10,000 to calculate the area in ha (Fig. 19).

This operation is carried out for both connected and isolated forests. The updated attribute tables are shown in Fig. 20: table for connected new forests on the left, and for unconnected new forests on the right.

Step 6

Of the various tools available to summarize the characteristics of the assessed patches, we use the *Basic statistics for fields* vector analysis tool. On the left of Fig. 21 we can see the various parameters that must be filled in, and on the right the log containing the sum of the areas of unconnected new patches of forest.

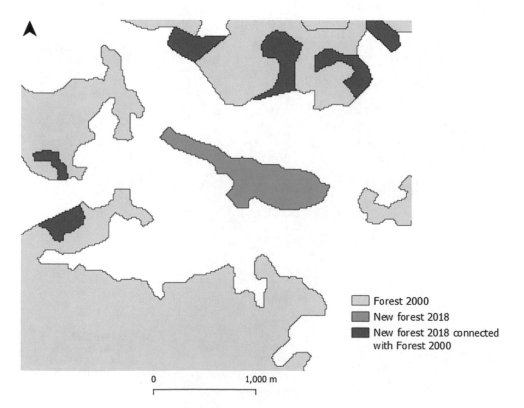

Fig. 16 Exercise 1. Step 3. An examople area of the resulting raster layer

Fig. 17 Exercise 1. Step 4. Extract by Location tool

Results and Comments

The results consist of update attribute tables and statistics, which appear in the log for the *Basic Statistics for Fields* function. After examining the attribute tables, we found that there were 74 contiguous and 2 isolated polygons representing new forests that did not appear on the map for the year 2000. Table 2 summarizes the basic statistics for both connected and unconnected new forest patches.

As can be seen in Table 2, almost all new forest patches (97.4%) are connected to forests that existed in 2000. These patches cover 92.94% of the total area of new forest. In addition, to better interpret these results, we have to bear in mind that most of the analysed territory is covered by forest; there are too few isolated patches of new forest to allow us to come to general conclusions; and changes take place more frequently on the edges of existing patches, especially for semi-natural dynamics like reforestation, than in new, separate areas of the landscape.

Fig. 18 Exercise 1. Step 4. An example area from the resulting raster layer showing not connected features

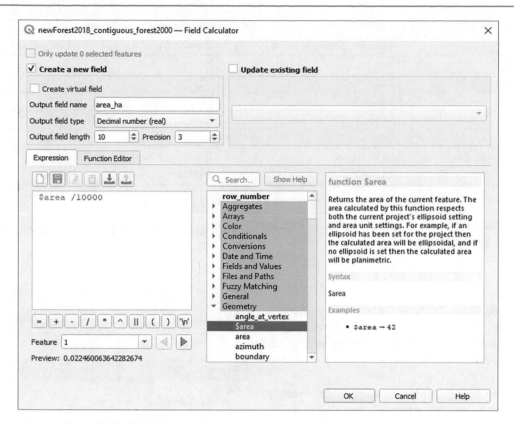

Fig. 19 Exercise 1. Step 5. Vector Table Field Calculator

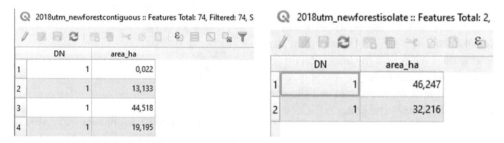

Fig. 20 Exercise 1. Step 5. Updated attribute tables

Fig. 21 Exercise 1. Step 6. Basic Statistics for Fields tool

Table 2 Results from Exercise 1. Spatial metrics for both, connected and not connected new forest patches

	Total new patches of forest	Connected patches	Unconnected patches
Number of patches	76	74	2
Minimum area (ha)	0.02	0.02	32.22
Maximum area (ha)	246.22	246.22	46.25
Mean area (ha)	14.63	13.97	39.23
Median area (ha)	5.61	4.40	39.23
Area standard deviation (ha)	31.34	31.48	7.02
Total area (ha)	1,112.00	1,033.54	78.46

3 Allocation Error Distance

Description

Allocation error distance refers to the distance between a wrongly allocated pixel compared to the closest object belonging to the same category on the reference map. It can be measured in different ways:

(a) The minimum distance from the edge of the wrongly allocated patch to the edge of the closest patch belonging to the same category on the reference map.

(b) The distance between the centroids of the two patches described in (a).

Allocation distance error can be expressed in terms of (i) individual pixels/patches, (ii) LUC classes (mean distance) or (iii) the mean distance for all the allocation errors. The mean allocation distance error can be usefully completed by calculating the minimum, maximum and standard deviation values when applied to several patterns (LUC class or whole map).

Utility

Exercises
1. To validate a simulation against a reference map (vector)
2. To validate a simulation against a reference map (raster)

Simulation accuracy can be measured in different ways, such as quantity agreement, allocation agreement, landscape structure agreement, etc. (Hagen-Zanker 2006; Paegelow et al. 2014) as described in Part III of this book. Generally, the indices and maps assessing allocation error tend to focus on the amount involved. Here we go further by measuring "how wrong" the simulation errors are. This analysis, which measures the individual (entity) or mean error distance (LUC class), is complementary to the cross-tabulation of maps at varying spatial resolution, often implemented by fuzzy logic.

QGIS Exercises

Available tools
• Raster
Raster Calculator
• Raster
Analysis
Proximity
• Processing Toolbox
GRASS
r.distance
r.grow.distance
• Processing Toolbox
SAGA
Distance

GRASS and SAGA toolboxes offer several algorithms for measuring the distance inside a raster grid (*r.grow.distance*; SAGA *distance*) or the minimum distance between pixels/patches belonging to two different grid layers (*r.distance*). Their use inside QGIS may be unstable.

Vector analysis tools require converting raster layers into vector format and then calculate the centroids of the polygons obtained. The *Distance to nearest centre (points)* tool creates a points layer whose table contains minimum distances between the points in one layer to the nearest point in the second layer.

Both tools (raster and vector) are used in the next two exercises because they provide complementary results.

Exercise 1. To validate a simulation against a reference map (vector)

Aim

To calculate the seriousness (degree) of allocation errors for a specific LUC category, expressed as the minimum mean distance between all the pixels wrongly allocated to this category in the simulation and the nearest patch belonging to the same category on the reference map.

Materials

CORINE Land Cover Map Val d'Ariège 2018
Simulation LCM Val d'Ariège 2018

Requisites

Maps can be raster or vector. They must have the same resolution, extent and projection. If using vector maps, readers can skip the first steps detailed in the execution.

Execution

Step 1

We extract real built-up areas in 2018 (Fig. 22) and the pixels wrongly allocated as built-up areas with the *Raster Calculator* (Fig. 23). They are areas wrongly simulated as built-up areas, which are not built up according to the reference map.

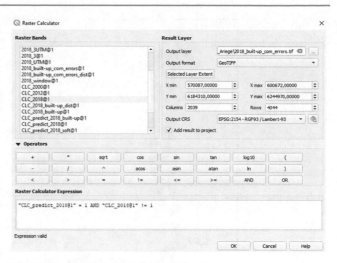

Fig. 23 Exercise 1. Step 1. Raster Calculator

The right map (A) in Fig. 24 is an overlay of real built-up areas (light grey) in 2018 (Corine Land Cover) and areas wrongly simulated as built-up (black). The left map in Fig. 24 represents the allocation errors that we will now go on to analyse.

Step 2

The two raster layers obtained in Step 1 are now polygonized into vector layers. This is done using the *Polygonize* function in the Raster—Conversion menu (Fig. 25).

The above map (Fig. 26) shows an overlay of the two vector layers: real built-up polygons in 2018 (reference map) and areas wrongly allocated as built-up (red) by the simulation. Results vary depending on whether or not diagonal connexions are allowed.

Step 3

We then calculate the centroids for each of these vector layers with the *Centroids* tool (Vector—Geometric tools) (Fig. 27).

Step 4

Once we have obtained the two centroids maps (built-up areas in 2018 and built-up allocation errors), we use the *Distance to nearest hub (points)* tool available in the Processing Toolbox (QGIS Vector). The source points layer is the point layer containing allocation errors and the destination hubs layer is the layer containing the built-up centroids from the reference map (Fig. 28). We measure the distance in metres and give the output point layer a name.

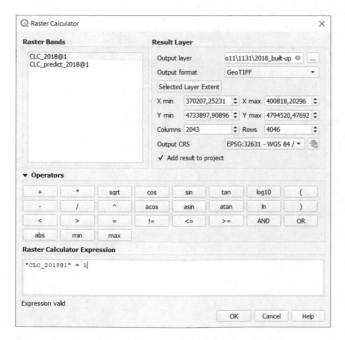

Fig. 22 Exercise 1. Step 1. Raster Calculator

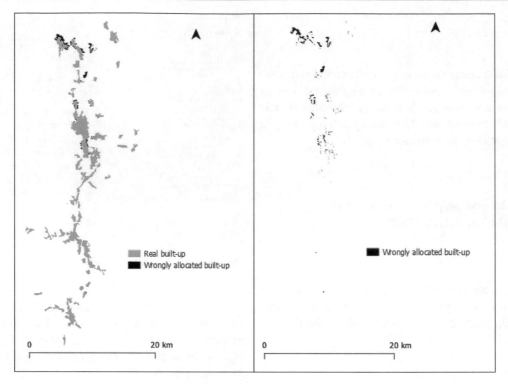

Fig. 24 Exercise 1. Step 1. Intermediate map showing the built-up areas correctly allocated in light gray (A) and the wrongly simulated built-up areas in black (B)

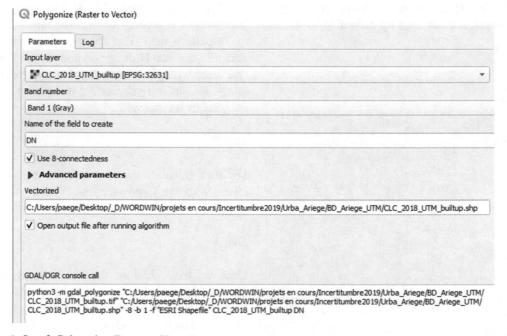

Fig. 25 Exercise 1. Step 2. Polygonize (Raster to Vector)

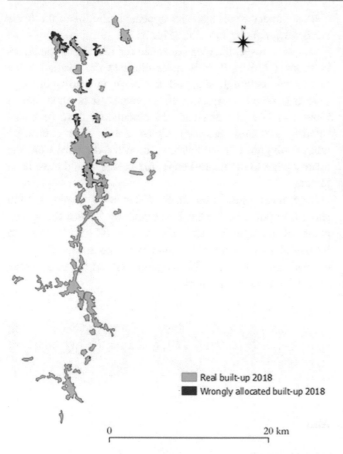

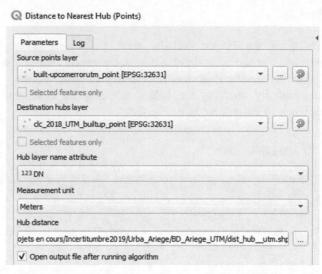

Fig. 28 Exercise 1. Step 4. Distance to Nearest Hub (Points)

Fig. 26 Exercise 1. Step 2. Intermediate map showing built-up areas correctly simulated in cyan and wrongly allocated built-up areas in red

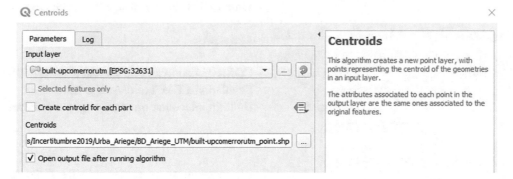

Fig. 27 Exercuse 1. Step 3. Centroids

Step 5

To obtain the desired statistics about the allocation error distance for wrongly simulated built-up areas, we use the *Basic statistics for fields* tool (Processing Toolbox, Vector—analysis) by selecting the field containing the calculated distance to the nearest hub (Fig. 29).

Fig. 29 Exercise 1. Step 5. Basic Statistics for Fields

Results and Comments

The resulting points layer contains the same number of points as the allocation error polygons at the same location. The corresponding table contains the minimum distance between each allocation error (centroid) and the nearest existing built-up area (centroid) on the reference map (Fig. 30).

	DN	HubName	HubDist
1	1	0	4052,723485740...
2	1	0	3506,743550374...
3	1	1	436,9969389902...
4	1	1	2646,064842092...
5	1	0	11715,63198073...

Fig. 30 Result from Exercise 1. Attribute table indicating (HubDist) the minimum distance between each allocation error and the nearest built-up area

A summary of the statistics appears in the log of the *Basic statistics for fields* function (Fig. 31).

As we can see, the mean distance for 132 allocation errors is about 1,236 m. This is quite close to the median value (1,119 m), although standard deviation is also quite high (775 m). When interpreting these values, it is important to remember how the distance was calculated: from centroids offering a one-dimensional representation of the built-up areas (polygons). If we had measured the distance from the nearest edge to the nearest edge, the values would have been lower.

The mean allocation error distance of about 1.2 km should be put into context by comparing it with the spatial extent of the layer, which is about 31 × 62 km. It may also be useful to compare this value with the mean allocation error distances for other LUC categories and the mean value for all the allocation errors.

Exercise 2. To validate a simulation against a reference map (raster)

Aim

To calculate the seriousness (degree) of allocation errors for a specific LUC category expressed as the minimum, individual and mean distance between wrongly allocated areas (simulation map) and the nearest patch belonging to the same LUC category on the reference map.

Materials

CORINE Land Cover Map Val d'Ariège 2018
Simulation LCM Val d'Ariège 2018
Built-up allocation error map (generated during Exercise 1)

Requisites

All maps must be rasters and have the same resolution, extent and projection.

Execution

Step 1

First, we compute a raster distance map up from built-up areas using the QGIS raster function *Proximity* (Fig. 32). If the built-up areas layer is not available, it must be extracted

```
Execution completed in 0.12 seconds
Results:
{'COUNT': 132,
'CV': 0.6272372130225321,
'EMPTY': 0,
'FILLED': 132,
'FIRSTQUARTILE': 608.3191480711826,
'IQR': 1188.9517389924517,
'MAJORITY': 35.285750182722104,
'MAX': 4031.2025628251654,
'MEAN': 1235.8905702081654,
'MEDIAN': 1118.7276322916427,
'MIN': 35.285750182722104,
'MINORITY': 35.285750182722104,
'OUTPUT_HTML_FILE': 'C:\\Users\\paege\\AppData\\Local\
\Temp/processing_1e206d02af6140b286011f29f1b8ac1c/
48165055bd1d4cdfa74fb6da4652f7a9/OUTPUT_HTML_FILE.html',
'RANGE': 3995.9168126424433,
'STD_DEV': 775.1965568581977,
'SUM': 163137.55526747784,
'THIRDQUARTILE': 1797.2708870636343,
'UNIQUE': 132}
```

Fig. 31   Result from Exercise 1. Log window from Basic Statistics for Fields tool

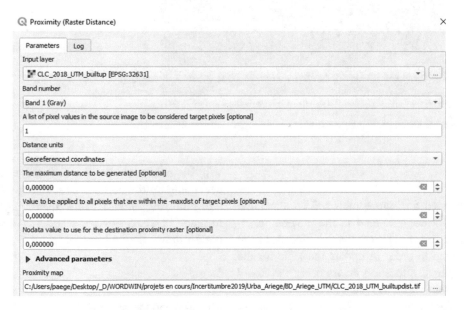

Fig. 32   Exercise 2. Step 1. Proximity (Raster Distance)

from the CLC_2018 layer using *Raster Calculator* (see Step 1 of the previous exercise).

In the *Proximity* tool, the input layer is built-up areas in 2018. We have to specify the target pixels (allocation errors = 1) and the fact that we want to calculate the distance in Coordinate Reference System (CRS) units (Fig. 32). The result is shown in Fig. 33. This map illustrates the distance between areas wrongly allocated to built-up (red) in the simulation and real built-up areas on the reference map (mapped in grey).

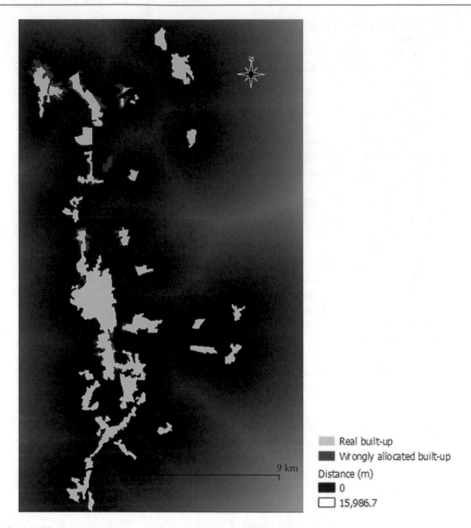

Real built-up
Wrongly allocated built-up
Distance (m)
0
15,986.7

Fig. 33 Exercise 2. Step 1. Distance map between wrongly simulated and real built-up areas

Step 2

Once you have obtained a distance map and an allocation error map in vector format (obtained in the previous exercise, Step 2), the next step involves extracting statistics from the raster distance map in order to update the table for the polygon (vector) layer of allocation errors using the *Zonal statistics* tool (Processing toolbox) (Fig. 34).

Open this function and choose the distance map to built-up areas 2018 (reference map) and the vector layer containing the allocation errors for the built-up category in 2018 (simulation). The table (Fig. 36) for the vector layer

will be enhanced by one or more additional columns depending on the number of statistics selected. In this case, the following values were measured: minimum, mean, median, standard deviation and maximum (Fig. 35). Figure 36 shows the updated table.

Step 3

The third and last step can be done on a spreadsheet. We will calculate the mean values (mean, median, standard deviation, minimum and maximum) for the individual distances extracted (Table 3).

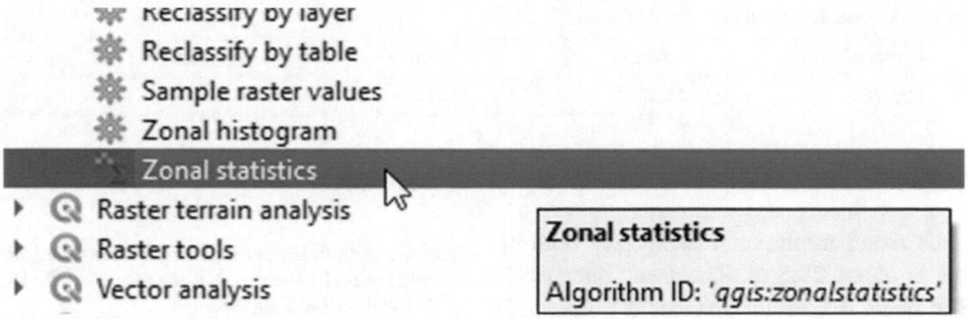

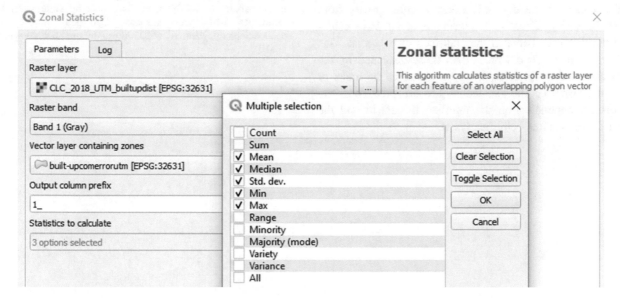

Fig. 34   Exercise 2. Step 2. Zonal Statistics

Fig. 35   Exercise 2. Step 2. Zonal Statistics

	DN	1_mean	1_median	1_stdev	1_min	1_max
1	1	71,33016204833...	74,91667175292...	34,18298759754...	14,98333358764...	134,8500061035...
2	1	37,23363629509...	33,50375366210...	20,49935032771...	14,98333358764...	91,14006042480...
3	1	58,41231830596...	54,02317810058...	34,33703505354...	14,98333358764...	156,4306030273...
4	1	21,56998003446...	21,18963241577...	7,267292783947...	14,98333358764...	33,50375366210...
5	1	82,94153703754...	74,91667175292...	55,17205590117...	14,98333358764...	239,7333374023...
6	1	39,63797190491...	29,96666717529...	23,99610053437...	14,98333358764...	89,90000152587...
7	1	14,98337658053...	18,08648300170...	4,388485586978...	14,98333358764...	21,18963241577...
8	1	14,98333358764...	14,98333358764...	0	14,98333358764...	14,98333358764...

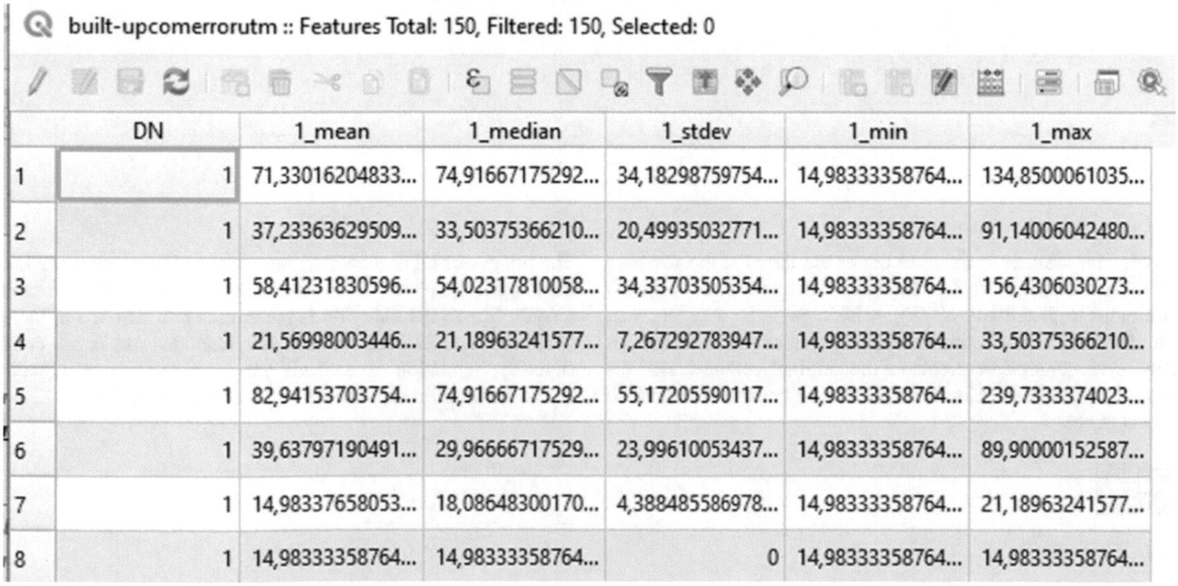

built-upcomerrorutm :: Features Total: 150, Filtered: 150, Selected: 0

Fig. 36   Exercise 2. Step 2. Updated attribute table

Table 3 Results from Exercise 2. Calculated statistics

	_mean	_median	_stdev	_min	_max
Mean	56.96	53.11	21.98	28.54	106.81

Results and Comments

As we can see, the mean minimum distance for built-up commission errors is about 28.5 m. The mean distance is close to 57 m. The mean maximum distance is quite small (106.81) and the standard deviation is low (21.98). This means that allocation errors affect small patches or are close to the right location.

The values obtained in this exercise differ greatly from those obtained in Exercise 1. During Exercise 1 we calculated the distances between the centroids of polygons. This may result in longer distances than those generated by the technique used in Exercise 2, which measures the mean or minimum distance. The two techniques can produce different results, depending on the number, the extent and the shape of the features being analysed.

References

Hagen- A (2006) Map comparison methods that simultaneously address overlap and structure. J Geogr Syst 8:165–185. https://doi.org/10.1007/s10109-006-0024-y

Hargrove WW, Hoffman FM, Hessburg PF (2006) Mapcurves: a quantitative method for comparing categorical maps. J Geogr Syst 8:187–208. https://doi.org/10.1007/s10109-006-0025-x

Paegelow M, Camacho Olmedo MT, Mas JF, Houet T (2014) Benchmarking of LUCC modelling tools by various validation techniques and error analysis. Cybergeo. https://journals.openedition.org/cybergeo/26610. https://doi.org/10.4000/cybergeo.26610

Geographically Weighted Methods to Validate Land Use Cover Maps

Ramón Molinero-Parejo

Abstract

One of the most commonly used techniques for validating Land Use Cover (LUC) maps are the accuracy assessment statistics derived from the cross-tabulation matrix. However, although these accuracy metrics are applied to spatial data, this does not mean that they produce spatial results. The overall, user's and producer's accuracy metrics provide global information for the entire area analysed, but shed no light on possible variations in accuracy at different points within this area, a shortcoming that has been widely criticized. To address this issue, a series of techniques have been developed to integrate a spatial component into these accuracy assessment statistics for the analysis and validation of LUC maps. Geographically Weighted Regression (GWR) is a local technique for estimating the relationship between a dependent variable with respect to one or more independent variables or explanatory factors. However, unlike traditional regression techniques, it considers the distance between data points when estimating the coefficients of the regression points using a moving window. Hence, it assumes that geographic data are non-stationary i.e., they vary over space. Geographically weighted methods provide a non-stationary analysis, which can reveal the spatial relationships between reference data obtained from a LUC map and classified data. Specifically, logistic GWR is used in this chapter to estimate the accuracy of each LUC data point, so allowing us to observe the spatial variation in overall, user's and producer's accuracies. A specific tool (*Local accuracy assessment statistics*) was specially developed for this practical exercise, aimed at validating a Land Use Cover map. The Marqués de Comillas region was selected as the study area for implementing this tool and demonstrating its applicability. For the calculation of the user's and producer's accuracy metrics, we selected the tropical rain forest category [50] as an example. Furthermore, a series of maps were obtained by interpolating the results of the tool, so enabling a visual interpretation and a description of the spatial distribution of error and accuracy.

Keywords

Geographically Weighted Regression • Overall accuracy • User's accuracy • Producer's accuracy

1 Overall, User's and Producer's Accuracy Through GWR

Description

Overall accuracy (OA), user's accuracy (UA) and producer's accuracy (PA) are assessment metrics obtained from the cross-tabulation matrix (see Sect. 5 in chapter "Metrics Based on a Cross-Tabulation Matrix to Validate Land Use Cover Maps"). Overall accuracy is expressed as the proportion of the map that has been correctly classified. User's accuracy indicates the probability that a pixel from a specific category on the classified map correctly represents the real situation on the ground or reference map. Producer's accuracy indicates the probability that a reference pixel belonging to a specific category has been correctly allocated to that category (Story and Congalton 1986). These last two metrics (user's and producer's accuracies) refer to commission and omission errors, respectively.

None of these accuracy assessment statistics produces spatially distributed information, i.e., they provide a single accuracy value for the entire study area or for each land use/land cover class. However, it is possible to explore how the error and accuracy of a classified map is spatially distributed with respect to reference data using Geographically Weighted Regression (GWR) methods.

R. Molinero-Parejo (✉)
Departamento de Geología, Geografía y Medio Ambiente,
Universidad de Alcalá, Alcalá de Henares, Spain
e-mail: ramon.molinero@uah.es

© The Author(s) 2022
D. García-Álvarez et al. (eds.), *Land Use Cover Datasets and Validation Tools*,
https://doi.org/10.1007/978-3-030-90998-7_13

GWR allow us to explore local spatial relationships between a dependent variable and a set of explanatory variables (Brunsdon et al. 1996; Fotheringham et al. 2002). In this chapter, we use the logistics version of the geographical weighting method (GWLR) to generate land use/land cover accuracy metrics with spatial variation, according to the proposal by Comber (2013), which was later developed in Comber et al. (2012), Comber et al. (2017) and Tsutsumida and Comber (2015).

GWR is a statistical technique in which regression points are estimated on the basis of the spatial distribution of data points. A moving window analyses the data points it collects to estimate the coefficients of the selected regression point. This window, or kernel, weights each data point according to the distance within the window and the assigned weighting function (*gaussian, exponential, bisquare, tricube, boxcar*). Its maximum weighting value is 1 and this decreases as the distance between the observation and calibration data points increases. The size of the kernel is defined by the bandwidth, which indicates the number of data points that will be included in the local calculation for each regression point. This can consider either a fixed or a variable number of reference data points. If a fixed number of points are considered, a specific number will be obtained, while in the case of a variable number, a distance value is given. The number of reference data points therefore varies according to their distribution. It is important to select a suitable bandwidth so as to minimise the cross-validation prediction error. According to Fotheringham et al. (2002), the GWR formula is

$$y_i = \beta_{0(u_i,v_i)} + \sum_n \beta_{n(u_i,v_i)} x_n$$

where β_0 is the intercept, β_n is the coefficient, x_n is the value of the explanatory variable, and u_i, v_i are the coordinates of the data point (Fig. 1).

This geographically weighted method was adapted for the calculation of local accuracy assessment statistics by Comber (2013). According to his proposal, the probability that a reference data point is correctly identified by a classified data point is given by

$$\text{Overall accuracy} \rightarrow P(A = 1) = \text{logit}(\beta_{0(u_i,v_i)})$$

where $P(A = 1)$ is the probability that the agreement between the classified data and the reference data is equal to 1. This value is 0 when there is no agreement and 1 when there is agreement.

To estimate user's accuracy, it is necessary to analyse the reference data against the classified data. This metric indicates the probability that the reference LUC class y_i and is correctly predicted by the classified data x_i.

$$\text{User's accuracy} \rightarrow P(y_i = 1)$$
$$= \text{logit}\left(\beta_{0(u_i,v_i)} + \beta_{1(u_i,v_i)} x_i\right).$$

To estimate producer's accuracy, it is necessary to analyse the classified data against reference data. This indicates the probability that the classified data x_i correctly represents reference LUC class y_i.

$$\text{Producer's accuracy} \rightarrow P(x_i = 1)$$
$$= \text{logit}\left(\beta_{0(u_i,v_i)} + \beta_{1(u_i,v_i)} y_i\right)$$

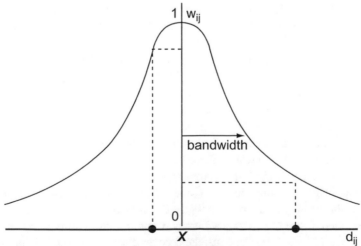

X regression point w_{ij} is the weight of data point *j* at regression point *i*

● data point d_{ij} is the distance between regression point *i* and data point *j*

Fig. 1 Spatial kernel. Regression point, data points and bandwidth are observed. The curve represents the Gaussian function that establishes the weighting of the data points for the regression point. Retrieved from Fotheringham et al. (2002)

Finally, in order to obtain the accuracy values, the coefficients have to be adjusted. To this end, the coefficients are added together, and an alogit function (inverse logit) is applied.

Utility

Exercises
1. To validate a map against reference data/map

Geographically Weighted methods can be used to validate single LUC maps by analysing spatial variations in the agreement between reference data and classified remotely sensed data, so enabling us to analyse the spatial non-stationarity of LUC data error and accuracy. They allow to explore the spatial relationships between the reference data and the classified data, exposing possible clusters of land cover errors, and reporting the values for each data point in contrast to global accuracy assessment statistics, which only provide a global value for the entire map.

This technique allows us not only to discover what proportion of the map has been correctly classified but also to estimate in which areas the classification fits best and to analyse possible trends that are only visible spatially. In this way, the spatial distribution of the overall, user's and producer's accuracy metrics can be visualized on a map so as to enable a better understanding of classification uncertainty.

QGIS Exercise

Available tools
▪ Processing Toolbox

    R
        Geographically weighted methods
            Local accuracy assessment statistics
        Interpolation
            IDW Interpolation
        GDAL
            Raster extraction
                Clip raster by mask layer

By default, there are no tools in QGIS that carry out a Geographically Weighted Methods analysis to estimate overall, user's and producer's accuracy values for local areas. We have therefore developed an R tool to calculate these local accuracy assessment metrics in QGIS, in which Geographically Weighted Methods are already implemented.

The *Local accuracy assessment statistics* script is based on the code developed by Professor Alexis J. Comber from the University of Leicester,[1] which was created using above all the "spgwr" R package.[2] The script provides overall, user's and producer's accuracy values for each data point, so allowing accuracy and error distribution areas to be generated by interpolation of the results obtained by the tool.

First, to estimate local OA values, the tool calculates internally, for each data point, the agreement between the reference data and the classified data, where 0 represents disagreement and 1 represents agreement. Agreement is automatically selected as dependent variable [y] and "1" is selected as independent variable [x], where $P(A = 1)$ is the probability that agreement is equal to 1.

To estimate local UA values, the tool generates a new data frame and obtains two columns. One column shows the presence (1)/absence (0) of the chosen category for the reference data, while the other column shows the same for the classified data. The reference data (RD) is selected as dependent variable [y], and the classified data (CD) is selected as independent variable [x], where $P(RD = 1|CD = 1)$. The procedure for producer's accuracy is very similar. The classified data for the chosen category is selected as dependent variable [y], and the reference data is selected as independent variable [x], where $P(CD = 1|RD = 1)$.

In order to ensure that the tool works correctly, various parameters must be configured. Selecting an appropriate bandwidth is therefore crucial. A small bandwidth would include too few data points in the local sample, making it unreliable for calibrating the model, while a large bandwidth would include too many data points, so reducing the local analysis capacity. A spatially distributed data sample is also required.

The fact that the parameters must be configured and the need for more in-depth knowledge to interpret the results could be considered a disadvantage when choosing these validation methods. Another important consideration is that using large data samples can lead to long runtimes.

Exercise 1. To validate a map against reference data/map

Aim

To assess the spatial variation of accuracy assessment measures (overall, user's and producer's accuracy) when

---

[1] The code is available at the personal repository of Professor Alexis J. Comber. https://github.com/lexcomber/AccuracyWorkshop2016.
[2] Full details of this R package and the functions it includes, may be found at https://cran.r-project.org/web/packages/spgwr/spgwr.pdf.

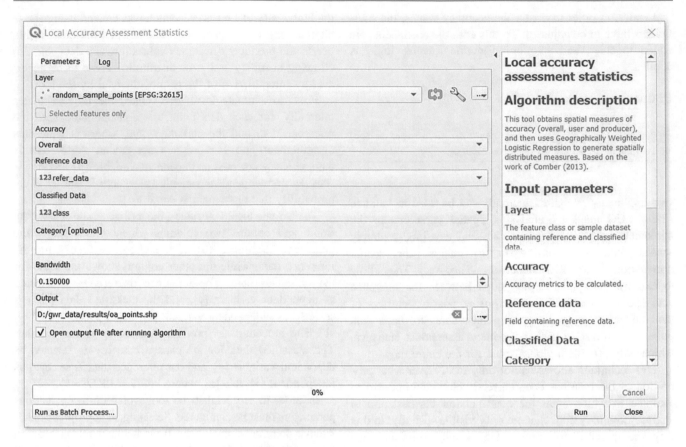

Fig. 2 Excersice 1. Step 1. Local accuracy assessment statistics (Overall accuracy)

validating the Marqués de Comillas LUC map against a reference set of points.

Materials

Marqués de Comillas random sample points from Mexico (2019)
Boundary of Marques de Comillas

Requisites

The data points must be projected in their corresponding reference system. The vector point file must include two attributes, one corresponding to reference LUC data and one to classified LUC data. It is recommended that the data points have an appropriate random distribution.

Sample size should not be overly large, as this could lead to long runtimes.

Execution

If necessary, install the Processing R provider plugin, and download the *Local accuracy assessment statistics.rsx* R script into the R scripts folder (processing/ rscripts). For more details, see chapter "About This Book" of this book.

Step 1

Open the *Local accuracy assessment statistics* function and fill in the required parameters (see Fig. 2). The input for this tool is the point layer containing the LUC random sample dataset. Select the type of accuracy assessment statistic to be obtained ("Overall"), and indicate the corresponding

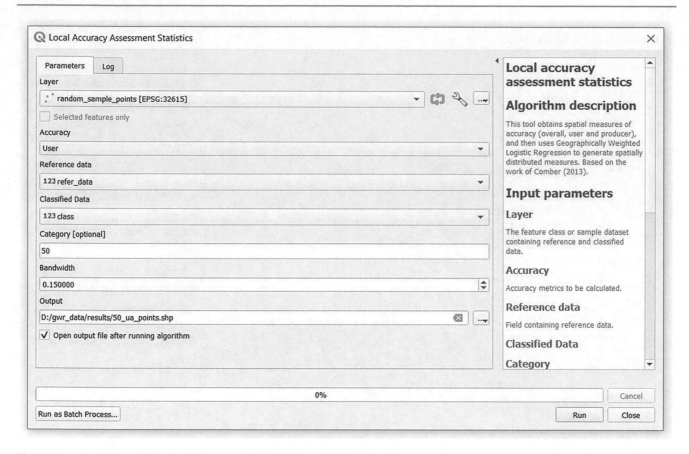

Fig. 3 Excersice 1. Step 2. Local accuracy assessment statistics (User's accuracy)

attribute table columns with the reference data and the classified data. The category can also be indicated, although this is only used to estimate the user's and producer's accuracy values. The remaining value to be set is the bandwidth, which in this exercise is 0.15. This means that 15% of the nearest neighbours will be used to estimate the coefficient for each regression point. The kernel is set internally in the tool by default with a *Gaussian* function.

Step 2

The parameter configuration for calculating User's Accuracy is very similar. Select the corresponding accuracy assessment statistic in the "Accuracy" option ("User") and the category you want to assess in the "Category" option, (see

Fig. 3). In this exercise, we will be using the tropical rain forest class [50] as an example.

Step 3

To estimate the producer's accuracy values, the same steps must be followed (see Fig. 4). Select the corresponding accuracy assessment statistic ("Producer"), and the tool will modify the internal inputs. The tropical rain forest class [50] will again be used as an example.

Step 4

Finally, the coefficients adjusted by the *Local accuracy assessment statistics* tool were interpolated using the Inverse

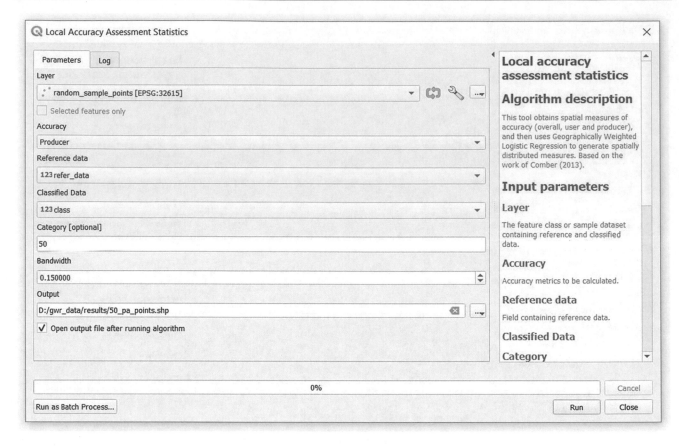

Fig. 4 Excersice 1. Step 3. Local accuracy assessment statistics (Producer's accuracy)

Distance Weighted method (*IDW interpolation* tool in QGIS) (see Fig. 5) to obtain a map showing the continuous variation in the spatial distribution of the accuracy measures, and to facilitate understanding in a more visual manner.

The names of the column or attribute obtained as a result of applying the tool and indicating the local overall, user's and producer's accuracy values are "g__SDF_", "coefs_u" and "coefs_p" respectively. This column must be specified in the "Interpolation attribute" option in line with the accuracy metric being analysed.

Step 5

As an additional, optional step, the raster images obtained by interpolation can be clipped by mask using the Marques de Comillas boundary (*Clip raster by mask layer* tool in QGIS) in order to provide a better visual representation. In addition,

a discrete colour scale using six classes was chosen in order to make interpretation of the data more straightforward.

Results and Comments

After the execution of the previous steps, we obtain a new attribute column with the estimated local values for OA, UA and PA respectively, and the interpolated distribution maps for these accuracy measures. Another output of the tool is a new layer that includes the estimated Overall Accuracy value for each data point. In addition, a summary of the local and overall values calculated is displayed in the log window (Fig. 6). It shows the minimum, first quantile, median, mean, third quartile, maximum and global overall accuracy values (Table 1).

The IDW interpolation method is used to generate an area that visually represents the distribution of the values obtained, offering a more detailed spatial representation of

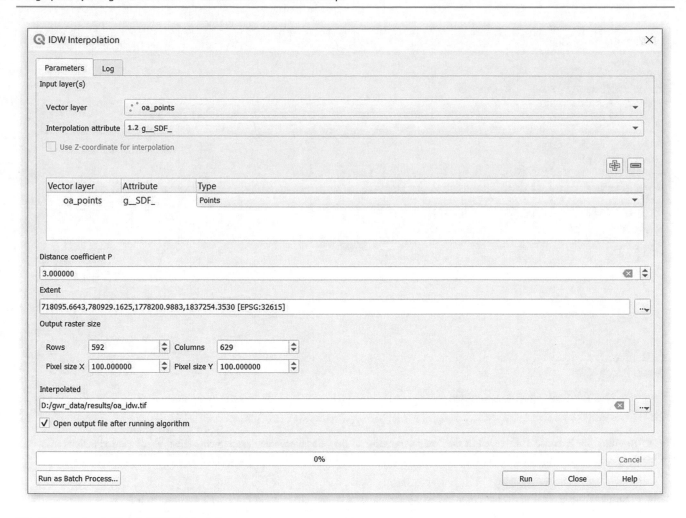

Fig. 5 Excersice 1. Step 4. IDW Interpolation

the distribution of accuracy and error than that provided by a single overall accuracy value. Figure 7 clearly shows a higher degree of accuracy in the north of the map, which decreases as it moves south and east.

The example category in this exercise is tropical rain forest (code 50). User's accuracy describes the commission errors in the tropical rain forest category. Its values range between 0.55 and 0.87, with a variation of 0.32, despite the overall value for the entire study area of 0.74 (Fig. 8).

Figure 9 represents the probability that a classified data point belonging to the tropical rain forest class is correctly represented by the reference data (User's accuracy). Values

are high through the centre and south of the region, but fall as we move away to the northeast.

The last part of this exercise focuses on Producer's Accuracy. In this case, it describes omission errors related to the tropical rain forest class. User's accuracy varies from 0.56 to 0.89 (variation of 0.33), despite the global value for the entire area of 0.74 (Fig. 10).

Figure 11 represents the probability that any reference data point is correctly classified (producer's accuracy). Most of the omission errors are concentrated in the north-east of our study area, while higher levels of producer's accuracy can be seen in the south-west.

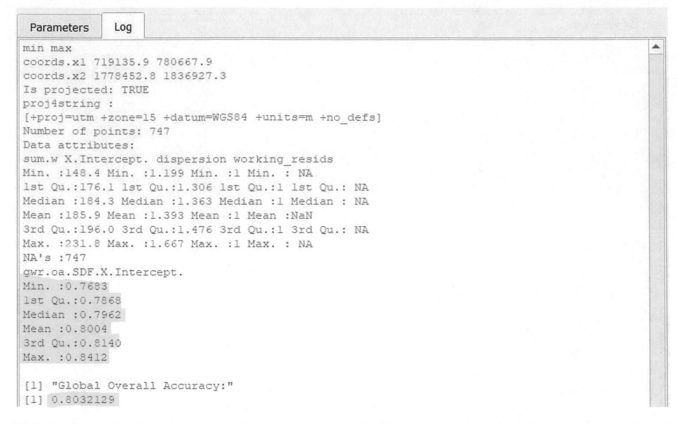

Fig. 6 Results from Exercise 1 displayed in the ``output' window of the ``Local accuracy assessment statistics' showing variations in overall accuracy

Table 1 Results from Exercise 1. Table summarizing the variations in Overall, User's and Producer's accuracy values

	Min	1st Qu	Median	Mean	3rd Qu	Max	Global
Overall Accuracy (OA)	0.7683	0.7868	0.7962	0.8004	0.8140	0.8412	0.8032
User's Accuracy (UA)[a]	0.5564	0.6702	0.7609	0.7432	0.8201	0.8777	0.7403
Producer's Accuracy (PA)[a]	0.5665	0.6424	0.6986	0.7168	0.7979	0.8905	0.7403

[a] These values are for the tropical rain forest class [50]

The values set out in Figs. 6, 8 and 10 are summarized in Table 1, which shows the variations in the accuracy of the classified data points with respect to the reference data points. The Overall accuracy value for the entire study area is 0.80. Nonetheless, it has been demonstrated that OA varies over space. The minimum value is 0.77 and the maximum is 0.84, which means that a variation of 0.07 is observed.

Producer's accuracy has the highest range of variation, with User's accuracy close behind. By contrast, Overall accuracy has a relatively small range, indicating low levels of spatial variation. Despite this, the maximum Overall accuracy value (0.84) is below the value proposed by Anderson (1971).

In conclusion, *Local accuracy assessment statistics* should be considered as a useful complement to the cross-tabulation matrix and its global accuracy statistics in that they provide more detailed information that can help improve classification techniques by locating possible error clusters with greater precision. It is also important to stress that a visual interpretation can enable better decisions to be taken when evaluating and validating LUC maps.

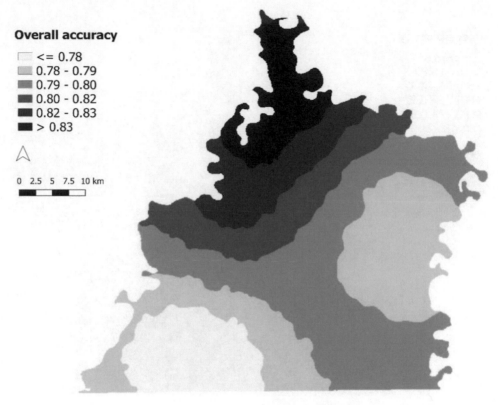

Fig. 7 Results from Exercise 1. Map showing the spatial distribution of overall accuracy values

```
Parameters    Log

min max
coords.xl 719135.9 780667.9
coords.x2 1778452.8 1836927.3
Is projected: TRUE
proj4string :
[+proj=utm +zone=15 +datum=WGS84 +units=m +no_defs]
Number of points: 747
Data attributes:
sum.w X.Intercept. df...x. dispersion working_resids
Min. :148.4 Min. :-4.005 Min. :3.397 Min. :1 Min. : NA
1st Qu.:176.1 1st Qu.:-3.662 1st Qu.:4.025 1st Qu.:1 1st Qu.: NA
Median :184.3 Median :-3.355 Median :4.467 Median :1 Median : NA
Mean :185.9 Mean :-3.375 Mean :4.490 Mean :1 Mean :NaN
3rd Qu.:196.0 3rd Qu.:-3.100 3rd Qu.:5.020 3rd Qu.:1 3rd Qu.: NA
Max. :231.8 Max. :-2.761 Max. :5.571 Max. :1 Max. : NA
NA's :747
coefs_ua
Min. :0.5564
1st Qu.:0.6702
Median :0.7609
Mean :0.7432
3rd Qu.:0.8201
Max. :0.8777

[1] "Global user accuracy:"
[1] 0.7402597
```

Fig. 8 Results from Exercise 1 displayed in the ``output' window of the ``Local accuracy assessment statistics' showing variations in user's accuracy

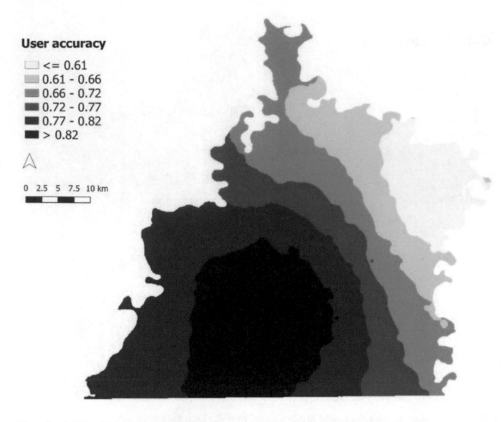

Fig. 9 Result from Exercise 1. Map showing the spatial distribution of user's accuracy values

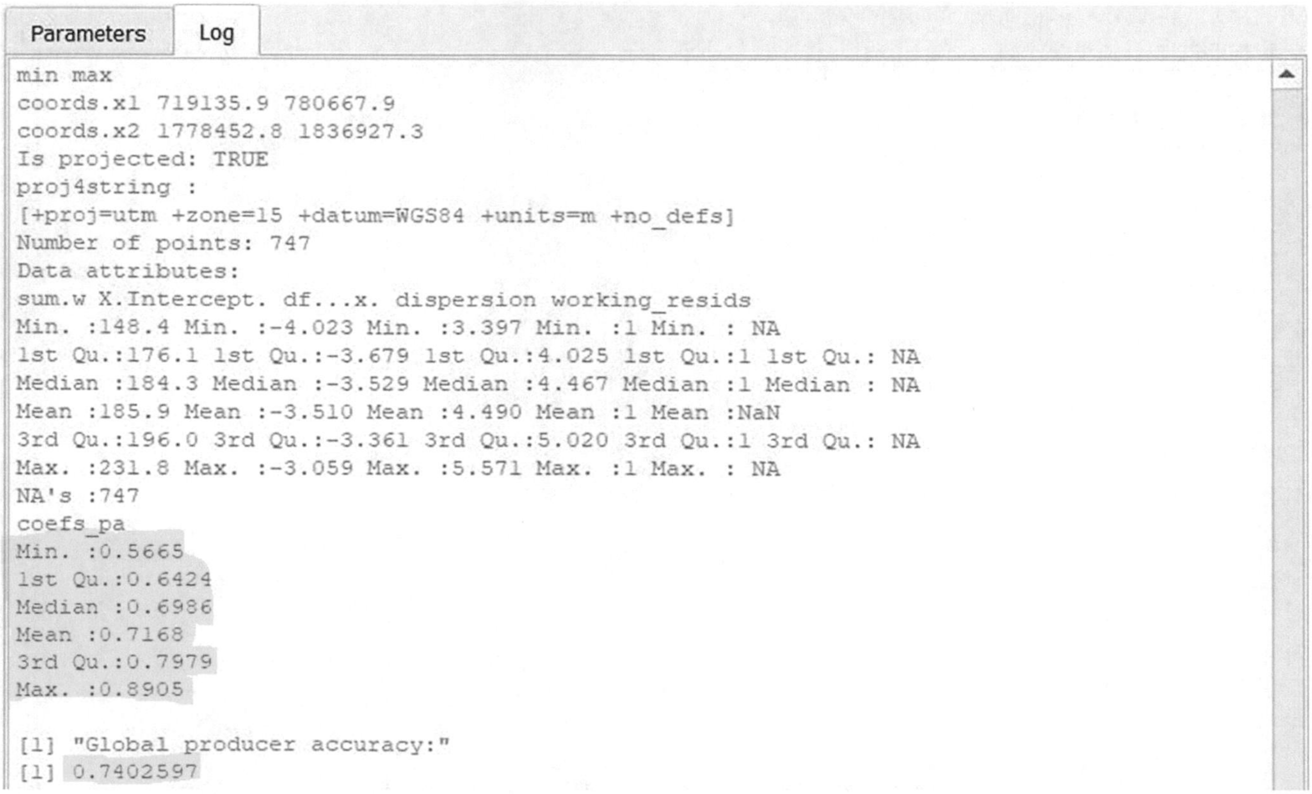

```
Parameters    Log

min max
coords.x1 719135.9 780667.9
coords.x2 1778452.8 1836927.3
Is projected: TRUE
proj4string :
[+proj=utm +zone=15 +datum=WGS84 +units=m +no_defs]
Number of points: 747
Data attributes:
sum.w X.Intercept. df...x. dispersion working_resids
Min. :148.4 Min. :-4.023 Min. :3.397 Min. :1 Min. : NA
1st Qu.:176.1 1st Qu.:-3.679 1st Qu.:4.025 1st Qu.:1 1st Qu.: NA
Median :184.3 Median :-3.529 Median :4.467 Median :1 Median : NA
Mean :185.9 Mean :-3.510 Mean :4.490 Mean :1 Mean :NaN
3rd Qu.:196.0 3rd Qu.:-3.361 3rd Qu.:5.020 3rd Qu.:1 3rd Qu.: NA
Max. :231.8 Max. :-3.059 Max. :5.571 Max. :1 Max. : NA
NA's :747
coefs_pa
Min. :0.5665
1st Qu.:0.6424
Median :0.6986
Mean :0.7168
3rd Qu.:0.7979
Max. :0.8905

[1] "Global producer accuracy:"
[1] 0.7402597
```

Fig. 10 Results from Exercise 1 displayed in the ``output' window of the ``Local accuracy assessment statistics' showing variations in producer's accuracy

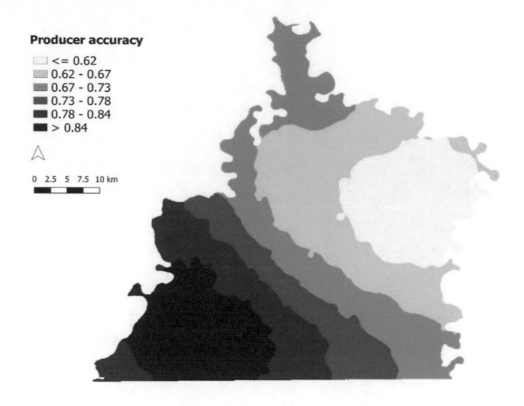

Producer accuracy

- ☐ <= 0.62
- ☐ 0.62 - 0.67
- ☐ 0.67 - 0.73
- ■ 0.73 - 0.78
- ■ 0.78 - 0.84
- ■ > 0.84

0 2.5 5 7.5 10 km

Fig. 11 Result from Exercise 1. Map showing the spatial distribution of producer's accuracy values

References

Brunsdon C, Fotheringham AS, Charlton ME (1996) Geographically weighted regression: a method for exploring spatial nonstationarity. Geogr Anal 28(4):281–298

Comber A, Brunsdon C, Charlton M, Harris P (2017) Geographically weighted correspondence matrices for local error reporting and change analyses: mapping the spatial distribution of errors and change. Remote Sens Lett 8(3):234–243. https://doi.org/10.1080/2150704X.2016.1258126

Comber A, Fisher P, Brunsdon C, Khmag A (2012) Spatial analysis of remote sensing image classification accuracy. Remote Sens Environ 127:237–246. https://doi.org/10.1016/j.rse.2012.09.005

Comber AJ (2013) Geographically weighted methods for estimating local surfaces of overall, user and producer accuracies. Remote Sens Lett 4(4):373–380. https://doi.org/10.1080/2150704X.2012.736694

Fotheringham AS, Brunsdon C, Charlton M (2002) Geographically weighted regression. The analysis of spatially varying relationships. In: Fotheringham AS, Brunsdon C, Charlton M (eds) Wiley

Story M, Congalton RG (1986) Accuracy assessment: a user's perspective. Photogramm Eng Remote Sens 52:397–399

Tsutsumida N, Comber AJ (2015) Measures of spatio-temporal accuracy for time series land cover data. Int J Appl Earth Obs Geoinf 41:46–55. https://doi.org/10.1016/j.jag.2015.04.018

Global General Land Use Cover Datasets with a Single Date

David García-Álvarez, Javier Lara Hinojosa,
and Jaime Quintero Villaraso

Abstract

Global general Land Use and Land Cover (LUC) datasets map all land uses and covers across the globe, without focusing on any specific use or cover. This chapter only reviews those datasets available for one single date, which have not been updated over time. Seven different datasets are described in detail. Two other ones were identified, but are not included in this review, because of its coarsens, which limits their utility: Mathews Global Vegetation/Land Use and GMRCA LULC. The first experiences in global LUC mapping date back to the 1990s, when leading research groups in the field produced the first global LUC maps at fine scales of 1 km spatial resolution: the UMD LC Classification and the Global Land Cover Characterization. Not long afterwards, in an attempt to build on these experiences and take them a stage further, an international partnership produced GLC2000 for the reference year 2000. These initial LUC mapping projects produced maps for just one reference year and were not continued or updated over time. Subsequent projects have mostly focused on the production of timeseries of global LUC maps, which allow us to study LUC change over time (see Chapter "Global General Land Use Cover Datasets with a Time Series of Maps"). As a result, there are relatively few single-date global LUC maps for recent years of reference. The latest projects and initiatives producing global LUC maps for single dates have focused on improving the accuracy of global LUC mapping and the use of crowdsourcing production strategies. The Geo-Wiki Hybrid and GLC-SHARE datasets built on the previous research in a bid to obtain more accurate global LUC maps by merging the data from existing datasets. OSM LULC is an ongoing test project that is trying to produce a global LUC map cheaply, using crowdsourced information provided by the Open Street Maps community. The other dataset reviewed here is the LADA LUC Map, which was developed for a specific thematic project (Land Degradation Assessment in Dryland). This dataset is not comparable to the others reviewed in this chapter in terms of its purpose and nature, as is clear from its coarse spatial resolution (5 arc minutes). We therefore believe that this dataset should not be considered part of initiatives to produce more accurate, more detailed land use maps at a global level.

Keywords

UMD LC Classification • GLCC 2.0 Global • GLC2000 • Geo-Wiki Hybrid • GLC-SHARE • LADA LUC Map • OSM Landuse/Landcover

D. García-Álvarez (✉)
Departamento de Geología, Geografía y Medio Ambiente,
Universidad de Alcalá, Alcalá de Henares, Spain
e-mail: David.garcia@uah.es

J. Lara Hinojosa · J. Quintero Villaraso
Departamento de Análisis Geográfico Regional y Geografía
Física, Universidad de Granada, Granada, Spain

© The Author(s) 2022
D. García-Álvarez et al. (eds.), *Land Use Cover Datasets and Validation Tools*,
https://doi.org/10.1007/978-3-030-90998-7_14

1 UMD LC Classification—University of Maryland Land Cover Classification

Product
LULC general
Dates
1992/93 (1 km) 1984 (8 km) 1987 (1°)
Formats
Raster
Pixel size
1 km, 8 km, 1°
Thematic resolution
15 Classes – 1 km products 1 (a), 1 (ag), 10 (v), 1 (m), 1 (na)[1]
Compatible legends
UMD, IGBP
Extent
Global
Updating
No
Change detection
No (only one date)
Overall accuracy
Expected to be >65%

Website of reference	Website Language English, Spanish, French, Arabic, Russian
https://daac.ornl.gov/ISLSCP_II/guides/umd_landcover_xdeg.html	

Download site
http://iridl.ldeo.columbia.edu/SOURCES/.UMD/.GLCF/.GLCDS/.lc/datafiles.html

Availability	Format(s)
Open Access	.lan, .img

Technical documentation
Hansen et al. (2000)

Other references of interest
DeFries and Townshend (1994), DeFries et al. (1995), Hansen and Reed (2000), McCallum et al. (2006)

---

[1] (a): artificial; (ag): agriculture; (v): vegetation; (m): mixed classes;
(na): no data.

Project

The Department of Geography of the University of Maryland hosted one of the first research groups to use the classification of satellite imagery for global LUC mapping. They initially produced an LUC map at a spatial resolution of 1 degree for the year of reference 1987. This was followed sometime later by the production of a finer map at 8 km for 1984. Finally, the project delivered a map at 1 km, which at that time was the finest resolution at which global LUC mapping had ever been carried out.

The Global Land Cover Facility that hosted all this data recently went offline. This means that there is currently no official website that supports the datasets and provides information about their particular specifications. The map at 1 km can however be downloaded from external sites. The earlier maps at coarser resolutions are no longer available.

Production method

The UMD LC was obtained through supervised classification with a decision tree algorithm of imagery captured by the AVHRR sensor. Urban and built-up areas were not mapped, nor were water covers. Instead, they were extracted from auxiliary sources. The classification obtained in this way was then improved in a post-classification stage by expert regional labelling, based on inconsistencies that were identified by the experts.

Product description

Users can download the UMD LC Classification in two formats (.lan, .img), which are available in the section "GIS-Compatible Formats". The download is not easy and does only include the raster file with LUC information.

Downloads

LAN file
– Raster file with LUC map

Legend and codification

Code	Label	Code	Label
0	Water	8	Closed Shrubland
1	Evergreen Needleleaf Forest	9	Open Shrubland
2	Evergreen Broadleaf Forest	10	Grassland
3	Deciduous Needleleaf Forest	11	Cropland
4	Deciduous Broadleaf Forest	12	Bare Ground
5	Mixed Forest	13	Urban and Built-up
6	Woodland	14	Unclassified
7	Wooded Grassland		

Practical considerations

There is no official website hosting this dataset, which makes it more difficult to access and understand. Users must bear in mind that this was one of the first global LUC datasets ever developed and it can therefore be considered outdated in technical terms.

Coarser versions of the 1 km map, resampled at 0.25, 0.5 and 1 degree of spatial resolution, are also available.[2]

---

[2] https://daac.ornl.gov/cgi-bin/dsviewer.pl?ds_id=969.

2   GLCC 2.0 Global—Global Land Cover Characterization 2.0

Product	
LULC general	
Dates	
1992 / 93	
Formats	
Raster	
Pixel size	
1 km	
Thematic resolution	
100 classes (Global ecosystems legend) 19 classes (IGBP legend): 1 (a), 1 (ag), 10 (v), 2 (m), 2 (na)	
Compatible legends	
Global Ecosystems, IGBP, USGS LULC system, SiB, SiB 2, BATS, Vegetation lifeforms	
Extent	
Global	
Updating	
No	
Change detection	
No (only one date)	
Overall accuracy	
Expected to be > 66%	

Website of reference	**Website Language** English

https://www.usgs.gov/centers/eros/science/usgs-eros-archive-land-cover-products-global-land-cover-characterization-glcc?qt-science_center_objects=0#qt-science_center_objects

Download site

https://earthexplorer.usgs.gov/

Availability	**Format(s)**
Open Access after registration	.tiff, .bil

Technical documentation

Belward et al. (1999), Brown et al. (1999), Loveland and Belward (1997), Loveland et al. (2000), Reed et al. (2000)

Other references of interest

Hansen and Reed (2000)

Project

The GLCC dataset was the result of collaboration between several international institutions: the U.S. Geological Survey (USGS), the Earth Resources Observation and Science (EROS) Center, the University of Nebraska-Lincoln (UNL) and the Joint Research Centre (JRC) of the European Commission. The project aimed to create a dataset of reference for global land monitoring. One of the LUC maps obtained from the project is usually referred to as the DISCover LUC map and follows the IGBP classification scheme.

The global LUC map was created by joining various continental LUC maps together, and the final product consisted of a generalized global map and a set of more detailed continental maps.

Two versions of the dataset have been produced so far, with the first being released in 1997. The second version (2.0) improved on the first by applying both the lessons learnt and user feedback. Version 1.2 of the product included the IGBP classification (DISCover LUC map).

Production method

The dataset was obtained through unsupervised classification (CLUSTER classifier) of AVHRR imagery at a spatial resolution of 1 km. The classification obtained was further refined with the help of auxiliary data from the Digital Elevation Model (DEM), Ecoregions data and other thematic maps specific for each region. Label-assignment for the spectral classes was based on expert interpretation.

The dataset production was split into different continents, according to their specific characteristics. A detailed LUC map was produced for each continent and these were then joined together to create the global LUC product.

Product description

Two GLCC maps are available for download: the global product and the specific LUC product for each continent. The continental LUC maps show more detail than the global one and have specific legends that disaggregate the complexity of the land uses and covers for each continent.

The data can be downloaded in two different formats (.bil, .tiff). The download for each format includes the LUC maps with all the various classification schemes, together with technical documentation about the product. The continental product also includes a specific binary raster which maps the built-up land cover.

The product is distributed in two different projections: the Goode projection and a geographic projection.

Downloads

Global land cover product—Goode projection ("glccgbg20_tif")
– Raster files with LUC maps for each of the 7 classification schemes included in the product
– PDF document with technical information about the product

European land cover product—Goode projection ("glcceag20_tif")
– Raster files with LUC maps for each of the classification schemes included in the product
– Raster file with urban land cover information (built-up/non built-up)
– PDF document with technical information about the product

Legend and codification

LUC maps for each continent include a specific regional classification scheme, which is not shown here. The global dataset also supports seven different classification schemes. The most detailed of these is the Global Ecosystems (GLCC) scheme. In this case, however, we will only display the IGBP Land Cover classification scheme (IGBP), because it is the most commonly used of all the schemes provided by the dataset.

Information about the codification and the meaning of all the other classification schemes can be found in the technical documentation included in the downloaded product, as well as in the documentation available on the project's website.[3]

IGBP Land Cover (IGBP) Legend

Code	Label	Code	Label
1	Evergreen Needleleaf Forest	11	Permanent Wetlands
2	Evergreen Broadleaf Forest	12	Croplands
3	Deciduous Needleleaf Forest	13	Urban and Built-Up
4	Deciduous Broadleaf Forest	14	Cropland/Natural Vegetation Mosaic

(continued)

IGBP Land Cover (IGBP) Legend

Code	Label	Code	Label
5	Mixed Forest	15	Snow and Ice
6	Closed Shrublands	16	Barren or Sparsely Vegetated
7	Open Shrublands	17	Water Bodies
8	Woody Savannas	99	Interrupted Areas (Goode's Homolosine Projection)
9	Savannas	100	Missing Data
10	Grasslands		

Practical considerations

For more information about the product, users are referred to its readme file,[4] which explains the project history, the dataset production workflow and all the characteristics of the product.

---

[3] https://www.usgs.gov/media/files/global-land-cover-characteristics-data-base-readme-version2.

[4] https://prd-wret.s3.us-west-2.amazonaws.com/assets/palladium/production/s3fs-public/atoms/files/GlobalLandCoverCharacteristicsDataBaseReadmeVersion2.pdf.

3 GLC2000—Global Land Cover 2000

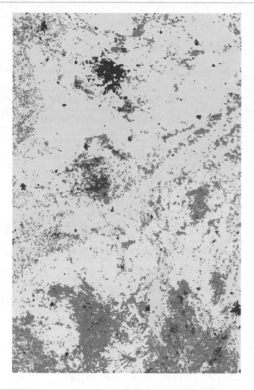

Product
LULC general
Dates
2000
Formats
Raster
Pixel size
1 km
Thematic resolution
23 classes: 1 (a), 1 (ag), 15 (v), 3 (m), 1 (na)
Compatible legends
FAO LCCS, IGBP
Extent
Global
Updating
No
Change detection
No (only one date)
Overall accuracy
Expected to be >68%

Website of reference	Website Language English
https://forobs.jrc.ec.europa.eu/products/glc2000/glc2000.php	

Download site

https://forobs.jrc.ec.europa.eu/products/glc2000/products.php

Availability	Format(s)
Open Access	.tiff, ESRI GRID, .img and Binary

Technical documentation

Hua et al. (2018), McCallum et al. (2006), Neumann et al. (2007), Pérez-Hoyos et al. (2012), Tchuenté et al. (2011)

Other references of interest

Bartholomé et al. (2002), Bartholomé and Belward (2005), Eva et al. (2004), Fritz et al. (2003)

Project

GLC2000 was a project run by the Joint Research Centre (JRC) of the European Commission in collaboration with regional teams across the globe. The objective of the project was to create a homogeneous, coherent global LUC map that was suitable for environmental monitoring. The reference year 2000 was chosen because of its particular significance for that purpose.

One of the most successful aspects of the project was the coordination of different teams across the globe to produce a global LUC map. To this end, GLC2000 provides a global dataset, together with a set of more detailed regional datasets adapted to the specificities of each territory.

Production method

GLC2000 was produced by different work teams across the globe. To this end, the world was split into 18 different regions, with each team mapping either a specific region or an area of special interest within a region.

A LUC map for each region was obtained through unsupervised classification of imagery captured by the VEGETATION sensor. The classifications obtained were then labelled by each regional team according to their local expertise in the area. Input for the classification varied in line with the particular characteristics of each region.

Regional LUC maps were merged into the global product, which is a coherent and homogeneous generalized mosaic of the set of regional maps. However, these regional maps provide more detail than the global one.

Product description

GLC2000 consists of two main products: the harmonized global LUC dataset covering the whole earth and the set of detailed regional LUC datasets. The Global LUC map can be downloaded in four different formats (ESRI, Binary, Tiff, Img), whereas the regional maps are only available in two (ESRI, Binary). The product for download includes a file to symbolize the raster LUC map as well as auxiliary information to interpret the legend.

Downloads

GLC2000 (Global)
– Raster file with LUC map
– Colormap file to symbolize the raster in ArcGIS (.clr)
– Excel spreadsheet with the map legend

GLurope)
– Folder with raster file of the regional LUC map (glc_eu_v2)
– Colormap file to symbolize the raster in ArcGIS (.clr)
– DBF file with the map legend

Legend and codification

Code	Label
1	Tree Cover, broadleaved, evergreen
2	Tree Cover, broadleaved, deciduous, closed
3	Tree Cover, broadleaved, deciduous, open
4	Tree Cover, needle-leaved, evergreen
5	Tree Cover, needle-leaved, deciduous
6	Tree Cover, mixed leaf type
7	Tree Cover, regularly flooded, fresh
8	Tree Cover, regularly flooded, saline, (daily variation)
9	Mosaic: Tree cover/Other natural vegetation
10	Tree Cover, burnt
11	Shrub Cover, closed-open, evergreen (with or without sparse tree layer)
12	Shrub Cover, closed-open, deciduous (with or without sparse tree layer)
13	Herbaceous Cover, closed-open
14	Sparse Herbaceous or sparse shrub cover
15	Regularly flooded shrub and/or herbaceous cover
16	Cultivated and managed areas
17	Mosaic: Cropland/Tree Cover/Other Natural Vegetation
18	Mosaic: Cropland/Shrub and/or Herbaceous cover
19	Bare Areas
20	Water Bodies (natural and artificial)
21	Snow and Ice (natural and artificial)
22	Artificial surfaces and associated area
23	No data

Practical considerations

Information about map metadata is easily available on the project's website together with technical documents describing the products. This information can help users gain a better understanding of the maps and all their specific characteristics, advantages and disadvantages. GLC2000 has also been widely analysed in the scientific literature. Users can find out more about the particular characteristics and the accuracy of the database by consulting some of the references of interest cited above.

4 Geo-Wiki Hybrid

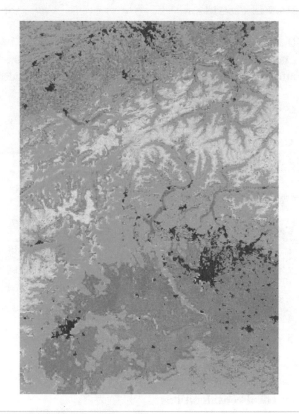

Product
LULC general
Dates
2000/05
Formats
Raster
Pixel size
300 m
Thematic resolution
10 classes: 1 (a), 1 (ag), 3 (v), 1 (m), 0 (na)
Compatible legends
FAO LCCS
Extent
Global
Updating
Not planned
Change detection
No (only one date)
Overall accuracy
Expected to be > 82% (87.9% for Hybrid Map 1 and 82.8% for Hybrid Map 2)

Website of reference	Website Language English
https://www.geo-wiki.org/	

Download site
https://application.geo-wiki.org/Application/index.php

Availability	Format(s)
Open Access after registration	.img

Technical documentation
See et al. (2015)

Other references of interest
Fritz et al. (2012)

Project

This project aimed to merge available global LUC maps to create a new, more accurate dataset, in a bid to enable more accurate global LUC mapping. Reference LUC data collected by the Geo-Wiki platform via crowdsourcing was employed in the fusion process, so pioneering a practice that has become more common in recent years. The dataset obtained in this way was one of the first, best-known examples of data fusion for global LUC mapping.

Production method

The hybrid map of the Geo-Wiki project was produced by merging three global LUC datasets: GLC2000, GlobCover and MODIS LC. Whereas GLC2000 shows the LUC state of the world for the reference year 2000, the other two sources provide LUC information for the reference year 2005. The spatial resolution of the hybrid map is the same as applied in the dataset with the highest resolution: GlobCover (300 m). The other two datasets, which had a spatial resolution of 1 km, were resampled to fit this resolution.

For each dataset, a probability layer was produced indicating the probability of that source representing the correct LUC class on the ground. These layers were obtained by regressing the datasets with validation points created through Geo-Wiki campaigns. A Geographically Weighted Regression (GWR) algorithm was employed to this end.

The probability layers were later merged in two different ways, delivering two LUC maps. For Hybrid Map 1, the LUC category from the dataset with the highest probability in the probability layers was selected. For Hybrid Map 2, when two LUC datasets agreed on a LUC category, this was selected. When the LUC datasets disagreed, the LUC category from the dataset with the highest probability in the probability layers was chosen.

Product description

Users can download the hybrid map in a compressed folder (.rar) which also contains the raster layers that store the LUC information. No other auxiliary information is provided.

Downloads

Geo-Wiki Hybrid (folder)
– A raster file with LUC information (.img)

Legend and codification

Code	Label	Code	Label
1	Tree cover	6	Flooded/wetland
2	Shrub cover	7	Urban
3	Herbaceous vegetation/Grassland	8	Snow and ice
4	Cultivated and managed	9	Barren
5	Mosaic of cultivated and managed/natural vegetation	10	Open water

Practical considerations

The Hybrid map is available online through the Geo-Wiki platform.[5] Although two hybrid maps were produced, only one was finally distributed. No information is provided as to which of these two maps is the one available online and for download.

---

[5] https://www.geo-wiki.org/.

5 LADA LUC Map—Land Degradation Assessment in Drylands

Product	
LULC general	
Dates	
2007	
Formats	
Raster	
Pixel size	
5 arc minutes	
Thematic resolution	
40 classes 1 (a), 7 (ag), 23 (v), 0 (m), 0 (na)	
Compatible legends	
–	
Extent	
Global	
Updating	
No	
Change detection	
No (only one date)	
Overall accuracy	
Not specified	

Website of reference	**Website Language** English
http://www.fao.org/land-water/land/land-governance/land-resources-planning-toolbox/category/details/en/c/1036360/	

Download site
https://data.apps.fao.org/map/catalog/srv/eng/catalog.search?currTab=simple&id=37139#/metadata/fc32c5de-440c-46aa-9cad-81f4c8b84c6a

Availability	**Format(s)**
Open Access	ESRI GRID, .tiff

Technical documentation
Nachtergaele and Petri (2013)

Other references of interest
–

Project

Land Degradation Assessment in Dryland (LADA) is a project led by the Food and Agriculture Organization (FAO) of the United Nations that aims to assess and map land degradation at different scales and levels, so as to understand its impact on land use. As part of the datasets created in the project, a map of the world's Land Use Systems (LUS) was developed. Many other datasets were also created within the framework of this project, which may be of interest to users.

Production method

The dataset was obtained after the interpretation of LUC units over a spatial dataset generated by the overlay of different spatial thematic layers: the GLC2000 LUC map, cropland LUC maps, livestock distribution data, ecosystem and ecological indicators and socioeconomic factors such as population density.

Product description

The LADA LUC map can be downloaded in two different formats (ESRI GRID or TIF). In each case, users download the raster files containing the LUC information, together with a layer style file to symbolize the dataset in a GIS.

Downloads

ESRI GRID folder
– Folder with raster files including LUC information ("lus")
– Folder with product metadata ("info")
– Layer style file for ArcGIS (.lyr)

TIF folder
– Raster file with LUC map (.tiff)
– Layer style file for ArcGIS (.lyr)

Legend and codification

Code	Label
1	Forest—Virgin
2	Forest—Protected
3	Forest—With agricultural activities
4	Forest—With moderate or high livestock density
5	Forest—Agroforestry
6	Forest—Plantations

(continued)

Code	Label
7	Grasslands—Unmanaged
8	Grasslands—Protected
9	Grasslands—Low livestock density
10	Grasslands—Moderate livestock density
11	Grasslands—High livestock density
12	Grasslands—Stable fed
13	Shrubs—Unmanaged
14	Shrubs—Protected
15	Shrubs—Low livestock density
16	Shrubs—Moderate livestock density
17	Shrubs—High livestock density
18	Shrubs—Stable fed
19	Agricultural land—Rainfed crops (subsistence/commercial)
20	Agricultural land—Crops and mod. Intensive livestock density
21	Agricultural land—Crops and intensive livestock density
22	Agricultural land—Crops with large scale irrigation and mod. Intensive or higher livestock density
23	Agricultural land—Large-scale irrigation (>25% pixel size)
24	Agricultural land—Protected
25	Urban land
26	Wetlands—Not used/not managed
27	Wetlands—Protected
28	Wetlands—Mangrove
29	Wetlands—With agricultural activities
30	Sparsely vegetated areas—Unmanaged
31	Sparsely vegetated areas—Protected
32	Sparsely vegetated areas—Low livestock density
33	Sparsely vegetated areas—With mod or higher livestock density
34	Barren areas—Unmanaged
35	Barren areas—Protected
36	Barren areas—Low livestock density
37	Barren areas—With mod. livestock density
38	Open water—Unmanaged
39	Open water—Protected
40	Open water—Inland fisheries

Practical considerations

The LADA LUC dataset is not a standard LUC map. It is a map of land use systems that was specifically created for the purposes of the LADA project, i.e. to study land degradation.

6 GLC-SHARE—Global Land Cover-SHARE

Product	
LULC general	
Dates	
Only one date, different for each part of the Earth	
Formats	
Raster	
Pixel size	
1 km	
Thematic resolution	
11 classes: 1 (a), 1 (ag), 6 (v), 0 (m), 0 (na)	
Compatible legends	
FAO LCCS	
Extent	
Global	
Updating	
None planned	
Change detection	
No (only one date)	
Overall accuracy	
Expected to be >80%	

Website of reference	**Website Language** English
http://www.fao.org/land-water/land/land-governance/land-resources-planning-toolbox/category/details/en/c/1036355/	

Download site

https://data.apps.fao.org/map/catalog/srv/eng/catalog.search?uuid=ba4526fd-cdbf-4028-a1bd-5a559c4bff38&currTab=distribution#/metadata/ba4526fd-cdbf-4028-a1bd-5a559c4bff38

Availability	**Format(s)**
Open Access	.tiff, .kml, WMS

Technical documentation

Latham et al. (2014)

Other references of interest

–

Project

GLC-SHARE was a project led by the Land and Water Division of the Food and Agriculture Organization (FAO), in collaboration with other institutions across the world. It aimed to create a global LUC map by mixing different sources of LUC information available at detailed scales. The objective was to improve the accuracy and quality of LUC information, so as to have a reliable source of global LUC information for policymaking.

Unlike other global LUC mapping projects, GLC-SHARE provides detailed LUC information in a single global product. Usually, this is only available in national, regional and local datasets.

Although the GLC-SHARE was produced in 2014, it was conceived as a living database that could integrate new LUC datasets as they were released or updated. Its production method has been made public, so enabling product replication.

As GLC-SHARE was produced by merging data from multiple databases, it has no specific date of reference. There are different dates for each part of the world, according to the main product that was used to map them.

Production method

GLC-SHARE was produced by merging and integrating high-quality LUC data for different areas of the world. LUC data at all scales (global, national, sub-national, regional) was used to produce the map.

In order to merge the various LUC datasets into a single product, their legends had to be harmonized. When different products were available for the same area, the one with the most detailed, most accurate data was chosen. If no products were available at detailed or national scales, global LUC datasets (Globcover 2009, MODIS VCF 2010 and Cropland database 2012) were used instead. The main areas not covered by high-resolution datasets included Latin America, West Africa, Indonesia and important parts of Asia, such as Thailand and the Arabian Peninsula.

An initial map for each of the 11 land cover classes that make up the classification legend of the GLC-SHARE was obtained. Each map shows the proportion that each land cover occupies in each pixel of the GLC-SHARE grid. Finally, from the 11 thematic rasters created, a general raster was obtained indicating the dominant land cover type in each pixel.

Product description

GLC-SHARE products can be downloaded in raster format or as a kml file to upload in Google Earth or any other GIS software. GLC-SHARE maps are also available through a WMS web service.

Users can download the global GLC-SHARE LUC map, which indicates the dominant land cover type in each pixel, or individual LUC rasters showing the proportions of each LUC type in each pixel. In these rasters, the pixel value refers to the proportion (0–100) at which each category is represented in the pixel. A pixel covered exclusively by artificial surfaces would have a value of 100 in the "GLC-Share – Artificial surfaces" raster.

Users can also download auxiliary information about the dataset from the website. This includes a technical report about the product (GLC-Share report) as well as a raster and an excel spreadsheet explaining which dataset was used to map each area of the world (GLC-Share—Sources).

Downloads

GLC-Share—Dominant land cover type

– Raster file with LUC map displaying the dominant land cover type
– Layer style file for ArcGIS (.lyr)
– Text document showing the classification legend for the dataset

---

GLC-Share—Sources

– Raster file with information about which LUC dataset was used to map each area of the world
– Layer style file for ArcGIS (.lyr)
– Excel spreadsheet with information about which LUC dataset was used to map each area of the world
– Text document explaining the downloaded product

GLC-Share—Artificial surfaces

– Raster file with information about the proportion of artificial surfaces in each pixel

Legend and codification

Code	Label	Code	Label
1	Artificial Surfaces	7	Mangroves
2	Cropland	8	Sparse vegetation
3	Grassland	9	Bare soil
4	Tree covered areas	10	Snow and glaciers
5	Shrubs covered areas	11	Water bodies
6	Herbaceous vegetation, aquatic or regularly flooded		

Practical considerations

GLC-SHARE is a single product with no information about changes in LUC over time. It was created in 2014, which may therefore be considered the reference year for the dataset. However, this date may vary a great deal between the different parts of the world. GLC-SHARE is therefore not recommended for studies or analyses of LUC change.

Although the dataset was conceived as a live map, it has not been further updated with the inclusion of new LULC datasets since 2014.

7 OSM Landuse/Landcover

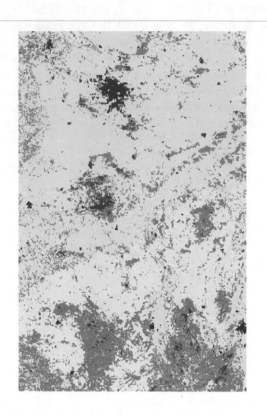

Product	
LULC general	
Dates	
Only one date, which cannot be specified	
Formats	
Raster	
Pixel size	
10 m	
Thematic resolution	
14 classes: 4 (a), 3 (ag), 2 (v), 3 (m), 1 (na)	
Compatible legends	
CLC	
Extent	
Global (with gaps) / Europe (full coverage)	
Updating	
Completion of the map is ongoing, although new editions of the map for different years of reference are not expected	
Change detection	
No (only one date)	
Overall accuracy	
Not specified	

Website of reference	Website Language English
https://data.osmlanduse.org	
Download site	
https://data.osmlanduse.org	

Availability	Format(s)
Under request (email to producers)	.tiff
Technical documentation	
Schultz et al. (2017)	
Other references of interest	
Fonte and Martinho (2017), Fonte et al. (2017a, b), Viana et al. (2019)	

Project

OSM Landuse/Landcover (LULC) is a LUC dataset created as part of the H2020 project "LandSense", which aims to engage citizens in the production of LUC information. The OSM LULC has been developed above all by the GIScience research group from Heidelberg University.

OSM LULC is an attempt to exploit the LUC information contained in the OpenStreetMaps (OSM) database. It is a test project and therefore cannot be regarded as a final product with full global coverage. Nevertheless, the project has developed a workflow to obtain LUC information from the OSM database as well as a methodology for obtaining an LUC map with full coverage over a specific test area (Heidelberg), filling the gaps in the OSM via classification of satellite imagery.

Production method

OSM LULC was produced using a very simple method. Authors downloaded the OSM database and translated the tags that define the features stored in the database into LUC terms (the legend for the Corine Land Cover (CLC) survey was used as a reference). An equivalence table between the OSM tags and the CLC level 2 legend was created.

The OSM LUC information, in vector, was generalized in a 30 m pixel side grid. In the event of feature overlap when aggregating information, preference was given to the smaller features.

Gap areas not covered by the OSM database were filled with the LUC information obtained by a supervised classification of Landsat imagery with the random forest classifier. This process was only carried out for a European test area, leaving important information gaps in the rest of the global map.

Due to the particular characteristics of the OSM database, LUC information is not provided for a single date. Each feature of the database has a different date. This makes it difficult to determine the date of reference for each pixel in the dataset.

Product description

The product was initially distributed in tiles. However, users can also request a specific file for their area of interest by email. These files contain the LUC map and an Excel spreadsheet with the pixel count for each category. They do not include the qualitative meaning of the category codes.

Downloads

OSM Landuse
– Raster file with LUC map (.tiff)
– Excel file with class codes and pixel count per class

Legend and codification

Code	Label	Code	Label
5	Water bodies	23	Pastures
11	Urban fabric	31	Forests
12	Industrial, commercial and transport units	32	Shrub and/or herbaceous vegetation associations
13	Mine, dump and construction sites	33	Open spaces with little or no vegetation
14	Artificial, non-agricultural vegetated areas	41	Inland wetlands
21	Arable land	42	Coastal wetlands
22	Permanent crops	NA	No data

Practical considerations

The website for this database includes a form for those who want to download the map. However, interested users are recommended to contact the map producers directly, as the first approach does not always work. Contact details for the map producers are available at the project's website.[6]

Users should be aware of the limitations of this dataset. As there is no single reference year for all the mapped areas, it may be difficult to use the map as a reference when analysing changes over time.

---

[6] https://osmlanduse.org/.

References

Bartholome E, Belward AS, Achard F et al (2002) GLC 2000. Global Land Cover mapping for the year 2000. Project status November 2002

Bartholomé E, Belward AS (2005) GLC2000: a new approach to global land cover mapping from earth observation data. Int J Remote Sens 26:1959–1977. https://doi.org/10.1080/01431160412331291297

Belward AS, Estes JE, Kline KD (1999) The IGBP-DIS global 1-km land-cover data set DISCover: a project overview. Photogramm Eng Remote Sens 65:1013–1020

Brown JF, Loveland TR, Ohlen DO, Zhu ZL (1999) The global land-cover characteristics database: the users' perspective. Photogramm Eng Remote Sens 65:1069–1074

DeFries RS, Townshend JRG (1994) NDVI-derived land cover classifications at a global scale. Int J Remote Sens 15:3567–3586. https://doi.org/10.1080/01431169408954345

DeFrics RS, Hansen MC, Townshend JRG, Sohlberg R (1995) Global land cover classification at 8 km spatial resolution: the use of training data derived from Landsat imagery in decision tree classifiers. Remote Sens Environ 19:3141–3168

Eva HD, Belward AS, De Miranda EE et al (2004) A land cover map of South America. Glob Chang Biol 10:731–744. https://doi.org/10.1111/j.1529-8817.2003.00774.x

Fonte CC, Martinho N (2017) Assessing the applicability of OpenStreetMap data to assist the validation of land use/land cover maps. Int J Geogr Inf Sci 1–19. https://doi.org/10.1080/13658816.2017.1358814

Fonte CC, Minghini M, Patriarca J et al (2017a) Generating up-to-date and detailed land use and land cover maps using OpenStreetMap and GlobeLand30. ISPRS Int J Geo Inf 6:1–22. https://doi.org/10.3390/ijgi6040125

Fonte CC, Patriarca JA, Minghini M et al (2017b) Using OpenStreetMap to create land use and land cover maps: development of an application. Volunt Geogr Inf Futur Geospatial Data i:113–137. https://doi.org/10.4018/978-1-5225-2446-5.ch007

Fritz S, Bartholomé E, Belward A et al (2003) Harmonisation, mosaicing and production of the Global Land Cover 2000 database (Beta Version). Accessed September 24, 2020. https://forobs.jrc.ec.europa.eu/data/products/glc2000/GLC2000_EUR_20849EN.pdf

Fritz S, McCallum I, Schill C et al (2012) Geo-Wiki: An online platform for improving global land cover. Environ Model Softw 31:110–123. https://doi.org/10.1016/j.envsoft.2011.11.015

Hansen MC, Defries RS, Townshend JRG, Sohlberg R (2000) Global land cover classification at 1 km spatial resolution using a classification tree approach. Int J Remote Sens 21:1331–1364. https://doi.org/10.1080/014311600210209

Hansen MC, Reed B (2000) A comparison of the IGBP DISCover and University of Maryland 1 km global land cover products. Int J Remote Sens 21:1365–1373. https://doi.org/10.1080/014311600210218

Hua T, Zhao W, Liu Y et al (2018) Spatial consistency assessments for global land-cover datasets: a comparison among GLC2000, CCI LC, MCD12. GLOBCOVER and GLCNMO. Remote Sens 10. https://doi.org/10.3390/rs10111846

Latham J, Cumani R, Rosati I, Bloise M (2014) Global Land Cover SHARE (GLC-SHARE) database Beta-Release Version 1.0. Accessed on 21 August 2020. http://www.fao.org/uploads/media/glc-share-doc.pdf

Loveland TR, Belward AS (1997) The IGBP-DIS global 1 km land cover data set, discover: first results. Int J Remote Sens 18:3289–3295. https://doi.org/10.1080/014311697217099

Loveland T, Reed B, Brown J, Ohlen D, Zhu J, Yang L, Merchant J (2000) Development of a global land cover characteristics database and IGBP DISCover from 1-km AVHRR Data. Int J Remote Sens 21(6/7):1303–1330

McCallum I, Obersteiner M, Nilsson S, Shvidenko A (2006) A spatial comparison of four satellite derived 1 km global land cover datasets. Int J Appl Earth Obs Geoinf 8:246–255. https://doi.org/10.1016/j.jag.2005.12.002

Neumann K, Herold M, Hartley A, Schmullius C (2007) Comparative assessment of CORINE2000 and GLC2000: spatial analysis of land cover data for Europe. Int J Appl Earth Obs Geoinf 9:425–437. https://doi.org/10.1016/j.jag.2007.02.004

Pérez-Hoyos A, García-Haro FJ, San-Miguel-Ayanz J (2012) Conventional and fuzzy comparisons of large scale land cover products: application to CORINE, GLC2000, MODIS and GlobCover in Europe. ISPRS J Photogramm Remote Sens 74:185–201. https://doi.org/10.1016/j.isprsjprs.2012.09.006

Reed BC, Brown JF, Ohlen O et al (2000) Development of a global land cover characteristics database and IGBP DISCover from 1-km AVHRR Data. Int J Remote Sens 21:1303–1330

Schultz M, Voss J, Auer M et al (2017) Open land cover from OpenStreetMap and remote sensing. Int J Appl Earth Obs Geoinf 63:206–213. https://doi.org/10.1016/j.jag.2017.07.014

See L, Schepaschenko D, Lesiv M et al (2015) Building a hybrid land cover map with crowdsourcing and geographically weighted regression. ISPRS J Photogramm Remote Sens 103:48–56. https://doi.org/10.1016/j.isprsjprs.2014.06.016

Tchuenté ATK, Roujean JL, de Jong SM (2011) Comparison and relative quality assessment of the GLC2000, GLOBCOVER, MODIS and ECOCLIMAP land cover data sets at the African continental scale. Int J Appl Earth Obs Geoinf 13:207–219. https://doi.org/10.1016/j.jag.2010.11.005

Viana CM, Encalada L, Rocha J (2019) The value of OpenStreetMap historical contributions as a source of sampling data for multi-temporal land use/cover maps. ISPRS Int J Geo-Information 8. https://doi.org/10.3390/ijgi8030116

Global General Land Use Cover Datasets with a Time Series of Maps

David García-Álvarez, Javier Lara Hinojosa,
Francisco José Jurado Pérez, and Jaime Quintero Villaraso

Abstract

General Land Use Cover (LUC) datasets provide a holistic picture of all the land uses and covers on Earth, without focusing specifically on any individual land use category. As opposed to the LUC maps which are only available for one date or year, reviewed in Chap. "Global General Land Use Cover Datasets with a Single Date", the maps with time series allow users to study LUC change over time. Time series of general LUC datasets at a global scale is useful for understanding global patterns of LUC change and their relation with global processes such as climate change or the loss of biodiversity. MCD12Q1, also known as MODIS Land Cover, was the first time series of LUC maps to be produced on a global scale. When it was first launched in 2002, there were already many organizations and researchers working on accurate, detailed global LUC maps, although these were all one-off editions for single years. The MCD12Q1 dataset continues to be updated today, providing a series of maps for the period 2001–2018. Since the launch of MCD12Q1, many other historical series of LUC maps have been produced, especially in the last decade. This has resulted in the LUC map series covering a longer time period at higher spatial resolution. Recent efforts have focused on producing consistent time series of maps that can track LUC changes over time with low levels of uncertainty. GLCNMO (500 m), GlobCover (300 m) and GLC250 (250 m) provide time series of LUC maps at similar spatial resolutions to MCD12Q1 (500 m), although for fewer reference years. GLCNMO provides information for the years 2003, 2008 and 2013, GlobCover for 2005 and 2009 and GLC250 for 2001 and 2010. GLASS-GLC is the dataset with the coarsest spatial resolution of all those reviewed in this chapter (5 km), even though it was released very recently, in 2020. Map producers have focused on this dataset's long timespan (1982–2015) rather than on its spatial detail. LC-CCI and CGLS-LC100 are the recently launched datasets providing a consistent series of LUC maps, which show LUC changes over time with lower levels of uncertainty. LC-CCI provides LUC information for one of the longest timespans reviewed here (1992–2018) at a spatial resolution of 300 m. CGLS-LC100 provides LUC information for a shorter period (2015–2019) but at a higher spatial resolution (100 m). In both cases, updates are scheduled. The datasets with the highest levels of spatial detail are FROM-GLC and GLC30. These were produced using highly detailed Landsat imagery, delivering time series of maps at 30 m. The FROM-GLC project even has a test LUC map at a spatial resolution of 10 m from Sentinel-2 imagery for the year 2017, making it the global dataset with the greatest spatial detail of all those reviewed in this book. Both FROM-GLC and GLC30 provide data for three different dates: the former for 2010, 2015 and 2017 and the latter for 2000, 2010 and 2020.

Keywords

Time Series • Land Use Cover Change • GLASS-GLC • LC-CCI • GLC30 • GLC250 • MCD12Q1 • GLCNMO • GlobCover • FROM-GLC • CGLS-LC100

D. García-Álvarez (✉)
Departamento de Geología, Geografía y Medio Ambiente, Universidad de Alcalá, Alcalá de Henares, Spain
e-mail: David.garcia@uah.es

J. Lara Hinojosa · F. J. Jurado Pérez · J. Quintero Villaraso
Departamento de Análisis Geográfico Regional y Geografía Física, Universidad de Granada, Granada, Spain

© The Author(s) 2022
D. García-Álvarez et al. (eds.), *Land Use Cover Datasets and Validation Tools*,
https://doi.org/10.1007/978-3-030-90998-7_15

1 GLASS-GLC—Global Land Surface Satellite-Global Land Cover

Product	
LULC general	
Dates	
1982–2015	
Formats	
Raster	
Pixel size	
5 km	
Thematic resolution	
8 classes: 0 (a), 1 (ag), 4 (v), 0 (m), 1 (na)[1]	
Compatible legends	
FROM-GLC	
Extent	
Global	
Updating	
Not planned	
Change detection	
Possible, although sources of uncertainty may arise	
Overall accuracy	
Expected to be >82%	

Website of reference	**Website Language** English
http://data.ess.tsinghua.edu.cn/	
Download site	
https://doi.pangaea.de/10.1594/PANGAEA.913496	
Availability	**Format(s)**
Open Access	.tiff
Technical documentation	
Liu et al. (2020)	
Other references of interest	
–	

[1] (a): artificial; (ag): agriculture; (v): vegetation; (m): mixed classes; (na): no data.

Project

GLASS-GLC is the result of the research activity on LUC mapping carried out by a group of Chinese researchers. It is part of the efforts led by the Tsinghua University to map global LUC information, which also includes the FROM-GLC project, reviewed later in this chapter.

The project has delivered a series of global LUC maps at coarse resolution (5 km). This spatial resolution may limit the applicability of the dataset as, for example, it does not include information on impervious areas.

Production method

GLASS-GLC was obtained after making a supervised classification of AVHRR satellite imagery with the Google Earth Engine cloud platform. Random forest was the selected classifier. Auxiliary data, such as a Vegetation Continuous Field layer or a Digital Elevation Model, were also used in the classification.

To ensure the consistency of the maps over time, the authors applied the "LandTrendr" method and a linear regression-based algorithm. These helped to detect the LUC changes in the imagery archive used to obtain the LUC maps.

Product description

GLASS-GLC can be downloaded as a single compressed file. This file includes all the LUC maps for each year in the map series (1982–2015), as well as auxiliary data to help users understand the product.

Downloads

GLASS-GLC

– A raster file with the LUC information for each available year (.tiff)
– Word document with a technical description of the product

Legend and codification

Code	Label	Code	Label
10	Cropland	70	Tundra
20	Forest	90	Barren land
30	Grass	100	Snow/ice
40	Shrubland	0	No data

2 LC-CCI—Land Cover-Climate Change Initiative

Product	
LULC general	
Dates	
1992–2018	
Formats	
Raster	
Pixel size	
300 m (150 m for water bodies and 500 m for snow condition) MMU Changes: 1 km	
Thematic resolution	
37 classes: 1 (a), 2 (ag), 26 (v), 4 (m), 1 (na)	
Compatible legends	
PFT, FAO LCCS	
Extent	
Global	
Updating	
Updated planned (no date)	
Change detection	
Yes	
Overall accuracy	
Expected to be >70%	

Website of reference	**Website Language** English
https://www.esa-landcover-cci.org/	

Download site	
http://maps.elie.ucl.ac.be/CCI/viewer/download.php	

Availability	**Format(s)**
Open Access after provision of name, institution and email	.tiff, .nc (NetCDF4)

Technical documentation	
ESA (2017)	

Other references of interest	
Bontemps et al. (2012), Hollmann et al. (2013), Hua et al. (2018), Mousivand and Arsanjani (2019), Vilar et al. (2019)	

Project

The Land Cover-Climate Change Initiative is a project run by the European Space Agency (ESA) that seeks to create LUC products that meet the requirements of the Global Climate Observing System (GCOS) for Essential Climate Variables (ECV) and the Climate Modelling Community (CMC). It builds on the lessons learned during the Glob-Cover project. It also takes into account the opinion and the needs of users working in the climate and global land cover research communities, who were consulted and engaged with during the project.

The purpose of the project is to deliver a time series of land cover data that is stable, dynamic, transparent and flexible. This means: first, obtaining a historical series of land cover maps that show the changes over time, with no technical errors or instability: second, the production of a LUC dataset with a wide range of applications; and third, the provision of all relevant information regarding the quality of the dataset.

The project was launched in 2009 and has been developed in different phases. The initial idea was to create a LUC product covering three time periods (1998–2002, 2003–2007 and 2008–2012). Later, an improved yearly LUC product for the period 1992–2015 was launched, which replaced the previous one. Recently, this latter product has been updated and now includes new LUC maps for the period 2016–2018 which are consistent with the previous series.

Apart from LUC maps, other interesting products have also been created as part of the Climate Change Initiative: weekly image composites of the AVHRR (1992–1999, 1 km), MERIS (2003–2012, 300 m and 1 km) and PROBA-V (2014–2015, 1 km) sensors; a static map of open water bodies; and three global land surface seasonality products characterizing the dynamics of vegetation greenness, snow and burnt areas.

Production method

The LC-CCI LUC map series is based on a single base LUC map that is progressively updated and backdated. The base LUC layer was created by classifying a series of composite MERIS imagery for the period 2003–2012. A different classification was carried out for each year of this period, and the map finally obtained was a combination of all these classifications. This allowed them to differentiate between land cover states (i.e. those land features that remain stable over time) and land cover seasonality (i.e. natural, seasonal variability of land cover features that do not imply a change in the cover itself).

The classification method combined the GlobCover unsupervised classification chain with a machine learning algorithm. During the classification process, a series of spectrotemporal classes were identified. These were later labelled to LUC classes with the help of experts. The classification was regionalized to account for regional diversity and local heterogeneity of land cover characteristics.

Change detection for updating and backdating the base map was carried out with imagery from different sensors (AVHRR, SPOT, MERIS and PROVA), according to image availability. Changes were detected at a spatial resolution of 1 km, and since 2013 have been delineated at 300 m. Previously, delineation of changes at finer spatial resolutions had been impossible due to the lack of available images.

As a general rule, the only changes studied were those between six wide categories, which are not semantically close to each other: agriculture, forest, grassland, wetland, settlement and others. These changes had to persist for at least two years to be considered. The purpose of these rules was to try to ensure the stability over time of the LUC map series, avoiding technical changes and noise.

Product description

The LC-CCI dataset is distributed in different ways. This gives users the flexibility to download the product that best suits their needs. A single LUC map in either GeoTIFF or NetCDF4 may be downloaded for each year of the period 1992–2015. For the most recent years (2016–2018), these are only available in NetCDF4 format. Additionally, the whole time series of maps for the period 1992–2015 can be downloaded as a single raster with multiple bands, in either of the two formats available.

When downloading the LUC maps, users only gain access to the rasters with LUC information. However, other supplementary information is available on the project's website. This includes a CSV file with the legend description; layer style files for displaying the rasters in common GIS software (ArcGIS, ENVI and QGIS); GeoTIFF files with information about the quality and uncertainty of the LUC maps time series (Quality flags); and a data package for users working with the Sen2Cor classification software.

Downloads

LC Map 2015

– Raster file with LUC map

LC maps full 1992–2015 series

– Raster file with LUC maps series

Legend and codification

Code	Label
0	No data
10	Cropland, rainfed
11	Herbaceous cover
12	Tree or shrub cover
20	Cropland, irrigated or post-flooding
30	Mosaic cropland (>50%)/natural vegetation (tree, shrub, herbaceous cover) (<50%)
40	Mosaic natural vegetation (tree, shrub, herbaceous cover) (>50%)/cropland (<50%)
50	Tree cover, broadleaved, evergreen, closed to open (>15%)
60	Tree cover, broadleaved, deciduous, closed to open (>15%)
61	Tree cover, broadleaved, deciduous, closed (>40%)
62	Tree cover, broadleaved, deciduous, open (15–40%)
70	Tree cover, needleleaved, evergreen, closed to open (>15%)
71	Tree cover, needleleaved, evergreen, closed (>40%)
72	Tree cover, needleleaved, evergreen, open (15–40%)
80	Tree cover, needleleaved, deciduous, closed to open (>15%)
81	Tree cover, needleleaved, deciduous, closed (>40%
82	Tree cover, needleleaved, deciduous, open (15–40%)
90	Tree cover, mixed leaf type (broadleaved and needleleaved)
100	Mosaic tree and shrub (>50%)/herbaceous cover (<50%)
110	Mosaic herbaceous cover (>50%)/tree and shrub (<50%)
120	Shrubland
121	Evergreen shrubland
122	Deciduous shrubland

(continued)

Code	Label
130	Grassland
140	Lichens and mosses
150	Sparse vegetation (tree, shrub, herbaceous cover) (<15%)
152	Sparse shrub (<15%)
153	Sparse herbaceous cover (<15%)
160	Tree cover, flooded, fresh or brackish water
170	Tree cover, flooded, saline water
180	Shrub or herbaceous cover, flooded, fresh/saline/brackish water
190	Urban areas
200	Bare areas
201	Consolidated bare areas
202	Unconsolidated bare areas
210	Water bodies
220	Permanent snow and ice

Practical considerations

The project is aimed at the climate change research community and therefore provides the LUC data in the NetCDF4 raster file format commonly used by this community. However, .nc files are much heavier than .tiff files.

LUC maps for single years are easily displayed in QGIS. However, raster files storing the whole series of LUC maps for the period 1992–2015 are very heavy and are difficult to display in QGIS without a computer with good processing power.

3 GLC30—GlobeLand30

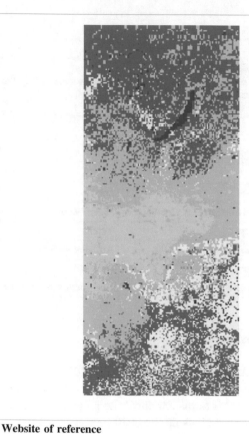

Product	
LULC general	
Dates	
2000, 2010, 2020	
Formats	
Raster	
Pixel size	
30 m	
Variable UMC depending on the category (3×3 to 10×10 pixels)	
Thematic resolution	
10 classes:	
1 (a), 1 (ag), 4 (v), 0 (m), 0 (na)	
Compatible legends	
GLC30	
Extent	
Global	
Updating	
Not planned	
Change detection	
Yes	
Overall accuracy	
Expected to be >78%	

Website of reference	**Website Language** English
http://www.globallandcover.com/home_en.html	

Download site

http://www.globallandcover.com/defaults_en.html?src=/Scripts/map/defaults/En/download_en.html&head=download&type=data

Availability	**Format(s)**
Open access under registration	.tiff

Technical documentation

Chen et al. (2010, 2011a, b, 2012, 2014, 2016), Tang et al. (2014), Xie et al. (2015), Zhu et al. (2010)

Other references of interest

Cao et al. (2014), Chen et al. (2013, 2017), Han et al. (2015), Jun et al. (2014), Manakos et al. (2018), Shi et al. (2016a, b), Wu et al. (2016), Yang et al. (2017)

Project

GlobeLand30 (GLC30) is a project funded and promoted by the Chinese government and the National Science Foundation of China. It aims to coherently map the land uses and covers on the world's surface at a detailed scale, using images from the Landsat satellite imagery archive.

The project initially focused on analysing the best methods and procedures to carry out such an ambitious task. It then produced a global LUC map at 30 m for the reference years 2000 and 2010. An update of the dataset for the year 2020 was recently released, in which Antarctica was mapped for the first time.

Production method

GLC30 was obtained after classifying Landsat imagery using a pixel-object-knowledge-based (POK-based) classification approach. Other sources of complementary imagery were also used for the reference years 2010 (HJ-1—China Environment and Disaster Reduction Satellite) and 2020 (GF-1—China High Resolution Satellite).

The classification was carried out independently for each of the mapped categories. Water bodies were mapped first, followed by wetlands, snow and ice, artificial surfaces, cultivated land, forest, scrubland, grassland, barren land and finally tundra. Once a LUC category had been classified, the pixels assigned to that category were masked for the following classifications.

Each category was classified according to a specific approach, adapted to the characteristics of the features being mapped. For most of the categories, the classification approach consisted of three main steps: a pixel-based classifier, image segmentation and knowledge-based verification. For this last step, different sources of auxiliary information were used via their integration in a web-based data platform.

Product description

GLC30 is distributed in tiles. Users can separately download a LUC map for each tile and year of reference. The download includes the LUC map in raster format, a metadata file and a vector file with information about the satellite imagery used to obtain the map.

Downloads

GLC30 2020
– Raster file with LUC map (.tiff)
– Shapefile file with information about the imagery source used in the LUC classification (.shp)
– Metadata file (.xls)

Legend and codification

Code	Label	Code	Label
10	Cultivated land	60	Water bodies
20	Forest	70	Tundra
30	Grassland	80	Artificial surfaces
40	Shrubland	90	Bareland
50	Wetland	100	Permanent snow and ice

Practical considerations

The GLC30 LUC maps for 2000, 2010 and 2020 can also be accessed online through the project website,[2] which also includes plenty of information about the project and various other datasets. These include the 2020 imagery used to map the latest update of the dataset and different sources of reference data used as auxiliary information in the mapping process.

There are no technical documents describing the latest update of the map for the year 2020. Methodological changes in the production of the map could have been implemented which could lead to errors when comparing with previous editions.

The project website is not always maintained. It has been unattended for many months over recent years. If the website is not maintained, it is possible that the dataset may be not accessible in the future.

---

[2] http://www.globallandcover.com/.

4   GLC250—Global Land Cover 250 m

Product
LULC general
Dates
2001, 2010
Formats
Raster
Pixel size
250 m
Thematic resolution
25 classes: 0 (a), 6 (ag), 7 (v), 1 (m), 0 (na)
Compatible legends
FAO-LCCS, IGBP
Extent
Global
Updating
Not expected
Change detection
Yes
Overall accuracy
Expected to be >75%

Website of reference	Website Language English
http://data.ess.tsinghua.edu.cn/	
Download site	
http://data.ess.tsinghua.edu.cn/	

Availability	Format(s)
Open Access	.tiff

Technical documentation
Wang et al. (2015)
Other references of interest
–

Project

This product forms part of the project led by Tsinghua University to effectively map land uses and covers across the world, which mainly focused on FROM-GLC and the production of thematic LUC databases. Several of these datasets were used in the production of GLC250. The classification legend for GLC250 was also taken from FROM-GLC.

Production method

GLC250 was obtained after the classification of MODIS imagery (MOD13Q1) with a random forest classifier fed with auxiliary data: slope, latitude, MODIS vegetation indexes. For each year of reference (2001, 2010), a classification was carried out for three different dates: the year of reference, the year before and the year after. For the year 2001, for example, images from 2000, 2001 and 2002 were classified.

The three probability maps obtained after the classification carried out for each year of reference were processed through a spatial–temporal consistency model (MAP-MRF) to improve the LUC classification. The final LUC map was improved in a post-classification phase through a rule-based label adjustment method using auxiliary data from MODIS Vegetation Continuous Fields (MOD44B), slope and Enhanced Vegetation Index series.

Product description

A map for each year of reference can be downloaded in a single compressed file. Each file contains all the raster files that make up the LUC map for each year of reference. To this end, the global map is split into multiple tiles following the MODIS tile grid.[3]

Downloads

GLC250—2010

– Raster files with a LUC map for each tile making up the MODIS tile grid (296 files)

Legend and codification

The GLC250 classification scheme is the same as that developed for FROM-GLC. It is a two-level classification scheme, which allows the LUC map to be displayed at two different levels of detail. Only the most detailed scheme (Level 2) is displayed here. Interested users can consult the correspondence between Level 2 and Level 1 classes on the project's website.[4]

Code	Label	Code	Label
11	Rice fields	42	Other shrublands
12	Greenhouse farming	61	Lake
13	Other croplands	62	Reservoir/pond
14	Seasonal croplands	63	River
15	Pastures	64	Ocean
21	Broadleaf forests	91	Dry salt flats
22	Needleleaf forests	92	Sandy areas
23	Mixed forests	93	Exposed bare rock
24	Orchards	94	Dry lake/river bottoms
31	Marshland	95	Tidal area
32	Herbaceous tundra	101	Snow
33	Other grasslands	102	Ice
41	Shrub and brush tundra		

---

[3] https://modis-land.gsfc.nasa.gov/MODLAND_grid.html.

[4] http://data.ess.tsinghua.edu.cn/.

5 MCD12Q1—MODIS/Terra + Aqua Land Cover Type

Product
LULC general
Dates
2001–2020
Formats
Raster
Pixel size
500 m, 1 km, 0.05°
Thematic resolution
18 classes (IGBP legend): 1 (a), 1 (ag), 10 (v), 2 (m), 1 (na)
Compatible legends
IGBP, UMD, LAI, BGC, PFT, FAO-LCCS
Extent
Global
Updating
Expected
Change detection
Not recommended
Overall accuracy
Expected to be >71%

Website of reference	Website Language English
https://lpdaac.usgs.gov/products/mcd12q1v006/	

Download site
https://lpdaac.usgs.gov/products/mcd12q1v006/

Availability	Format(s)
Open access under registration	.hdf

Technical documentation
Friedl et al. (2002, 2010), Friedl and Sulla-Menashe (2019), Sulla-Menashe et al (2019)

Other references of interest
Fritz and See (2005), Giri et al. (2005), Hao and Gen-Suo (2014), Tchuenté et al. (2011)

Project

MCD12Q1, also known as MODIS Land Cover type, dates back to 2002, after the launch into space of the TERRA satellite carrying the MODIS sensor. The MODIS sensor provided a new source of imagery for global LUC mapping. This led to the appearance of the MODIS Land Cover project, which aimed to produce a yearly series of LUC maps that could satisfy the demands of different communities interested in climate and environmental monitoring at global or very coarse scales. At the time the dataset was launched, only a few global LUC datasets were available, usually at coarser resolutions.

MODIS Land Cover was created by a team led by the University of Boston. Since 2002, six versions of the product have been developed. The latest is MODIS Land Cover Collection 6, which has included the most important changes in the production method of the dataset since its early developments.

A complementary product at coarser resolution has been developed as part of the same project: MCD12C1 (0.05 Deg).

Production method

MCD12Q1 was obtained by means of supervised classification (Random Forests) of MODIS imagery for the period 2001–2020. Once the classification had been obtained for each year, it was adjusted with the aid of auxiliary data: C5 MCD12Q1, C6 MODIS Land Water mask, C5 MODIS Vegetation Continuous Fields (VCF), WorldClim dataset, a global urban layer and global crop type information compiled from census data.

As a result of the classification, class probability rasters were obtained for each LUC category. These inform about the probability of each pixel belonging to a specific LUC category. These probability layers provided a base on which to map LUC covers according to six different classification schemes: IGBP, UMD, LAI, BGC, PFT and FAO-LCCS. In order to ensure the consistency of the classification over time, a hidden Markov model (HMM) was applied to the adjusted classification to reduce spurious changes over time.

Product description

MCD12Q1 may be downloaded through different servers or tools: AppEEARS, Data Pool, NASA Earthdata Search, USGS EarthExplorer, OPeNDAP, DAAC2Disk Utility and LDOPE. Depending on the server or tool chosen, users can download the product as a single file for each year of reference or in tiles for specific areas of interest.

The download includes the raster file with LUC data in six different classification schemes and PDF documents with the technical specifications for the product.

Legend and codification

MCD12Q1 is distributed for six different, widely used classification schemes. The only one displayed here is the IGBP scheme, which is one of the most commonly used. However, more information about the codes and class descriptions for the other classification legends is available in the user guide for this dataset (Sulla-Menashe et al. 2019).

MCD12Q1—IGBP (International Geosphere-Biosphere Programme)

Code	Label	Code	Label
1	Evergreen needleleaf forests	10	Grasslands
2	Evergreen broadleaf forests	11	Permanent wetlands
3	Deciduous needleleaf forests	12	Croplands
4	Deciduous broadleaf forests	13	Urban and built-up lands
5	Mixed forests	14	Cropland/natural vegetation mosaics
6	Closed shrublands	15	Permanent snow and ice
7	Open shrublands	16	Barren
8	Woody savannas	17	Water bodies
9	Savannas	18	Unclassified

Practical considerations

Users can consult the dataset online through the Web Map Service (WMS) available here.[5] The dataset is also available at a spatial resolution of 0.05 : MCD12C1 (0.05 Deg).[6]

This dataset is not recommended for the study of LUC change, because of the high technical variability in LUC covers from one year to the next.

---

[5] https://lpdaacgis.cr.usgs.gov/arcgis/rest/services/WMS?f=pjson.
[6] https://lpdaac.usgs.gov/products/mcd12c1v006/.

6 GLCNMO—Global Land Cover by National Mapping Organization

Product	
LULC general	
Dates	
2003, 2008, 2013	
Formats	
Raster	
Pixel size	
1 km (2003) 500 m (2008, 2013)	
Thematic resolution	
20 classes: 1 (a), 3 (ag), 11 (v), 2 (m), 0 (na)	
Compatible legends	
FAO LCCS	
Extent	
Global	
Updating	
Not planned	
Change detection	
No	
Overall accuracy	
Expected to be >75%	

Website of reference	**Website Language** English
https://globalmaps.github.io/glcnmo.html	

Download site
https://globalmaps.github.io/glcnmo.html

Availability	**Format(s)**
Open Access	.tiff

Technical documentation
Kobayashi et al. (2017), Tateishi et al. (2011, 2014)

Other references of interest
Hua et al. (2018)

Project

GLCNMO is a project promoted by the International Steering Committee for Global Mapping (ISCGM) in collaboration with the Geospatial Information Authority of Japan (GSI), Chiba University and national mapping organizations from different participant countries. It is part of a wider effort to create global datasets on different subjects, including land cover and land use.

The project has delivered three global LUC maps. Each one was produced at a different time and various methodological changes were introduced between the production of each map. The most evident one was the change in spatial resolution after the 2003 map. Another important difference was the number of countries taking part in each edition of the map: 40 countries took part in the production of the 2003 map, 14 in the 2008 map and 22 in the map for 2013.

The ISCGM was wound up in 2016 and its data was transferred to the Geospatial Information Section in the United Nations. We, therefore, do not expect any updates on this project.

Production method

The three LUC maps were produced at the continental level using a mixture of different methods. The maps for each continent were prepared by separate groups, with national experts providing assistance for each case.

Most of the categories (14 in 2003 and 2008 and 11 in 2013) were obtained through supervised classification of MODIS imagery. The training samples for the classifier were selected with great care using photointerpretation from sources like Google Earth and other auxiliary data. Different classifiers were used for the different maps. Whereas the map for 2003 was produced using a maximum likelihood classifier, the ones for 2008 and 2013 were based on a decision tree classifier.

The remaining categories that were not classified using the method described above were individually mapped according to different procedures adapted to the specific needs of each category. These were urban, tree open, mangrove, wetland, snow/ice and water in 2003 and 2008. In addition to those, herbaceous areas, forests and agricultural areas were also mapped in this way in 2013. The strategies used to map these categories also varied in the different editions of the map, mainly involving specific classification methods of MODIS imagery, as well as the use of additional information, such as population density datasets, thematic MODIS products and other global LUC maps.

Product description

The GLCNMO LUC map is distributed individually for each available year. The map for each year is split into four tiles, which can be downloaded in different zipped files. No other additional information is provided, except for the scientific papers presenting each map.

Downloads

GLCNMO (version 3)
– Raster LUC map covering North America, the north of South America and the west of Europe and Africa (1_1)
– Raster LUC map covering Europe, the east of Africa and Asia (1_2)
– Raster LUC map covering South America and the west of the Antarctic (2_1)
– Raster LUC map covering Africa, the south of Asia and Oceania (2_2)

Legend and codification

Code	Label	Code	Label
1	Broadleaf evergreen forest	11	Cropland
2	Broadleaf deciduous forest	12	Paddy field
3	Needleleaf evergreen forest	13	Cropland/other vegetation mosaic
4	Needleleaf deciduous forest	14	Mangrove
5	Mixed forest	15	Wetland
6	Tree open	16	Bare area, consolidated (gravel, rock)
7	Shrub	17	Bare area, unconsolidated (sand)
8	Herbaceous	18	Urban
9	Herbaceous with sparse tree/shrub	19	Snow/ice
10	Sparse vegetation	20	Water bodies

Practical considerations

As there are no auxiliary datasets or documentation, users who require more detailed information about the characteristics of the dataset should consult the scientific papers cited above (14.6 Technical Documentation).

7 GlobCover

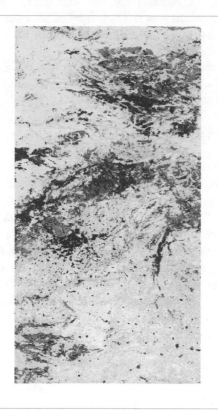

Product
LULC general
Dates
2005, 2009
Formats
Raster
Pixel size
300 m
Thematic resolution
23 classes: 1 (a), 2 (ag), 14 (v), 4 (m), 1 (na)
Compatible legends
FAO LCCS
Extent
Global
Updating
Not planned
Change detection
Not recommended
Overall accuracy
Expected to be >78.0%

Website of reference	Website Language English
http://due.esrin.esa.int/page_globcover.php	

Download site
http://due.esrin.esa.int/page_globcover.php

Availability	Format(s)
Open Access	.tiff

Technical documentation
Bicheron et al. (2008), Bontemps et al. (2011)

Other references of interest
Defourny et al. (2010)

Project

GlobCover is a project run by the European Space Agency (ESA) in collaboration with the Joint Research Centre (JRC) of the European Commission, the European Environment Agency, the FAO, the UN Environment Programme (UNEP), the Global Observations of Forest Cover Land-use Dynamics (GOFC–GOLD) programme and the International Geosphere-Biosphere Programme (IGBP). It started in 2005 and produced two global LUC maps for the reference years 2005 and 2009. The Université Catholique de Louvain (UCL) also contributed to the 2009 edition of the map.

The aim of the project was to develop global LUC maps using images from the MERIS sensor onboard the ENVISAT satellite. At the time it was launched, the 2005 Glob-Cover map was the first global LUC map at a spatial resolution of 300 m.

Based on the results of GlobCover, the ESA launched a new project called GlobCorine in which two new LUC maps compatible with the Corine Land Cover classification legend were created for Europe from the same imagery. The LC-CCI project from the ESA (see Sect. 2) builds on the progress made and the lessons learnt during the GlobCover project.

Production method

GlobCover maps were obtained by classifying imagery captured by the MERIS sensor. Urban and wetland areas, which are not well represented, were classified using a supervised classifier. The remaining categories were classified in a series of spectro-temporal classes through an unsupervised classifier. Once classified, the spectro-temporal classes were labelled automatically according to the information provided by the reference datasets. For the 2005 map, the reference datasets were the GLC2000 global LUC map (see Sect. 3 in Chap. "Global General Land Use Cover Datasets with a Single Date" Global General Land Use Cover Datasets with a Single Date) and other high-quality regional LUC maps. For the 2009 map, the GlobCover 2005 map was used as a reference.

The area for classification was divided into different regions, to account for the ecological and reflectance diversity of the world. Once labelled after classification, the LUC map was finally edited to account for inaccuracies in the representation of certain features.

For the 2005 version, regional maps with a more detailed legend were also produced following the same classification procedure.

Product description

A zipped file is available for each GlobCover map. It contains the raster layer with the LUC information and all the auxiliary data that users may need to correctly interpret the dataset. This includes the classification legend, technical and data quality information, and files with the layer style of the map to automatically symbolize the raster in GIS software. A complementary raster detailing the source of the LUC information for each pixel (MERIS sensor classification (value = null) or a land cover database (value = 1)) is also provided. In a separate file, users can also download a raster for a coloured version of the LUC map.

Downloads

GlobCover

– Raster file with LUC map
 ("GLOBCOVER_L4_200901_200912_V2.3")
– Raster file with quality information
 ("GLOBCOVER_L4_200901_200912_V2.3_CLA_QL")
– Preview image of the product
– Excel sheet with the map legend ("Globcover2009_Legend")
– Layer style files for ArcGIS (.lyr) and ENVI (.dsr)
– PDFs with technical information about the product

GlobCover coloured

– Raster file with coloured version of LUC map

Legend and codification

Code	Label
11	Post-flooding or irrigated croplands (or aquatic)
14	Rainfed croplands
20	Mosaic cropland (50–70%)/vegetation (grassland/shrubland/forest) (20–50%)
30	Mosaic vegetation (grassland/shrubland/forest) (50–70%)/cropland (20–50%)
40	Closed to open (>15%) broadleaved evergreen or semi-deciduous forest (>5 m)
50	Closed (>40%) broadleaved deciduous forest (>5 m)
60	Open (15–40%) broadleaved deciduous forest/woodland (>5 m)
70	Closed (>40%) needleleaved evergreen forest (>5 m)
90	Open (15–40%) needleleaved deciduous or evergreen forest (>5 m)
100	Closed to open (>15%) mixed broadleaved and needleleaved forest (>5 m)
110	Mosaic forest or shrubland (50–70%)/grassland (20–50%)
120	Mosaic grassland (50–70%)/forest or shrubland (20–50%)
130	Closed to open (>15%) (broadleaved or needleleaved, evergreen or deciduous) shrubland (<5 m)
140	Closed to open (>15%) herbaceous vegetation (grassland, savannas or lichens/mosses)

(continued)

Code	Label
150	Sparse (<15%) vegetation
160	Closed to open (>15%) broadleaved forest regularly flooded (semi-permanently or temporarily)—fresh or brackish water
170	Closed (>40%) broadleaved forest or shrubland permanently flooded—saline or brackish water
180	Closed to open (>15%) grassland or woody vegetation on regularly flooded or waterlogged soil—fresh, brackish or saline water
190	Artificial surfaces and associated areas (urban areas >50%)
200	Bare areas
210	Water bodies
220	Permanent snow and ice
230	No data (burnt areas, clouds...)

Practical considerations

Eleven regional maps with more detailed classification schemes were developed as part of the GlobCover Project for 2005. These maps were produced using the same methodology as the global GlobCover, but provided more thematic detail. Unfortunately, they are currently unavailable for download.

8   FROM-GLC—Finer Resolution Observation and Monitoring of Global Land Cover

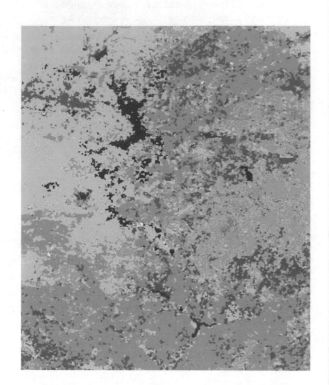

Product	
LULC general	
Dates	
2010, 2015, 2017	
Formats	
Raster	
Pixel size	
250 m, 500 m, 1 km, 5 km, 25 km, 50 km, 100 km (2010) 30 m (2010, 2015, 2017) 10 m (2017)	
Thematic resolution	
8 classes (2017): 1 (a), 1 (ag), 3 (v), 0 (m), 0 (na)	
Compatible legends	
–	
Extent	
Global	
Updating	
Not planned	
Change detection	
Not recommended	
Overall accuracy	
Expected to be >65%	

Website of reference	**Website Language** English
http://data.ess.tsinghua.edu.cn/	

Download site
http://data.ess.tsinghua.edu.cn/

Availability	**Format(s)**
Open Access	.tiff

Technical documentation
Chen et al. (2019), Gong et al. (2013), Yu et al. (2013, 2014)

Other references of interest
Ji et al. (2015), Xu et al. (2019)

Project

FROM-GLC was a project funded by Chinese research and innovation programmes that was led by Tsinghua University. It brought together researchers from Chinese and other international institutions.

The goal of this project was to produce global LUC datasets at medium to high spatial resolution. When the project started, there were no global LUC maps available at a resolution of 30 m using images from the Landsat archive. Maps at that resolution are useful for different user communities working in cross-regional and cross-national areas at that level of detail. The aim of FROM-GLC was therefore to provide new sources of data for modelling communities that required detailed global datasets. Global LUC maps at detailed scales are also useful for countries for which no other detailed LUC datasets are available.

Three global LUC maps at three different time points (2010, 2015 and 2017) were created as part of this project. Three LUC maps are available for the year 2010. The original (FROM-GLC) was successively improved by changes in the production method, producing maps known as FROM-GLC-egg and FROM-GLC-agg, the latter being the final, most updated version. It is available at the original (30 m) and 7 other spatial resolutions: 250 m, 500 m, 1 km, 5 km, 25 km, 50 km and 100 km. Unlike the maps for 2010 and 2015, the one for 2017 was produced at two spatial resolutions: 10 and 30 m.

The research team involved in the production of FROM-GLC has also taken part in related projects to produce other national, regional and thematic LUC maps, most of them at fine spatial resolutions. These maps can be accessed through the project website and include national maps of China or Chile, thematic maps about water covers and other global LUC datasets.

Production method

Each FROM-GLC map was produced using a different method. The maps for 2010, 2015 and 2017 at 30 m were produced using a supervised classification of Landsat imagery.

Four different classifiers were compared in the production of FROM-GLC for 2010. The first improved version of FROM-GLC, known as FROM-GLC-egg, included an image-segmentation method in the classification process and used two different classifiers. In addition, impervious surfaces were individually mapped. For its part, FROM-GLC-agg was obtained by combining the previous two LUC maps (FROM-GLC, FROM-GLC-egg) using a decision tree algorithm. Impervious surfaces were remapped according to the information provided by the Nighttime Light Impervious Surface Area (NL-ISA) and the MODIS urban extent (MODIS-urban) datasets. Once the FROM-GLC-agg map had been obtained at 30 m, it was then aggregated at seven other spatial resolutions through majority aggregation and proportion aggregation approaches.

The map for 2017 at 10 m was obtained through a supervised classification of Sentinel-2 imagery with a random forest classifier in the Google Earth Engine.

Product description

The FROM-GLC LUC maps are not provided as a single global file. To facilitate downloading of the product, the world is split into different tiles. Users can download the tile corresponding to their area of interest according to its latitude and longitude values.

FROM-GLC products for the year 2010 can also be downloaded through an assisted kmz layer. When uploading it in Google Earth, users can visualize their area of interest and automatically download the map corresponding to that area.

Downloads

FROM-GLC-agg (2010)

– Raster file with LUC map

FROM-GLC-agg hierarchy (2010)

– Raster file with LUC map at 30 m
– Raster file with LUC map at 250 m obtained by majority aggregation
– Raster file with LUC map at 500 m obtained by majority aggregation
– Raster file with LUC map at 1 km obtained by majority aggregation
– Raster file with LUC map at 5 km obtained by majority aggregation
– Raster file with LUC map at 5 km obtained by proportion aggregation
– Raster file with LUC map at 10 km obtained by majority aggregation
– Raster file with LUC map at 10 km obtained by proportion aggregation
– Raster file with LUC map at 25 km obtained by majority aggregation
– Raster file with LUC map at 25 km obtained by proportion aggregation
– Raster file with LUC map at 50 km obtained by majority aggregation
– Raster file with LUC map at 50 km obtained by proportion aggregation
– Raster file with LUC map at 100 km obtained by majority aggregation
– Raster file with LUC map at 100 km obtained by proportion aggregation

FROM-GLC (2015)

– Raster file with LUC map

FROM-GLC 30 m (2017)

– Raster file with LUC map

FROM-GLC 10 m (2017)

– Raster file with LUC map

Legend and codification

A specific two-level classification scheme legend was initially developed for the FROM-GLC project in 2010. This
was updated with various changes for the FROM-GLC map
for 2015. The map for 2017 has the simplest, least detailed
classification legend (Level 1). In each case, we include the
most detailed classification scheme available for each year.
Users can consult the correspondence between level 2 and
level 1 of the classification scheme for the years 2010 and
2015 at the project website.[7]

FROM-GLC (2010)

Code	Label	Code	Label
11	Rice	62	Pond
12	Greenhouse	63	River
13	Other	64	Sea
39	*Crop in urban*	*69*	*Water in urban*
21	Broadleaf	71	Shrub
22	Needleleaf	72	Grass
23	Mixed	81	High albedo
24	Orchard	82	Low albedo
29	*Forest in urban*	91	Saline-Alkali
31	Managed	92	Sand
32	Nature	93	Gravel
39	*Grass in urban*	94	Bare Cropland
40	Shrubland	95	Dry river/lake bed
49	*Shrub in urban*	96	Other
51	Grass	*99*	*Bareland in urban*
52	Silt	101	Snow
59	*Wetland in urban*	102	Ice
61	Lake	120	Cloud

* Categories only available in the FROM-GLC-Hierarchy product are
shown in italics

FROM-GLC (2015)

Code	Label	Code	Label
11	Rice paddy	41	Shrubland, leaf-on
12	Greenhouse	42	Shrubland, leaf-off
13	Other	51	Marshland
14	Orchard	52	Mudflat
15	Bare farmland	53	Marshland, leaf-off
21	Broadleaf, leaf-on	60	Water
22	Broadleaf, leaf-off	71	Shrub and brush tundra
23	Needleleaf, leaf-on	72	Herbaceous tundra
24	Needleleaf, leaf-off	80	Impervious surface
25	Mixed leaf, leaf-on	90	Bareland
26	Mixed leaf, leaf-off	92	Bareland
31	Pasture	101	Snow
32	Natural grassland	102	Ice
33	Grassland, leaf-off	120	Cloud

FROM-GLC (2017)

Code	Label	Code	Label
1	Cropland	6	Water
2	Forest	8	Impervious
3	Grass	9	Bareland
4	Shrubland	10	Snow/ice

Practical considerations

The project website, where all the information is stored and
available for download, is not user-friendly. It is not easy to
find the information the user is looking for. Users may also
struggle to download datasets for their area of interest
according to latitude and longitude information. When
available, we recommend using the kmz file with Google
Earth for this purpose.

There is little additional information. For a complete
description of the characteristics of the different maps, we
recommend users to read the scientific papers cited in the
introduction to this dataset above (14.8. Technical
Documentation).

---

[7] http://data.ess.tsinghua.edu.cn/.

9 CGLS-LC100—Copernicus Global Land Service Dynamic Land Cover Map

Product	
LULC general	
Dates	
2015–2019	
Formats	
Raster	
Pixel size	
100 m	
Thematic resolution	
24 classes: 1 (a), 1 (ag), 18 (v), 2 (m), 1 (na)	
Compatible legends	
FAO LCCS	
Extent	
Global	
Updating	
Yes, every year	
Change detection	
Possible, although sources of uncertainty may arise	
Overall accuracy	
Expected to be >80%	

Website of reference	**Website Language** English
https://land.copernicus.eu/global/products/lc	

Download site	
https://lcviewer.vito.be/download	

Availability	**Format(s)**
Open Access	.tiff

Technical documentation	
Buchhorn et al. (2020a, b, c), Tsendbazar et al. (2019, 2020)	

Other references of interest	
–	

Project

CGLS-LC100 is one of the deliverables produced as part of the Copernicus Global Land Service (CGLS), which aims to provide a series of bio-geophysical products to monitor land surface at a global scale. In addition to this LUC package, the programme produces other relevant variables, such as the Leaf Area Index (LAI), the Fraction of Absorbed Photosynthetically Active Radiation (FAPAR), the Land Surface Temperature, soil moisture and other vegetation indices.

The first version of CGLS-LC100 was released in 2017, mapping LUC for Africa. Since then, several updates of the product have improved the production methodology and extended its temporal and geographical coverage. The last version of the product (Collection 3), released in 2021, covers the whole world for the period 2015–2019. It includes a method for detecting land cover change that addresses the main sources of technical uncertainty when studying change in a time series of LUC maps.

In addition to the LUC map described here, the product also includes a series of continuous field layers or "fraction maps" for the basic LUC classes mapped. Future updates of the product are expected on an annual basis, using the imagery provided by the Sentinel satellite missions.

Production method

The Copernicus Global Land Service Dynamic Land Cover map is produced through a multistep processing framework. First, PROBA-V satellite images are pre-processed and merged following a Sentinel-2 tiling grid to create a 3-year epoch mosaic for each reference year. Second, a series of metrics (spectral and textural metrics, descriptive statistics) are extracted from each epoch mosaic. Third, imagery for all the epochs is classified using a regression algorithm, which delivers a cover fraction layer for each basic LUC class and reference year, and a supervised classification algorithm, which delivers a LUC map for each reference year.

Various auxiliary data sources are used in the classification phase, i.e. seven different data masks and three extra datasets: biome clusters, water cover fractions and built-up cover fractions.

In order to ensure the temporal consistency of the LUC map series, it was decided to include a temporal postprocessing phase in the production of the dataset. This consists of a BFAST break detection algorithm and a Hidden Markov Model. The former is used to detect changes in an independent time series of MODIS NIRv imagery, while the latter is used to rule out technical changes in the classified epoch images.

Product description

CGLS-LC100 is distributed in tiles, following the Sentinel-2 tiling grid (110 × 110 km). For each tile, users can download many different layers: the discrete classification containing the LUC map for the selected area; a layer with the classification probability; layers of cover fractions for each of the basic LUC classes mapped; a layer showing the level of confidence for the change measured between the different years in each pixel; and two extra layers: forest types and input data density.

The download of the LUC map only includes the raster file with the LUC information. Each reference year must be downloaded separately.

Downloads

Land Cover classification—discrete classification
– Raster file with LUC map

Cover fractions—bare and sparse vegetation
– Raster file with the cover fraction for the land cover under consideration

Land Cover changes—change confidence
– Raster file indicating the reliability of the change in the discrete class

Others—forest types
– Raster file indicating for all pixels with a cover fraction >1% the type of forest represented in the pixel

Legend and codification

Land Cover classification–discrete classification

Code	Label	Code	Label
0	No input data	113	Closed forest, deciduous needle leaf
20	Shrubs	114	Closed forest, deciduous broad leaf
30	Herbaceous vegetation	115	Closed forest, mixed
40	Cultivated and managed vegetation/agriculture (cropland)	116	Closed forest, unknown
50	Urban/built up	121	Open forest, evergreen needle leaf

(continued)

Land Cover classification–discrete classification

Code	Label	Code	Label
60	Bare/sparse vegetation	122	Open forest, evergreen broad leaf
70	Snow and ice	123	Open forest, deciduous needle leaf
80	Permanent water bodies	124	Open forest, deciduous broad leaf
90	Herbaceous wetland	125	Open forest, mixed
100	Moss and lichen	126	Open forest, unknown
111	Closed forest, evergreen needle leaf	200	Open sea
112	Closed forest, evergreen, broad leaf	113	Closed forest, deciduous needle leaf

Cover fractions—bare and sparse vegetation

Code	Meaning
0–100	Percentage of the pixel (0–100%) covered by the land cover under consideration
200	Masked sea

Land Cover changes—change confidence

Code	Change confidence	Code	Change confidence
0	No change	2	Medium confidence
1	Potential change	3	High confidence

Others—forest types

Code	Forest type	Code	Forest type
0	Unknown	3	Deciduous, needle leaf forest (DNF)
1	Evergreen, needle leaf forest (ENF)	4	Deciduous, broad leaf forest (DBF)
2	Evergreen, broad leaf forest (EBF)	5	Mixed

Practical considerations

Because of the large number of datasets available through this project, users are encouraged to make use of the different layers of LUC information available. This will give them a better understanding of the uncertainties and limitations of the product.

Users can download the product covering the whole globe, which is distributed through the files in the Zenodo repository.[8]

----

[8] 2015: https://doi.org/10.5281/zenodo.3939038; 2016: https://doi.org/10.5281/zenodo.3518026; 2017: https://doi.org/10.5281/zenodo.3518036; 2018: https://doi.org/10.5281/zenodo.3518038; 2019: https://doi.org/10.5281/zenodo.3939050.

References

Bicheron P, Defourny P, Brockmann C, et al (2008) GLOBCOVER. Products description and validation report. http://due.esrin.esa.int/files/GLOBCOVER_Products_Description_Validation_Report_I2.1.1.pdf. Accessed 20 Aug 2020

Bontemps S, Herold M, Kooistra L et al (2012) Revisiting land cover observation to address the needs of the climate modeling community. Biogeosciences 9:2145–2157. https://doi.org/10.5194/bg-9-2145-2012

Bontemps S, Defourny P, Van Bogaert E, et al (2011) GLOBCOVER 2009 products description and validation report. http://due.esrin.esa.int/files/GLOBCOVER_Products_Description_Validation_Report_I2.1.1.pdf. Accessed 20 Aug 2020

Buchhorn et al., 2020aBuchhorn M, Lesiv M, Tsendbazar NE, et al (2020a) Copernicus global land cover layers-collection 2. Remote Sens 12. https://doi.org/10.3390/rs12061044

Buchhorn M, Smets B, Bertels L, et al (2020b) Copernicus global land service: land cover 100 m: version 3 globe 2015–2019: product user manual. https://land.copernicus.eu/global/sites/cgls.vito.be/files/products/CGLOPS1_PUM_LC100m-V3_I3.3.pdf. Accessed 28 Dec 2020b

Buchhorn M, Bertels L, Smets B, et al (2020c) Copernicus global land service: land cover 100 m: version 3 globe 2015–2019: algorithm theoretical basis document. https://land.copernicus.eu/global/sites/cgls.vito.be/files/products/CGLOPS1_ATBD_LC100V3_I3.3.pdf. Accessed 28 Dec 2020c

Cao X, Chen J, Chen L et al (2014) Preliminary analysis of spatiotemporal pattern of global land surface water. Sci China Earth Sci 57:2330–2339. https://doi.org/10.1007/s11430-014-4929-x

Chen J, Zhu X, Imura H, Chen X (2010) Consistency of accuracy assessment indices for soft classification: simulation analysis. ISPRS J Photogramm Remote Sens 65:156–164. https://doi.org/10.1016/j.isprsjprs.2009.10.003

Chen J, Chen X, Cui X, Chen J (2011a) Change vector analysis in posterior probability space: a new method for land cover change detection. IEEE Geosci Remote Sens Lett 8:317–321. https://doi.org/10.1109/LGRS.2010.2068537

Chen J, Zhu X, Vogelmann JE et al (2011b) A simple and effective method for filling gaps in Landsat ETM+ SLC-off images. Remote Sens Environ 115:1053–1064. https://doi.org/10.1016/j.rse.2010.12.010

Chen X, Chen J, Shi Y, Yamaguchi Y (2012) An automated approach for updating land cover maps based on integrated change detection and classification methods. ISPRS J Photogramm Remote Sens 71:86–95. https://doi.org/10.1016/j.isprsjprs.2012.05.006

Chen J, Wu H, Li S et al (2013) Temporal logic and operation relations based knowledge representation for land cover change web services. ISPRS J Photogramm Remote Sens 83:140–150. https://doi.org/10.1016/j.isprsjprs.2013.02.005

Chen J, Chen J, Liao A et al (2014) Global land cover mapping at 30 m resolution: a POK-based operational approach. ISPRS J Photogramm Remote Sens 103:7–27. https://doi.org/10.1016/j.isprsjprs.2014.09.002

Chen F, Chen J, Wu H et al (2016) A landscape shape index-based sampling approach for land cover accuracy assessment. Sci China Earth Sci 59:2263–2274. https://doi.org/10.1007/s11430-015-5280-5

Chen J, Li S, Wu H, Chen X (2017) Towards a collaborative global land cover information service. Int J Digit Earth 10:356–370. https://doi.org/10.1080/17538947.2016.1267268

Chen B, Xu B, Zhu Z, et al (2019) Stable classification with limited sample: transferring a 30 m resolution sample set collected in 2015 to mapping 10 m resolution global land cover in 2017. Sci Bull

Defourny P, Bontemps S, Obsomer V, et al (2010) Accuracy assessment of global land cover maps: lessons learnt from the

GlobCover and GlobCorine experiences. In: SP-686 ESA living planet symposium Bergen, Norway, 2010-06-28/2010-07-02

ESA (2017) Land cover CCI. Product user guide. Version 2.0. http://maps.elie.ucl.ac.be/CCI/viewer/download/ESACCI-LC-Ph2-PUGv2_2.0.pdf. Accessed 19 Aug 2020

Friedl MA, McIver DK, Hodges JCF et al (2002) Global land cover mapping from MODIS: algorithms and early results. Remote Sens Environ 83:287–302. https://doi.org/10.1016/S0034-4257(02)00078-0

Friedl MA, Sulla-Menashe D, Tan B et al (2010) MODIS collection 5 global land cover: algorithm refinements and characterization of new datasets. Remote Sens Environ 114:168–182. https://doi.org/10.1016/j.rse.2009.08.016

Friedl MA, Sulla-Menashe D (2019) User guide to collection 6 MODIS land cover (MCD12Q1 and MCD12C1) product. https://lpdaac.usgs.gov/documents/101/MCD12_User_Guide_V6.pdf. Accessed 26 Sept 2020

Fritz S, See L (2005) Comparison of land cover maps using fuzzy agreement. Int J Geogr Inf Sci 19:787–807. https://doi.org/10.1080/13658810500072020

Giri C, Zhu Z, Reed B (2005) A comparative analysis of the global land cover 2000 and MODIS land cover data sets. Remote Sens Environ 94:123–132. https://doi.org/10.1016/j.rse.2004.09.005

Gong P, Wang J, Yu L et al (2013) Finer resolution observation and monitoring of global land cover: first mapping results with Landsat TM and ETM+ data. Int J Remote Sens 34:2607–2654. https://doi.org/10.1080/01431161.2012.748992

Han G, Chen J, He C et al (2015) A web-based system for supporting global land cover data production. ISPRS J Photogramm Remote Sens 103:66–80. https://doi.org/10.1016/j.isprsjprs.2014.07.012

Hao G, Gen-Suo J (2014) Assessing MODIS land cover products over China with probability of interannual change. Atmos Ocean Sci Lett 7:564–570. https://doi.org/10.1080/16742834.2014.11447225

Hollmann R, Merchant CJ, Saunders R et al (2013) The ESA climate change initiative: satellite data records for essential climate variables. Bull Am Meteorol Soc 94:1541–1552. https://doi.org/10.1175/BAMS-D-11-00254.1

Hua T, Zhao W, Liu Y, et al (2018) Spatial consistency assessments for global land-cover datasets: a comparison among GLC2000, CCI LC, MCD12, GLOBCOVER and GLCNMO. Remote Sens 10. https://doi.org/10.3390/rs10111846

Ji L, Gong P, Geng X, Zhao Y (2015) Improving the accuracy of the water surface cover type in the 30 m FROM-GLC product. Remote Sens 7:13507–13527. https://doi.org/10.3390/rs71013507

Jun C, Ban Y, Li S (2014) Open access to earth land-cover map. Nature 514:434–434. https://doi.org/10.1038/514434c

Kobayashi T, Tateishi R, Alsaaideh B et al (2017) Production of global land cover data—GLCNMO2013. J Geogr Geol 9:1. https://doi.org/10.5539/jgg.v9n3p1

Liu H, Gong P, Wang J et al (2020) Annual dynamics of global land cover and its long-term changes from 1982 to 2015. Earth Syst Sci Data 12:1217–1243. https://doi.org/10.5194/essd-12-1217-2020

Manakos I, Tomaszewska M, Gkinis I, et al (2018) Comparison of global and continental land cover products for selected study areas in South Central and Eastern European Region. Remote Sens 10. https://doi.org/10.3390/rs10121967

Mousivand A, Arsanjani JJ (2019) Insights on the historical and emerging global land cover changes: the case of ESA-CCI-LC datasets. Appl Geogr 106:82–92. https://doi.org/10.1016/j.apgeog.2019.03.010

Shi X, Nie S, Ju W, Yu L (2016a) Climate effects of the GlobeLand30 land cover dataset on the Beijing climate center climate model

simulations. Sci China Earth Sci 59:1754–1764. https://doi.org/10.1007/s11430-016-5320-x

Shi X, Nie S, Ju W, Yu L (2016b) Application and impacts of the GlobeLand30 land cover dataset on the Beijing climate center climate model. In: IOP conference series: earth and environmental science. pp 12–32

Sulla-Menashe D, Gray JM, Abercrombie SP, Friedl MA (2019) Hierarchical mapping of annual global land cover 2001 to present: the MODIS collection 6 land cover product. Remote Sens Environ 222:183–194. https://doi.org/10.1016/j.rse.2018.12.013

Tang P, Zhang H, Zhao Y et al (2014) Practice and thoughts of the automatic processing of multispectral images with 30 m spatial resolution on the global scale. J Remote Sens 18:231–253. https://doi.org/10.11834/jrs.20143287

Tateishi R, Uriyangqai B, Al-Bilbisi H et al (2011) Production of global land cover data—GLCNMO. Int J Digit Earth 4:22–49. https://doi.org/10.1080/17538941003777521

Tateishi R, Hoan NT, Kobayashi T et al (2014) Production of global land cover data—GLCNMO2008. J Geogr Geol 6:99–122. https://doi.org/10.5539/jgg.v6n3p99

Tchuenté ATK, Roujean JL, de Jong SM (2011) Comparison and relative quality assessment of the GLC2000, GLOBCOVER, MODIS and ECOCLIMAP land cover data sets at the African continental scale. Int J Appl Earth Obs Geoinf 13:207–219. https://doi.org/10.1016/j.jag.2010.11.005

Tsendbazar NE, Herold M, Tarko A, et al (2019) Copernicus global land service: land cover 100 m: version 2: validation report. https://land.copernicus.eu/global/sites/cgls.vito.be/files/products/CGLOPS1_VR_LC100m-V2.0_I1.00.pdf. Accessed 28 Dec 2020

Tsendbazar NE, Tarko A, Linlin L, et al (2020) Copernicus global land service: land cover 100 m: version 3 globe 2015–2019: validation report. https://land.copernicus.eu/global/sites/cgls.vito.be/files/products/CGLOPS1_VR_LC100m-V3.0_I1.00.pdf. Accessed 28 Dec 2020

Vilar L, Garrido J, Echavarría P et al (2019) Comparative analysis of CORINE and climate change initiative land cover maps in Europe: implications for wildfire occurrence estimation at regional and local scales. Int J Appl Earth Obs Geoinf 78:102–117. https://doi.org/10.1016/j.jag.2019.01.019

Wang J, Zhao Y, Li C et al (2015) Mapping global land cover in 2001 and 2010 with spatial-temporal consistency at 250 m resolution. ISPRS J Photogramm Remote Sens 103:38–47. https://doi.org/10.1016/j.isprsjprs.2014.03.007

Wu H, Chen J, Xing H, et al (2016) Pragmatics driven land cover service composition utilizing behavior-intention model. In: ISPRS—international archives of the photogrammetry, remote sensing and spatial information sciences. pp 1319–1325

Xie H, Tong X, Meng W et al (2015) A multilevel stratified spatial sampling approach for the quality assessment of remote-sensing-derived products. IEEE J Sel Top Appl Earth Obs Remote Sens 8:4699–4713. https://doi.org/10.1109/JSTARS.2015.2437371

Xu H, Wei Y, Liu C, et al (2019) A scheme for the long-term monitoring of impervious-relevant land disturbances using high frequency Landsat archives and the Google Earth Engine. Remote Sens 11. https://doi.org/10.3390/rs11161891

Yang Z, Dong J, Liu J et al (2017) Accuracy assessment and inter-comparison of eight medium resolution forest products on the Loess Plateau. China. ISPRS Int J Geo-Information 6:152. https://doi.org/10.3390/ijgi6050152

Yu L, Wang J, Gong P (2013) Improving 30 m global land-cover map FROM-GLC with time series MODIS and auxiliary data sets: a segmentation-based approach. Int J Remote Sens 34:5851–5867. https://doi.org/10.1080/01431161.2013.798055

Yu L, Wang J, Li X et al (2014) A multi-resolution global land cover dataset through multisource data aggregation. Sci China Earth Sci 57:2317–2329. https://doi.org/10.1007/s11430-014-4919-z

Zhu X, Chen J, Gao F et al (2010) An enhanced spatial and temporal adaptive reflectance fusion model for complex heterogeneous regions. Remote Sens Environ 114:2610–2623. https://doi.org/10.1016/j.rse.2010.05.032

General Land Use Cover Datasets for Europe

David García-Álvarez, Javier Lara Hinojosa,
Francisco José Jurado Pérez, and Jaime Quintero Villaraso

Abstract

The land uses and covers of Europe are the most systematically mapped in the world today, and their associated datasets offer the greatest spatial and thematic detail. Thanks to the work done within the Copernicus Land Monitoring programme run by the European Environmental Agency (EEA) and the Joint Research Centre (JRC) of the European Commission, there are many general LUC datasets covering most of the European continent. These general datasets map all land uses and covers on the ground, without focusing on any specific type. However, whereas some cover the whole of Europe, others only map specific local areas of interest, such as urban or coastal areas, riparian zones or spaces protected under the Nature 2000 network. CORINE Land Cover (CLC) is the flagship European LUC mapping programme and a reference worldwide. It has provided consistent LUC information at a detailed scale (1:100,000) every 6 years since 1990. This is the result of a high degree of coordination between many different organizations and institutions across Europe. The Copernicus programme also includes other European datasets such as Urban Atlas, N2K, Riparian Zones and Coastal Zones, which provide very detailed LUC information at higher levels of spatial detail (scale 1:10,000) for specific geographical area types: Functional Urban Areas, the Natura 2000 network, riparian zones from Strahler level 2–8 rivers and areas 10 km away from the coastline. However, these projects do not cover the same long timeframe as CLC. In addition, their long-term future is far from clear in that updates are only planned for Urban Atlas and Coastal Zones. PELCOM, GlobCorine and the Annual Land Cover Product are the European projects that most resemble the LUC maps available at global and supra-national scales for other parts of the world. They were obtained through classification of satellite imagery. PELCOM and GlobCorine are only available for a few dates and at quite coarse spatial resolutions: 1 km and 300 m respectively. The Annual Land Cover Product consists of a series of LUC maps for the period 2000–2019 at a highly detailed spatial resolution (30 m). It offers information for a large number of different points in time. However, it makes a separate classification of land uses each year, which means that change analysis with this dataset is more uncertain than with CLC or other Copernicus Land Monitoring products. HILDA and S2GLC 2017 are LUC datasets produced within the framework of different research projects, which can be considered reference products in their respective fields. HILDA provides one of the largest time series of LUC maps currently available, spanning the period from 1900 to 2010. S2GLC 2017 is one of the most spatially detailed LUC mapping experiences at a supra-national scale, with a spatial resolution of 10 m.

Keywords

Europe • European Union • HILDA • CORINE Land Cover • PELCOM • Annual Land Cover Product • GlobCorine • Urban Atlas • N2K • Riparian Zones • Coastal Zones • S2GLC 2017

D. García-Álvarez (✉)
Departamento de Geología, Geografía y Medio Ambiente, Universidad de Alcalá, Alcalá de Henares, Spain
e-mail: David.garcia@uah.es

J. Lara Hinojosa · F. J. Jurado Pérez · J. Quintero Villaraso
Departamento de Análisis Geográfico Regional y Geografía Física, Universidad de Granada, Granada, Spain

D. García-Álvarez et al. (eds.), *Land Use Cover Datasets and Validation Tools*,
https://doi.org/10.1007/978-3-030-90998-7_16

1   HILDA

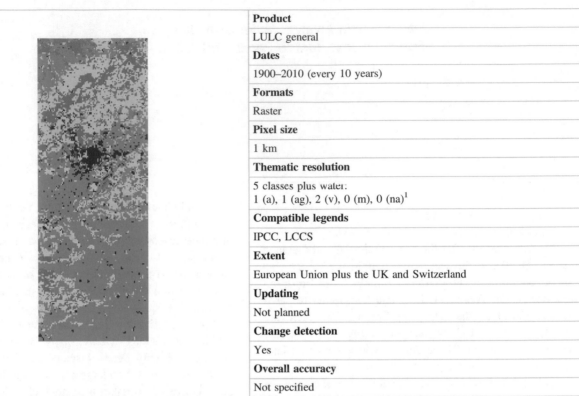

Product
LULC general
Dates
1900–2010 (every 10 years)
Formats
Raster
Pixel size
1 km
Thematic resolution
5 classes plus water: 1 (a), 1 (ag), 2 (v), 0 (m), 0 (na)[1]
Compatible legends
IPCC, LCCS
Extent
European Union plus the UK and Switzerland
Updating
Not planned
Change detection
Yes
Overall accuracy
Not specified

Website of reference	Website Language English

https://www.wur.nl/en/Research-Results/Chair-groups/Environmental-Sciences/Laboratory-of-Geo-information-Science-and-Remote-Sensing/Models/Hilda.htm

Download site

https://www.wur.nl/en/Research-Results/Chair-groups/Environmental-Sciences/Laboratory-of-Geo-information-Science-and-Remote-Sensing/Models/Hilda/HILDA-data-downloads.htm

Availability	Format(s)
Open Access, after providing personal data	ESRI Grid, .tiff, .ascii

Technical documentation

Fuchs et al. (2013,2015a, b)

Other references of interest

Fuchs (2015)

---

[1] (a): artificial; (ag): agriculture; (v): vegetation; (m): mixed classes; (na): no data.

Project

HIstoric Land Dynamics Assessment (HILDA) is a project aimed at reconstructing historic land cover/use and LUC changes in Europe. Unlike other LUC reconstruction projects and datasets, it allows us to study LUC changes over time. The recently launched HILDA + project takes the original project one step further by mapping historical LUC changes at a global scale for the period 1960–2019.

The reconstruction of historic LUC landscapes and changes is carried out using a model maintained and developed by the Department of Geoinformation Science and Remote Sensing of Wageningen University. The model allocates non-spatial historic LUC information on the ground.

Production method

Historic LUC maps for the HILDA project were obtained through an extensive workflow involving various steps. First, gross and net LUC changes per decade were obtained for the period 1950–2010 from a set of sources providing historic LUC information: UNFCCC national reporting data, CORINE Land Cover, Historisch Grondgebruik Nederland (HGN) for the Netherlands, FAO-RSS data and BioPress data with classified aerial photographs of 73 sample sites across Europe. Later, LUC data was spatially allocated by the HILDA model. Four categories were spatially allocated at this stage. A fifth category (other land) remained static throughout the time series. Water was a subclass of the "other land" category, which was only separated in the final maps for visualization purposes.

The model allocates the LUC categories using a series of probability maps. A specific probability map for each category was created on the basis of historical LUC maps and a range of socioeconomic and physical (soil properties, climate and terrain) factors. The categories were allocated hierarchically according to their socioeconomic value: settlements were allocated first, followed by croplands, forest and grasslands.

Once the model had been run for the 1950–2010 timeframe, four extra maps were obtained for the period 1900–1950 based on historical LUC statistics and an extrapolation of the change matrix. The pre-1950 maps therefore assume stable transition rates for the period 1950–2010. This could be an important source of uncertainty in these maps.

Product description

The product is delivered in four different packages, two of which include the series of LUC maps (1900–2010). Of these, one considers the net changes over the course of each decade, while the other considers the gross changes. The other two packages detail the specific transitions that take place between the different categories, one charting net changes and the other gross changes.

Each package can be downloaded in three different file formats (ESRI Grid, TIFF, ASCII). Each download includes a raster with LUC information for each decade and a supplementary file with the technical description of the product.

Downloads

Gross land changes
– Raster files with LUC maps for each decade
– Text document with technical information and the legend

Net land changes
– Raster files with LUC maps for each decade
– Text document with technical information and the legend

Transitions maps (for gross and net)
– Raster files with LUC maps for each decade
– Text document with technical information and the legend

Legend and codification

HILDA gross and net maps

Code	Label	Code	Label
111	Settlement	444	Grassland
222	Cropland	555	Other land
333	Forest	666	Water

HILDA gross and net transitions maps (1900–2000)

Code	Label	Code	Label
112	Cropland to settlement	242	Cropland to grassland
113	Forest to settlement	252	Cropland to other land
114	Grassland to settlement	262	Cropland to water

(continued)

HILDA gross and net transitions maps (1900–2000)

Code	Label	Code	Label
115	Other land to settlement	334	Grassland to forest
116	Water to settlement	335	Other land to forest
121	Settlement to cropland	336	Water to forest
131	Settlement to forest	343	Forest to grassland
141	Settlement to grassland	353	Forest to other land
151	Settlement to other land	363	Forest to water
161	Settlement to water	445	Other land to grassland
223	Forest to cropland	446	Water to grassland
224	Grassland to cropland	454	Grassland to other land
225	Other land to cropland	464	Grassland to water
226	Water to cropland	556	Water to other land
232	Cropland to forest	565	Other land to Water

HILDA gross and net transitions maps (2000–2010)

Code	Label	Code	Label
112	Settlement to cropland	242	Grassland to cropland
113	Settlement to forest	252	Other land to cropland
114	Settlement to grassland	262	Water to cropland
115	Settlement to other land	334	Forest to grassland
116	Settlement to water	335	Forest to other land
121	Cropland to settlement	336	Forest to water
131	Forest to settlement	343	Grassland to forest

(continued)

HILDA gross and net transitions maps (2000–2010)

Code	Label	Code	Label
141	Grassland to settlement	353	Other land to forest
151	Other land to settlement	363	Water to forest
161	Water to settlement	445	Grassland to other land
223	Cropland to forest	446	Grassland to water
224	Cropland to grassland	454	Other land to grassland
225	Cropland to other land	464	Water to grassland
226	Cropland to water	556	Other land to Water
232	Forest to cropland	565	Water to other land

Practical considerations

This is a valuable dataset because of the rich historic LUC information it provides. There are very few long, dense historical series of LUC maps that measure LUC change over time. Nonetheless, users should be aware of the uncertainties associated with this dataset. The maps prior to 1950 were created by extrapolating the patterns of change for the period 1950–2010. This could introduce a high degree of uncertainty.

An online visualization of the maps for the years 1900 and 2010 is available, together with other auxiliary information, at http://www.geo-informatie.nl/fuchs003/#.

To study global historical LUC change at a similar level of detail, users should refer to the associated HILDA+ project.

2 CLC—CORINE Land Cover

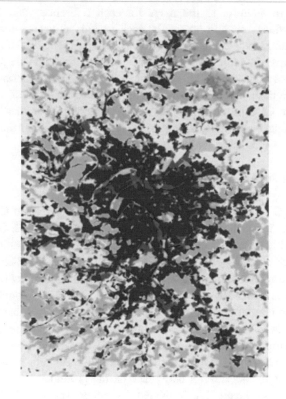

Product	
LULC general	
Dates	
1990, 2000, 2006, 2012, 2018	
Formats	
Vector and raster	
Scale/Pixel size	
Photointerpretation scale: 1:100,000 Minimum Mapping Unit: 25 ha/5 ha for changes Minimum Mapping Width: 100 m Pixel size (raster): 100 m	
Thematic resolution	
44 classes: 11 (a), 8 (ag), 8 (v), 6 (m), 3 (na)	
Compatible legends	
CLC	
Extent	
Europe, with an increasing number of countries taking part in the project each year (39 in CLC18)	
Updating	
Scheduled updates every 6 years	
Change detection	
Yes, through the layer of changes	
Overall accuracy	
Expected to be >85%	

Website of reference	**Website Language** English, German and French
https://land.copernicus.eu/pan-european/corine-land-cover	

Download site

https://land.copernicus.eu/pan-european/corine-land-cover

Availability	**Format(s)**
Open Access previous registration	.tiff, .gdb, .gpkg

Technical documentation

Bossard et al. (2000), Büttner et al. (2002, 2011, 2012, 2014) European Environment Agency (1994, 2006a, b, 2007), Jaffrain et al (2017), Kosztra et al. (2019), Soukup et al. (2017)

Other references of interest

Bach et al. (2006), Bielecka and Jenerowicz (2019), Büttner (2014), European Environment Agency (2006c), Feranec et al. (2010, 2016), Gallego (2001), García-Álvarez and Camacho Olmedo (2017), Neumann et al. (2007)

Project

CORINE Land Cover (CLC) is a European project monitoring Land Use and Cover that dates back to 1985. It aims to map land uses and land covers across the whole continent according to the same rules. It is currently part of the land monitoring efforts of the Copernicus programme.

The number of countries taking part in the project has been increasing since its inception, from the initial group of 26 countries that created the CLC 1990 to the 39 countries that participated in the most recent edition[2]. In the meantime, the production of CLC has undergone several technical and methodological changes. The fact that CLC is produced at a national level means that methods vary from one country to the next.

Because of its long life, detail, consistency and wide range of applications, CLC is one of the most renowned LUC mapping initiatives worldwide. Various European countries have developed national LUC products based on CLC. In some cases, these products are new CLC layers with an extended legend, adapted to the specificities of the country. In other cases, they are new CLC layers for different dates to those used in the main Europe-wide project.

Production method

The production of CLC is coordinated by the European Environment Agency (EEA). Each participant country is responsible for mapping its own territory according to the general guidelines developed by the EEA.

The method of production may vary from country to country. Initially, CLC was mapped at national scales based on the photointerpretation of Landsat imagery. In the following editions, most of the countries decided to stick to this method, using different satellite imagery according to EEA prescriptions: Landsat, SPOT; ITS P6, RapidEye, LISS III, Sentinel. In the latest editions, the production method has varied in some cases. A few countries, like Germany or Spain, produce the CLC database by generalizing national LUC databases at finer scales. This has introduced important changes in the way land uses and covers are mapped over time for these countries. For both production methods, photointerpretation and map generalization, the CLC map obtained is then subject to expert review to ensure its consistency and validity.

The first CLC map was produced for the reference year 1990 and the subsequent editions have been updates of this initial map. The national teams do not draw a new map for each new reference year. Instead, they map the changes for the analysed period (e.g. 1990–2000) and then update the base map for the new reference year. In this updating process, any errors detected in the base map are also corrected. If important changes have been made in the CLC production method, the base map is also updated according to the new method.

In addition to the maps for each reference year, CLC produces change layers for each period between reference years: 1990–2000, 2000–2006, 2006–2012, 2012–2018. The maps showing changes do not follow the same mapping rules as the base CLC maps and show more information than the base layers for the reference years (MMU of 5ha). The CLC production team therefore recommends that LUC changes be studied using these change layers, rather than by cross-tabulating and comparing base CLC maps.

Product description

CLC is made up of two spatial layers: a Land Use Cover map for each reference year (1990, 2000, 2006, 2012, 2018) and a layer of Land Use Cover changes for each analysis period (1990–2000, 2000–2006, 2006–2012, 2012–2018). The reference map for each year provides Land Use Cover information for the total area of the participant countries. The map of changes only accounts for the changes that took place in the period under consideration. Rather than comparing two reference maps, the CLC layer of changes maps all changes bigger than 5ha and discards all technical changes that did not take place on the ground.

CLC layers are provided in either vector (ESRI or GeoPackage databases) or raster (.tiff) formats. As might be expected, the vector data is much heavier than the raster data, because of its higher definition.

Together with the LUC layers, the CLC product includes all the auxiliary information required to understand the LUC information provided by the CLC layers: a style layer for the raster, the legend description, technical information and other relevant metadata. LUC maps for the French overseas departments (Guadeloupe, French Guinea, Martinique, Mayotte and Reunion) are also provided in auxiliary layers.

Downloads

The base layers with LUC maps for each reference year (CLC) have the same structure and group of files, as do the change layers for each period of analysis (CHA). This is why we only describe the file structure once for each type of format.

CLC 2018 (Geodatabase)/CHA 2012–2018 (Geodatabase)

– Geodatabase files with CLC vector layers (*DATA* folder)
– Folder with CLC vector data for French overseas departments
– Layer style files for ArcGIS (.lyr), QGIS (.qml) and any other GIS software (.sld) (*Legend* folder)

(continued)

---

[2] https://land.copernicus.eu/pan-european/corine-land-cover.

CLC 2018 (Geodatabase)/CHA 2012–2018 (Geodatabase)

- Excel presenting the CLC legend, including information about the RGB colours for each class (*Legend* folder)
- Text documents describing the CLC legend, including information about the RGB colours for each class (*Legend* folder)
- Folder with metadata files (.xml)
- PDF and Excel sheet with information about CLC country coverage (*Documents* folder)
- A Word document explaining how to use the CLC files for the product in QGIS (*Documents* folder)
- Three text documents with technical information about the CLC layers (*Documents* folder)

CLC 2018 (GeoPackage)/CHA 2012–2018 (GeoPackage)

- GeoPackage file with CLC vector layers (*DATA* folder)
- Layer style files for ArcGIS (.lyr), QGIS (.qml) and any other GIS software (.sld) (*Legend* folder)
- Excel presenting the CLC legend, including information about the RGB colours for each class (*Legend* folder)
- Text documents describing the CLC legend, including information about the RGB colours for each class (*Legend* folder)
- Folder with metadata files (.xml)
- PDF and Excel sheet with information about CLC country coverage (*Documents* folder)

CLC 2018 (GeoPackage)/CHA 2012–2018 (GeoPackage)

- A Word document explaining how to use the CLC files for the product in QGIS (*Documents* folder)
- PDFs and text documents with technical information about the CLC layers (*Documents* folder)

CLC 2018 (Raster)/CHA 2012–2018 (Raster)

- Raster file with CLC map (*DATA* folder)
- Folder with CLC raster data for French Overseas Departments (*DATA* folder)
- Layer style files for ArcGIS (.lyr) and QGIS (.qml) (*Legend* folder)
- Layer style files for ArcGIS (.lyr) and QGIS (.qml) for French Overseas Departments (*French_DOMs* folder)
- Text document describing the CLC legend, including information about the RGB colours for each class (*Legend* folder)
- Folder with metadata files (.xml)
- PDF and Excel sheet with information about CLC country coverage (*Documents* folder)
- A Word document explaining how to use the CLC files for the product in QGIS (*Documents* folder)
- PDFs and text documents with technical information about the CLC layers (*Documents* folder)

(continued)

Database

CLC 2018

OBJECTID	Code_18	Remark	Area_Ha	ID	Shape_Length	Shape_Area	c18
1	111		130,86365376999143	EU_1	10902,412470996784	1308636,5376999143	111
2	111		53,7445236800477	EU_2	6329,456389055609	537445,236800477	111
3	111		30,719103909971782	EU_3	3371,7747186580314	307191,0390997178	111

- OBJECTID: Unique identifier for each polygon.

- Code_18: LUC code for the year 2018.

- Remark

- Area_Ha: Area of the polygon, in hectares.

- ID: Unique identifier for each polygon.

- Shape_Length: Perimeter of the polygon, in metres.

- Shape_Area: Area of the polygon, in square metres.

- C18: LUC code for the year 2018.

CHA 2012–2018

OBJECTID	Change	ID	Code_12	Code_18	Chtype	Remark	Area_Ha	Shape_Length	Shape_Area
1	111-112	EU-1	111	112	R		8,049594709980532	1774,991620734559	80495,94709980533
2	111-112	EU-2	111	112	R		18,185012579975673	2282,695093131555	181850,12579975673
3	111-112	EU-3	111	112	R		12,21255290507005	2012,3640670179973	122125,5290507005

– OBJECTID: Unique identifier for each polygon

– Change: Change code made up of the CLC code for the oldest year (on the right) and the CLC code for the most recent year on the left (2018)

– ID:

– Code_12: LUC code for the year 2012

– Code_18: LUC code for the year 2018

– Chtype

– Remark

– AREA_HA: Area of the polygon, in hectares

– Shape_Length: Perimeter of the polygon, in metres

– Shape_Area: Area of the polygon, in square metres

Legend and codification

Code	Label	Code	Label
111	Continuous urban fabric	313	Mixed forest
112	Discontinuous urban fabric	321	Natural grasslands
121	Industrial or commercial units	322	Moors and heathland
122	Road and rail networks and associated land	323	Sclerophyllous vegetation
123	Port areas	324	Transitional woodland-shrub
124	Airports	331	Beaches, dunes, sands
131	Mineral extraction sites	332	Bare rocks
132	Dump sites	333	Sparsely vegetated areas
133	Construction sites	334	Burnt areas
141	Green urban areas	335	Glaciers and perpetual snow
142	Sport and leisure facilities	411	Inland marshes
211	Non-irrigated land	412	Peat bogs
213	Rice fields	421	Salt marshes
221	Vineyards	422	Salines
222	Fruit trees and berry plantations	423	Intertidal flats
223	Olive groves	511	Water courses
231	Pastures	512	Water bodies
241	Annual crops associated with permanent crops	521	Coastal lagoons
242	Complex cultivation patterns	522	Estuaries

(continued)

Code	Label	Code	Label
243	Land principally occupied by agriculture, with significant areas of natural vegetation	523	Sea and ocean
244	Agro-forestry areas	999	NO DATA
311	Broad-leaved forest	990	UNCLASSIFIED LAND SURFACE
312	Coniferous forest	995	UNCLASSIFIED WATER BODIES

Practical considerations

CLC was originally mapped in vector format. This format provides higher precision and detail and is therefore recommended when working at local and regional scales. At national and supranational scales, raster data can be more suitable, as vector data is too heavy and may be difficult to handle in desktop computers with insufficient processing power.

Users can download the vector CLC to rasterize the database to the spatial resolution they require. The 100 m offered is the reference resolution provided by the EEA, but it is not the only one at which the map could be used.

Users should be aware that different mapping methodologies were used in different countries, and in some countries, at different times. This could result in significant differences in the way the landscape is mapped and conceptualised, which could introduce important sources of uncertainty in our studies and analyses. The same category could be interpreted differently in different countries, and even within the same country, a particular category could be mapped differently at different times if the production method changes. Those wishing to analyse LUC change should therefore use the change layers rather than the maps.

3   PELCOM—Pan-European Land Use and Land Cover Monitoring

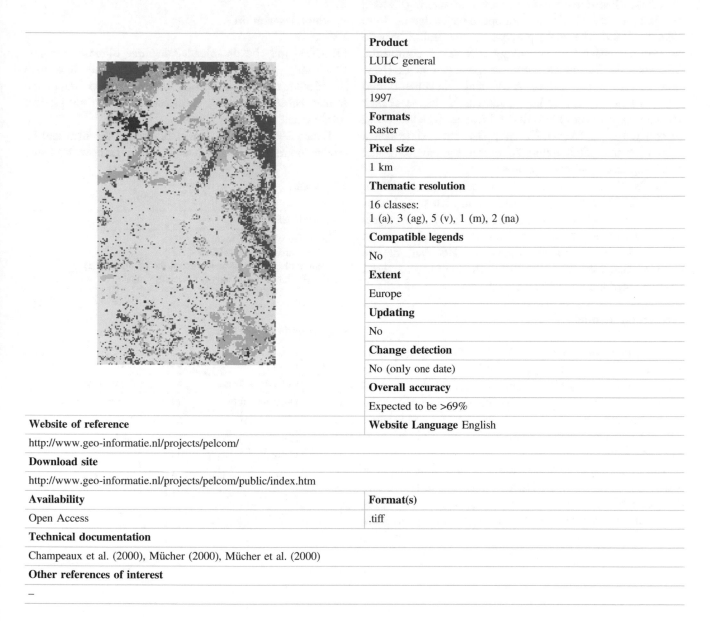

Product
LULC general
Dates
1997
Formats
Raster
Pixel size
1 km
Thematic resolution
16 classes: 1 (a), 3 (ag), 5 (v), 1 (m), 2 (na)
Compatible legends
No
Extent
Europe
Updating
No
Change detection
No (only one date)
Overall accuracy
Expected to be >69%

Website of reference	**Website Language** English
http://www.geo-informatie.nl/projects/pelcom/	

Download site
http://www.geo-informatie.nl/projects/pelcom/public/index.htm

Availability	**Format(s)**
Open Access	.tiff

Technical documentation
Champeaux et al. (2000), Mücher (2000), Mücher et al. (2000)

Other references of interest
–

Project

PELCOM (Pan-European Land Cover Monitoring) was a research project funded by the European Union that ran from 1996 to 1999. The main purpose of the project was to develop a consistent methodology to create a continental LUC map for Europe from remote sensing sources. Users were consulted about their needs and requirements and revealed that they would like to have LUC data at coarser and finer spatial resolutions than CLC, and that CLC could be updated more frequently. They also made clear that a dataset of this kind would be useful for environmental modelling and monitoring purposes.

At the time the project was launched, no consistent continental LUC maps were available at high spatial resolution (at least 1 km). The map created through the project sought to provide a high-resolution continental LUC dataset that could later be updated frequently. However, despite these original intentions, the PELCOM map has not been updated since the project came to an end.

Production method

The classification carried out for the PELCOM map was based on AVHRR imagery and NDVI composites from the DLR archive of the JRC. An improved stratified, integrated classification methodology was specifically developed by the creators of this map. To this end, Europe was divided into different strata according to similarities in LULC patterns and phenology.

The classification process consisted of several steps, in which users played an important role. Both supervised and unsupervised classifiers were employed. Some classes (forest, water bodies, urban areas) were mapped through specific workflows, using masks and other strategies, to improve the uncertainty and errors associated with their classification.

Product description

PELCOM may be downloaded in three different formats: ESRI-grid, ERDAS-Image and ENVI. The download includes the raster with the LUC map and, depending on the format chosen, auxiliary information about the product (readme and symbology files).

Detailed technical documentation about the map and its production method is also available from the download site.

Downloads

PELCOM ESRI-grid
– Raster file with LUC map
– Preview image of the product
– Readme file with information about the product (.doc)
– File with raster symbology for ArcGIS (.avl)

Legend and codification

Code	Label	Code	Label
11	Coniferous forest	60	Barren land
12	Deciduous forest	70	Permanent Ice & Snow
13	Mixed forest	80	Wetlands
20	Grassland	91	Inland waters
31	Rainfed arable land	92	Sea
32	Irrigated arable land	100	Urban areas
40	Permanent crops	110	Data gaps
50	Shrubland	111	Out of scope

4 Annual Land Cover Product

Product	
LULC general	
Dates	
2000–2019	
Formats	
Raster	
Pixel size	
30 m	
Thematic resolution	
33 classes: 8 (a), 7 (ag), 7 (v), 1 (m), 0 (na)	
Compatible legends	
LUCAS, CLC	
Extent	
Europe	
Updating	
Not planned	
Change detection	
Not recommended	
Overall accuracy	
Evaluation in process	

Website of reference	**Website Language** English
https://medium.com/swlh/europe-from-above-space-time-machine-learning-reveals-our-changing-environment-1b05cb7be520	

Download site

https://maps.opendatascience.eu/

Availability	**Format(s)**
Open Access	.tiff

Technical documentation

Not published yet

Other references of interest

–

Project

An open annual land cover dataset for Europe has been produced in the context of the "Geo-harmonizer: EU-wide automated mapping system for harmonization of Open Data based on FOSS4G and Machine Learning", a project coordinated by the Czech Technical University in Prague. This project is part of the Connecting Europe Facility (CEF) in Telecom, which aims to deploy digital service infrastructures (DSIs) that can facilitate cross-border interaction between public administrations, businesses and citizens.

The Geo-harmonizer project has developed a web-based system (Open Data Science Europe) that hosts open European thematic geospatial layers, including one on land cover. They were specifically created for the project from other data sources for the period 2000–2020 using modelling techniques. These harmonized European layers overcome the limitations resulting from the use of national datasets that were created with different parameters and have different characteristics.

Apart from a layer on land cover, Open Data Science Europe hosts data on subjects such as the environment, terrain, clime, soils or vegetation. These data are complementary to the datasets provided by the Copernicus Land Monitoring Service, also at continental level.

The project has the same values and approach as other Open Science projects in the geospatial field, such as Open Land Map and Open Street Map.

Production method

Open Data Science Europe's Annual Land Cover Product is obtained by producing a series of probability layers for each of the 33 LUC categories that were mapped. The land cover with the highest probability for each year and pixel according to these layers was the one finally selected to create the general LUC maps.

Probability layers were obtained through a set of three Machine Learning (ML) models: Random Forest, XGBoost and Artificial Neural Network. The models were trained with reference data obtained from CLC and LUCAS and input Landsat imagery (LANDSAT ARD), night lights data (VIIRS/SUOMI NPP), Global surface water frequency and an EU DTM.

A final probability layer for each LUC category was obtained after running a Logistic regression classifier on the results of three ML models. The uncertainty of the probability layers for each LUC category was also calculated as the standard deviation of the three predicted probabilities from the ML models.

Product description

The dataset can be individually downloaded for each available year of the period 2000–2019 from the Open Data Science Europe viewer. The download contains the raster file with the LUC information, but offers no other auxiliary data. Nonetheless, a layer style file to symbolize the dataset in QGIS[3] can be downloaded separately.

Downloads

Annual Land Cover Product 2019
– Raster file with LUC map (.tiff)

Legend and codification

Code	Label	Code	Label
111	Urban fabric	321	Natural grasslands
122	Road and rail networks and associated land	322	Moors and heathland

(continued)

---

[3] http://s3.eu-central-1.wasabisys.com/eumap/lcv/lcv_landcover.hcl_lucas.corine.rf_p_30m_0..0cm_2000_eumap_epsg3035_v0.1.qml.

Code	Label	Code	Label
123	Port areas	323	Sclerophyllous vegetation
124	Airports	324	Transitional woodland-shrub
131	Mineral extraction sites	331	Breaches, dunes, sands
132	Dump sites	332	Bare rocks
133	Construction sites	333	Sparsely vegetated areas
141	Green urban areas	334	Burnt areas
211	Non-irrigated arable land	335	Glaciers and perpetual snow
212	Permanently irrigated arable land	411	Inland wetlands
213	Rice fields	421	Maritime wetlands
221	Vineyards	511	Water courses
222	Fruit trees and berry plantations	512	Water bodies
223	Olive groves	521	Coastal lagoons
231	Pastures	522	Estuaries
311	Broad-leaved forest	523	Sea and ocean
312	Coniferous forest		

Practical considerations

The dataset is currently only available for download at the Opendatascience website. However, it will soon be uploaded to public repositories, where users will be able to access all data from the project, including layers of uncertainty. Information about the dataset production procedure will also be published in the coming months together with other relevant information.

The dataset can also be accessed through a WFS[4] and a file service (Cloud-Optimized GeoTIFFs)[5] in QGIS or other common GIS software. The map producers also provide information about how to access the data through GDAL, R and Python. This can be found by clicking on the *About* tab in the Opendatascience website.

---

[4] https://geoserver.opendatascience.eu/geoserver/wfs.
[5] http://s3.eu-central-1.wasabisys.com/eumap/lcv/lcv_landcover.hcl_ lucas.corine.rf_p_30m_0..0cm_2019_eumap_epsg3035_v0.1.tif.

5 GlobCorine

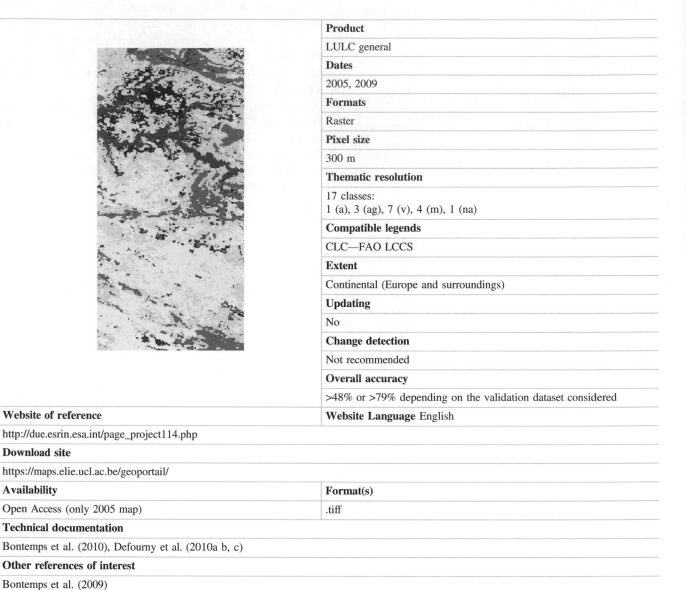

Product	
LULC general	
Dates	
2005, 2009	
Formats	
Raster	
Pixel size	
300 m	
Thematic resolution	
17 classes: 1 (a), 3 (ag), 7 (v), 4 (m), 1 (na)	
Compatible legends	
CLC—FAO LCCS	
Extent	
Continental (Europe and surroundings)	
Updating	
No	
Change detection	
Not recommended	
Overall accuracy	
>48% or >79% depending on the validation dataset considered	

Website of reference	Website Language English
http://due.esrin.esa.int/page_project114.php	
Download site	
https://maps.elie.ucl.ac.be/geoportail/	
Availability	**Format(s)**
Open Access (only 2005 map)	.tiff
Technical documentation	
Bontemps et al. (2010), Defourny et al. (2010a b, c)	
Other references of interest	
Bontemps et al. (2009)	

Project

Based on earlier efforts in GlobCover, the ESA launched the GlobCorine project in collaboration with the European Environment Agency (EEA) and the Université Catholique de Louvain (UCL). The aim was to create a new LUC product for the European continent that was compatible with the Corine Land Cover (CLC) classification and built on the work already carried out as part of the GlobCover project.

Production method

GlobCorine was produced by classifying the same MERIS imagery used for GlobCover. The same production method was used in the two LUC maps available and was similar to the one already used for GlobCover. It consisted of a series of supervised and unsupervised classification routines to identify spectro-temporal classes. These were later automatically labelled with the information provided by auxiliary datasets, mainly Corine Land Cover (CLC) and GlobCover. For classification purposes, the world was divided into different regions according to their ecological and reflectance characteristics.

An extra classification was carried out for mixed categories. The final LUC maps were then corrected and improved in a post-classification phase with the help of auxiliary data and expert knowledge.

Product description

Only one of the two GlobCorine maps is currently available for download: the map for the reference year 2005. The download includes the raster with the LUC map, the legend, a file to symbolize it in GIS software and all relevant technical information explaining the characteristics of the dataset.

Downloads

GlobCorine 2005

- Raster file with LUC map (GLOBCORINE_LC)
- Preview image of the product (GLOBCORINE_LC)
- Layer style files for ArcGIS (.lyr) and ENVI (.dsr) (GLOBCORINE_LC)

(continued)

GlobCorine 2005

- Excel sheet with the map legend ("GlobCorine_legend") (GLOBCORINE_LC)
- PDFs with technical information about the product (Documentation)
- PDF with a description of the downloaded product (README)

Legend and codification

Code	Label	Code	Label
10	Urban areas and associated areas	100	Complex cropland
20	Rainfed cropland	110	Mosaic cropland (50–80%) / natural vegetation (20–50%)
30	Irrigated cropland	120	Mosaic natural vegetation (50–80%) / cropland (20–50%)
40	Forest	130	Mosaic herbaceous (50–80%) / shrub-trees (20–50%)
50	Shrubland	140	Mosaic shrub-trees (50–80%) / herbaceous (20–50%)
60	Grassland	200	Water bodies
70	Sparsely vegetated areas	210	Permanent snow and ice
80	Vegetated low-lying areas on regularly flooded soil	220	No Data
90	Bare areas		

Practical considerations

The product is no longer available for download from the official website of the EEA. The only edition that can still be obtained is the map for 2005, which is available through the geoportal of the Université Catholique de Louvain, one of the producers of the dataset. The map can also be consulted online at the same website, without having to download it.

While the GlobCorine classification legend focuses particularly on land use, GlobCover centres on land cover. GlobCorine can therefore be regarded as a complementary dataset to GlobCover.

6 Urban Atlas

Product
LULC general
Dates
2006, 2012, 2018
Formats
Vector
Scale
Photointerpretation scale: 1:10,000 Minimum Mapping Unit: 0.25ha in urban areas and 1ha in rural areas 0.1ha for urban changes and 0.25ha for rural/natural changes Minimum Mapping Width: 10 m
Thematic resolution
29 classes: 17 (a), 4 (ag), 2 (v), 2 (m), 2 (na)
Compatible legends
CLC
Extent
Europe (39 countries)
Updating
Every 6 years
Change detection
Through map of changes
Overall accuracy
Expected to be > 80%

Website of reference	Website Language English, German and French

https://land.copernicus.eu/local/urban-atlas

Download site

https://land.copernicus.eu/local/urban-atlas

Availability	Format(s)
Open Access under registration	.gpkg

Technical documentation

Copernicus Programme (2020), Gallaun (2017), Hirschmugl et al. (2018), Silva et al. (2013, 2016)

Other references of interest

Barranco et al. (2014), European Commission and OECD (2012), Jaffrain et al. (2016), Montero et al. (2014), Petrişor and Petrişor (2015), Prastacos et al. (2011), Seifert (2009)

Project

Urban Atlas is part of the Copernicus programme and provides very detailed LUC information for Functional Urban Areas (FUA) in Europe. A Functional Urban Area (as defined by the European Commission and the OECD) is an urban space that joins the core areas of cities with their surrounding commuter belts.

The Urban Atlas aims to contribute to the study of urban areas and their dynamics, in line with the needs of the European Commission and other European initiatives, such as ESPON and INTERREG. It therefore has a clear goal to inform policy-making.

Three editions of the Urban Atlas have so far been published, with more FUAs participating in each one. 319 FUAs were mapped for reference year 2006, 785 for 2012 and 788 for 2018. New updates of the Urban Atlas are expected every 6 years.

For each edition, a detailed LUC map of the FUAs is provided, as well as a map of the changes that have taken place over the period under consideration (e.g. 2006–2012). A Street Tree Layer map is also provided for the 2012 and 2018 editions. The 2012 Urban Atlas includes a building height map for core areas (not FUA) of European capitals in the EEA39. Polygons of the 2012 Urban Atlas also include population estimates.

Production method

Urban Atlas is obtained through automatic classification and manual photointerpretation of high-resolution satellite imagery: the optical VHR coverage of the Copernicus programme, at a spatial resolution of 2–4 m.

First, the imagery is automatically segmented and classified, differentiating between basic land cover classes. Later, the detailed interpretation of land cover classes is carried out visually. A range of auxiliary data are applied in this process: topographic maps, the High Resolution Layer for impervious surfaces, road networks from COTS (Commercial Off-The-Shelf) navigation data and OSM as well as other data sources depending on the class under consideration (e.g. Google Earth, local city maps, cadastral data or very high resolution imagery, at a spatial resolution of up to 1 m).

Change detection for each period is carried out independently, based on the Urban Atlas map for the previous year and a combination of both automatic and manual approaches for change detection. In the change detection process, misclassifications for the previous year of reference are corrected.

There are certain exceptions to the Minimum Mapping Units and Minimum Mapping Widths, depending on the characteristics and pattern of the class being analysed. However, no features are mapped below the 0.5ha threshold.

Product description

Urban Atlas is distributed in single files for each FUA. There is no single common file that hosts all the FUAs together. A different file must be downloaded for each year and for each available change layer.

Downloads include the vector layers in Geopackage format with the LUC information, the boundaries of the FUAs and their urban cores, a metadata file and layer style files to symbolize the vector layers in GIS.

For reference years 2012 and 2018, the Street Tree Layer can be downloaded for each FUA. This layer represents contiguous rows or patches of trees covering at least 0.5ha. For the reference year 2012, a building height model in raster format can also be downloaded.

Downloads

Urban Atlas 2018 (Madrid)

- GeoPackage file with Urban Atlas vector layers: Urban Atlas 2018, Urban Core and Boundary (*DATA* folder)
- Layer style files for ArcGIS (.lyr), QGIS (.qml) and any other GIS software (.sld) (*Legend* folder)
- Metadata file (.xml) (*Metadata* folder)

Urban Atlas changes 2012–2018 (Madrid)

- GeoPackage file with Urban Atlas Change vector layers: Urban Atlas Change 2012–2018, Urban Core and Boundary (*DATA* folder)
- Layer style files for ArcGIS (.lyr), QGIS (.qml) and any other GIS software (.sld) (*Legend* folder)
- Metadata file (.xml) (*Metadata* folder)

Street tree layer 2012 – STL (Madrid)

- Vector file with STL layer
- Vector file with FUA boundary

Building height 2012 (Madrid)

- Raster file with building heights (*DATA* folder)
- PDFs with technical information about the product (*DOC* folder)
- Metadata file (.xml) (*Metadata* folder)

Database

Urban area 2018 (Madrid)

	fid	country	fua_name	fua_code	code_2018	class_2018	prod_date	identifier	perimeter	area	comment
1	1	ES	Madrid	ES001L3	11210	Discontinuous ...	2020-02	20215-ES001L3	288,108889117289	4195,51178273653	
2	2	ES	Madrid	ES001L3	11220	Discontinuous ...	2020-02	34093-ES001L3	338,42589076045	7376,70547635699	
3	3	ES	Madrid	ES001L3	11240	Discontinuous ...	2020-02	43827-ES001L3	266,701707505638	4077,81816957153	

– FID: Unique identifier for each polygon
– Country: Country code
– FUA_name: Name of the Functional Urban Area
– FUA_code: Code for the Functional Urban Area
– Code_2018: LUC code for the year 2018
– Class_2018: LUC description for the year 2018
– Prod_date: Map production year
– Identifier: Unique identifier for each polygon
– Perimeter: Perimeter of the polygon, in metres
– Area: Area of the polygon, in square metres
– Comment: Extra field for additional comments about the mapped features

Urban area change 2012–2018 (Madrid)

	fid	country	fua_name	fua_code	code_2018	class_2018	prod_date	identifier	perimeter	area	comment	code_2012	class_2012
1	3334	ES	Madrid	ES001L3	14100	Green urban ar...	2020-02	84414-ES001L3	509,613513775099	16445,1738405536		23000	Pastures
2	3333	ES	Madrid	ES001L3	13400	Land without c...	2020-02	84413-ES001L3	545,957445830923	11287,5073455729		23000	Pastures
3	3332	ES	Madrid	ES001L3	11230	Discontinuous I...	2020-02	64723-ES001L3	216,313014959852	2768,58596393346		13300	Construction si...

– FID: Unique identifier for each polygon
– Country: A two-letter code to identify each country
– FUA_name: Name of the Functional Urban Area
– FUA_code: Code for the Functional Urban Area
– Code_2018: LUC code for the year 2018
– Class_2018: LUC description for the year 2012
– Prod_date: Map production year
– Identifier: Unique identifier for each polygon
– Perimeter: Perimeter of the polygon, in metres
– Area: Area of the polygon, in square metres
– Comment: Extra field for additional comments about the mapped features
– Code_2012: LUC code for the year 2012
– Class_2012: LUC description for the year 2012

Street tree layer 2012—STL (Madrid)

	COUNTRY	CITIES	FUA_OR_CIT	STL	Shape_Leng	Shape_Area
1	ES	Madrid	ES001L2	1	236,03710343800	2115,64219083000
2	ES	Madrid	ES001L2	1	135,50042612100	679,89521552300
3	ES	Madrid	ES001L2	1	94,47833948390	565,27038061200

– COUNTRY: A two-letter code for each different country
– CITIES: Name of the Functional Urban Area
– FUA_OR_CIT: Code of the Functional Urban Area
– STL: Street Tree Layer code
– Shape_Leng: Perimeter of the polygon, in metres
– Shape_Area: Area of the polygon, in square metres

Legend and codification

Urban Atlas

Code	Label	Code	Label
11100	Continuous urban fabric (S.L. > 80%)	14100	Green urban areas
11210	Discontinuous dense urban fabric (S.L. 50–80%)	14200	Sports and leisure facilities
11220	Discontinuous medium-density urban fabric (S.L. 30–50%)	21000	Arable land (annual crops)
11230	Discontinuous low-density urban fabric (S.L. 10–30%)	22000	Permanent crops
11240	Discontinuous very low-density urban fabric (S.L. < 10%)	23000	Pastures
11300	Isolated structures	24000	Complex and mixed cultivation
12100	Industrial, commercial, public, military and private units	25000	Orchards
12210	Fast transit roads and associated land	31000	Forests
12220	Other roads and associated land	32000	Herbaceous vegetation associations
12230	Railways and associated land	33000	Open spaces with little or no vegetation

(continued)

Urban Atlas

Code	Label	Code	Label
12300	Port areas	40000	Wetlands
12400	Airports	50000	Water
13100	Mineral extraction and dump sites	91000	No data (Clouds and shadows)
13300	Construction sites	92000	No data (Missing imagery)
13400	Land without current use		

Street tree layer

Code	Land cover
1	Tree cover

Practical considerations

LUC change must be analysed using the change layer. Comparing Urban Atlases for different years of reference will highlight many technical changes that did not actually happen on the ground.

The Urban Atlas product can be also consulted online at the download webpage.

7 N2K—Natura 2000

Product	
LULC general	
Dates	
2006, 2012, 2018	
Formats	
Vector	
Scale	
Photointerpretation scale: 1:5,000–1:10,000 Minimum Mapping Unit: 0.5 ha Minimum Mapping Width: 10 m	
Thematic resolution	
48 classes: 8 (a), 6 (ag), 13 (v), 9 (m), 0 (na)	
Compatible legends	
Urban Atlas, Riparian Zones, Coastal Zone product	
Extent	
Europe (29 countries)	
Updating	
Not planned	
Change detection	
Yes	
Overall accuracy	
Expected to be >80%	
Website of reference	**Website Language** English, German and French
http://land.copernicus.eu/local/natura	
Download site	
http://land.copernicus.eu/local/natura	
Availability	**Format(s)**
Open Access under registration	.gdb, .gpkg
Technical documentation	
Buck and Büscher (2018)	
Other references of interest	
–	

Project

N2K was developed as part of the Copernicus Land Monitoring programme. It maps land uses and covers in the areas that form part of the Natura 2000 network, plus a 2 km buffer zone around their perimeters. Natura 2000 is a network that protects natural areas with rare and threatened species or with rare types of natural habitat.

The dataset first appeared in 2015. A reviewed edition was issued in 2017 with a new classification legend that made it compatible with other European local reference LUC datasets: Riparian Zones, N2K and the Coastal Zone product.

Production method

N2K is obtained by photointerpretation of high-resolution imagery. Various auxiliary datasets are used in the photointerpretation process, namely CORINE Land Cover, Urban Atlas, High Resolution Layers, topographic maps, national WMS services and COTS navigation data. The changes are also photointerpreted by comparing satellite images at two different points in time.

Database

Product description

N2K is distributed as a single vector file covering all mapped Nature 2000 areas. Two formats are available: ESRI Geodatabase and Geopackage. Downloads include the layers with LUC information, a style file to symbolize the layers in GIS and a pdf with the product classification scheme.

Downloads

N2K 2012 (Geodatabase)

– Geodatabase files with N2K vector layers
– Layer style files for ArcGIS (.lyr), QGIS (.qml) and any other GIS software (.sld) (*Legend* folder)
– Metadata file (.xml) (*Metadata* folder)
– PDF with nomenclature guidelines

N2K 2012 (GeoPackage)

– GeoPackage files with N2K vector layers
– Layer style files for ArcGIS (.lyr), QGIS (.qml) and any other GIS software (.sld) (*Legend* folder)
– Metadata file (.xml) (*Metadata* folder)
– PDF with nomenclature guidelines

Product: N2K 2000

	OBJECTID	ID	UID	SITECODE	GRASSTYPE	MAES_1_12
1	1	4114	4114_1			4
2	2	4114	4114_2			4
3	3	4114	4114_3			4

– OBJECTID: Unique identifier for each polygon
– ID
– UID
– SITECODE
– GRASSTYPE
– MAES_1_12: MAES class Level 1 for 2012
– MAES_2_12: MAES class Level 2 for 2012
– MAES_3_12: MAES class Level 3 for 2012
– MAES_4_12: MAES class Level 4 for 2012
– COMMENT_12: Comments on the 2012 mapping
– NODATA_12: Objects with no data in 2012
– MAES_1_06: MAES class Level 1 for 2006
– MAES_2_06: MAES class Level 2 for 2006
– MAES_3_06: MAES class Level 3 for 2006
– MAES_4_06: MAES class Level 4 for 2006
– COMMENT_06: Comments on the 2006 mapping
– NODATA_06: Objects with no data in 2006
– CHANGECODE: 2006–2012 changes
– AREA_HA: Area of the polygon, in hectares
– ID: unique identifier for each polygon

Legend and codification

N2K was produced according to a hierarchical classification legend made up of four different levels, the most detailed of which is provided here (MAES L3). Information about the other levels of classification and their codes can be found in the technical documents accompanying the dataset.

Code	Label	Code	Label
111	Urban fabric (predominantly public and private units)	4211	Semi-natural grassland with woody plants (C. C.D. ≥ 30%)
112	Industrial, commercial and military units	4212	Semi-natural grassland without woody plants (C. C.D. ≤ 30%)
121	Road networks and associated land	422	Alpine and sub-alpine natural grassland
122	Railways and associated land	511	Heathland and Moorland
123	Port areas and associated land	512	Other scrub land
124	Airports and associated land	621	Beaches and dunes
131	Mineral extraction, dump and construction sites	622	River banks
132	Land without current use	631	Bare rocks and rock debris
211	Arable land	632	Burn areas (except burnt forest)
212	Greenhouses	633	Glaciers and perpetual snow
221	Vineyards, fruit trees and berry plantations	721	Exploited peat bog
222	Olive groves	722	Unexploited peat bog
231	Annual crops associated with permanent crops	811	Coastal salt marshes

(continued)

Code	Label	Code	Label
232	Complex cultivation patterns	812	Salines
233	Land principally occupied by agriculture with significant areas of natural vegetation	813	Intertidal flats
234	Agro-forestry	821	Coastal lagoons
311	Natural and semi-natural broadleaved forest	822	Estuaries
312	Highly artificial broadleaved plantations	911	Interconnected water courses
321	Natural and semi-natural coniferous forest	912	Highly modified water courses and canals
322	Highly artificial coniferous plantations	913	Separated water bodies belonging to the river system
331	Natural and semi-natural mixed forest	921	Natural water bodies
332	Highly artificial mixed plantations	922	Artificial standing water bodies
341	Transitional woodland and scrub	923	Intensively managed fish ponds
342	Lines of threes and scrubs	924	Standing water bodies of extractive industrial sites

Practical considerations

N2K files are very heavy (over 2gb), which means that they may be difficult to use for those without powerful computers. The map can also be consulted online in a viewer included in the download website of the product.

8 Riparian Zones Land Cover/Land Use—Riparian Zones (RZ)

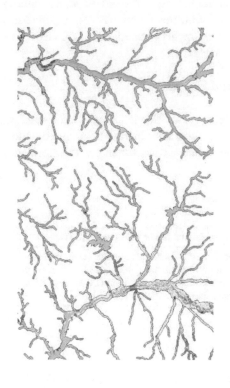

Product	
LULC general	
Dates	
2012, 2018	
Formats	
Vector	
Scale	
Photointerpretation scale: 1:10,000 Minimum Mapping Unit: 0.5 ha Minimum Mapping Width: 10 m	
Thematic resolution	
56 classes: 11 (a), 6 (ag), 16 (v), 10 (m), 0 (na)	
Compatible legends	
Urban Atlas, N2K, Coastal Zone product	
Extent	
Europe (39 countries)	
Updating	
Not expected	
Change detection	
No (only one date)	
Overall accuracy	
Expected to be >85%	

Website of reference	Website Language English, French and German

https://land.copernicus.eu/local/riparian-zones

Download site

https://land.copernicus.eu/local/riparian-zones/land-cover-land-use-lclu-image?tab=download

Availability	Format(s)
Open Access after registration	.shp

Technical documentation

Tamame et al. (2018), Vandeputte et al. (2018), Weissteiner et al. (2016)

Other references of interest

Piedelobo et al. (2019), Ugille (2019)

Project

Riparian Zones (RZ) is one of the local datasets produced as part of the Copernicus Land Monitoring Programme. It focuses on riparian areas (i.e. transitional areas between land and freshwater ecosystems with very specific characteristics) associated with Strahler level 2–8 rivers.

This product was created to support the Mapping and Assessment of Ecosystems and their Services (MAES) within the context of the EU Biodiversity Strategy for 2020. It is also intended for use in relation to the Habitats, Birds and Water Framework Directives.

The Riparian Zones dataset was initially launched in 2015, with an extension in 2017/18 to include riparian areas from Strahler 2 rivers. In 2017 the classification scheme was adapted to make it compatible with other local products developed under the Copernicus Land Monitoring framework.

Together with the LUC map of riparian zones, two extra complementary products are also provided: a delineation of Riparian Zones based on a fuzzy modelling approach and an inventory of the Green Linear Elements (hedgerows and lines of trees) growing in those riparian areas.

Production method

The RZ map was obtained through semi-automatic classification of very high-resolution imagery captured by the SPOT and Pleiades satellites (1.5–2.5 m). This classification was later refined with the aid of visual interpretation and intersected with the following auxiliary datasets: CORINE Land Cover, Imperviousness HRL, Tree Cover Density HRL and Urban Atlas.

Product description

A different vector file is provided for each riparian area mapped. Downloads include the vector file with LUC information and pdf documents with information about the product.

Downloads

Riparian zones land cover land use 2012 (vector)

– Vector files with LUC information (*Data* folder)
– PDF with nomenclature guidelines (*Documents* folder)
– PDF with product specifications (*Documents* folder)
– Metadata files (.xml) (*Metadata* folder)

Database

Riparian zones land cover land use 2012—LCLU (vector)

	ID	DU_ID	MAES_1	MAES_2	MAES_3	MAES_4	UA	AREA_HA	NODATA	COMMENT
1	1	DU020A	1	11	111	1111		0,73018853235	0	
2	2	DU020A	1	11	111	1111		0,90258671155	0	
3	3	DU020A	1	11	111	1111		0,64638451960	0	

– ID: Unique identifier for each polygon
– DU_ID: Mapped area Code
– MAES_1: MAES class Level 1
– MAES_2: MAES class Level 2
– MAES_3: MAES class Level 3
– MAES_4: MAES class Level 4
– UA
– AREA_HA: Area of the polygon, in hectares
– NODATA: Unclassifiable areas due to clouds, shadows, snow, haze or missing data
– COMMENT: Comment field for additional information

Legend and codification

The Riparian Zones dataset was produced following a hierarchical classification legend made up of three different levels, the most detailed of which is provided here. Information about the other levels of classification of LUC categories can be found in the technical documents accompanying the dataset.

Code	Label	Code	Label
1111	Continuous Urban Fabric (IM.D $\geq$ 80%)	41	Managed grassland
1112	Dense Urban Fabric (IM.D $\geq$ 30–80%)	421	Semi-natural grassland
1113	Low Density Fabric (IM.D <30%)	422	Alpine and sub-alpine natural grassland
112	Industrial, commercial and military units	511	Heathland and Moorland
121	Road networks and associated land	512	Other scrub land
122	Railways and associated land	52	Sclerophyllous vegetation
123	Port areas and associated land	61	Sparsely vegetated areas
124	Airports and associated land	621	Beaches and dunes
131	Mineral extraction, dump and construction sites	622	River banks
132	Land without current use	631	Bare rocks and rock debris
14	Green urban, sports and leisure facilities	632	Burnt areas (except burnt forest)
211	Arable land	633	Glaciers and perpetual snow
212	Greenhouses	71	Inland marshes
221	Vineyards, fruit trees and berry plantations	721	Exploited peat bog
222	Olive groves	722	Unexploited peat bog

(continued)

Code	Label	Code	Label
231	Annual crops associated with permanent crops	811	Coastal salt marshes
232	Complex cultivation patterns	812	Salines
233	Land principally occupied by agriculture with significant areas of natural vegetation	813	Intertidal flats
234	Agro-forestry	821	Coastal lagoons
311	Natural and semi-natural broadleaved forest	822	Estuaries
312	Highly artificial broadleaved plantations	911	Interconnected water courses
321	Natural and semi-natural coniferous forest	912	Highly modified water courses and canals
322	Highly artificial coniferous plantations	913	Separated water bodies belonging to the river system
331	Natural and semi-natural mixed forest	921	Natural water bodies
332	Highly artificial mixed plantations	922	Artificial standing water bodies
341	Transitional woodland and scrub	923	Intensively managed fish ponds
342	Lines of trees and scrub	924	Standing water bodies of extractive industrial sites
35	Damaged forest	10	Sea and ocean

Practical considerations

The map can also be consulted online in a viewer available on the download site (see link above).

9   Coastal Zones

Product
LULC general
Dates
2012, 2018
Formats
Vector
Scale
1:10,000 Minimum Mapping Unit: 0.5 ha Minimum Mapping Width: 10 m
Thematic resolution
71 classes: 19 (a), 6 (ag), 17 (v), 9 (m), 0 (na)
Compatible legends
Urban Atlas, N2K, Riparian Zones
Extent
Coastlines of EEA member states (39 countries)
Updating Yes
Change detection
Yes, through the change layer
Overall accuracy
Expected to be >85%

Website of reference	Website Language English, German and French
https://land.copernicus.eu/local/coastal-zones	

Download site
https://land.copernicus.eu/local/coastal-zones

Availability	Format(s)
Open Access after registration	.gdb, .gpkg

Technical documentation
European Environment Agency (2021)

Other references of interest
–

Project

The Coastal Zones Land Cover/Land Use dataset is produced by the European Environment Agency (EEA) as part of the Copernicus Land Monitoring Service (CLMS). The dataset has been developed in collaboration with the Copernicus Marine Environment Monitoring Service (CMEMS) and representatives from the potential community of users.

It is specifically intended for monitoring coastal areas and provides an important source of information for all EU policies dealing with coastal management and maritime spatial planning.

The dataset maps, at very detailed scale, the land uses and covers in coastal areas in the 39 countries belonging to the EEA. The coastal area mapped is defined by a 10 km inland buffer zone and the Corine Land Cover (CLC) seawards buffer zone. Relevant estuaries, coastal lowlands and nature reserves that extend beyond the buffer zone have also been included.

The dataset's classification legend has been specifically designed to fit the needs of its user community. It is based on the Mapping and Assessment of Ecosystems and their Services (MAES) ecosystem typology and makes the product compatible with other CLMS local monitoring datasets, such as Urban Atlas, Riparian Zones and N2K.

The dataset is composed of two LUC maps for the reference years 2012 and 2018, plus a change layer for the period 2012–2018. The dataset will be updated every 6 years, in accordance with the CLC production timeline.

Production method

The Coastal Zones Land Cover/Land Use dataset is produced via computer-assisted photointerpretation of very high spatial resolution (1.5–4 m) imagery from a wide variety of missions: SPOT, Pléiades, WorldView, SuperView, KOMPSat, Planet Dove, Deimos and TripleSat. A variable photointerpretation scale (1:5,000–1:10,000) was selected depending on the mapped landscape and feature characteristics. The following auxiliary datasets were also used in support of the photointerpretation process: CLC, Urban Atlas, HRL, Bing Maps and different imagery sources (DWH_MG2_CORE_01 Coverage, Sentinel-2, Landsat-8, national aerial imagery, Google Earth).

Product description

Users can download the Coastal Zones dataset in two different formats: Geodatabase and GeoPackage. Different download files are available for each year of reference (2012, 2018) as well as for the change layer (2012–2018). All downloads include the same information: layers with LUC information, a style file for their symbolization in GIS and auxiliary data.

Downloads

Coastal zones 2018 (Geodatabase)

– Geodatabase files with Coastal Zones vector layers
– Layer style files for ArcGIS (.lyr), QGIS (.qml) and any other GIS software (.sld) (*Symbology* folder)
– Metadata file (.xml, .gfs) (*Metadata* folder)

Database

Coastal zones 2018 (Geodatabase)

	fid	ID ▲	DU	CODE_1_18	CODE_2_18	CODE_3_18	CODE_4_18	CODE_5_18	COMMENT_18	NODATA_18	AREA_HA	Shape_Length	Shape_Area
1	1	1	4	1	11	111	1111	11110	*NULL*	0	0.907902455860...	448.5645642872...	9079.024558604...
2	2	2	4	1	11	111	1111	11110	*NULL*	0	2.709284033416...	664.5918333262...	27092.84033416...
3	3	3	4	1	11	111	1112	11120	*NULL*	0	1.804702394212...	565.373960242318	18047.02394212...
4	4	4	4	1	11	111	1111	11110	*NULL*	0	2.832704905509...	674.9293361490...	28327.0490550978

– fid: Identifier for each polygon
– ID: Unique identifier for each polygon
– DU:
– CODE_1_18: LUC category for the Level 1 classification legend
– CODE_2_18: LUC category for the Level 2 classification legend
– CODE_3_18: LUC category for the Level 3 classification legend
– CODE_4_18: LUC category for the Level 4 classification legend
– CODE_5_18: LUC category for the Level 5 classification legend
– COMMENT_18: Comments on the mapping
– NODATA_18: Objects with no data in 2018
– AREA_HA: Area of the polygon, in hectares
– Shape_Length: Perimeter of the polygon, in metres
– Shape_Area: Area of the polygon, in square metres

Coastal zones change 2012–2018 (Geodatabase)

	OBJECTID ▲	ID	DU	CODE_1_12	CODE_2_12	CODE_3_12	CODE_4_12	CODE_5_12	CODE_1_18
1	1	1	4	1	11	111	1111	11110	1
2	2	2	4	1	11	111	1112	11120	1
3	3	3	4	1	11	111	1112	11120	1
4	4	4	4	1	11	111	1112	11120	4

– fid: Identifier for each polygon
– ID: Unique identifier for each polygon
– DU:
– CODE_1_12: LUC category for the Level 1 classification legend in 2012
– CODE_2_12: LUC category for the Level 2 classification legend in 2012
– CODE_3_12: LUC category for the Level 3 classification legend in 2012
– CODE_4_12: LUC category for the Level 4 classification legend in 2012
– CODE_5_12: LUC category for the Level 5 classification legend in 2012
– CODE_1_18: LUC category for the Level 1 classification legend in 2018
– CODE_2_18: LUC category for the Level 2 classification legend in 2018
– CODE_3_18: LUC category for the Level 3 classification legend in 2018
– CODE_4_18: LUC category for the Level 4 classification legend in 2018
– CODE_5_18: LUC category for the Level 5 classification legend in 2018
– COMMENT: Comments on the mapping
– NODATA_12: Objects with no data in 2012
– NODATA_18: Objects with no data in 2018
– AREA_HA: Area of the polygon, in hectares
– Shape_Length: Perimeter of the polygon, in metres
– Shape_Area: Area of the polygon, in square metres

Legend and codification

The Coastal Zones dataset was produced following a hierarchical classification legend made up of five different levels, the most detailed of which is provided here. Information about the full classification scheme, including the five different levels, can be found in the technical documentation accompanying the dataset.

Code	Label	Code	Label
11110	Continuous urban fabric (IMD $\geq$ 80%)	36000	Damaged forest
11120	Dense urban fabric (IMD $\geq$ 30–80%)	41000	Managed grassland
11130	Low-density fabric (IMD < 30%)	42100	Semi-natural grassland
11210	Industrial, commercial, public and military units (other)	42200	Alpine and sub-alpine natural grassland
11220	Nuclear energy plants and associated land	51000	Heathland and moorland
12100	Road networks and associated land	52000	Alpine scrub land
12200	Railways and associated land	53000	Sclerophyllous scrubs

(continued)

Code	Label	Code	Label
12310	Cargo port	61100	Sparse vegetation on sands
12320	Passenger port	61200	Sparse vegetation on rocks
12330	Fishing port	62111	Sandy beaches
12340	Naval port	62112	Shingle beaches
12350	Marinas	62120	Dunes
12360	Local multi-functional harbours	62200	River banks
12370	Shipyards	63110	Bare rocks and outcrops
12400	Airports and associated land	63120	Coastal cliffs
13110	Mineral extraction sites	63200	Burnt areas (except burnt forest)
13120	Dump sites	63300	Glaciers and perpetual snow
13130	Construction sites	71100	Inland marshes
13200	Land without current use	71210	Exploited peat bogs

(continued)

Code	Label	Code	Label
14000	Green urban, sports and leisure facilities	71220	Unexploited peat bogs
21100	Arable irrigated and non-irrigated land	72100	Salt marshes
21200	Greenhouses	72200	Salines
22100	Vineyards, fruit trees and berry plantations	72300	Intertidal flats
22200	Olive groves	81100	Natural & semi-natural water courses
23100	Annual crops associated with permanent crops	81200	Highly modified water courses and canals
23200	Complex cultivation patterns	81300	Seasonally connected water courses (oxbows)
23300	Land principally occupied by agriculture with significant areas of natural vegetation	82100	Natural lakes
23400	Agro-forestry	82200	Reservoirs
31100	Natural & semi-natural broadleaved forest	82300	Aquaculture ponds
31200	Highly artificial broadleaved plantations	82400	Standing water bodies of extractive industrial sites
32100	Natural & semi-natural coniferous forest	83100	Lagoons
32200	Highly artificial coniferous plantations	83200	Estuaries
33100	Natural & semi-natural mixed forest	83300	Marine inlets and fjords
33200	Highly artificial mixed plantations	84100	Open sea
34000	Transitional woodland and scrub	84200	Coastal waters
35000	Lines of trees and scrub		

Practical considerations

Coastal Zones files are very heavy (above 3gb), which means that the dataset may be difficult to use for those without powerful computers. The dataset can also be consulted online using the viewers available when downloading the different layers.

10 S2GLC 2017—Sentinel-2 Global Land Cover 2017

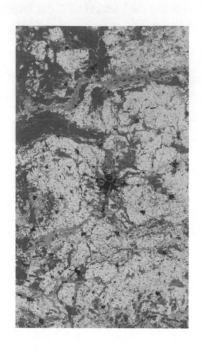

Product
LULC general
Dates
2017
Formats
Raster
Pixel size 10 m
Thematic resolution
13 classes: 1 (a), 2 (ag), 5 (v), 0 (m), 0 (na)
Compatible legends
CLC
Extent
Europe, except Russia
Updating
Not expected
Change detection
No (only one date)
Overall accuracy
Expected to be >86%

Website of reference	Website Language English
http://s2glc.cbk.waw.pl/extension	

Download sites
http://s2glc.cbk.waw.pl/extension https://finder.creodias.eu/

Availability	Format(s)
Open Access	.tiff

Technical documentation
Gromny et al. (2019a, b), Kukawska et al. (2017), Malinowski et al. (2019), Nowakowski et al. (2017)

Other references of interest
–

Project

Sentinel-2 Global Land Cover (S2GLC) was a project funded by the European Space Agency (ESA) in order to create an automatic methodology to globally map LUC at high resolution from Sentinel-2 imagery. The project was led by the Space Research Centre of the Polish Academy of Sciences (CBK PAN). Its main output is the S2GLC 2017 map.

The project was developed in two phases. In the first phase, the proposed methodology was tested in five prototype sites: Germany, Italy, China, Columbia and Namibia. In the second phase, the methodology was adjusted to map LUC for the whole of Europe, except Russia, Belarus and Ukraine.

Production method

S2GLC was obtained by classifying Sentinel-2 imagery. Each Sentinel-2 scene was individually classified using a set of multi-temporal images through a random forest classifier. Training data was automatically extracted from existing datasets, such as CORINE Land Cover. A set of probability rasters were obtained from the random forest classifier, and the class finally selected for each pixel was the one with the highest probability over the whole time series. A post-classification step was applied for those pixels with low probabilities.

Product description

S2GLC 2017 can be downloaded as a single file or in tiles. In the first case, users can choose to download the raster LUC file, either symbolized (RGB GeoTiff file) or not (GeoTiff file). Users who opt to download a tile from the map will automatically download both types of rasters.

Downloads

European land cover map (single-band file, RGB file)
– Raster file with LUC map
– TXT file with the product legend
– PDF with technical information about the product (tiles decomposition)

Single tile
– Raster file with LUC map
– Coloured raster file with LUC map
– Preview image of the product
– TXT file with the product legend

Legend and codification

Code	Label	Code	Label
62	Artificial surfaces and constructions	104	Sclerophyllous vegetation
73	Cultivated areas	105	Marshes
75	Vineyards	106	Peatbogs
82	Broadleaf tree cover	121	Natural material surfaces
83	Coniferous tree cover	123	Permanent snow-covered surfaces
102	Herbaceous vegetation	162	Water bodies
103	Moors and Heathland		

Practical considerations

The map can be consulted online using the CREODIAS Browser application (https://browser.creodias.eu/). Single file download options involve very heavy files (8–16.2 gb), for which a powerful computer will be required.

References

Bach M, Breuer L, Frede HG et al (2006) Accuracy and congruency of three different digital land-use maps. Landsc Urban Plan 78:289–299. https://doi.org/10.1016/j.landurbplan.2005.09.004

Barranco RR, Silva FBE, Marin Herrera M, Lavalle C (2014) Integrating the MOLAND and the urban Atlas geo-databases to analyze urban growth in European cities. J Map Geogr Libr 10:305–328. https://doi.org/10.1080/15420353.2014.952485

Bielecka J (2019) Intellectual structure of CORINE land cover research applications in web of science: a Europe-wide review. Remote Sens 11:2017. https://doi.org/10.3390/rs11172017

Bontemps S, Defourny P, Van Bogaert E et al (2009) GlobCorine—a joint EEA-ESA project for operational land dynamics monitoring at pan-European scale. In: The 33rd international symposium on remote sensing of environment. Stresa, Italy

Bontemps S, Defourney P, Van Bogaert E et al (2010) GlobCover 2009. Description and validation report. http://due.esrin.esa.int/files/p114/GLOBCORINE2009_DVR_2.1.pdf. Accessed 21 Aug 2020

Bossard M, Feranec J, Otahel J (2000) CORINE land cover technical guide: addendum 2000. EEA Technical Report No. 40. https://land.copernicus.eu/user-corner/technical-library/tech40add.pdf. Accessed 19 Aug 2020

Buck O, Büscher O (2018) NOMENCLATURE and MAPPING GUIDELINE copernicus land monitoring service local component : Natura 2000 mapping. https://land.copernicus.eu/user-corner/technical-library/N2K_Nomenclature_Guidelines.pdf. Accessed 30 Sept 2020

Büttner G (2014) CORINE land cover and land cover change products. In: Manakos I, Braun M (eds) Land use and land cover mapping in Europe: practices & trends. Springer, Dordrecht, Heidelberg, New York, London, pp 55–74

Büttner G, Feranec J, Jaffrain G (2002) Corine land cover update 2000—technical guidelines. http://www.eea.europa.eu/publications/technical_report_2007_17. Accessed 19 Aug 2020

Büttner G, Kosztra B (2011) Manual of CORINE land cover changes. https://land.copernicus.eu/user-corner/technical-library/manual_of_changes_final_draft.pdf. Accessed 19 Aug 2020

Büttner G, Kosztra B, Maucha G, Pataki R (2012) Implementation and achievements of CLC 2006. https://land.copernicus.eu/user-corner/technical-library/CLCFinalrep_revised_finaldraft.pdf. Accessed 19 Aug 2020

Büttner G, Soukup T, Kosztra B (2014) CLC2012. Addendum to CLC2006 technical guidelines. https://land.copernicus.eu/user-corner/technical-library/Addendum_finaldraft_v2_August_2014.pdf. Accessed 19 Aug 2020

Champeaux JL, Mucher CA, Steinnocher K et al (2000) PELCOM project: a 1-km pan-European land cover database for environmental monitoring and use in meteorological models. Int Geosci Remote Sens Symp 5:1915–1917. https://doi.org/10.1109/igarss.2000.858165

Copernicus Programme (2020) Mapping Guide v6.1 for an European Urban Atlas. https://land.copernicus.eu/user-corner/technical-library/urban_atlas_2012_2018_mapping_guide_v6-1.pdf. Accessed 29 Sept 2020

Defourny P, Bontemps S, Van Bogaert E (2010a) GlobCorine. Product description manual. http://due.esrin.esa.int/files/p114/GLOBCORINE_LC_PDM_2.2.pdf. Accessed 21 Aug 2020

Defourny P, Bontemps S, Van Bogaert E et al (2010b) GlobCorine 2009. Product description manual. http://due.esrin.esa.int/files/p114/GLOBCORINE2009_DVR_2.1.pdf. Accessed 21 Aug 2020

Defourny P, Bontemps S, Van Bogaert E et al (2010c) GlobCorine. Validation report. http://due.esrin.esa.int/files/p114/GLOBCORINE_VR_2.1_2010c0406.pdf. Accessed 21 Aug 2020

European Commission, OECD (2012) Cities in Europe: the new OECD-EC definition. https://land.copernicus.eu/user-corner/technical-library/oecd-definition-of-functional-urban-area-fua. Accessed 29 Sept 2020

European Environment Agency (1994) corine land cover. ISSN 92-826-2579-6. https://www.eea.europa.eu/publications/COR0-landcover. Accessed 19 Aug 2020

European Environment Agency (2006a) CORINE land cover nomenclature illustrated guide. https://land.copernicus.eu/user-corner/technical-library/Nomenclature.pdf. Accessed 19 Aug 2020

European Environment Agency (2006b) The thematic accuracy of Corine land cover 2000. Assessment using LUCAS (land use/cover area frame statistical survey). EEA Technical Report No. 7/2006b ISSN 1725-2237. https://land.copernicus.eu/user-corner/technical-library/technical_report_7_2006b.pdf. Accessed 19 Aug 2020

European Environment Agency (2006c) Land accounts for Europe 1990–2000. EEA Report No 11/2006c. https://www.eea.europa.eu/publications/eea_report_2006c_11. Accessed 20 Aug 2020

European Environment Agency (2007) CLC2006 technical guidelines. EEA Technical Report No. 17/2007 ISSN 1725–2237. https://land.copernicus.eu/user-corner/technical-library/techrep89.pdf. Accessed 19 Aug 2020

European Environment Agency (2021). Copernicus land monitoring service—ocal component: coastal zones monitoring nomenclature guideline. https://land.copernicus.eu/user-corner/technical-library/coastal-zones-nomenclature-and-mapping-guideline.pdf. Accessed 25 March 2021

Feranec J, Jaffrain G, Soukup T, Hazeu G (2010) Determining changes and flows in European landscapes 1990–2000 using CORINE land cover data. Appl Geogr 30:19–35. https://doi.org/10.1016/j.apgeog.2009.07.003

Feranec J, Soukup T, Hazeu G, Jaffrain G (2016) European landscape dynamics: CORINE land cover data. CRC Press

Fuchs R (2015) A data-driven reconstruction of historic land cover/use change of Europe for the period 1900 to 2010. PhD thesis

Fuchs R, Herold M, Verburg PH, Clevers JGPW (2013) A high-resolution and harmonized model approach for reconstructing and analysing historic land changes in Europe. Biogeosciences 10:1543–1559. https://doi.org/10.5194/bg-10-1543-2013

Fuchs R, Herold M, Verburg PH et al (2015a) Gross changes in reconstructions of historic land cover/use for Europe between 1900 and 2010. Glob Chang Biol 21:299–313. https://doi.org/10.1111/gcb.12714

Fuchs R, Verburg PH, Clevers JGPW, Herold M (2015b) The potential of old maps and encyclopaedias for reconstructing historic European land cover/use change. Appl Geogr 59:43–55. https://doi.org/10.1016/j.apgeog.2015.02.013

Gallaun H (2017) Urban Atlas 2012 validation report. https://land.copernicus.eu/user-corner/technical-library/ua-2012-validation-report. Accessed 29 Sept 2020

Gallego J (2001) Fine scale profile of CORINE land cover classes with LUCAS data. In: Building agri-environmental indicators: focussing on the European area frame survey LUCAS, pp 121–136

García-Álvarez D, Camacho Olmedo MT (2017) Changes in the methodology used in the production of the Spanish CORINE: uncertainty analysis of the new maps. Int J Appl Earth Obs Geoinf 63:55–67

Gromny E, Lewiński S, Rybicki M et al (2019a) Creation of training dataset for Sentinel-2 land cover classification. In: Proceedings of SPIE 11176, photonics applications in astronomy, communications, industry, and high-energy physics experiments 2019a. Wiga, Poland

Gromny E, Lewiński S, Rybicki M et al (2019b) Post-processing tools for land cover classification of Sentinel-2. In: Proceedings of SPIE 11176, photonics applications in astronomy, communications, industry, and high-energy physics experiments 2019b. Wiga, Poland

Hirschmugl M, Pennec A, Lhernould A et al (2018) Urban Atlas 2012 extension to Western Balkan and Turkey validation report. https://

land.copernicus.eu/user-corner/technical-library/urban-atlas-2012-extension-to-western-balkan-and-turkey-validation-report. Accessed 29 Sept 2020

Jaffrain G (2017) CORINE land cover 2012. Final validation report. https://land.copernicus.eu/user-corner/technical-library/clc-2012-validation-report-1. Accessed 19 Aug 2020

Jaffrain G, Sannier C, Feranec J (2016) Monitoring of urban fabric classes and their validation in selected European cities (Urban Atlas). In: Feranec J, Soukup T, Hazeu G, Jaffrain G (eds) European landscape dynamics. CORINE land cover data. CRC Press, Boca Raton, pp 141–156

Kosztra B, Büttner G, Hazeu G, Arnold S (2019) Updated CLC illustrated nomenclature guidelines. https://land.copernicus.eu/user-corner/technical-library/corine-land-cover-nomenclature-guidelines/docs/pdf/CLC2018_Nomenclature_illustrated_guide_20190510.pdf. Accessed 19 Aug 2020

Kukawska E, Lewinski S, Krupinski M et al (2017) Multitemporal Sentinel-2 data—remarks and observations. In: 2017 9th international workshop on the analysis of multitemporal remote sensing images, MultiTemp 2017

Malinowski R, Lewinski S, Rybicki M et al (2019) Deliverable 1.2. Final report. http://users.cbk.waw.pl/~mkrupinski/S2GLC_Phase2_FinalReport.pdf. Accessed 26 Sept 2020

Montero E, Van Wolvelaer J, Garzón A (2014) The European urban atlas. In: Manakos I, Braun M (eds) Land use and land cover mapping in Europe: practices & trends. Springer, Dordrecht, Heidelberg, New York, London, pp 115–124

Mücher CA (2000) PELCOM. Final report. http://www.geo-informatie.nl/projects/pelcom/download/FINALREP.zip. Accessed 26 Sept 2020

Mücher S, Steinnocher K, Champeaux J-L, Griguolo S, Wester K, Heunks C, Van Katwijk V (2000) Establishment of a 1-km pan-European land cover database for environmental monitoring. Int Archiv Photogram Remote Sens Spat Inf Sci ISPRS Arch 33

Neumann K, Herold M, Hartley A, Schmullius C (2007) Comparative assessment of CORINE2000 and GLC2000: spatial analysis of land cover data for Europe. Int J Appl Earth Obs Geoinf 9:425–437. https://doi.org/10.1016/j.jag.2007.02.004

Nowakowski A, Rybicki M, Kukawska E et al (2017) Aggregation of Sentinel-2 time series classifications as a solution for multitemporal analysis. In: Proceedings of SPIE 10427, image and signal processing for remote sensing XXIII, 104270B

Petrişor A-I, Petrişor LE (2015) Assessing microscale environmental changes: CORINE vs. the urban Atlas. Present Environ Sustain Dev 9:95–104. https://doi.org/10.1515/pesd-2015-0027

Piedelobo L, Taramelli A, Schiavon E et al (2019) Assessment of green infrastructure in Riparian zones using copernicus programme. Remote Sens 11. https://doi.org/10.3390/rs11242967

Prastacos P, Chrysoulakis N, Kochilakis G (2011) Urban Atlas, land use modelling and spatial metric techniques. In: ERSA conference papers

Seifert F (2009) Improving Urban Monitoring toward a European Urban Atlas. In: Gamba P, Herold M (eds) Global mapping of human settlement. experiences, datasets, and prospects. CRC Press, Boca Raton, p 231

Silva FB, Poelman H, Martens V, Lavalle C (2013) Population estimation for the urban Atlas polygons. https://ec.europa.eu/jrc/en/publication/eur-scientific-and-technical-research-reports/population-estimation-urban-atlas-polygons. Accessed 29 Sept 2020

Silva FB, Poleman H (2016) Mapping population density in functional urban areas. https://publications.europa.eu/en/publication-detail/-/publication/8568b1b3-b864-11e6-9e3c-01aa75ed71a1. Accessed 29 Sept 2020

Soukup T, Sousa A, Langanke T (2017) CLC2018 Technical guidelines. https://land.copernicus.eu/user-corner/technical-library/clc2018technicalguidelines_final.pdf. Accessed 19 Aug 2020

Tamame M, Lorenzo A, Lindmayer A, Richter R, Ramminger G, Köpp W, Martinez A et al (2018) Riparian zones nomenclature guideline. https://land.copernicus.eu/user-corner/technical-library/rz_nomenclature_guideline_v1_4_19-10-2018.pdf. Accessed 30 2020

Ugille J-P (2019) Potential of Copernicus riparian layers to assess riparian zones integrity with landscape metrics. Retrieved from https://matheo.uliege.be/handle/2268.2/7958. Accessed 11 June 2021. https://converges.eu/wp-content/uploads/2019/10/report_copernicus_ugille.pdf

Vandeputte R, Lhernould A, Pennec A et al (2018) Riparian zones land cover/land use extension to Strahler 2 validation report. https://land.copernicus.eu/user-corner/technical-library/riparian-zones-land-cover-land-use-validation-report. Accessed 30 Sept 2020

Weissteiner CJ, Ickerott M, Ott H et al (2016) Europe's green arteries—a continental dataset of riparian zones. Remote Sens 8. https://doi.org/10.3390/rs8110925

General Land Use Cover Datasets for Africa

David García-Álvarez and Javier Lara Hinojosa

Abstract

Several general Land Use Cover (LUC) datasets are available for Africa. They provide a general picture of the land uses and covers in more than one African country, rather than focusing on any specific type. In this chapter, we review six datasets of this kind. Only one (CCI LAND COVER – S2 PROTOTYPE, 30 m) covers the whole continent, while the others map certain specific regions of Africa. All these datasets have been produced within the context of specific projects, usually sponsored by international organizations such as the European Space Agency (ESA), the Food and Agriculture Organization (FAO) or the National Aeronautics and Space Administration (NASA). Once these projects come to an end, no new updates of the maps were published, which limits the potential and the temporal resolution of the available datasets. For Africa, only the West Africa Land Use Land Cover (2 km) and the SERVIR-ESA (30 m) provide a time series of LUC maps. The first provides maps for three reference years (1975, 2000, 2013), while in the second the number of maps available and their respective reference years vary from country to country: from 2 to 4 different editions issued between 1990 and 2015. AFRICOVER (1:200,000) and the Congo Basin Vegetation Types dataset (300 m) provide LUC information for just one reference year, although they were created from imagery covering a long time-span: 1994–2001 for AFRICOVER and 2000–2007 for Congo Basin Vegetation Types. The SADC Land Cover Database (1:250,000) was obtained by merging and harmonizing national and regional LUC datasets. As a result, the reference year varies from one country to the next, always between 1990 and 1997. The CCI LAND COVER – S2 PROTOTYPE was produced at the highest spatial resolution of all the datasets reviewed in this chapter (30 m). It also provided the most comprehensive, most updated LUC image of Africa, with information for the year 2015/16.

Keywords

Africa • West Africa Land Use Land Cover • SERVIR-ESA • SADC Land Cover Database • AFRICOVER • CCI LAND COVER – S2 PROTOTYPE • Congo Basin Vegetation Types

D. García-Álvarez (✉)
Departamento de Geología, Geografía y Medio Ambiente, Universidad de Alcalá, Alcalá de Henares, Spain
e-mail: David.garcia@uah.es

J. Lara Hinojosa
Departamento de Análisis Geográfico Regional y Geografía Física, Universidad de Granada, Granada, Spain

© The Author(s) 2022
D. García-Álvarez et al. (eds.), *Land Use Cover Datasets and Validation Tools*,
https://doi.org/10.1007/978-3-030-90998-7_17

1   West Africa Land Use Land Cover

Product
LULC general
Dates
1975, 2000, 2013
Formats
Raster
Pixel size
2 km
Thematic resolution
30 classes: 2 (a), 5 (ag), 12 (v), 3 (m), 3 (na)[1]
Compatible legends
None
Extent
West Africa and Cape Verde
Updating
Not expected
Change detection
Yes
Overall accuracy
Not specified

Website of reference	Website Language English
https://eros.usgs.gov/westafrica/	

Download site
https://eros.usgs.gov/westafrica/data-downloads https://www.sciencebase.gov/catalog/item/5deffc05e4b02caea0f4f3fc

Availability	Format(s)
Open Access	.tiff

Technical documentation
CILSS (2016)

Other references of interest
Cotillon (2017), Cotillon and Mathis (2017)

---

[1] (a): artificial; (ag): agriculture; (v): vegetation; (m): mixed classes; (na): no data.

Project

West Africa Land Use Dynamics was a project led by the AGRHYMET Regional Centre in collaboration with the Sahel Institute (INSAH), the USGS Earth Resources Observation and Science (EROS) and the US Agency for International Development (USAID). 17 different countries took part: Benin, Burkina Faso, Cape Verde, Chad, Ivory Coast, Gambia, Ghana, Guinea, Guinea-Bissau, Liberia, Mali, Mauritania, Niger, Nigeria, Senegal, Sierra Leone and Togo.

As a result of the project, a LUC map series was created to monitor natural and environmental trends in the West Africa region. The dataset is part of a wider effort to create an atlas about landscape and environmental changes in West Africa.

Production method

West Africa Land Use Land Cover was obtained through photointerpretation of Landsat imagery with the Rapid Land Cover Mapper (RLCM) tool at a spatial resolution of 2 km. Gambia was photointerpreted at a spatial resolution of 2 km and Cape Verde at 500 m. Photointerpretation guidelines were developed specifically for the task.

Product description

Users can download a separate edition of the West Africa Land Use Land Cover dataset for each year of reference. In each case, the download includes the raster file with the LUC map as well as a file to symbolize the raster in GIS. An Excel file with the legend can also be downloaded from the website together with the detailed metadata files for each LUC map.

Downloads

West Africa Land Use Land Cover (2013)
– Raster file with LUC map
– File to symbolize the raster in GIS (.clr)

Legend and codification

Code	Label	Code	Label
0	No data	15	Gallery forest and riparian forest
1	Forest	16	Shrub and tree savanna
2	Savanna	21	Degraded forest
3	Wetland – floodplain	22	Bowé
4	Steppe	23	Thicket
5	Oasis	24	Agriculture in bottomlands and flood recessional
6	Plantation	25	Woodland
7	Mangrove	27	Cropland and fallow with oil palms
8	Agriculture	28	Swamp forest
9	Water bodies	29	Sahelian short grass savanna
10	Sandy area	31	Herbaceous savanna
11	Rocky land	32	Shrubland
12	Bare soil	78	Open mine
13	Settlements	98	Cloud shadow
14	Irrigated agriculture	99	Cloud

2   SERVIR-ESA—SERVIR Eastern and Southern Africa

Prcoduct	
LULC general	
Dates	
1990, 2000, 2010 (Malawi) 1990, 2000, 2010, 2015 (Rwanda) 2000, 2010 (Botswana, Namibia, Tanzania, Zambia) 2000, 2014 (Lesotho, Uganda) 2003, 2008 (Ethiopia)	
Formats	
Raster	
Pixel size	
30 m	
Thematic resolution	
7 classes: 1 (a), 1 (ag), 2 (v), 0 (m), 1 (na)	
Compatible legends	
IPCC	
Extent	
Eastern and Southern Africa	
Updating	
No updating confirmed	
Change detection	
Yes, but potential uncertainties have not been specified	
Overall accuracy	
Expected to be >63%	

Website of reference	**Website Language** English, Spanish, French
https://www.servirglobal.net/ServiceCatalogue/details/ 5bd052d451ebdcae79683375	

Download site
http://geoportal.rcmrd.org/layers/?limit=100&offset=0

Availability	**Format(s)**
Open Access	.tiff

Technical documentation
Oduor et al. (2016)

Other references of interest
Al-Hamdan et al. (2017), Searby et al. (2019)

Project

SERVIR is an initiative led by the National Aeronautics and Space Administration (NASA) and the United States Agency for International Development (USAID) that aims to help developing countries to produce geospatial information suitable for climate risks and land use management. SERVIR operates in West Africa, Eastern and Southern Africa, Hindu Kush Himalaya, the Lower Mekong, South America and Mesoamerica.

In Eastern and Southern Africa in 2008, SERVIR started a project in partnership with the Kenya-based Regional Centre for Mapping of Resources for Development (RCMRD). Training, geospatial tools and geospatial datasets were developed as part of the project, including a dataset specifically aimed at LUC monitoring. Six countries were initially mapped (Botswana, Malawi, Namibia, Rwanda, Tanzania, and Zambia), with three more countries participating since 2014/15 (Ethiopia, Uganda, and Lesotho).

As a result of this project, a LUC map covering all 9 countries was developed. National LUC maps, with detailed national legends, were also provided as part of the project.

Production method

SERVIR-ESA was produced by aggregating LUC maps created at the national level according to the same 7-class classification scheme. For each country, a map with a legend adapted to the country's specificities was also developed following the same general guidelines.

The maps were obtained through supervised classification of Landsat imagery through a Maximum likelihood classifier. Auxiliary spatial and non-spatial data were also used in the classification. Settlements were manually photointerpreted from Google Earth imagery.

Errors and uncertainties in the classification resulting from this process were corrected in a post-classification step, which included expert review.

Product description

The SERVIR-ESA LUC map is distributed at national level. For each country, users can download the harmonized map for all Eastern and Southern Africa (Scheme I) or the specific LUC map with a detailed legend for the selected country (Scheme II).

Downloads

Scheme I product/Scheme II product
– Raster file with coloured LUC map

Legend and codification

In this description, we only include the general 7-class legend adopted for all the LUC maps. However, a specific legend is available for each national map, which can be consulted online.

Scheme I legend			
Code	Label	Code	Label
0	Non data	4	Wetland
1	Forestland	5	Settlement
2	Grassland	6	Other land
3	Cropland		

Practical considerations

The maps for each country were usually produced at different dates, so making inter-country comparison difficult.

3   SADC Land Cover Database

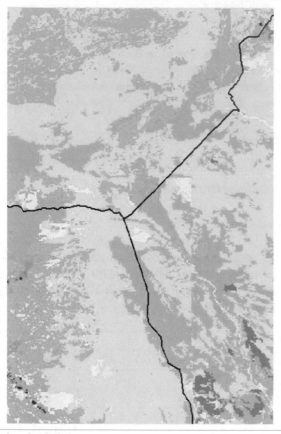

Product
LULC general
Dates
1990 / 91 (Malawi) 1997 (Tanzania, Zimbabwe) 1999 (Mozambique, South Africa, Lesotho, and Swaziland)
Formats
Vector
Scale
1:250,000
Thematic resolution
13 classes: 1 (a), 2 (ag), 5 (v), 0 (m), 1 (na)
Compatible legends
None
Extent
Southern African Development Community (Lesotho, Malawi, Mozambique, South Africa, Swaziland, Tanzania, Zimbabwe)
Updating
No
Change detection
No (only one date)
Overall accuracy
Not specified

Website of reference	Website Language English
http://gsdi.geoportal.csir.co.za/projects	

Download site
http://gsdi.geoportal.csir.co.za/projects

Availability	Format(s)
Open Access	.shp

Technical documentation
–

Other references of interest
–

Project

The SADC Land Cover Database is fruit of a project funded by the South African Department of Arts, Culture, Science and Technology (DACST) through the Regional Science and Technology Programme. It was coordinated by the Council for Scientific and Industrial Research (CSIR) in South Africa, with the participation of organizations from the different countries being mapped.

The objective of the project was to deliver a coherent Land Use Cover map covering the Southern African Development Community (SADC) region. The project builds on earlier LUC mapping work carried out at national and regional scales for each of the mapped countries.

The map covers those SADC countries that already had a LUC dataset available for their territory: Lesotho, Malawi, Mozambique, South Africa, Swaziland, Tanzania, Zimbabwe. The other countries in the region are not included in the map.

Production method

The SADC Land Cover Database was obtained by harmonizing and fusing the different national and regional LUC datasets. All the datasets were originally obtained by classification or photointerpretation of Landsat imagery, although the reference years vary from country to country.

The maps were combined by resampling to a spatial resolution of 1 km, before being reclassified according to the same classification system. This reduced the detail of the original maps, a deliberate action to avoid copyright and commercialisation issues.

Product description

The dataset is downloaded as a single compressed file (.zip), which includes the vector LUC map, a metadata file and a complete map (i.e. with colours, graphics, scale and legend) in jpg format that is ready to print out.

Downloads

SADC
– Vector file with LUC map (.shp)
– Edited map in a non-modifiable format (.jpg)
– Metadata file (.html)

Legend and codification

Label	Label	Label
Forest	Bare ground	Open water
Woodland	Plantation	Wetland
Bushland	Cultivation	Ice-cap/Snow
Low shrubland	Built-on	Not classified
Grassland		

Practical considerations

A detailed description of the map categories is available in the dataset's metadata. The map's production method entails certain limitations and uncertainties, in that each country has been mapped by a different team, using different sources of imagery for different reference years. Inconsistencies may therefore arise when comparing information between countries.

4 AFRICOVER

Product	
LULC general	
Dates	
1994/01 (the reference year varies according to the country)	
Formats	
Vector	
Scale	
1:200,000	
Thematic resolution	
8 classes: 2 (a), 2 (ag), 2 (v), 0 (m), 0 (na)	
Compatible legends	
FAO-LCCS	
Extent	
Africa	
Updating	
No	
Change detection	
No (only one date)	
Overall accuracy	
Expected to be > 80%	

Website of reference	**Website Language** English
http://www.fao.org/geospatial/projects/detail/en/c/1035404/	
Download site	
http://www.fao.org/geospatial/projects/detail/en/c/1035404/	
Availability	**Format(s)**
Open Access	.shp
Technical documentation	
Di Georgio and Jansen (1996), FAO (1997)	
Other references of interest	
Di Gregorio (2009), Kalensky (1998), Latham et al (2002)	

Project

AFRICOVER was a project led and coordinated by the Food and Agriculture Organization (FAO) of the United Nations, which aimed to create georeferenced data for the African continent. The FAO helped the different countries and regions to develop their reference maps, establishing the standards for the final product. Twelve countries participated in the project (Burundi, Democratic Republic of Congo, Egypt, Eritrea, Kenya, Rwanda, Somalia, Sudan, Tanzania, Uganda, Libya and Malawi), which therefore required extensive coordination of many national and regional teams across Africa.

A keystone of the project was the production of LUC maps for Africa. In addition to LUC maps, other georeferenced data were created for a range of themes: hydrology, geomorphology, demography...

Production method

The production of AFRICOVER was decentralised at a national and regional level. Although the FAO defined the guidelines and standards for the product, national and regional teams from each country were responsible for its execution. This meant that although a set of common characteristics regarding the production of AFRICOVER had been established for all the countries involved, certain specificities could also arise.

AFRICOVER LUC maps were mainly obtained through photointerpretation of satellite imagery, of which Landsat was the main source. The photointerpretation scale was 1:200,000. When drawing LUC polygons, the FAO LCSS classification scheme was followed. The FAO provided national and regional teams with specific software and training to carry out LUC mapping according to this approach.

Product description

AFRICOVER LUC maps are distributed at a national level. A compressed file can be downloaded for each country. This includes the vector LUC map and a legend description to help users interpret it.

Downloads

Land cover folder
– Vector file with LUC map (.shp)
– PDFs describing the classification legend
– Excel file with the classification legend

Legend and codification

Label	Label
Cultivated Terrestrial Areas and Managed Lands	Artificial Surfaces and Associated Areas
Natural and Seminatural Terrestrial Vegetation	Bare Areas
Cultivated Aquatic or Regularly Flooded Areas	Artificial Waterbodies
Natural and Seminatural Aquatic Vegetation	Inland Waterbodies

Practical considerations

AFRICOVER LUC maps have been created following the FAO LCSS classification scheme. This means that each LUC polygon is described through a specific code that identifies the general cover of the polygon and characterizes it through a series of labels. Users may find this system difficult to understand, as it does not follow a common hierarchical classification legend in which each polygon is defined by a single category.

5   CCI LAND COVER – S2 PROTOTYPE

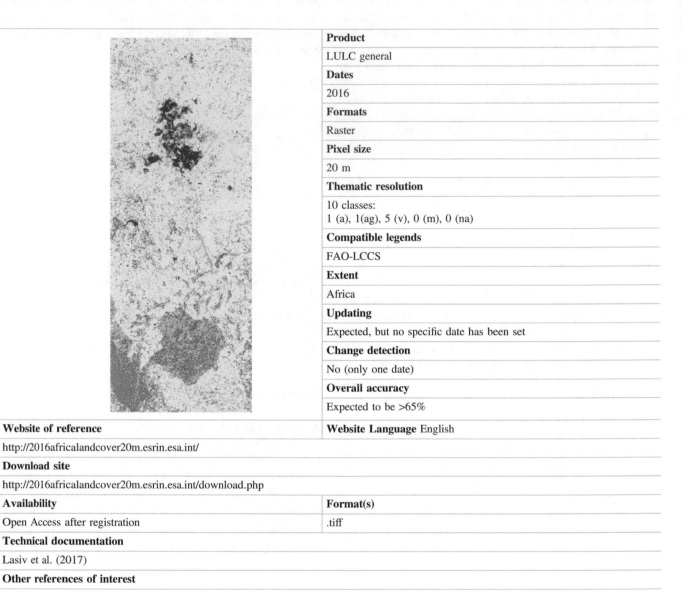

Product
LULC general
Dates
2016
Formats
Raster
Pixel size
20 m
Thematic resolution
10 classes: 1 (a), 1(ag), 5 (v), 0 (m), 0 (na)
Compatible legends
FAO-LCCS
Extent
Africa
Updating
Expected, but no specific date has been set
Change detection
No (only one date)
Overall accuracy
Expected to be >65%

Website of reference	Website Language English
http://2016africalandcover20m.esrin.esa.int/	

Download site
http://2016africalandcover20m.esrin.esa.int/download.php

Availability	Format(s)
Open Access after registration	.tiff

Technical documentation
Lasiv et al. (2017)

Other references of interest
–

Project

The CCI LAND COVER – S2 PROTOTYPE map is part of the Land Cover – Climate Change Initiative led by the European Space Agency (ESA). The purpose of this initiative is to deliver Land Cover products that meet the requirements of the climate change research community.

The map was created as a prototype to collect feedback from users for future improvements. At the time it was released, it was the highest spatial resolution LUC map covering the whole African continent and one of the few products providing consistent LUC coverage for all of Africa.

Production method

The map was obtained after classification of Sentinel-2A imagery for the reference year 2016. Two different classifications were carried out, through random forest and machine learning classifiers. The final map is a combination of the two classifications. Auxiliary datasets were used to map the "open water" (extracted from the Global Surface Water product) and "urban areas" (extracted from Global Human Settlement Layer and the Global Urban Footprint) categories.

Product description

CCI LAND COVER – S2 PROTOTYPE is distributed as a single compressed file, including the raster with LUC information and a style layer to symbolize the map in GIS software. The legend is described in two auxiliary files, in Excel and pdf.

Downloads

S2 PROTOTYPE LC at 20 m AFRICA 2016
– Raster file with LUC map ("ESACCI-LC-L4-LC10-Map-20 m-P1Y-2016-v1.0")
– Layer style file for GIS software (.qml)
– Excel sheet with the map legend
– PDF describing the map legend

Legend and codification

Code	Label	Code	Label
1	Tree cover areas	6	Lichens and mosses
2	Shrubs cover areas	7	Bare areas
3	Grassland	8	Built up areas
4	Cropland (rainfed or irrigated)	9	Snow and/or Ice
5	Vegetation aquatic or regularly flooded	10	Open water

Practical considerations

The map is distributed as a single, very heavy file (6 Gb). Users with limited computer and internet capacities may find it difficult to download and work with this product. Nonetheless, a preview tool is available online for any user wishing to consult the map.

6 Congo Basin Vegetation Types

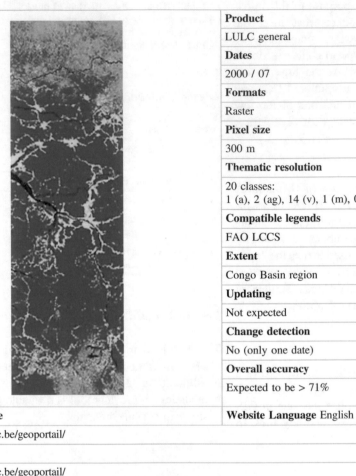

Product
LULC general
Dates
2000 / 07
Formats
Raster
Pixel size
300 m
Thematic resolution
20 classes: 1 (a), 2 (ag), 14 (v), 1 (m), 0 (na)
Compatible legends
FAO LCCS
Extent
Congo Basin region
Updating
Not expected
Change detection
No (only one date)
Overall accuracy
Expected to be > 71%

Website of reference	Website Language English
http://maps.elie.ucl.ac.be/geoportail/	

Download site
http://maps.elie.ucl.ac.be/geoportail/

Availability	Format(s)
Open access	.tiff

Technical documentation
Verhegghen et al. (2012)

Other references of interest
–

Project

The Congo Basin Vegetation Types map was produced by a team of experts from the Université Catholique de Louvain, the Joint Research Centre (JRC) of the European Commission and the Observatory for the Forests of Central Africa (OFAC).

The map was produced in an attempt to aid forest and vegetation monitoring in Central Africa. It provided a spatially coherent dataset for all the Congo Basin region with improved spatial discrimination with respect to previous datasets of similar nature.

Production method

The Congo Basin Vegetation Types was obtained by unsupervised classification of imagery composites created from the images provided by the MERIS and VEGETATION sensors.

To account for the regional disparities of the mapped area and its different cloud coverage, the Congo Basin was split into four different zones: North, South, Western Centre and Eastern Centre. Seasonal imagery composites were created for each specific season in the northern and southern regions. In addition, an annual composite was generated for the whole mapped area.

A different classification exercise was performed for each mapped zone based on a cluster k-means algorithm. The resulting clusters were labelled on the basis of the information provided by reference maps when LUC information on these sources covered at least 50% of the identified cluster. The rest of the clusters were manually labelled on the basis of visual interpretation and expert knowledge.

Product description

A compressed file (.zip) containing the raster layer with the LUC data can be downloaded, together with other auxiliary information to interpret and symbolize the map content.

Downloads

Congo Basin Vegetation Types map

– Raster file with LUC map (.tif)
– Layer style files for ArcGIS (.lyr)
– Excel file with the map legend (.xls)
– Text file with the metadata for the product (.txt)

Legend and codification

Code	Label	Code	Label
1	Dense moist forest	11	Grassland
2	Submontane forest	12	Aquatic grassland
3	Mountain forest	13	Swamp grassland
4	Edaphic forest	14	Sparse vegetation
5	Mangrove	15	Mosaic cultivated areas/ vegetation
6	Forest/savanna mosaic	16	Agriculture
7	Rural complex (forest area)	17	Irrigated agriculture
8	Closed to open deciduous woodland	18	Bare areas
9	Savanna woodland/tree savanna	19	Artificial surfaces and associated areas
10	Shrubland	20	Water bodies

Practical considerations

Users can consult the LUC map online on the Université Catholique de Louvain website (http://maps.elie.ucl.ac.be/geoportail/).

References

Al-Hamdan MZ, Oduor P, Flores AI et al (2017) Evaluating land cover changes in Eastern and Southern Africa from 2000 to 2010 using validated Landsat and MODIS data. Int J Appl Earth Obs Geoinf 62:8–26. https://doi.org/10.1016/j.jag.2017.04.007

CILSS (2016) Landscapes of West Africa—a window on a changing world. Accessed September 28, 2020. https://eros.usgs.gov/westafrica/atlas-document-downloads

Cotillon SE (2017) West Africa land use and land cover time series. Fact Sheet 2017–3004. Accessed September 28, 2020. https://pubs.er.usgs.gov/publication/fs20173004

Cotillon SE, Mathis ML (2017) Mapping land cover through time with the Rapid Land Cover Mapper—Documentation and user manual

Di Georgio A, Jansen LJM (1996) Part I: Technical documentation on the Africover Land Cover Classification Scheme. 4–33, 63–76, in FAO (1997). Africover Land Cover Classification

Di Gregorio A (2009) AFRICOVER Land Cover classification and mapping project. In: Verheye WH (ed) Land use, land cover and soil sciences - Volume I: Land cover, land use and the global change. EOLSS Publishers, pp 236–251

FAO (1997) AFRICOVER land cover classification. Accessed September 24, 2020. http://www.fao.org/3/a-bd854e.pdf

Kalensky ZD (1998) AFRICOVER land cover database and map of africa. Can J Remote Sens 24:292–297. https://doi.org/10.1080/07038992.1998.10855250

Lasiv M, Fritz S, McCallum I, Tsendbazar N, Herold M, Pekel J-F, Buchhorn M, Smets B, Van De Kerchove R (2017) Evaluation of ESA CCI prototype land cover map at 20 m. Working paper available at http://pure.iiasa.ac.at/14979/1/WP-17-021.pdf (accessed on 8 September 2020)

Latham JS, He C, Alinovi L et al (2002) FAO methodologies for land cover classification and mapping. Link People, Place, Policy 283–316. https://doi.org/10.1007/978-1-4615-0985-1_13

Oduor P, Ababu J, Mugo R et al (2016) Land cover mapping for green house gas inventories in Eastern and Southern Africa using landsat and high resolution imagery: approach and lessons learnt. In: Hossain F (ed) Earth science satellite applications. Springer, pp 85–116

Searby ND, Irwin D, Kim T (2019) Servir: leveraging the expertise of a space agency and a development agency to increase impact of earth observation in the developing world. In: Proceedings of the international astronautical congress, IAC, pp 1–9

Verhegghen A, Mayaux P, De Wasseige C, Defourny P (2012) Mapping Congo Basin vegetation types from 300 m and 1 km multi-sensor time series for carbon stocks and forest areas estimation. Biogeosciences 9:5061–5079. https://doi.org/10.5194/bg-9-5061-2012

General Land Use Cover Datasets for America and Asia

David García-Álvarez and Javier Lara Hinojosa

Abstract

In this chapter we review some examples of general Land Use Cover (LUC) mapping at a supra-national level in America and Asia. These datasets provide a general overview of the land uses and covers in specific American or Asian regions, without focusing on any particular land use or cover. For Asia, we have only identified one dataset mapping the Himalayan region, whereas for America five different datasets were identified. Only three of these are reviewed here, as the other two (SERENA, South America 30 m) are not available for download. The most ambitious project of all those reviewed is NALCMS, which coordinates the production of a LUC map for the whole of North America (Canada, Mexico, USA) at detailed scales (30–250 m) and using the same classification legend. It is the only dataset of all those reviewed that provides a time series of LUC maps (2005, 2010 and 2015). The Himalaya Regional Land Cover database is a vector-based map that provides information on LUC changes over the period 1970/80–2007 at a scale of 1:350,000. The other two American datasets—LBA-ECO LC-08 (1 km, 1987/91) and MER-ISAM2009 (300 m, 2008/10)—are raster-based and only available for one date, therefore making change detection impossible.

Keywords

America • Asia • LBA-ECO LC-08 • NALCMS • MERISAM2009 • Himalaya Regional Land Cover database

D. García-Álvarez (✉)
Departamento de Geología, Geografía y Medio Ambiente, Universidad de Alcalá, Alcala de Henares, Spain
e-mail: David.garcia@uah.es

J. Lara Hinojosa
Departamento de Análisis Geográfico Regional y Geografía Física, Universidad de Granada, Granada, Spain

1   LBA-ECO LC-08—Land Cover Map of South America

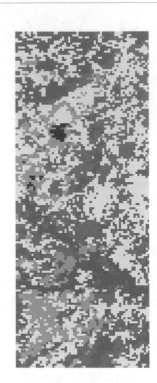

Product	
LULC general / LULC thematic (vegetation)	
Dates	
1987 / 91	
Formats	
Raster	
Pixel size 1 km	
Thematic resolution	
42 classes: 1 (a), 1 (ag), 27 (v), 7 (m), 3 (na)[1]	
Compatible legends	
None	
Extent	
South America	
Updating	
Not expected	
Change detection	
No (only one date)	
Overall accuracy	
Depending on the class. Expected to be >90% for 24 classes covering 85% of the map. Classes with an accuracy of <75% only cover 6.5% of the map	

Website of reference	**Website Language** English
https://daac.ornl.gov/LBA/guides/LC08_EOS_Maps.html#references	

Download site
https://daac.ornl.gov/cgi-bin/dsviewer.pl?ds_id=1155

Availability	**Format(s)**
Open Access after registration	.tiff, .nc, .asc, .nitf, .img

Technical documentation
Stone et al. (1994)

Other references of interest
–

---

[1] (a): artificial; (ag): agriculture; (v): vegetation; (m): mixed classes; (na): no data.

Project

The Large-Scale Biosphere-Atmosphere Experiment in the Amazon (LBA) was an international project launched by the Brazilian scientific community in 1993. The main objectives were to study Amazonia and its role in the earth's ecosystem as well as to understand LUC changes in the area and their environmental consequences.

As part of the project, a global LUC map covering South America was produced from imagery and data of the period 1987/91. Vegetation and soil maps for Brazil were also digitalized on the basis of previous resources. These maps are also available for any interested user as part of the same dataset.

Production method

The LBA LUC map was produced after unsupervised classification of AVHRR imagery, postprocessing and labelling of the classification results. Different sources of auxiliary data were used in the production of the dataset to overcome the limitations of the imagery, including a Global Vegetation Index (GVI) layer, the UNESCO's Vegetation Map of South America, the Hueck's Vegetationsskarte Von Sudamerika and a potential vegetation map of South America based on the Holdridge bioclimatic scheme.

Production description

Users can download the LUC map as a single raster file including the LUC information or as part of a data package including all the products produced within the LBA project. As part of these, we find different vegetation and soil maps for Brazil. In all cases, the download only includes the raster files and no auxiliary information is provided.

Downloads

RAR folder with all products

– Raster file with LUC map
– 3 raster files with Brazil vegetation maps at different levels of thematic resolution
– 3 raster files with Brazil soil maps at different levels of thematic resolution

SA_lc_Map_41class.tif

– Raster file with LUC map

Legend and codification

Code	Label	Code	Label
0	Off Map (Fill Value)	21	Secondary seasonal forest with agriculture
1	Tropical moist and semi-deciduous forest	22	Urban and degraded lands
2	Cleared tropical moist Forest	23	Degraded tropical seasonal forest
3	Unclassified	24	Mixed pine forest with secondary forest and agriculture
4	Water	25	Xerophytic scrubland
5	Savanna/Grasslands	26	Xerophytic littoral vegetation
6	Wet vegetation/Mixed	27	Montane grassland
7	Unclassified	28	Montane woodlands
8	Mangroves	29	Montane forests
9	Seasonally deciduous Woodlands.	30	Degraded montane grasslands
10	Forest (Bamboo dominated?)	31	Degraded montane woodlands
11	Secondary tropical moist forest with agriculture	32	Degraded montane forests
12	Pantanal grassland (seasonally flooded)	33	Cool deciduous shrublands
13	Tropical seasonal or deciduous forest	34	Bare soil/Rock
14	Agriculture	35	Cool deciduous woodlands
15	Gallery forests	36	Cool deciduous forests
16	Tropical open forests (mixed)	37	Snow/Rock
17	Cerrado (woodlands) degraded	38	Salt marsh community
18	Grasslands or Savanna with agriculture	39	Desert
19	Xerophytic woodlands with agriculture	40	Degraded temperate deciduous forest
20	Degraded xerophytic woodlands	41	Temperate deciduous forests

2   NALCMS—North American Land Change Monitoring System

Product
LULC general
Dates
2005, 2010, 2015
Formats
Raster
Pixel size
30 m (2010, 2015) 250 m (2005, 2010)
Thematic resolution
19 classes: 1 (a), 1 (ag), 13 (v), 1 (m), 0 (na)
Compatible legends
FAO-LCCS
Extent
North America
Updating
Unknown
Change detection
Through change layers
Overall accuracy
Expected to be >79.9%

Website of reference	**Website Language** English, Spanish, French
http://www.cec.org/north-american-land-change-monitoring-system/	
Download site	
http://www.cec.org/north-american-land-change-monitoring-system/	

Availability	**Format(s)**
Open Access	.tiff, .img, .mxd

Technical documentation
Colditz et al. (2012, 2014a, b, c), Gebhardt et al. (2014), Homer et al. (2015), Jin et al. (2013, 2019), Latifovic et al. (2012, 2017)
Other references of interest
Yang et al (2018)

Project

The NALCMS project started in 2006 fruit of the collaboration between the following Canadian, American and Mexican institutions: the Natural Resources Canada/Canada Centre for Remote Sensing (NRCan/CCRS), the United States Geological Survey (USGS) and the Mexican National Institute of Statistics and Geography (INEGI), National Commission for the Knowledge and Use of Biodiversity (CONABIO) and the National Forestry Commission of Mexico (CONAFOR). The project is also supported by the Commission for Environmental Cooperation (CEC), a body comprising all three North American countries.

The objective of the project was to create a homogeneous, coherent LUC dataset for North America that could be used for environmental monitoring at a continental scale, and which also addressed the needs and requirements of scientific and policy-making communities. Each country produced its own LUC map according to its needs and requirements. The purpose of the project was to coordinate the homogenization and harmonization of these national maps to create a single map of the whole North America.

Since it was launched in 2006, three LUC maps have been produced. Important improvements have been made over time. The most significant change was the improved spatial resolution of 30 m applied in the latest maps, compared to 250 m in the first edition.

Production method

There is no single production methodology for NALCMS. Each country is responsible for producing its own LUC map, according to its particular needs and interests.

The first edition of the product for 2005 was created via a classification of MODIS imagery at 250 m following a similar workflow for the three countries. In 2010, the initial map for 2005 at 250 m was revised, mapping only the LUC changes that happened over the period 2005–2010. LUC changes for Hawaii were not mapped in this update. Mapped changes were individually distributed through a specific change layer at 250 m for the period 2005–2010.

For 2015, Canada and the USA obtained their respective LUC maps after classification of Landsat imagery, while Mexico obtained its map via the classification of RapidEye (5 m) imagery resampled at 30 m. Whereas for Canada and Mexico the imagery mostly dates from 2015, most of the imagery used in the US map was from the year 2016. For 2010, the three countries obtained the map at 30 m from the classification of Landsat imagery. However, whereas most of the imagery for Canada and Mexico was captured in 2010, the images used to map USA were taken in 2011.

A change layer at 30 m for the period 2010–2015 was obtained by comparing the base LUC maps at the two different dates for Canada and USA. In Mexico, because different imagery sources had been used for the different reference years, the changes were individually extracted from Landsat imagery based on an independent change detection algorithm.

Product description

NALCMS can be separately downloaded for each of the reference years. A change layer for each mapped period is also available: 2005–2010 and 2010–2015. For those years for which more than one spatial resolution is available, users can download a separate product at each resolution.

The datasets at 250 m can be downloaded in different formats: GeoTIFF, ERDAS Imagine (.img), Map Exchange Document (.mxd) and as a georeferenced PDF file (GeoPDF). Datasets at 30 m are downloaded in a compressed file (.zip) in GeoTIFF. They can be downloaded for the whole of North America or individually for each of the mapped countries.

Different auxiliary information is provided with each downloaded product. Nonetheless, the metadata for all the available products can be downloaded separately from the dataset's website.

Downloads

Land Cover, 2005–2010 (MODIS, 250 m), TIFF

– Raster files with North America and Hawaii LUC maps (.tiff)
– Metadata file (.doc)
– Definitions of the different classes (.doc) [Only 2010 map]
– Press release presenting the product (.doc)
– Terms of use of the product (.doc)

Land Cover Change, 2005–2010 (MODIS, 250 m), TIFF

– Raster files with LUC changes (.tiff)
– Layer style file for ArcGIS (.lyr)
– Cross tabulation matrixes of change (in ha, percent and pixels) at two different classification schemes (.xlsx)
– Metadata file (.doc)—Press release presenting the product (.doc)
– Terms of use of the product (.doc)

Land Cover, 2010–2015 (Landsat, 30 m), North America

– Raster file with LUC map
– Layer style files for ArcGIS (.lyr) in English, French and Spanish
– Metadata file (.doc)

Land Cover Change, 2010–2015 (Landsat, 30 m), North America

– Raster files with gains and losses for the Forest, Shrubland, Grassland, Wetland, Cropland, Barren Land, Urban and Built-up, Water and Snow and Ice categories (.tiff)
– Raster file with LUC changes
– Metadata file (.doc)
– Text document with a description of the dataset (.txt)

Legend and codification

The change layers include a qualitative description of the classes at the two different points in time. In addition, the pixel values are formed by combining the class code for the land use at point 1 in time with the class code for the new land use at point 2. e.g. the code 1011 refers to a pixel that was *Temperate or sub-polar grassland* (10) on the first date assessed and had changed to *Sub-polar or polar shrubland-lichen-moss* (11) on the second.

Code	Label	Code	Label
1	Temperate or sub-polar needleleaf forest	11	Sub-polar or polar shrubland-lichen-moss
2	Sub-polar taiga needleleaf forest	12	Sub-polar or polar grassland-lichen-moss
3	Tropical or sub-tropical broadleaf evergreen forest	13	Sub-polar or polar barren-lichen-moss
4	Tropical or sub-tropical broadleaf deciduous forest	14	Wetland
5	Temperate or sub-polar broadleaf deciduous forest	15	Cropland
6	Mixed forest	16	Barren lands
7	Tropical or sub-tropical shrubland	17	Urban

(continued)

Code	Label	Code	Label
8	Temperate or sub-polar shrubland	18	Water
9	Tropical or sub-tropical grassland	19	Snow and Ice
10	Temperate or sub-polar grassland		

Practical considerations

Maps at 30 m and 250 m were obtained following a different workflow and are not comparable. The maps for Mexico for 2010 and 2015 were obtained from different imagery sources, which means that changes cannot be calculated by subtracting one map from the other and should only be studied using the change layer distributed by the production team.

No information is offered about the uncertainty of the change layers. They may be subject to important sources of uncertainty and may include a lot of technical or spurious changes that did not actually happen on the ground.

NALCMS is one of the products in the North American Environmental Atlas. Users can consult the different NALCMS layers online, together with a lot of other relevant geospatial information for North America, as part of the Atlas website at http://www.cec.org/files/atlas/. Users can also download any of the displayed layers, including the LUC maps, from the same website.

3 MERISAM2009—MERIS MAP 2009/2010 South America

Product	
LULC thematic	
Dates	
2009/10	
Formats	
Raster, Vector	
Pixel size	
300 m	
Thematic resolution	
11 classes: 0 (a), 3 (ag), 5 (v), 5 (m), 1 (na)	
Compatible legends	
None	
Extent	
South America	
Updating	
Not expected	
Change detection	
No (only one date)	
Overall accuracy	
Not specified	

Website of reference	Website Language English
Not available	
Download site	
Not available	
Availability	**Format(s)**
On request	.img, .shp
Technical documentation	
Hojas-Gascon et al. (2012)	
Other references of interest	
–	

Project

MERISAM is a map developed by the Joint Research Centre (JRC) of the European Commission as part of the regional LUC mapping efforts for South America. With the production of MERISAM, the JRC team aimed to overcome some of the limitations encountered during the production of GlobCover for South America. These referred mainly to spatial and thematic inaccuracies due to the limited number of MERIS images acquired and the method followed to produce the imagery mosaic required to carry out the classification.

The MERISAM dataset was used to assess LUC change in the first decade of the 21st century by comparing it with the GLC2000 dataset.

Production method

MERISAM was obtained after unsupervised classification of MERIS imagery for the period 2008–2010 using the ISO-DATA classification algorithm, which identified 100 different spectral classes. These were manually assigned to 6 LUC categories based on the information provided by auxiliary datasets, such as national vegetation maps and Google Earth imagery. FAPAR data, which provide information on the photosynthetic activity of the vegetation, were also used as auxiliary information to disaggregate the initial set of LUC categories.

Product description

Interested users can access this dataset by contacting the JRC team that produced it. The dataset includes the LUC map in two formats: raster (.img) and vector (.shp). The vector file was obtained by vectorizing the original raster file.

Downloads

MERISAM2009
– Raster file with LUC map (.img)
– Vector file with LUC map (.shp)
– Two versions of the scientific paper presenting the dataset (.pdf)

Legend and codification

Here are the codes used to produce the raster version of the map.

Code	Label	Code	Label
1	Evergreen forest	6	Sparse and barren
2	Dry forest and shrubs	10	Inland water
3	Dry open forest and shrubs	41	Grasslands and shrubs
4	Grasslands	51	Agriculture mosaic
5	Agriculture and pasture	52	Agriculture intensive
0	Background		

Practical considerations

This dataset is not directly available for download. Users wishing to access it must contact the JRC team that produced it (Hugh.EVA@ec.europa.eu, Rene.BEUCHLE@ec.europa.eu).

Although the dataset has been used to assess LUC changes by comparing it with GLC2000, this exercise has many limitations and uncertainties and is therefore not recommended.

4 The Himalaya Regional Land Cover Database

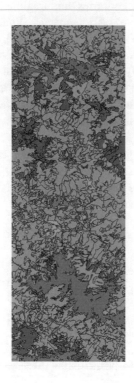

Product	
LULC general	
Dates	
2000 (base LUC map) 1970–2007 (LUC changes)	
Formats	
Vector	
Scale	
1:350,000	
Thematic resolution	
35 classes: 1 (a), 7(ag), 15 (v), 7 (m), 0 (na)	
Compatible legends	
LCCS	
Extent	
Himalaya region	
Updating	
No	
Change detection	
Yes, through the change layer	
Overall accuracy	
Not specified	

Website of reference	**Website Language** English
http://www.fao.org/geonetwork/srv/en/main.home?uuid=46d3c2ef-72c3-4f96-8e32-40723cd1847b	
Download site	
http://www.fao.org/geonetwork/srv/en/main.home?uuid=46d3c2ef-72c3-4f96-8e32-40723cd1847b	
Availability	**Format(s)**
Open Access	.shp
Technical documentation	
–	
Other references of interest	
–	

Project

The Himalaya Regional Land Cover database was developed within the context of the Global Land Cover Network— Regional Harmonization Programme, promoted by the Food and Agriculture Organization of the United Nations (FAO) and UN Environment in collaboration with the Geographic Information for Sustainable Development (GISD) global partnership. The programme aimed to produce reliable, harmonized global land cover information, providing guidance and methodologies for the production of LUC information at national, regional and global levels.

Production method

The database was obtained by automatic segmentation of Landsat imagery for the reference year 2000 plus visual interpretation. The initial classification was refined by interpreting high resolution imagery from Google Earth.

A layer of LUC changes was obtained by assessing the base map (2000) against historical imagery for the periods 1970–80, 1990 and 2007. No maps for the other years of reference are available, but only the respective layers of changes.

Product description

The database is distributed at regional level in vector format for each of the countries and regions that make up the Himalayan region: Afghanistan, Bhutan, China-Yunnan Sheng, China-Xizang Zizhiqu, India, Nepal, Pakistan, Aksai Chin, Arunachal Pradesh, China/India, Jammu Kashmir and Myanmar. An additional vector layer with LUC changes for the period 1970–2007 is also included. The downloaded products consist solely of the vector layers with LUC data. No other auxiliary information is provided with the downloaded file.

A detailed legend for the product can be downloaded separately in Excel or mdb formats. A layer with the boundaries of the region and its administrative units is also available for download.

Downloads

Land Cover map (country/region)
– Vector file with Land Cover map (.shp)

Land change Himalaya region
– Vector file with map of Land Cover changes (.shp) – Vector file with boundaries of the Himalaya region (.shp)

Database

Himalaya Regional Land Cover Database

	Z007CODE	Z007USLB	Z007PERC	HECTARES	AREA	AGG	ZONE	CODE1	CODE2	BOOLEAN1	BOOLEAN2	LCCSMAIN1	LCCSMAIN2	AUTO_ID ▲
1	20059-122...	2HS//6BR	100	97714.744	977147...	BS	Zone 44	2HS//6BR	NULL	A2A14B4-B13 /...	NULL	2//6	NULL	221550
2	20377-120...	2SOd	100	51050.075	510500...	H&S	Zone 44	2SOd	NULL	A4A11B3XXXXX...	NULL	2	NULL	221580
3	20059-122...	2HS//6BR	100	48652.191	486521...	BS	Zone 44	2HS//6BR	NULL	A2A14B4-B13 /...	NULL	2//6	NULL	222894

– Z007CODE: LUC Code
– Z007USLB: LUC User Label
– Z007PERC: Percentage of the LUC(s) making up the polygon
– HECTARES: Area of the polygon, in hectares
– AREA: Area of the polygon, in square meters
– AGG
– ZONE: UTM Zone
– CODE 1: Code LUC 1
– CODE 2: Code LUC 2
– BOOLEAN1: LUC Label 1
– BOOLEAN 2: LUC Label 2
– LCCSMAIN1: Main LUC 1
– LCCSMAIN2: Main LUC 2
– AUTO_ID: Unique identifier for each polygon

Legend and codification

Code	Label
1H	Herbaceous Crops
1HI	Irrigated Herbaceous Crops
1T	Tree Crop
1S	Tea Crop
1HSs	Small Herbaceous Crops in sloping land
1HLMv	Large to Medium Herbaceous Crops in valley floor
1HSv	Small Herbaceous Crops in valley floor
2HCO	Closed to Open Herbaceous
2HS	Sparse Herbaceous
2HS//6BR	Sparse Herbaceous OR Bare Rock
2HCO//1H	Closed to Open Herbaceous OR Rainfed Herbaceous Crops
2SCO	Closed to Open Shrubs
2SS	Sparse Shrubs with Sparse Herbaceous
2SSd	Sparse Dwarf Shrubs with Sparse Herbaceous
2SOd	Open Dwarf Shrubs with Sparse Herbaceous
2TCOne//2TCObe	Closed to Open Needleleaved Trees OR Closed to Open Broadleaved Trees
2TCOne	Closed to Open Needleleaved Trees
2TCObe	Closed to Open Broadleaved Trees
2TSne//2TSbe	Sparse Needleleaved Trees OR Sparse Broadleaved Trees
2TSne	Sparse Needleleaved Trees
2TSbe	Sparse Broadleaved Trees
4HCOp	Closed to Open Permanently Flooded Herbaceous
4SCOs	Closed to Open Seasonally Flooded Shrubs
5UI	Urban and Industrial Areas
6BR	Bare Rock
6S	Bare Soil
6GR	Rock Debris
8ICE	Glacier
8ICEr	Rocky Glacier
8SN	Perennial Snow
8SNs	Seasonal Snow
8WNP	Non-Perennial Lakes
8WBS	Bare Soil in seasonally flooded area
8WP	Lakes
8WF	Rivers

References

Colditz RR, Llamas RM, Ressl RA (2014a) Detecting change areas in Mexico between 2005 and 2010 using 250 m MODIS images. IEEE J Sel Top Appl Earth Obs Remote Sens 7:3358–3372. https://doi.org/10.1109/JSTARS.2013.2280711

Colditz RR, Llamas RM, Ressl RA (2014c) Annual land cover monitoring using 250M MODIS data for Mexico. Int Geosci Remote Sens Symp: 4664–4667. https://doi.org/10.1109/IGARSS.2014b.6947533

Colditz RR, López Saldaña G, Maeda P et al (2012) Generation and analysis of the 2005 land cover map for Mexico using 250m MODIS data. Remote Sens Environ 123:541–552. https://doi.org/10.1016/j.rse.2012.04.021

Colditz RR, Pouliot D, Llamas RM et al (2014c) Detection of North American land cover change between 2005 and 2010 with 250m MODIS Data. Photogram Eng Remote Sens 80:918–924

Gebhardt S, Wehrmann T, Ruiz MAM et al (2014) MAD-MEX: automatic wall-to-wall land cover monitoring for the mexican REDD-MRV program using all landsat data. Remote Sens 6:3923–3943. https://doi.org/10.3390/rs6053923

Hojas-Gascon L, Eva HD, Gobron N, Simonetti D, Fritz S (2012) The application of medium-resolution MERIS satellite data for continental land-cover mapping over South America results and caveats. Remote Sens Land Use Land Cover Princip Appl. https://doi.org/10.1201/b11964-27

Homer C, Dewitz J, Yang L et al (2015) Completion of the 2011 national land cover database for the conterminous United States—representing a decade of land cover change information. Photogramm Eng Remote Sensing 81:345–354. https://doi.org/10.1016/S0099-1112(15)30100-2

Jin S, Homer C, Yang L et al (2019) Overall methodology design for the United States national land cover database 2016 products. Remote Sens 11. https://doi.org/10.3390/rs11242971

Jin S, Yang L, Danielson P, Homer C, Fry J, Xian G (2013) A comprehensive change detection method for updating the National Land Cover Database to circa 2011. Remote Sens Environ 132:159–175. https://doi.org/10.1016/j.rse.2013.01.012

Latifovic R, Homer C, Ressl R et al (2012) North American land change monitoring system. In: Giri C (ed) Remote sensing of land use and land cover: principles and applications. CRC Press, pp 303–324

Latifovic R, Pouliot D, Olthof I (2017) Circa 2010 land cover of Canada: local optimization methodology and product development. Remote Sens 9. https://doi.org/10.3390/rs9111098

Stone TA, Schlesinger P, Houghton RA, Woodwell GM (1994) A map of the vegetation of South America based on satellite imagery. Photogram Eng Remote Sensing 60:541–551

Yang L, Jin S, Danielson P et al (2018) A new generation of the United States National Land Cover Database: requirements, research priorities, design, and implementation strategies. ISPRS J Photogram Remote Sens 146:108–123. https://doi.org/10.1016/j.isprsjprs.2018.09.006

Global Thematic Land Use Cover Datasets Characterizing Vegetation Covers

David García-Álvarez and Javier Lara Hinojosa

Abstract

Vegetation covers were one of the first land covers to receive special attention when thematic Land Use Cover (LUC) maps first appeared. Interest in this subject has remained strong since then because of the valuable information that these datasets provide for monitoring forests, deforestation and climate change, among other issues. A wide variety of thematic LUC datasets characterizing vegetation covers are currently available. In this chapter, we review eleven of these datasets, most of which provide long series of LUC maps, so permitting the study of LUC change. In thematic terms, most of the maps provide information on the vegetation or tree cover fraction per pixel, so characterizing the vegetation covers on Earth in great detail. A specific dataset has been found that maps mangrove distribution across the globe at 30 m for one date (1997/00). It is not included in this review because of its high specificity, which means it is only of interest to certain communities of users. Of all the products reviewed here, the World's Forests 2000 is probably the most basic, providing information about three wooded cover categories for the year 1995/96 at a spatial resolution of 1 km. SYNMAP is a very specific thematic map designed to meet the needs of the carbon cycle and vegetation modelling community, which was produced at a spatial resolution of 1 km and with a legend of 48 categories. Among the maps providing information on the fraction of vegetation cover per pixel, the Hybrid Forest Mask 2000 (1 km) and the PTC Global Version (500 m– 1 km) offer relatively coarse resolutions and few points in time: just one date in the former (2000) and two in the latter (2003, 2008). The Forests of the World 2010 is also available for just one year (2010), albeit at a more detailed spatial resolution (250 m). Various datasets provide information on the cover fraction for long periods of time at medium and high spatial resolutions. FCover provides the longest time series (1999-present) at 1 km, although since 2014 this dataset is also available at 300 m. Modis VCF also offers a long data series (2000–2019) at a spatial resolution of 250 m. MEaSUREs Vegetation Continuous Fields (VCF) is another thematic LUC dataset providing information on the tree cover fraction of the earth surface for a very long time period: 1982–2016. However, it is not reviewed here because of its coarse spatial resolution (around 5.6 km at the Equator). At very detailed spatial resolutions, GFCC30TC Landsat VCF (30 m) provides data on the cover fraction for four different points in time, between 2000 and 2015. It also gives information on forest change for two periods (1990–2000/2000–2005) through the associated GFCC30FCC dataset. The Hansen forest map (30 m) also provides one of the longest time series, from 2000 to 2019. Global FNF is the dataset with the highest resolution (25 m) of all those reviewed. It is available for two periods of time: 2007–2010 and 2015–2017. In thematic terms, however, this dataset is less detailed, in that it only differentiates between forest and non-forest covers. TanDEM-X Forest/Non-Forest also provides information on the forest extent at high spatial resolution (50 m). However, the map is only available for one point in time. Like Global FNF, it was also obtained from the classification of radar data.

Keywords

Vegetation • Wood • Tree cover • World's Forests 2000 • FCover • Hybrid Forest Mask 2000 • SYNMAP • GFCC30TC Landsat VCF • GFCC30FCC • Hansen Forest Map • MODIS VCF • PTC Global Version • Global FNF • Forests of the World 2010 • TanDEM-X Forest/Non-Forest Map

D. García-Álvarez (✉)
Departamento de Geología, Geografía y Medio Ambiente, Universidad de Alcalá, Alcalá de Henares, Spain
e-mail: David.garcia@uah.es

J. Lara Hinojosa
Departamento de Análisis Geográfico Regional y Geografía Física, Universidad de Granada, Granada, Spain

D. García-Álvarez et al. (eds.), *Land Use Cover Datasets and Validation Tools*,
https://doi.org/10.1007/978-3-030-90998-7_19

1   The World's Forests 2000

Product
LULC thematic
Dates
1995 / 96
Formats
Raster
Pixel size
1 km
Theme
3 forest categories out of 6
Extent
Global
Updating
No
Change detection
No (only one date)
Overall accuracy
Expected to be >80%

Website of reference	Website Language English
http://www.fao.org/forest-resources-assessment/past-assessments/fra-2000/en/	

Download site
http://www.fao.org/geonetwork/srv/en/main.home?uuid=b9f2ee20-88fd-11da-a88f-000d939bc5d8

Availability	Format(s)
Open Access	.adf

Technical documentation
FAO (2000, 2001)

Other references of interest
–

Project

The World's Forests 2000 map was one of the products generated within the context of the Global Forest Resources Assessment (FRA) for the year 2000. FRA is a project run by the Food and Agriculture Organization (FAO) that dates back to the year 1946. A new edition is issued every five years on average.

The project, which is carried out in collaboration with the different countries that form part of the FAO, aims to assess the state of the world's forests and understand the changes that they undergo over time. Satellite imagery and remote sensing techniques were used for the first time in the FRA2000 survey. A global map of forests was produced as part of the project. The U.S. Geological Survey (USGS) EROS Data Center (EDC) was in charge of map production. Two extra maps were also produced as part of the project: an ecological zoning map and a map of protected forests.

Production method

The World's Forests 2000 map was produced in two stages. In the first stage, closed forest and open or fragmented forest categories were mapped on the basis of a classification of AVHRR imagery for the period 1995–1996. A complex methodology based on a mixture analysis model and a geographical stratification to account for regional variation in the mapped features was employed to calculate the fraction cover per pixel. The two LUC categories were extracted from these layers based on the tree cover percentages defined by the FAO: 40–100% for closed forest and 10–40% for open or fragmented forest.

In the second stage, the Global Land Cover Characteristics Database (GLCC), obtained from a classification of AVHRR imagery for the period 1992/93, was used to map the remaining categories: other wooded land, other land cover and water. The fact that the different input data (AVHRR and GLCC) had different reference dates led to temporal inconsistency between forest and non-forest categories.

Some auxiliary datasets were also used in the production of the map, such as ecoregion maps and digital elevation models. These helped to merge and split the different categories being mapped.

Product description

The map can be downloaded as a zipped file containing the raster with the LUC information and other auxiliary information. The download includes two versions of the LUC map, one classifying the land covers in a range of values from 1 to 6 and the other classifying the land covers in a range of values from 100 to 600.

Downloads

The World's Forests 2000
– Raster file with LUC map (for_2000)
– Raster file with LUC map (info, forest)
– Preview image of the product
– ArcGIS file (.avl) with symbology for the raster

Legend and codification

Code	Label	Code	Label
1/100	Closed forest	4/400	Other land cover
2/200	Open or fragmented forest	5/500	Water
3/300	Other wooded land	6/600	Undefined

2 FCover—Fraction of Green Vegetation Cover

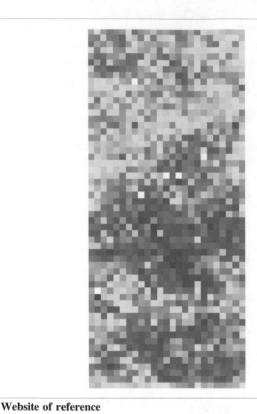

Product	
LULC thematic	
Dates	
Every 10 days from 1999 to 2020 (1 km) Every 10 days from 2014 to the present (300 m)	
Formats	
Raster	
Pixel size	
300 m,	
1 km	
Theme	
Percentage of vegetation cover	
Extent	
Global	
Updating	
Expected, but no specific date	
Change detection	
Supported via specific layers of forest change	
Overall accuracy	
Not specified	

Website of reference	Website Language English
https://land.copernicus.eu/global/products/fcover	

Download site

https://land.copernicus.vgt.vito.be/PDF/portal/Application.html#Browse;Root=512260;Collection=1000061;Time=NORMAL,NORMAL,-1,,,-1,,, (300 m)
https://land.copernicus.vgt.vito.be/PDF/portal/Application.html#Browse;Root=512260;Collection=1000081;Time=NORMAL,NORMAL,1, JANUARY,2014,31,DECEMBER,2020;isReserved=false (1 km)

Availability	Format(s)
Open Access under registration	.nc

Technical documentation

Baret et al. (2016), Jolivet (2020), Lacaze et al. (2020), Martínez-Sánchez and Sánchez-Zapero (2020), Ramon et al. (2020), Sánchez-Zapero et al. (2018), Smets et al. (2018), Toté and Tansey (2020), Verger (2020), Wolfs et al. (2020)

Other references of interest

–

Project

The Fraction of Vegetation Cover (FCover) is a product developed as part of the Copernicus programme, which is led and coordinated by the European Commission. The Copernicus Global Land Service (CGLS) aims to provide bio-geophysical land information to monitor the status and evolution of land surface across the globe. FCover provides information on the fraction of the ground surface that is covered by green vegetation.

FCover is jointly produced with two other products, which also help to characterize the vegetation cover on Earth: the Leaf Area Index (LAI) and the Fraction of Absorbed Photosynthetically Active Radiation (FAPAR). All three were initially produced at a spatial resolution of 1 km, although a finer version of the product has recently been developed at 300 m. There are two versions of the 1 km product. The second version is an improved version of the first.

Production method

FCover is obtained after processing satellite imagery using a neuronal networks method, which has been successively improved in the different versions of the product.

PROVA-V imagery is used to create the product with a spatial resolution of 300 m. The product at 1 km also makes use of imagery from the VEGETATION sensor to increase the coverage over time. In both cases, various different techniques (smoothing, gap filling and temporal compositing) are applied to ensure the temporal consistency of the product time series.

Product description

The different versions of FCover at spatial resolutions of 1 km and 300 m can be downloaded from the same website.

In all cases, the product is distributed in single files covering the whole world for each period of 10 days.

The product is delivered in the same format regardless of the particular version and/or spatial resolution chosen. It contains a raster with the LUC information, a preview picture of the product and technical information regarding the creation process. The raster includes information on the vegetation cover fraction, plus a series of technical parameters: uncertainty on the FCover, a quality flag, etc.

Downloads

FCover 300 m/1 km
– Raster file with LUC map in netCDF4 format (.nc) – A metadata file (.xml) – Preview image of the product (.tiff) – PDFs with technical information about the product

Legend and codification

Code	Label
0–100	Vegetation fraction cover (0–1.0)

Practical considerations

This is a thematically rich, complex product that some users may find hard to understand at first glance. Nonetheless, the product's website includes all the relevant information to enable users to apply the product correctly and understand its characteristics. We therefore recommend users to visit the website before taking a look at the technical documents.

3 Hybrid Forest Mask 2000

Product
LULC thematic
Dates
2000
Formats
Raster
Pixel size
1 km
Theme
Percentage of forest cover
Extent
Global
Updating
Not expected
Change detection
No (only one date)
Overall accuracy
Expected to be >=85% and up to 93%

Website of reference	Website Language English
Not available	

Download site

https://application.geo-wiki.org/Application/index.php

Availability	Format(s)
Open Access under registration	.tiff,.img

Technical documentation

Schepaschenko et al. (2015)

Other references of interest

FAO (2010)

Project

Researchers from several institutions across the world joined this project to produce a forest mask for the reference year 2000 by data fusion. The purpose was to create a new LUC map that charted the extent of forests at a global level and outperformed previous maps of a similar nature. The resulting map is consistent with FAO national forest statistics.

This is one of many projects that have benefited from the Geo-Wiki platform through which crowdsourced data were collected for use in the production of the map.

Production method

The forest map was produced by merging different LUC databases at global (GLC2000, GLCNMO, GlobCover, MODIS LC, MODIS VCF, Landsat VCF, Hansen Forest map) and regional (Congo Basin forest types map, Brazil PRODES forest mask, ALUM, Pan-European Forest/Non-Forest Map, NLCD 2006, Land cover of Russia, Forest mask for European Russia) scales. Although the reference year for the Hybrid Forest Mask is 2000, many of the input maps refer to different years.

The input maps were combined using a Geographical Weighted Regression (GWR) algorithm that produced two intermediate layers: a map of forest probability and a map of percentage forest cover. Reference points collected through crowdsourcing campaigns were used to train the GWR algorithm and validate the maps obtained.

From the two intermediate layers obtained, three maps were finally created. The first map indicates the percentage of forest cover in pixels with a probability of being forest of more than 0.5. For the second map, the pixels with the highest probability of being forest were selected until the number of pixels determined according to the FAO FRA national statistics were reached. The third map was obtained by repeating the same procedure using regional statistics.

Product description

Each of the three maps produced by this project can be independently downloaded. In all cases, the download contains just one file about the LUC layer, with no auxiliary information.

Downloads

Hybrid Forest mask 2000–Best guess/FAO FRA national statistics/FAO FRA regional statistics
– Raster file with information on tree canopy cover for the year 2000

Legend and codification

Code	Label
0–100	Forest Coverage (0–100%)
128	Non forest cover

Practical considerations

The maps can be accessed online through the viewer included in the Geo-Wiki platform. Users should be aware that although the reference year for the product is 2000, it was obtained by merging products with different reference years. This map is therefore unsuitable for land change analysis.

4 SYNMAP Global Potential Vegetation

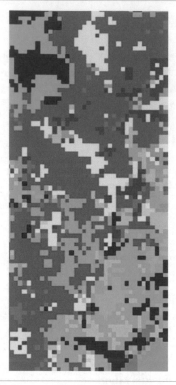

Product
LULC thematic
Dates
2000
Formats
Raster
Pixel size
1 km
Theme
43 vegetation categories out of 48
Compatible legends
GLCC, GLC2000, MODIS
Extent
Global
Updating
Not expected
Change detection
No (only one date)
Overall accuracy
Not specified

Website of reference	Website Language English, Spanish
https://databasin.org/datasets/112a942ec4294e5284e63d5e6bf14b29	

Download site

https://webmap.ornl.gov/wcsdown/dataset.jsp?ds_id=10024

Availability	Format(s)
Open Access under registration	.nc, .tiff, .xyz, .nitf, .img, .asc

Technical documentation

Jung et al. (2006)

Other references of interest

–

Project

SYNMAP is a dataset produced by German researchers from the University of Jena. It was developed to meet the requirements of carbon cycle and vegetation models. To this end, all the classes in the dataset were defined in terms of plant functional type mixtures, with information about the type of tree leaf and its longevity. The dataset was obtained by merging data from existing global LUC products.

Production method

SYNMAP was obtained by merging GLCC, MODIS Land Cover and GLC2000. From GLCC and MODIS Land Cover, two different classification schemes were used: USGS and IGBPP for GLCC and PFT and IGBP for MODIS Land Cover. The tree classes obtained after merging the previous maps were complemented with information about leaf type and phenology from AVHRR-CFTC (Continuous Fields of Tree Cover).

A specific legend adapted to the requirements of the carbon cycle and vegetation modelling communities was developed for SYNMAP. Each class in the new map was linked with each class in the input datasets through three affinity scores: one for life forms, one for leaf type and one for leaf longevity. AVHRR-CFTC provided auxiliary data regarding leaf attributes. The different maps were combined using fuzzy agreement to define the classes for the new map.

Product description

SYNMAP can be downloaded in multiple formats via a web application. Users must select the product corresponding to their geographical area of interest. The product is downloaded in the form of a raster file with LUC information.

Downloads

SYNMAP
– Raster file with LUC map

Legend and codification

Code	Label	Code	Label
0	Water	24	Mixed-broadleaf-trees and grasses
1	Evergreen-needle-trees	25	Evergreen-mixed-trees and grasses
2	Deciduous-needle-trees	26	Deciduous-mixed-trees and grasses

(continued)

Code	Label	Code	Label
3	Mixed-needle–trees	27	Mixed-trees and grasses
4	Evergreen-broadleaf-trees	28	Evergreen-needle-trees and crops
5	Deciduous-broadleaf-trees	29	Deciduous-needle-trees and crops
6	Mixed-broadleaf–trees	30	Mixed-needle-trees and crops
7	Evergreen-mixed-trees	31	Evergreen-broadleaf-trees and crops
8	Deciduous-mixed-trees	32	Deciduous-broadleaf-trees and crops
9	Mixed–trees	33	Mixed-broadleaf-trees and crops
10	Evergreen-needle-trees and shrubs	34	Evergreen-mixed-trees and crops
11	Deciduous-needle-trees and shrubs	35	Deciduous-mixed-trees and crops
12	Mixed-needle-trees and shrubs	36	Mixed-trees and crops
13	Evergreen-broadleaf-trees and shrubs	37	Shrubs
14	Deciduous-broadleaf-trees and shrubs	38	Shrubs and grasses
15	Mixed-broadleaf-trees and shrubs	39	Shrubs and crops
16	Evergreen-mixed-trees and shrubs	40	Shrubs and barren
17	Deciduous-mixed-trees and shrubs	41	Grasses
18	Mixed-trees and shrubs	42	Grasses and crops
19	Evergreen-needle-trees and grasses	43	Grasses and barren
20	Deciduous-needle-trees and grasses	44	Crops
21	Mixed-needle-trees and grasses	45	Barren
22	Evergreen-broadleaf-trees and grasses	46	Urban
23	Deciduous-broadleaf-trees and grasses	47	Snow and ice

Practical considerations

SYNMAP was designed to satisfy the needs of a very specific community: carbon cycle and vegetation modellers. The dataset can be consulted online via a web application.[1]

---

[1] https://databasin.org/maps/new#datasets=112a942ec4294e5284e63 d5e6bf14b29.

5 GFCC—Global Forest Cover Change (GFCC30TC and GFCC30FCC)

Product	
LULC thematic	
Dates	
2000, 2005, 2010, 2015 (tree cover) 1990–2000, 2000–2005 (forest change)	
Formats	
Raster	
Pixel size	
30 m MMU Forest change: 0.27 ha	
Theme	
Percentage of tree cover and forest gains / losses	
Extent	
Global	
Updating	
Expected, but no date specified	
Change detection	
Yes, by comparing tree cover layers or though layer of forest changes	
Overall accuracy	
Expected to be >88–90%	

Website of reference	**Website Language** English

https://lpdaac.usgs.gov/products/gfcc30tcv003/
https://lpdaac.usgs.gov/products/gfcc30fccv001/

Download site

https://lpdaac.usgs.gov/products/gfcc30tcv003/
https://lpdaac.usgs.gov/products/gfcc30fccv001/

Availability	**Format(s)**
Open Access	.tiff

Technical documentation

Sexton et al. (2013, 2016a, b)

Other references of interest

–

Project

Global Forest Cover Change (GFCC) is a suite of products at 30 m providing information about tree cover, forest cover change, water cover and surface reflectance. The last two products are auxiliary datasets used in the production of the first two: the GFCC Tree Cover Multi-Year (GFCC30TC) and the GFCC Forest Cover Change Multi-Year (GFCC30FCC).

These datasets were developed by the Department of Geographical Sciences of the University of Maryland and form part of the NASA Making Earth System Data Records for Use in Research Environments (MEaSUREs). They aim to provide reference information for environmental monitoring and forest assessment at a global scale.

The aim of GFCC was to overcome the limitations imposed by the coarse resolution of the MODIS VCF dataset, as many forest changes take place at finer scales than 250 m. To this end, GFCC rescales at 30 m the information provided by the MODIS VCF dataset, which is described later on in this chapter.

The Tree Cover layer is also known as the Landsat Vegetation Continuous Fields (VCF) and was initially launched in 2013, with updates continuing until 2016. It describes the state of changes in the tree cover. Forest Cover Change focuses on forest covers and their changes. It was created from the Tree Cover layer, and there is only one edition.

Production method

The GFCC Tree Cover Multi-Year Global 30 m (GFCC30TC) was obtained by applying a model to Landsat reflectance imagery to rescale the MODIS VCF Tree Cover Layer at 30 m. The model consisted of a piecewise linear function of surface reflectance and temperature. Although Landsat imagery was available prior to the year 2000, the Tree Cover layer is only available for the reference years 2000, 2005, 2010 and 2015. This is because of the timeframe covered by MODIS VCF (2000–2019), which is essential for producing the dataset.

In the latest version of the product, the entire Landsat imagery archive was employed to obtain the dataset, whereas in the initial versions the Landsat Global Land Survey collection was used. In addition, a water mask, specifically created from Landsat imagery through a classification-tree model, was used in a post-classification step as an auxiliary dataset for generating the Tree Cover layer.

The layers of forest change (GFCC30FCC) were independently produced for each of the periods available (1990–2000 and 2000–2005) from the Tree Cover layer. First, forest areas were extracted by applying a specific threshold to the Tree Cover Layer. Then, four change categories were defined for the period 2000–2005 based on changes in the Tree Cover layer: stable forest, stable non-forest, forest gain and forest loss. To calculate the change for the period 1900–2000, a specific forest cover layer was obtained for 1990 from Landsat imagery based on a classification-tree algorithm.

Product description

GFCC30TC and GFCC30FCC are distributed as two independent products. Users can download the two datasets through four different servers or tools: Data Pool,[2] NASA Earthdata Search,[3] USGS EarthExplorer[4] and DAAC2Disk Utility.[5]

The datasets are distributed in tiles. Users must therefore download the tiles that cover their area of interest. The online viewers provided in the NASA Earthdata Search and USGS EarthExplorer tools are very useful for this purpose. The Data Pool option also includes a preview image of the tile as part of the download.

---

[2] https://lpdaac.usgs.gov/tools/data-pool/.
[3] https://lpdaac.usgs.gov/tools/earthdata-search/.
[4] https://lpdaac.usgs.gov/tools/usgs-earthexplorer/.
[5] https://lpdaac.usgs.gov/tools/daac2diskscripts/.

Downloads

GFCC30TC

– Raster file with the tree cover percentage per pixel
– Raster file with information about the LUC map error

GFCC30FCC

– Raster file with classes of forest change
– Raster file with forest change probability

Legend and codification

GFCC30TC-Tree Cover

Code	Label	Code	Label
0–100	Percent of pixel area covered by tree cover (0–100)	211	Shadow
200	Water	220	Fill Value
210	Cloud		

GFCC30FCC-Forest Cover Change Map

Code	Label	Code	Label
0	No Data	11	Persistent Forest
2	Shadow	19	Forest Loss
3	Cloud	91	Forest Gain
4	Water	99	Persistent Non-forest

GFCC30FCC-Forest Cover Change Probability

Code	Label
0–100	Probability (0–100%) of forest change

GFCC30TC-Tree Cover

Code	Label	Code	Label
0–100	Percent of pixel area covered by tree cover (0–100)	211	Shadow
200	Water	220	Fill Value
210	Cloud		

GFCC30FCC-Forest Cover Change Map

Code	Label	Code	Label
0	No Data	11	Persistent Forest
2	Shadow	19	Forest Loss
3	Cloud	91	Forest Gain
4	Water	99	Persistent Non-forest

GFCC30FCC-Forest Cover Change Probability

Code	Label
0–100	Probability (0–100%) of forest change

6 Hansen Forest Map—Global Forest Change 2000–2019

Product	
LULC thematic	
Dates	
2000–2019	
Formats	
Raster	
Pixel size	
30 m	
Theme	
Percentage of tree cover and forest gains / losses	
Extent	
Global	
Updating	
Expected, but no date specified	
Change detection	
Supported through specific layers of forest gains and losses	
Overall accuracy	
Not specified	

Website of reference	Website Language English
https://earthenginepartners.appspot.com/science-2013-global-forest/download_v1.7.html	

Download site
https://earthenginepartners.appspot.com/science-2013-global-forest/download_v1.7.html

Availability	Format(s)
Open Access	.tiff

Technical documentation
Hansen et al. (2013)

Other references of interest
Hansen et al. (2014)

Project

The Hansen forest map was named after the researcher leading the project that produced the dataset: Matthew Hansen, from the University of Maryland. Notwithstanding this, the project is the result of collaboration between scientists from various US institutions, including the USGS.

The database was initially released in 2013. Since then, it has been revised and improved on several occasions. The latest published version of the product is Version 1.7, which included significant improvements on the previous version. This is expected to be the first step towards the creation of Version 2.0 of the product.

Production method

Landsat imagery was pre-processed and classified using the Google Earth Engine to create the Hansen forest map. A decision tree classifier was used to independently produce the base forest map and the yearly maps of forest lost. For classification purposes, all vegetation taller than 5 m in height was considered to be a tree. Forest loss was defined as a stand-replacement disturbance.

Product description

The Hansen Global Forest Change dataset is made up of multiple layers. The base layer (treecover2000) provides information on forest cover across the world for the year 2000. Two other layers (gain, lossyear) help to interpret the changes in forest cover since 2000 by identifying both the areas where new forest cover has appeared during this period and the areas in which forest cover has been lost. Forest cover losses are disaggregated per year.

The product also includes an auxiliary layer which identifies the mapped areas, the water bodies and the areas with no data. Cloud-free composites of Landsat imagery for the product's first and last years (2000 and 2019) are also provided together with the LUC layers.

The map is distributed in tiles. For this purpose, the world is divided into equal-size areas of 10 × 10 degrees.

Downloads

Tree canopy cover for year 2000 (treecover 2000)
– Raster file with information on tree canopy cover for the year 2000

Global forest cover gain 2000–2012 (gain)
– Raster file with information about gains in forest cover

Year of gross forest cover loss event (lossyear)
– Raster file with information about the loss of forest cover

Data mask (datamask)
– Raster file indicating the areas with no data, water surfaces and mapped land surface

Legend and codification

Tree canopy cover for year 2000 (treecover 2000)	
Code	Label
0–100	Tree cover area density (1–100)

Global forest cover gain 2000–2012 (gain)			
Code	Label	Code	Label
0	Forest no gain	1	Forest gain

Year of gross forest cover loss event (lossyear)			
Code	Label	Code	Label
0	No forest loss	10	Forest loss in 2010
1	Forest loss in 2001	11	Forest loss in 2011
2	Forest loss in 2002	12	Forest loss in 2012
3	Forest loss in 2003	13	Forest loss in 2013
4	Forest loss in 2004	14	Forest loss in 2014
5	Forest loss in 2005	15	Forest loss in 2015
6	Forest loss in 2006	16	Forest loss in 2016
7	Forest loss in 2007	17	Forest loss in 2017
8	Forest loss in 2008	18	Forest loss in 2018
9	Forest loss in 2009	19	Forest loss in 2019

Data mask (datamask)	
Code	Label
0	No data
1	Mapped land surface
2	Water bodies

Practical considerations

The dataset can be easily visualized and consulted through a web-based visualization tool.[6] For those who want to work with data for the whole Earth rather than for specific areas of the world (tiles), the producers provide txt files with a full list of download links for each of the 6 layers that make up the product.

Landsat 8 imagery enabled better detection and mapping of forest disturbance. Some uncertainties may therefore emerge when comparing forest losses before and after the inclusion of Landsat 8 imagery.

---

[6] http://earthenginepartners.appspot.com/science-2013-global-forest.

7 MODIS Vegetation Continuous Fields—MOD44B

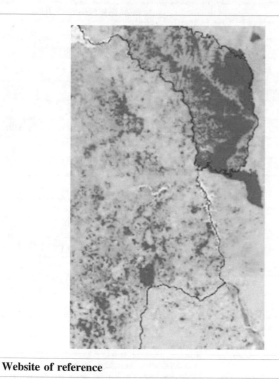

Product	
LULC thematic	
Dates	
2000–2019	
Formats	
Raster	
Pixel size	
250 m	
Theme	
Percentage of tree cover	
Extent	
Global	
Updating	
Expected	
Change detection	
Yes	
Overall accuracy	
Not specified	

Website of reference	**Website Language** English
https://lpdaac.usgs.gov/products/mod44bv006/ https://modis.gsfc.nasa.gov/data/dataprod/mod44.php	

Download site

https://lpdaac.usgs.gov/products/mod44bv006/

Availability	**Format(s)**
Open Access under registration	.hdf

Technical documentation

Hansen et al. (2003a, b), Townshend et al. (2011)

Other references of interest

Amarnath et al. (2017), Hansen et al. (2005), Jeganathan et al. (2009)

Project

The MODIS Vegetation Continuous Fields (VCF), also known as MOD44B, is a thematic LUC database developed by the Department of Geographical Sciences of the University of Maryland. This dataset was created in order to overcome the limitations of categorical LUC data for which it was necessary to define specific thresholds when characterizing vegetation cover. The team from the University of Maryland later applied Landsat imagery to produce a VCF product at finer spatial resolutions, so improving the quality of the information provided by this dataset.

The dataset was initially launched in 2003. Since then, several versions of the product have been produced, each making an improvement on its predecessors. The last version of the product was launched in 2015 (v6). Versions 1 to 3 of the dataset were produced at a spatial resolution of 500 m. Subsequent versions were produced at 250 m.

Production method

MODIS VCF was obtained from MODIS imagery and other MODIS-related products, such as the MODIS Global 250 m Land/Water Map. A regression tree model was applied to the imagery to obtain the MODIS VCF dataset. The model was applied through open-access and other software customized for the production of the dataset.

Product description

MOD44B can be downloaded from different servers or tools, including AppEEARS, Data Pool, Nasa Earthdata Search, USGS EarthExplorer and OPeNDAP. In all cases, the product is distributed in tiles. Users must select their area of interest.

The download consists of a single raster file made up of multiple bands, each one showing different information: percent of tree cover, percent of non-tree vegetation, percent of non-vegetation covers, and three extra bands with technical and quality information about the product.

Downloads

Single mosaic

– Raster file with multiple bands, including LUC and data quality information

Legend and codification

Percent Tree Cover

Code	Label
0–100	Percent tree cover (0–100)
200	Water
253	Fill/Outside of projection

Percent Non-tree vegetation

Code	Label
0–100	Percent non-tree vegetation of each pixel (1–100)
200	Water
253	Fill/Outside of projection

Percent Non-vegetation cover

Code	Label
0–100	Percent with no vegetation of each pixel (1–100)
200	Water
253	Fill/Outside of projection

Percent Tree Cover Standard Deviation (SD)

Code	Label
0–10,000	Percent with standard deviation as regards Percent Tree Cover layer (1–10,000)

Percent Non-vegetation Standard Deviation (SD)

Code	Label
0–10,000	Percent with standard deviation as regards Percent Non-vegetation (1–10,000)

Practical considerations

Users must bear in mind that although the dataset is distributed as a single raster file, this includes multiple layers with different, complementary information. Nonetheless, the core of the product is the band storing information about the percentage of tree cover. The dataset can be also consulted online through a Web Map Service (WMS).[7]

---

[7] https://lpdaacgis.cr.usgs.gov/arcgis/rest/services/WMS?f=pjson.

8 PTC Global Version—Percent Tree Cover Global Version

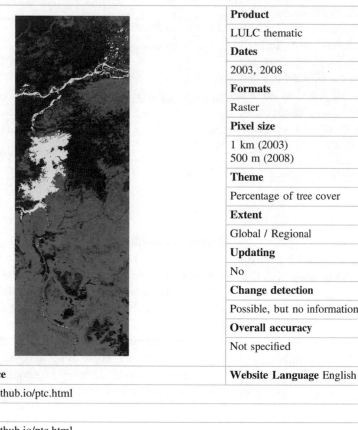

Product
LULC thematic
Dates
2003, 2008
Formats
Raster
Pixel size
1 km (2003)
500 m (2008)
Theme
Percentage of tree cover
Extent
Global / Regional
Updating
No
Change detection
Possible, but no information is available regarding its uncertainty
Overall accuracy
Not specified

Website of reference	**Website Language** English
https://globalmaps.github.io/ptc.html	

Download site
https://globalmaps.github.io/ptc.html

Availability	**Format(s)**
Open Access	.tiff

Technical documentation
–

Other references of interest
–

Project

The Percent Tree Cover Global version is a dataset created within the context of the Global Mapping Project, which aimed to create a global reference database of geospatial information. The project was promoted by the International Steering Committee for Global Mapping (ISCGM) in cooperation with National Geospatial Information Authorities (NGIAs) from different countries and regions across the world. It came to an end in 2016, when the ISCGM decided to wind up the project and transfer all the data to the Geospatial Information Section of the United Nations.

The PTC map was generated by a group of researchers from the Geospatial Information Authority of Japan (GSI) and Chiba University. Two versions of the map were produced: one for the reference year 2003 and another for the reference year 2008.

Production method

The map was obtained via the classification of MODIS imagery. No other information is available about how the PTC Global version was produced.

Product description

A single download containing the map for the entire globe is available for the year 2003. For the year 2008, the map is distributed in 12 different tiles. Each tile covers an area of 90 degrees of latitude and 60 degrees of longitude. The downloads only include the raster files with LUC information. There are no auxiliary data.

Downloads

PTC Global 2003/2008
– Raster file with global tree cover

Legend and codification

Code	Label
0–100	Tree Coverage (0–100%)
254	Water bodies
255	No data

Practical considerations

This dataset lacks auxiliary and technical information about specific characteristics and possible limitations, including data about its accuracy. It must therefore be used with caution.

General information about the Global Mapping Project can be found at https://www.gsi.go.jp/kankyochiri/gm_report_e.html. More information about the project within which the dataset was created can be found at this website.

9 FNF—Global Forest Non-Forest Map

Product
LULC thematic
Dates
2007, 2008, 2009, 2010, 2015, 2016, 2017
Formats
Raster
Pixel size
25 m, 100 m, 1 km, 0.25°
Theme
Forest extent
Extent
Global
Updating
Expected
Change detection
Possible, but no information available about its uncertainty
Overall accuracy
Expected to be > 84%

Website of reference	Website Language English
https://www.eorc.jaxa.jp/ALOS/en/dataset/fnf_e.htm	

Download site
https://www.eorc.jaxa.jp/ALOS/en/palsar_fnf/registration.htm

Availability	Format(s)
Open Access under registration	.hdr

Technical documentation
JAXA and EORC (2019), Shimada et al (2014)

Other references of interest
Altunel et al. (2020)

Project

The Global Forest Non-Forest map (FNF) is one of the datasets produced by the Earth Observation Research Center (EORC) and the Japan Aerospace Exploration Agency (JAXA) as part of the ALOS-2/ALOS Science Project. The project is responsible for the ALOS satellites (ALOS and ALOS-2) and the datasets obtained from them.

The FNF map aims to provide a reference dataset for the study of deforestation and forest degradation. As the map is obtained from imagery captured by Synthetic Aperture Radar (SAR) sensors (PALSAR and PALSAR-2), it can monitor forest changes regardless of the weather conditions, which is especially useful when monitoring tropical forests.

Production method

The main source of information for the FNF map is imagery from the PALSAR and PALSAR-2 sensors, on board the ALOS and ALOS-2 satellites. As these sensors are radar sensors, image classification is based on backscattering intensity values. Different parameters for classification are used depending on the region under consideration and its characteristics.

The original map is produced at 25 m and later generalized at coarser resolutions: 100 m, 1 km and 0.25°. Following the FAO definition, those areas of more than 0.5 ha covered by trees with a canopy cover of over 10% are considered to be forest.

Product description

Users can download the FNF map for each available year at different spatial resolutions. However, the map at 25 m is the only one available for all the different years covered by the product.

Whereas the maps at 1 km and 0.25° can be downloaded as a single file covering all the globe, the FNF map at higher resolutions (25 m, 100 m) is split into different tiles to facilitate downloading. Users can download the tile for their particular area of interest. All downloads include the FNF map for the selected area as well as the satellite imagery used to obtain it.

Downloads

Global Forest Non-Forest map (FNF)—25 m/100 m/1 km/0.25°
– Raster file with LUC map
– Raster files with satellite imagery

Legend and codification

Global Forest Non-Forest map (FNF)—25 m

Code	Label	Code	Label
0	No Data	2	Non-forest
1	Forest	3	Water

Global Forest Non-Forest map (FNF)—100 m

Code	Label	Code	Label
1	Water	5	Forest (26–50%)
3	Non-forest (0–9%)	6	Forest (51–75%)
4	Forest (10–25%)	7	Forest (76–100%)

Global Forest Non-Forest map (FNF)—1 km / 0.25°

Code	Label
0–100	Forest Coverage (0–100%)
200	Water
255	No Data

10 Forests of the World 2010

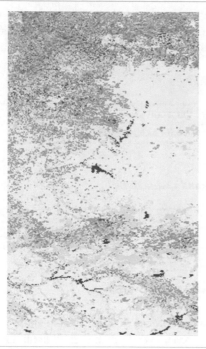

Product
LULC thematic
Dates
2010
Formats
Raster
Pixel size
250 m
Theme
Percentage of tree cover
Extent
Global
Updating
No
Change detection
No (only one date)
Overall accuracy
Not specified

Website of reference	Website Language English
http://www.fao.org/geonetwork/srv/en/main.home?uuid=063720fb-79b5-44e5-832b-1c03f6b845ac	

Download site
http://www.fao.org/geonetwork/srv/en/main.home?uuid=063720fb-79b5-44e5-832b-1c03f6b845ac

Availability	Format(s)
Open Access	.adf

Technical documentation
–

Other references of interest
FAO (2010); Ridder (2007) FAO et al. (2009)

Project

The Food and Agriculture Organization (FAO) carries out the Global Forest Resources Assessment (FRA) on average once every five years. The first LUC map produced for this project was the World's Forests 2000, described above. For the 2015 edition of the FRA, a new map for the reference year 2010 was produced.

The Forests of the World 2010 map was produced within the framework of the FRA 2010 and 2015 Global Remote Sensing Surveys. These surveys aimed to provide complementary information using remote sensing techniques and Landsat imagery, in addition to the data that was normally collected and analysed through the different FRA projects.

The FRA Global Remote Sensing Surveys, carried out by the FAO in collaboration with the Joint Research Centre (JRC) of the European Commission, provided systematic alphanumerical information on the dynamics of forest covers and uses for four dates (1990, 2000, 2005, 2010) at three different scales: regional, ecozone and global.

A new participatory global remote sensing survey is currently ongoing as part of the FRA 2020 project.

Production method

The Forests of the World 2010 map is partially based on the MODIS/Terra Vegetation Continuous Fields (VCF) product. Other auxiliary datasets were also employed in its production: water data from the Shuttle Radar Topography Mission (SRTM) and the MODIS global water mask; a Digital Elevation Model from the SRTM; the Global Administrative Unit Layer (GAUL); and a dataset of Global ecological zones. No information is available about the procedure followed to merge this information.

Product description

The map is downloaded as a single zip file, which contains the LUC raster and a series of auxiliary files that do not, however, provide any extra information to the user.

Downloads

Forests of the world 2010
– Raster file with LUC map (fao_fra2010)

Legend and codification

Code	Label
1–100	Percent of pixel area covered by tree cover (0–100)

Practical considerations

No technical information is available about the way the map was produced, which makes it difficult to understand its characteristics and potential disadvantages. As this map was created on the basis of information provided by the MODIS VCF map (see Sect. 7), there may be high correlation between the two maps.

When downloading the data, users will find many files making up the LUC map. To represent the map in QGIS they can open any of the files in the "fao_fra2010" folder.

11   TanDEM-X Forest/Non-Forest Map

Product	
LULC thematic	
Dates	
2011 / 15	
Formats	
Raster	
Pixel size	
50 m	
Theme	
Forest extent	
Extent	
Global	
Updating	
No	
Change detection	
No (only one date)	
Overall accuracy	
Expected to be >90%	

Website of reference	**Website Language** English
https://www.dlr.de/hr/en/desktopdefault.aspx/tabid-12538/21873_read-50027/	

Download site

https://download.geoservice.dlr.de/FNF50/

Availability	**Format(s)**
Open Access	.tiff

Technical documentation

Bueso Bello et al. (2019), Martone et al. (2016, 2018a, b)

Other references of interest

–

Project

The TanDEM-X Forest/Non-Forest Map is a dataset produced by the Microwaves and Radar Institute of the German
Aerospace Center (DLR). It aims to provide useful information for environmental assessment and forest monitoring.
Together with the Global Forest Non-Forest map, described
earlier in this chapter, it was one of the first projects to use
radar data for forest mapping at a global scale. Radar overcomes some of the limitations associated with forest mapping
using optical sensors, in that it can provide accurate LUCC
information regardless of the weather or daylight conditions.

The dataset was produced within the context of the
TanDEM-X mission. It makes use of TanDEM-X bistatic
interferometric synthetic aperture radar (InSAR) data,
mainly captured to produce a very precise Digital Elevation
Model (DEM) at a global scale.

Production method

The TanDEM-X Forest/Non-Forest Map was obtained by
classifying and processing interferometric synthetic aperture
radar (InSAR) data acquired by the TanDEM-X mission
over the period 2011–2015. The original data at 3 m was
resampled at 50 m for the classification. It includes two full
coverages of the Earth's surface.

Different factors in the InSAR data were used in the
classification of forest and non-forest areas. The most
important of these was the volume correlation factor. It
quantifies the amount of decorrelation caused by multiple
scattering within a volume, which is usually due to the
presence of vegetation. The other factors employed in the
classification process were bistatic coherence, calibrated
amplitude and DEM height information.

All this information was provided as input for a fuzzy
multi-clustering classification process at the scene level.
Specific parameters were used for different forest types
(tropical, temperate and boreal forest) due to differences in
forest structure, density and tree height.

Once the classification had been carried out for all the
available scenes, a Forest/Non-Forest Map was obtained by
mosaicking all the classification results. In a post-
classification stage, the accuracy of the map was improved
using auxiliary layers that provide information about urban
areas, water bodies, deserts and the tree line, i.e. the virtual
line marking the altitudes above which trees do not grow.

Product description

The TanDEM-X Forest/Non-Forest Map is distributed in
1 × 1° tiles. Users can select those within their area of
interest via the online viewer available at the download
website (see above). The files are also available through an
HTTPS Web browser: https://download.geoservice.dlr.de/
FNF50/files/. In the latter case, users must input the latitude
and longitude values for their specific area of interest when
downloading the files.

The download includes the forest/non-forest map plus
three auxiliary layers providing technical information about
the classification. Interested users can also download the
product's metadata as a separate file from the download
website.

Downloads

TanDEM-X Forest/Non-Forest Map
– Raster file with forest/non-forest map
– Raster file with coverage information (number of mosaicked acquisitions per pixel)
– Raster file with the number of reliable super pixels in input
– Raster file with the date of the most recent super pixels
– Text file with information about the data acquisition process
– PDF files with the product's license agreements in English and German
– Image preview of the product

Legend and codification

Code	Label	Code	Label
0	Invalid pixels and settlements	2	Non-forested areas
1	Forested areas	3	Water bodies

Practical considerations

This dataset was produced by means of a complex production method that is difficult to understand for those without
specialist knowledge of radar data. Those wishing to find out
more about this dataset should read the guide cited in the
specifications above and other information about the dataset
available at https://geoservice.dlr.de/web/dataguide/fnf50/.

References

Altunel AO, Akturk E, Altunel T (2020) Examining the PALSAR-2 Global forest/non-forest maps through Turkish afforestation practices. Int J Remote Sens 41:6071–6088. https://doi.org/10.1080/01431161.2020.1760397

Amarnath G, Babar S, Murthy MSR (2017) Evaluating MODIS-vegetation continuous field products to assess tree cover change and forest fragmentation in India–a multi-scale satellite remote sensing approach. Egypt J Remote Sens Sp Sci 20:157–168. https://doi.org/10.1016/j.ejrs.2017.05.004

Baret F, Weiss M, Verger A, Smets B (2016) ATBD for LAI, FAPAR and FCOVER from PROBA-V products at 300M resolution (GEOV3). https://land.copernicus.eu/global/sites/cgls.vito.be/files/products/ImagineS_RP2.1_ATBD-FCOVER300m_I1.73.pdf. Accessed 11 Feb 2021

Bueso Bello JL, González C, Martone M, Rizzoli P (2019) TanDEM-X. Forest/Non-Forest map product description. https://geoservice.dlr.de/web/dataguide/fnf50/pdfs/TD-GS-PS-0206_Forest_Non-Forest_Map_Product_Description.pdf. Accessed 12 July 2021

FAO (2000) Global forest resources assessment 2000. Main report. http://www.fao.org/3/Y1997E/Y1997E00.htm. Accessed 9 Feb 2021

FAO (2001) FRA 2000. Global forest cover mapping. Final report. Accessed February 9, 2021. http://www.fao.org/3/ad679e/ad679e00.htm#TopOfPage. Accessed 9 Feb 2021

FAO (2010) Global forest resources assessment 2010. Rome. http://www.fao.org/3/i1757e/i1757e.pdf. Accessed 12 Jan 2021

FAO, JRC, SDSU, UCL (2009) The FRA 2010 Remote sensing survey: an outline of objectives, data, methods and approach. http://www.fao.org/3/a-k7023e.pdf. Accessed 10 Feb 2021

Hansen MC, DeFries RS, Townshend JRG et al (2003a) Global percent tree cover at a spatial resolution of 500 meters: first results of the modis vegetation continuous fields algorithm. Earth Interact 7:1–15. https://doi.org/10.1175/1087-3562(2003)007%3c0001:gptcaa%3e2.0.co;2

Hansen MC, DeFries RS, Townshend JRG et al (2003b) Development of 500 meter vegetation continuous field maps using MODIS data. Int Geosci Remote Sens Symp 1:264–266. https://doi.org/10.1109/igarss.2003.1293745

Hansen MC, Townshend JRG, DeFries RS, Carroll M (2005) Estimation of tree cover using MODIS data at global, continental and regional/local scales. Int J Remote Sens 26:4359–4380. https://doi.org/10.1080/01431160500113435

Hansen MC, Potapov P V., Moore R, et al (2013) High-resolution global maps of 21st-century forest cover change. Science 80 (342):850–853. https://doi.org/10.1126/science.1244693

Hansen MC, Potapov P, Margono B et al (2014) Response to comment on High-resolution global maps of 21st-century forest cover change. Science 80(344):981–981. https://doi.org/10.1126/science.1248817

JAXA, EORC (2019) Global 25m Resolution PALSAR-2 / PALSAR Mosaic and Forest/Non-Forest Map (FNF). Dataset description. https://www.eorc.jaxa.jp/ALOS/en/palsar_fnf/DatasetDescription_PALSAR2_Mosaic_FNF_revI.pdf. Accessed 24 Sept 2020

Jeganathan C, Dadhwal VK, Gupta K, Raju PLN (2009) Comparison of MODIS vegetation continuous field-based forest density maps with IRS-LISS III derived maps. J Indian Soc Remote Sens 37:539–549. https://doi.org/10.1007/s12524-009-0050-6

Jolivet D (2020) Copernicus global land operations. Vegetation and energy (CGLOPS-1) Quality assessment report atmospheric correction for Sentinel-3 OLCI and SLSTR products (version 1.0). https://land.copernicus.eu/global/sites/cgls.vito.be/files/products/CGLOPS1_QAR_S3-AC_I1.00.pdf. Accessed 11 Feb 2021

Jung M, Henkel K, Herold M, Churkina G (2006) Exploiting synergies of global land cover products for carbon cycle modeling. Remote Sens Environ 101:534–553. https://doi.org/10.1016/j.rse.2006.01.020

Lacaze R, Bauer-Marschallinger B, Jolivet D et al (2020) Copernicus global land operations. vegetation and energy (CGLOPS-1) Product quality assurance document. https://land.copernicus.eu/global/sites/cgls.vito.be/files/products/CGLOPS1_PQAD_I2.00.pdf. Accessed 11 Feb 2021

Martínez-Sánchez E, Sánchez-Zapero J (2020) Copernicus global land operations: vegetation and energy (CGLOPS-1: scientific quality evaluation) LAI, FAPAR, FCOVER. collection 300 M (Version 1). https://land.copernicus.eu/global/sites/cgls.vito.be/files/products/CGLOPS1_SQE2019_LAI300m-V1_I1.00.pdf. Accessed 11 Feb 2021

Martone M, Rizzoli P, Wecklich C, González C, Bueso-Bello JL, Valdo P, Schulze D, Zink M, Krieger G, Moreira A (2018a) The global forest/non-forest map from TanDEM-X interferometric SAR data. Remote Sens Environ 205:352–373. https://doi.org/10.1016/j.rse.2017.12.002

Martone M, Rizzoli P, Wecklich C, González C, Bueso-Bello JL, Valdo P, Schulze D, Zink M, Krieger G, Moreira A (2018b) The global forest/non-forest map from TanDEM-X interferometric SAR data. Remote Sens Environ 205:352–373. https://doi.org/10.1016/j.rse.2017.12.002

Martone M, Rizzoli P, Krieger G (2016) Volume decorrelation effects in TanDEM-X interferometric SAR data. IEEE Geosci Remote Sens Lett 13(12):1812–1816. https://doi.org/10.1109/LGRS.2016.2614103

Ramon D, Jolivet D, Compiègne M (2020) Copernicus global land operations, vegetation and energy (CGLOPS-1) Algorithm theoretical basis document. Atmospheric correction for sentinel-3 OLCI and SLSTR Products (Version 1.0) https://land.copernicus.eu/global/sites/cgls.vito.be/files/products/CGLOPS1_ATBD_S3-AC-v1_I1.10.pdf. Accessed 11 Feb 2021

Ridder RM (2007) Global forest resources assessment 2010. Options and recommendations for a global remote sensing survey of forests. http://www.fao.org/forestry/20809-0b4ad8bc5f3c694d537d0d66ae8298008.pdf. Accessed 10 Feb 2021

Sánchez-Zapero J, Fuster B, Camacho F (2018) Copernicus global land operations, vegetation and energy (CGLOPS-1) Quality assessment report. LAI, FAPAR, FCOVER, collection 300 M (version 1). https://land.copernicus.eu/global/sites/cgls.vito.be/files/products/CGLOPS1_QAR_FCOVER300-V1_I2.00.pdf. Accessed 11 Feb 2021

Schepaschenko D, See L, Lesiv M et al (2015) Development of a global hybrid forest mask through the synergy of remote sensing, crowdsourcing and FAO statistics. Remote Sens Environ 162:208–220. https://doi.org/10.1016/j.rse.2015.02.011

Sexton JO, Feng M, Channan S, Song XP, Kim DH, Noojipady P, Song D, Huang C, Annand A, Collins K, Vermote EF, Wolfe R, Masek J, Townshend JRG (2016a) Earth science data records of global forest cover and change. User guide. https://lpdaac.usgs.gov/documents/145/GFCC_User_Guide_V1.pdf. Accessed 12 July 2021

Sexton JO, Feng M, Channan S, Song XP, Kim DH, Noojipady P, Song D, Huang C, Annand A, Collins K, Vermote EF, Wolfe R, Masek J, Townshend JRG (2016b) Earth science data records of global forest cover and change. Algorithm theoretical basis document. https://lpdaac.usgs.gov/documents/146/GFCC_ATBD.pdf. Accessed 12 July 2021

Sexton JO, Song XP, Feng M et al (2013) Global, 30-m resolution continuous fields of tree cover: landsat-based rescaling of MODIS vegetation continuous fields with lidar-based estimates of error. Int J Digit Earth 6:427–448. https://doi.org/10.1080/17538947.2013.786146

Shimada M, Itoh T, Motooka T et al (2014) New global forest/non-forest maps from ALOS PALSAR data (2007–2010). Remote Sens Environ 155:13–31. https://doi.org/10.1016/j.rse.2014.04.014

Smets B, Jacobs T, Verger A (2018) Gio global land component-Lot I operation of the global land component. Product user manual leaf area index (lai). Fraction of Photosynthetically Active Radiation (FAPAR), Fraction of vegetation cover (fcover). Collection 300M. (Version 1). https://land.copernicus.eu/global/sites/cgls.vito.be/files/products/GIOGL1_PUM_FCOVER300m-V1_I1.60.pdf. Accessed 11 Feb 2021

Toté C, Tansey K (2020) Copernicus global land operations, vegetation and energy (CGLOPS-1) evaluation report of OLCI and SLSTR cloud, cloud shadow and snow detection. https://land.copernicus.eu/global/sites/cgls.vito.be/files/products/CGLOPS1_QAR_S3-CloudMask_I1.00.pdf. Accessed 11 Feb 2021

Townshend JRG, Hansen MC, Carroll M, et al (2011) User guide for the MODIS vegetation continuous fields product, collection 5 (version 1). http://modis-land.gsfc.nasa.gov/pdf/VCF_C5_UserGuide_Feb2013.doc. Accessed 25 Sept 2020

Verger A (2020) Copernicus global land operations. Vegetation and energy (CGLOPS-1) Algorithm theoretical basis document, Fraction of Absorbed Photosynthetically Active Radiation (FAPAR). Fraction of green vegetation Cover (FCover), collection 300m (version 1.1). https://land.copernicus.eu/global/sites/cgls.vito.be/files/products/CGLOPS1_ATBD_FCOVER300m-V1.1_I1.00.pdf. Accessed 11 Feb 2021

Wolfs D, Verger A, Van der Goten R (2020) Copernicus global land operations, vegetation and energy (CGLOPS-1) product user manual. Leaf Area Index (LAI), Fraction of Absorbed Photosynthetically Active Radiation (FAPAR), Fraction of green Vegetation Cover (FCover), collection 300m (version 1.1). https://land.copernicus.eu/global/sites/cgls.vito.be/files/products/CGLOPS1_PUM_FCOVER300m-V1.1_I1.00.pdf. Accessed 11 Feb 2021

Global Thematic Land Use Cover Datasets Characterizing Agricultural Covers

David García-Álvarez and Javier Lara Hinojosa

Abstract

There is a wide variety of global thematic Land Use Cover (LUC) datasets characterizing agricultural covers. Most of them focus on cropland areas, providing information on their extent or the percentage of cropland cover on the ground. In some cases, the focus is more specific and they provide information on cropland irrigation practices. In other cases, specific maps charting the extension of different crops are also available. In this chapter, we review 8 different datasets with a spatial resolution of at least 1 km. There are many other datasets characterizing agricultural covers at coarser resolutions, such as the Historic Croplands Dataset, GMRCA or GIAM. Their coarse resolution hampers their potential application in practice, which is why they are not described in detail in this chapter. Nor do we analyse FROM-GC, a dataset mapping the extent of global cropland at 30 m, because it is not currently accessible. GFSAD30 has the highest resolution of all the datasets reviewed (30 m). It also provides some of the most up-to-date information (2015). However, it only charts the extent of cropland. As part of an associated project, GFSAD1KCD and GFSAD1KCM characterize cropland areas in 9 and 7 categories respectively at 1 km for 2010. They provide information on the irrigation status of the crops. GFSAD1KCD and GFSAD1KCM were obtained from data fusion. This method is commonly used in the production of many of the cropland datasets reviewed: IIASA-IFPRI cropland map, Global Synergy Cropland Map, Unified Cropland Layer (UCL) and ASAP Land Cover Masks. The IIASA-IFPRI (2005) and ASAP maps provide information on the proportion of cropland at a spatial resolution of 1 km. ASAP also includes a map on rangeland covers, and as such is the only dataset described in this chapter that maps a cover other than croplands. The Global Synergy Cropland Map (2010) and the Unified Cropland Layer (2014) also map cropland proportions, although they have been produced at higher spatial resolutions: 500 and 250 m respectively. The Global Cropland Extent product maps the extent of cropland at 250 m based on imagery from 2000-2008. Although thematically limited, this dataset is less affected by time variability, as it is based on imagery taken over a long period (8 years). Finally, GRIPC maps the extent of three types of cropland area (irrigated, rainfed and paddy crops) at 500 m for 2005.

Keywords

Agriculture • Cropland • Pastureland • Global Cropland Extent • IIASA-IFPRI Cropland Map • GRIPC • GFSAD1KCM • GFSAD1KCD • Global Synergy Cropland Map • UCL • GFSAD30 • ASAP Land Cover Masks

D. García-Álvarez (✉)
Departamento de Geología, Geografía y Medio Ambiente, Universidad de Alcalá, Alcalá de Henares, Spain
e-mail: David.garcia@uah.es

J. Lara Hinojosa
Departamento de Análisis Geográfico Regional y Geografía Física, Universidad de Granada, Granada, Spain
e-mail: jlarahinojosa@ugr.es

© The Author(s) 2022
D. García-Álvarez et al. (eds.), *Land Use Cover Datasets and Validation Tools*,
https://doi.org/10.1007/978-3-030-90998-7_20

1   Global Cropland Extent

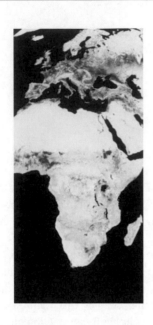

Product
LULC thematic
Dates
2000 / 08
Formats
Raster
Pixel size
250 m
Theme
Cropland extent
Extent
Global
Updating
Not expected
Change detection
No (only one date)
Overall accuracy
Not specified

Website of reference	Website Language English
https://glad.umd.edu/projects/croplands/globalindex.html	

Download site
https://glad.umd.edu/projects/croplands/dataindex.html

Availability	Format(s)
Open Access	.tiff

Technical documentation
Pittman et al. (2010)

Other references of interest
–

Project

The Global Cropland Extent was a map developed for the Global Agriculture Monitoring Project (GLAM). The project, promoted by NASA, the USDA, and Maryland and South Dakota State universities, aimed to take advantage of the new generation of NASA satellite observations to enhance the agricultural monitoring and crop-production estimation work carried out by the USDA Foreign Agriculture Service (FAS). At the time it was produced, Global Cropland Extent was the highest resolution cropland map at global scale produced using synoptic inputs.

Production method

The Global Cropland Extent map was obtained after thresholding a crop probability layer obtained from 16-day composites of MODIS imagery for the period 2000–2008. The probability layer was generated by averaging the results from multiple decision-tree classifications. They were trained with sub-pixel data obtained from multiple sources: GeoCover, AfriCover, USDA, Cropland Data Layer, NLCD, Agriculture and Agri-Food Canada, South Africa State of the Environment and CLC.

The selected threshold for differentiating between cropland and non-cropland areas in the probability layer was decided on the basis of information from the FAS Production, Supply and Distribution (PSD) database. The database provided, per country, the median harvested area of production field crops (barley, corn, cotton, oats, rice, rye, sorghum, soybeans and wheat) for the period 2000–2008. The pixels with the highest cropland probability were then considered cropland until those area thresholds were met. In the European Union, the threshold was defined for the whole EU area rather than at country level.

Product description

The Global Cropland Extent map is distributed in tiles following the MODIS tile grid.[1] To identify the file or files that fall within their area of interest, users must know the horizontal and vertical tile numbers that identify each area. The download only includes the raster file with the cropland information and no additional data is provided.

The Cropland probability layer can also be downloaded following the same procedure. In addition, the project provides a global mosaic at a spatial resolution of 1 km, merging all the tiles in one file.

Downloads

Global Cropland Extent h17v04
– Raster file with cropland extent (.tiff)

Global Cropland Probability h17v04
– Raster file with cropland probability (.tiff)

Legend and codification

Global Cropland Extent	
Code	Label
0	Cropland
1	No cropland
254	Water

Global Cropland Probability	
Code	Label
0	Water
1–100	Cropland probability (1–100%)

Practical considerations

According to the accuracy analyses carried out by the production team, the Global Cropland Extent map shows important accuracy differences when mapping cropland areas. Intensive broadleaf crop regions (corn and soybean) are the best mapped, while wheat-growing regions and, especially, rice production regions, present low levels of accuracy. The dataset also has problems mapping cropland areas in regions without intensive agriculture, like Africa.

Because of the 8-year timespan of the MODIS imagery used as an input for the production of the Global Cropland Extent, the dataset can be considered insensitive to inter-annual variability of cropland covers.

---

[1] The MODIS tile grid is available at https://modis-land.gsfc.nasa.gov/MODLAND_grid.html.

2   IIASA-IFPRI Cropland Map

Product
LULC thematic
Dates
2005
Formats
Raster
Pixel size
1 km
Theme
Percentage of cropland cover
Extent
Global
Updating
Not expected
Change detection
No (only one date)
Overall accuracy
Expected to be > 82%

Website of reference	Website Language English
https://geo-wiki.org/Application/index.php	

Download site
https://geo-wiki.org/Application/index.php

Availability	Format(s)
Open Access after registration	.img

Technical documentation
Fritz et al. (2015)

Other references of interest
Fritz et al. (2011)

Project

The IIASA-IFPRI Cropland Map was produced by an international consortium of researchers led by the International Institute for Applied Systems Analysis (IIASA) and the International Food Policy Research Institute (IFPRI). The project builds on the experience and the method proposed by Fritz et al. (2011) for mapping cropland areas in sub-Saharan Africa. It is part of a broader plan to provide better LUC mapping for food security studies and policies.

The aim of the project was to improve the spatial representation of cropland areas by fusing existing datasets. Unlike previous efforts, the focus was on cropland percentage instead of cropland extent. In addition, the project delivered the first ever global field-size map.

Production method

The IIASA-IFPRI Cropland Map was obtained by merging the cropland cover information provided by global (GLC2000, MODIS 2005, GlobCover), regional (CLC, AFRICOVER, Cropland mask for Africa) and national (14 countries) datasets. The datasets with a spatial resolution finer than 1 km were resampled and combined in a common grid at a spatial resolution of 1 km. For those datasets that do not provide information about the percentage of cropland, and merely inform about its presence or absence, minimum, average and maximum percentages of cropland cover were assigned according to the definition of the cropland categories.

Once all the input information had been homogenized, the different datasets were combined in a synergy layer. The synergy layer defines the cropland areas according to the agreement of the input datasets. The combination of datasets was hierarchical, according to their accuracy, which was determined by reference data collected through the Geo-Wiki platform. Together with the synergy layer, three other layers stating the minimum, average and maximum cropland percentage cover were obtained by averaging the minimum, average and maximum cropland percentage values from the input maps.

The final IIASA-IFPRI Cropland Map was obtained by combining the synergy and average cropland percentage layers with national cropland statistics provided by FAO. The areas with the highest probability of being cropland according to the synergy layer were selected until the total surface area for cropland according to FAO statistics for each country was reached. The specific area of cropland allocated to each pixel (e.g. 70 ha of cropland) was determined based on the average cropland percentage cover layer.

Finally, a visual verification with Google Earth imagery was carried out at the national level to correct possible omission errors.

Product description

The dataset can be downloaded as a single compressed file (.zip), including the raster with the LUC information and an auxiliary file with a brief technical description of the raster file.

Downloads

IIASA-IFPRI Cropland map
– Raster file with cropland percentage (.img)
– Text file with technical information about the raster

Legend and codification

Code	Label
0–100	Cropland Coverage (0-100%)

Practical considerations

The IIASA-IFPRI Cropland Map can be accessed online via the Geo-Wiki platform. The associated field-size map can be very useful for researchers studying food security and other aspects of cropland uses and practices. The field-size map can be downloaded and visualized at the same website as the Cropland map.

3 GRIPC—Global Rainfed, Irrigated, and Paddy Croplands

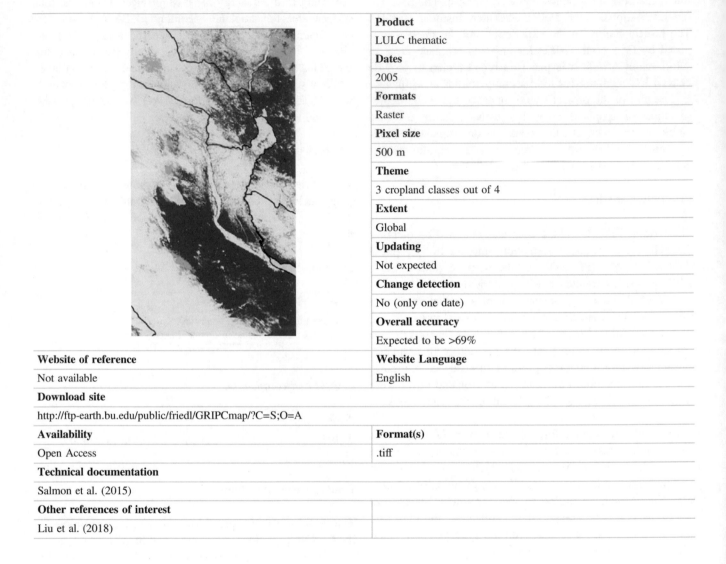

Product	
LULC thematic	
Dates	
2005	
Formats	
Raster	
Pixel size	
500 m	
Theme	
3 cropland classes out of 4	
Extent	
Global	
Updating	
Not expected	
Change detection	
No (only one date)	
Overall accuracy	
Expected to be >69%	

Website of reference	Website Language
Not available	English

Download site	
http://ftp-earth.bu.edu/public/friedl/GRIPCmap/?C=S;O=A	

Availability	Format(s)
Open Access	.tiff

Technical documentation	
Salmon et al. (2015)	

Other references of interest	
Liu et al. (2018)	

Project

Global Rainfed, Irrigated and Paddy Croplands (GRIPC) is a map developed by researchers from German and American universities, who aimed to overcome some of the limitations of previous datasets focusing on irrigated croplands. At the time it was released, the dataset offered an up-to-date representation of irrigated croplands across the world at the highest spatial resolution available. It could be useful for those studying agricultural productivity, agricultural hydrology and food security in general.

Production method

The GRIPC map is made up of 4 different categories. Uncropped areas were extracted from the non-cropland categories of the MODIS Land Cover database for the period 2004–2006. Paddy croplands were independently mapped from different sources, such as crop inventories, due to the challenges involved in classifying cloudy imagery in the tropics. Rainfed and irrigated cropland were mapped using a decision-tree classification algorithm (C4.5) and the "boosting" machine learning technique.

MODIS imagery was used as the input for the classification. Climate and agroecozones data were also used as auxiliary datasets. Probability layers obtained from the classification were combined with information from national and subnational cropland inventory-based datasets to finally map the rainfed and irrigated cropland areas. The information from these datasets served to define the probabilities of each category occupying a pixel. Then, the classification results were combined with these probabilities using a Bayes' rule to obtain the final map.

Product description

GRIPC is distributed in 273 tiles, according to the MODIS tile grid.[2] Users must consult the tiles that correspond to their area of interest. A lower-resolution version of the product, at 5 arc minutes, and a file with the main technical characteristics of the dataset, are also available for download.

Downloads

GRIPC h17v04
– Raster file with cropland information (.tiff)

Legend and codification

Code	Label	Code	Label
1	Rainfed cropland	3	Paddy cropland
2	Irrigated cropland	4	No cropland

Practical considerations

GRIPC does not map various important irrigated cropland categories, such as deficit irrigation (irrigation occurring less than once a year), permanent crops (orchards and vineyards) and unharvested pastures. As there is no official website describing the GRIPC and its characteristics, users wishing to find out more about this dataset should consult the scientific paper in which it was presented (Pittman et al. 2010).

---

[2] The MODIS tile grid is available at https://modis-land.gsfc.nasa.gov/MODLAND_grid.html.

4   GFSAD1KCM and GFSAD1KCD

Product
LULC thematic
Dates
2010
Formats
Raster
Pixel size
1 km Minimum mapping unit: 0.81 ha
Theme
5 cropland classes out of 7, focusing on cropland extent (GFSAD1KCM) 8 cropland classes out of 10, focusing on crop dominance (GFSAD1KCD)
Extent
Global
Updating
Not expected
Change detection
No (only one date)
Overall accuracy
Expected to be >70%

Website of reference	**Website Language** English

https://lpdaac.usgs.gov/products/gfsad1kcmv001/
https://lpdaac.usgs.gov/products/gfsad1kcdv001/

Download site

https://lpdaac.usgs.gov/products/gfsad1kcmv001/
https://lpdaac.usgs.gov/products/gfsad1kcdv001/

Availability	**Format(s)**
Open Access after registration	.tiff

Technical documentation

Teluguntla et al. (2020), USGS EROS (2017)

Other references of interest

Friedl et al. (2010), Pittman et al. (2010), Portmann et al. (2010), Ramankutty et al. (2008), Thenkabail and Lyon (2009), Thenkabail et al. (2009), Thenkabail et al. (2010), Thenkabail et al. (2011), Thenkabail et al. (2012), Yadav and Congalton (2018), Yu et al. (2013)

Project

The GFSAD1KCM and GFSAD1KCD datasets were created by NASA and the USGS within the context of the MEaSUREs (Making Earth System Data Records for Use in Research Environments) programme. MEaSUREs is one of the competitive programmes of the Earth Science Data Systems (ESDS), which aims to take full scientific advantage of NASA missions.

MEaSUREs projects make use of data from NASA satellites to produce innovative products that meet the needs of the research community, inform policy-making and provide a better understanding of the planet. GFSAD (Global Food Security Support Analysis Data) is a specific MEaSUREs project focused on mapping agricultural areas to contribute to global food security policies. The project aims to improve global cropland mapping, by providing a methodology that can map cropland areas across the world quickly, consistently and accurately.

As part of the GFSAD projects, cropland maps have been produced at three different spatial resolutions (1 km, 250 m and 30 m). The maps at 1 km and 30 m cover the whole globe. Various different supranational datasets are available at 250 m for Africa, Australia and South Asia at different years of reference. A similar dataset at 250 m is also available yearly for the United States from 2001 to 2013.

For the product at 1 km, two complementary maps were generated: GFSAD1KCM, mapping the extent of cropland at a global level, and GFSAD1KCD, which maps crop dominance across the world. The map at 30 m is described later in this chapter.

Production method

GFSAD1KCM and GFSAD1KCD were produced separately by aggregating different existing products. The input maps were first resampled at the same resolution (1 km) and later overlaid.

GFSAD1KCM was created by aggregating the maps produced by Thenkabail et al. (2009, 2011), Pittman et al. (2010), Yu et al. (2013), and Friedl et al. (2010). Cropland extent was obtained by agreement of these four maps. Other information and indicators, such as irrigation status, irrigation or rainfed dominance, were obtained from the map developed by Thenkabail et al. (2009, 2011).

GFSAD1KCD was created by combining the global irrigated and rainfed cropland area map produced by the International Water Management Institute with the maps of dominant global crop-types produced by Ramankutty et al. (2008), Monfreda et al. (2008), and Portmann et al. (2010). In both cases, the maps were obtained from data for the period 2007–2012.

Product description

GFSAD1KCM and GFSAD1KCD can be downloaded from various different servers or tools, such as Data Pool, NASA Earthdata Search, USGS EarthExplorer and the DAAC2Disk Utility. In all cases, users download a raster file with the cropland information. Downloads from Data Pool also include a metadata file and a preview image of the product.

Downloads

GFSAD1KCDv001
– Raster file with crop dominance information

GFSAD1KCMv001
– Raster file with cropland extent

Legend and codification

GFSAD1KCD

Code	Label
0	Ocean or Water areas
1	Irrigated (Wheat and Rice)
2	Irrigated Mixed Crops 1 (Wheat, Rice, Barley, Soybeans)
3	Irrigated Mixed Crops 2 (Wheat, Rice, Cotton, Orchards)
4	Rainfed (Wheat, Rice, Soybeans, Sugarcane, Corn, Cassava)
5	Rainfed (Wheat, Barley)
6	Rainfed (Corn, Soybeans)
7	Rainfed Mixed Crops (Wheat, Corn, Rice, Barley, Soybeans)
8	Fractions of Mixed Crops (Wheat, Maize, Rice, Barley, Soybeans)
9	Non-cropland areas

GFSAD1KCM

Code	Label	Code	Label
0	Ocean or Water areas	4	Croplands, Rainfed, Minor Fragments
1	Croplands, Irrigation Major	5	Croplands, Rainfed, Very Minor Fragments
2	Croplands, Irrigation Minor	9	Non-Cropland areas
3	Croplands, Rainfed		

Practical considerations

GFSAD1KCM and GFSAD1KCD were produced independently for different purposes and cannot therefore be compared. Although GFSAD1KCD provides information on crop dominance, it can also be used to study cropland extent.

According to the authors, data about cropping intensity can be obtained from this product using a time-series of Normalized Difference Vegetation Index (NDVI) data.

5 Global Synergy Cropland Map

Product
LULC thematic
Dates
2010
Formats
Raster
Pixel size
500 m
Theme
Percentage of cropland cover
Extent
Global
Updating
Not expected
Change detection
No (only one date)
Overall accuracy
Expected to be >90%

Website of reference	Website Language English
https://dataverse.harvard.edu/dataset.xhtml?persistentId=doi:10.7910/DVN/ZWSFAA	

Download site
https://dataverse.harvard.edu/dataset.xhtml?persistentId=doi:10.7910/DVN/ZWSFAA

Availability	Format(s)
Open Access	.tiff
Technical documentation	
Lu et al. (2020) .tiff	
Other references of interest	
Yu et al. (2020)	

Project

The Global Synergy Cropland Map is a dataset created within the framework of the Spatial Production Allocation Model (SPAM), which maps agriculture production across the world. It is a joint effort involving different institutions and universities across the world: AGRIRS, IFPRI Chinese Academy of Agricultural Sciences and Victoria University of Wellington.

The project team aimed to create a more accurate cropland dataset that would be useful for agricultural monitoring and food security policies and studies. The obtained map is a critical input of SPAM.

Production method

A self-adapting statistics allocation model (SASAM) is used to generate the Global Synergy Cropland Map, using LUC datasets at global, supranational and national scales as input, as well as FAO agricultural statistics at national and subnational levels.

Two layers were generated by the model. Firstly, an agreement layer, which shows the level of agreement of all the datasets regarding the location of cropland areas, and secondly, an average cropland percentage layer, obtained by calculating the average of all the input maps. For the agreement layer, datasets with a higher accuracy are given more weight. This accuracy is based on the agreement between each input dataset and the FAO statistics. For the cropland percentage layer, the cropland category definitions in the input maps were translated into cropland percentages.

The final cropland map was obtained after executing the SASAM model, which allocated cropland in the areas with the highest probability in the agreement layer until the total surface area for cropland according to FAO statistics for each country was reached.

Product description

The raster file showing the cropland percentage can be downloaded separately. However, we recommend the full download, which also contains additional information about the dataset, such as its level of confidence.

Downloads

Global synergy cropland map (full download)
– Raster file with cropland percentage (.tiff)
– Raster file with information about the confidence level of the cropland map (.tiff)
– A text file with information about the downloaded product

Legend and codification

Code	Label
0-1	Cropland extent percent (0–100%)

Practical considerations

More information about the associated SPAM project is available at www.mapspam.info. The website includes all the spatial datasets about agricultural production generated as part of the project. These complement the information provided by the cropland map reviewed here.

6   UCL—Unified Cropland Layer

Product	
LULC thematic	
Dates	
2014	
Formats	
Raster	
Pixel size	
250 m	
Theme	
Percentage of cropland cover, although for some areas it only informs about the extent	
Extent	
Global	
Updating	
Not planned	
Change detection	
No (only one date)	
Overall accuracy	
Expected to be >83%	

Website of reference	**Website Language** English
https://figshare.com/articles/dataset/ucl_2014_v2_0_tif/2066742	
Download site	
https://figshare.com/articles/dataset/ucl_2014_v2_0_tif/2066742	
Availability	**Format(s)**
Open Access	.tiff
Technical documentation	
Waldner et al. (2016)	
Other references of interest	
–	

Project

The Unified Cropland Layer (UCL) is one of the results of the SIGMA (Stimulating Innovation for Global Monitoring of Agriculture and its Impact on the Environment in support of GEOGLAM) project. SIGMA was a European funded project that sought to improve agricultural monitoring and forecasting tools, using earth observation data. The project was made up of 22 renowned international institutions, many of which were experts in agricultural monitoring. In addition, the project was part of the European contribution to the Global Agricultural Geo-Monitoring (GEOGLAM) initiative.

12 of the 22 institutions involved in this project took part in the production of the UCL. Its aim was to enhance the global mapping of cropland areas, contributing to studies and activities assessing the current situation of cropland areas across the world, assessing crop land changes and providing new data for the production of cropland statistics. The UCL uses the definition of cropland proposed by the Joint Experiment of Crop Assessment and Monitoring (JECAM).

Production method

The UCL was obtained by combining the best available LUC cropland datasets for each area of the world. To this end, up to 49 different LUC datasets at global, regional and national scales were reviewed and assessed. They were resampled at a spatial resolution of 250 m and, when several dates were available, the closest to 2014 was selected.

The best dataset was selected on the basis of a multi-criteria analysis considering 4 different criteria: (i) match between the legend and the definition of cropland used by the UCL; (ii) match between the spatial resolution and the cropland pattern in each area; (iii) the timeliness of the datasets regarding the UCL year of reference (2014); and (iv) the confidence level of each dataset.

Each input source was scored according to the four criteria. The scores were later reviewed by experts on the topic. After this review, the scores were combined to create a single indicator. The dataset with the highest score in this indicator was selected for each pixel. When the input datasets provided information on the proportion of cropland, this information was maintained. In all other cases, the UCL only differentiates binarily between cropland and non-cropland areas.

Product description

The UCL download includes the raster file with the cropland information, as well as a preview image of the product and the technical paper describing the map. Each file can also be downloaded independently.

Downloads

Unified Cropland Layer
– Raster file with cropland information (.tiff)
– Preview image of the map (.png)
– Paper describing the map

Legend and codification

Code	Label
0-100	Cropland proportion (0-100%)

7 GFSAD30 Cropland Extent

Product	
LULC thematic	
Dates	
2015 (2010 for North America)	
Formats	
Raster	
Pixel size	
30 m	
Theme	
Extent of Cropland	
Extent	
Global	
Updating	
Not expected	
Change detection	
No (only one date)	
Overall accuracy	
Expected to be > 91%	

Website of reference	Website Language English
https://www.usgs.gov/centers/wgsc/science/global-food-security-support-analysis-data-30-m-gfsad	

Download site
https://croplands.org/   https://croplands.org/downloadLPDAAC

Availability	Format(s)
Open Access	.tiff

Technical documentation
Gumma et al. (2020), Oliphant et al. (2019), Phalke et al. (2020), Teluguntla et al. (2018), Xiong et al. (2017)

Other references of interest
Teluguntla et al. (2015)

Project

Global Food Security-support Analysis Data 30 metre (GFSAD30) was a project aimed at producing high-resolution cropland maps to inform global food and water security studies and policies. The project sought to overcome some of the limitations presented by previous cropland datasets, such as sources of uncertainty, insufficient precision in the allocation of cropped areas, and a lack of information regarding the intensity and irrigation status of cropland areas.

GFSAD30 was the continuation of earlier projects (the GFSAD1KCM and GFSAD1KCD datasets described above) with similar purposes. They all formed part of the MEaSUREs (Making Earth System Data Records for Use in Research Environments) programme, which promotes the use of data from NASA missions to produce innovative products that are useful for research and policy-making.

Various different US institutions (USGS, BAER Institute, U.S. Department of Agriculture, U.S. Environmental Protection Agency) and universities (New Hampshire, California, Wisconsin, Northern Arizona) took part in the project, together with Google and institutions from other countries (ICRISAT, IAARD).

A global map of cropland extent at a spatial resolution of 30 m for the reference year 2015 was delivered as part of the project. The global map was obtained after merging different maps that had been independently produced for seven different regions across the world. The map for North America was produced for the reference year 2010, instead of 2015.

Production method

GFSAD30 is made up of 7 datasets which were independently produced for different regions across the world: Europe, Middle East, Russia and Central Asia; Africa; Australia, New Zealand, China, and Mongolia; Southeast and Northeast Asia; North America; and South America. Each dataset was produced following a specific production method, although they all share certain common features.

The same imagery source (Landsat) was used for all 7 datasets. Sentinel-2 imagery was also used to map the extent of cropland in Africa. Other auxiliary data, such as elevation data from the SRTM radar, were used for the production of several datasets. In all cases, the extent of cropland was computed using the Google Earth Engine (GEE) platform.

The classification workflow varies in each case. The most frequent classification method was the random forest algorithm. For some datasets, like Africa, additional classifiers

(support vector machines, an object-based classifier) were also used. In addition, in order to take the geographical variability within the mapped area into account, producers usually split the classification into agro-ecological zones (AEZs).

Product description

GFSAD30 is distributed in tiles with a 10° edge for each of the mapped regions. Datasets are available from different servers or tools, including Data Pool, NASA Earthdata Search, USGS EarthExplorer and the DAAC2Disk Utility. We recommend users to download the dataset through NASA Earthdata Search and USGS EarthExplorer, on which the geographical coverage of each tile can be visualized.

In most cases, the download only includes a raster file with the extent of cropland in .tiff format. Nonetheless, the download from the Data Pool server also includes a metadata file and a preview image of the product.

Downloads

GFSAD30AFCE v001
– Raster file with cropland extent

Legend and codification

Code	Label
0	Water
1	Non-Cropland
2	Cropland

Practical considerations

The global map obtained after merging the 7 GFSAD30 datasets can be consulted online at the project's website.[3] The website also includes other important products for mapping cropland at coarser scales (250 m, 1 km), as well as datasets about irrigated/rainfed cropland areas for South Asia, Iran, Afghanistan and Australia. Users can also download a dataset validating the product (GFSAD30VAL).[4]

In addition to the technical documentation published as reports and papers in journals, other interesting technical documents are also available on the website.[5]

---

[3] www.croplands.org.
[4] https://lpdaac.usgs.gov/products/gfsad30valv001/.
[5] https://www.croplands.org/documents.

8 ASAP Land Cover Masks

Product
LULC thematic
Dates
2019
Formats
Raster
Pixel size
1 km (resampled from 250 m original resolution)
Theme
Percentage of cropland/rangeland covers
Extent
Global
Updating
Not planned
Change detection
No (only one date)
Overall accuracy
Not specified

Website of reference	Website Language English
https://mars.jrc.ec.europa.eu/asap/index.php	

Download site

https://mars.jrc.ec.europa.eu/asap/download.php

Availability	Format(s)
Open Access	.tiff

Technical documentation

Meroni et al. (2019)

Other references of interest

Pérez-Hoyos et al. (2017a), Pérez-Hoyos et al. (2017b), Rembold et al. (2019), Vancutsem et al. (2013)

Project

Anomaly hot Spots of Agricultural Production (ASAP) is an online decision support system developed and maintained by the Monitoring Agricultural Resources unit (MARS) of the Joint Research Centre (JRC) of the European Commission to monitor anomalies in global agricultural production. The system supports early warnings and assessments on food security, so providing a useful tool for many international organizations working in this field.

Two land cover maps charting global crop and rangeland cover fractions were specifically produced for ASAP and are accessible to any interested user. These layers are required to compute anomalies based on rainfall and vegetation index data, which are later translated into timely warnings about potential food security problems.

The maps rely on previous work carried out for similar purposes by the JRC. In their studies of Africa, the maps follow a similar approach to that proposed by Vancutsem et al. (2013) and further refined by Pérez Hoyos (2017a).

Production method

The cropland and rangeland cover maps for ASAP were produced by combining the best available LUC data for each country. To select the best available source for each case, different criteria were employed depending on the country or geographical area. The selected data sources for each map (cropland, rangeland) also varied.

For Africa and part of Asia (Bangladesh, Indonesia, Laos, Myanmar, Thailand, Timor-Leste, Philippines and Vietnam), 8 global LUC datasets (CGLS-LC100, GLC2000, GLCNMO, GlobCover, GLC30, LC-CCI, MODISLC, S2 Prototype Land Cover) were compared according to different criteria. In the African case, the most suitable dataset was selected on the basis of timeliness, spatial resolution, agreement with FAO statistics, accuracy and expert knowledge. In the Asian case, only accuracy and agreement with FAO statistics were considered.

For the rest of the countries, when a suitable regional dataset was available, this was the one selected. In the cases when a suitable dataset was not available, the global LUC dataset with the highest spatial resolution was chosen. If this was not considered valid when assessed against Google Earth imagery, the FAO-GLCshare dataset was selected in its place.

The maps were initially produced at 250 m and later resampled at 1km in line with the requirements of the ASAP system.

Product description

The raster files containing the cropland and rangeland cover information can be downloaded from the ASAP website. No auxiliary information is available for these datasets.

Downloads

ASAP crop mask
– Raster file with cropland percentage (.tiff)

ASAP rangeland mask
– Raster file with rangeland percentage (.tiff)

Legend and codification

ASAP crop mask	
Code	Label
0-100	Cropland Coverage (0–100%)

ASAP rangeland mask	
Code	Label
0-100	Rangeland Coverage (0–100%)

Practical considerations

Although not directly available for download, access to the original map at a spatial resolution of 250m is possible on request to the members of the ASAP Team.[6] Previous versions of the dataset for Africa developed by Vancutsem et al. (2013) and Pérez Hoyos (2017a) can also be accessed in the same way.

---

[6] https://mars.jrc.ec.europa.eu/asap/about.php.

References

Friedl MA, Sulla-Menashe D, Tan B et al (2010) MODIS Collection 5 global land cover: Algorithm refinements and characterization of new datasets. Remote Sens Environ 114:168–182. https://doi.org/10.1016/j.rse.2009.08.016

Fritz S, See L, Mccallum I et al (2015) Mapping global cropland and field size. Glob Chang Biol 21:1980–1992. https://doi.org/10.1111/gcb.12838

Fritz S, You L, Bun A, et al (2011) Cropland for sub-Saharan Africa: A synergistic approach using five land cover data sets. Geophys Res Lett 38. https://doi.org/10.1029/2010GL046213

Gumma MK, Thenkabail PS, Teluguntla PG et al (2020) Agricultural cropland extent and areas of South Asia derived using Landsat satellite 30-m time-series big-data using random forest machine learning algorithms on the Google Earth Engine cloud. Giscience Remote Sens 57:302–322. https://doi.org/10.1080/15481603.2019.1690780

Liu Y, Wu W, Li H, et al (2018) Intercomparison on four irrigated cropland maps in Mainland China. Sensors (Switzerland) 18.https://doi.org/10.3390/s18041197

Lu M, Wu W, You L et al (2020) A cultivated planet in 2010 - Part 1: The global synergy cropland map. Earth Syst Sci Data 12:1913–1928. https://doi.org/10.5194/essd-12-1913-2020

Meroni M, Rembold F, Urbano F, et al (2019) The warning classification scheme of ASAP – Anomaly hot Spots of Agricultural Production, v4.0. Accessed 20 April, 2021. https://mars.jrc.ec.europa.eu/asap/files/asap_warning_classification_v_4_0.pdf

Monfreda C, Ramankutty N, Foley JA (2008) Farming the planet: 2. Geographic distribution of crop areas, yields, physiological types, and net primary production in the year 2000. Global Biogeochem Cycles 22:n/a-n/a. https://doi.org/10.1029/2007GB002947

Oliphant AJ, Thenkabail PS, Teluguntla P et al (2019) Mapping cropland extent of Southeast and Northeast Asia using multi-year time-series Landsat 30-m data using a random forest classifier on the Google Earth Engine Cloud. Int J Appl Earth Obs Geoinf 81:110–124. https://doi.org/10.1016/j.jag.2018.11.014

Pérez-Hoyos A, Rembold F, Kerdiles H, Gallego J (2017a) Comparison of global land cover datasets for cropland monitoring. Remote Sens 9.https://doi.org/10.3390/rs9111118

Pérez-Hoyos A, Rembold F, Gallego J, et al (2017b) Development of a new harmonized land cover/land use dataset for agricultural monitoring in Africa. ESA World Cover Conf 14-17 March 2017a Frascati, Rome

Phalke AR, Özdoğan M, Thenkabail PS et al (2020) Mapping croplands of Europe, Middle East, Russia, and Central Asia using Landsat, Random Forest, and Google Earth Engine. ISPRS J Photogramm Remote Sens 167:104–122. https://doi.org/10.1016/j.isprsjprs.2020.06.022

Pittman K, Hansen MC, Becker-Reshef I et al (2010) Estimating global cropland extent with multi-year MODIS data. Remote Sens 2:1844–1863. https://doi.org/10.3390/rs2071844

Portmann FT, Siebert S, Döll P (2010) MIRCA2000-Global monthly irrigated and rainfed crop areas around the year 2000: A new high-resolution data set for agricultural and hydrological modeling. Glob Biogeochem Cycles 24. https://doi.org/10.1029/2008GB003435

Ramankutty N, Evan AT, Monfreda C, Foley JA (2008) Farming the planet: 1. Geographic distribution of global agricultural lands in the year 2000. Glob Biogeochem Cycles 22. https://doi.org/10.1029/2007GB002952

Rembold F, Meroni M, Urbano F et al (2019) ASAP: A new global early warning system to detect anomaly hot spots of agricultural production for food security analysis. Agric Syst 168:247–257. https://doi.org/10.1016/j.agsy.2018.07.002

Salmon JM, Friedl MA, Frolking S, Wisser D, Douglas EM (2015) Global rain-fed, irrigated, and paddy croplands: A new high resolution map derived from remote sensing, crop inventories and climate data. Int J Appl Earth Obs Geoinf 38:321–334. https://doi.org/10.1016/j.jag.2015.01.014

Teluguntla P, Thenkabail P, Oliphant A et al (2018) A 30-m landsat-derived cropland extent product of Australia and China using random forest machine learning algorithm on Google Earth Engine cloud computing platform. ISPRS J Photogramm Remote Sens 144:325–340

Teluguntla P, Thenkabail PS, Xiong J et al (2015) Global cropland area database (GCAD) derived from remote sensing in support of food security in the twenty-first century : current achievements and future possibilities. Remote Sens Handb II:1–45

Teluguntla P, Thenkabail PS, Xiong J et al (2020) Global food security support analysis data at nominal 1 km (GFSAD1km) derived from remote sensing in support of food security in the twenty-first century: current achievements and future possibilities. In: Thenkabail PS (ed) Land resources monitoring, modeling, and mapping with remote sensing. CRC Press, Boca Raton, pp 131–160

Thenkabail P, Lyon J (2009) Remote sensing of global croplands for food security. CRC Press, Boca Raton

Thenkabail PS, Biradar CM, Noojipady P et al (2009) Global irrigated area map (GIAM), derived from remote sensing, for the end of the last millennium. Int J Remote Sens 30:3679–3733. https://doi.org/10.1080/01431160802698919

Thenkabail PS, Hanjra MA, Dheeravath V, Gumma M (2010) A holistic view of global croplands and their water use for ensuring global food security in the 21st century through advanced remote sensing and non-remote sensing approaches. Remote Sens 2:211–261. https://doi.org/10.3390/rs2010211

Thenkabail PS, Knox JW, Ozdogan M et al (2012) Assessing future risks to agricultural productivity, water resources and food security: How can remote sensing help? Photogramm Eng Remote Sensing 78:773–782

Thenkabail PS, Lyon GJ, Huete A (2011) Advances in hyperspectral remote sensing of vegetation. In: Thenkabail P, Lyon GJ, Huete A (eds) Hyperspectral remote sensing of vegetation. CRC Press, Boca Raton, pp 3–38

USGS EROS (2017) NASA Making Earth System Data Records for Use in Research Environments (MEaSUREs) Global Food Security-support Analysis Data (GFSAD) 1 km datasets. Accessed February 18, 2021. https://lpdaac.usgs.gov/documents/172/GFSAD1K_User_Guide_V1.pdf

Vancutsem C, Marinho E, Kayitakire F et al (2013) Harmonizing and combining existing land cover/land use datasets for cropland area monitoring at the African continental scale. Remote Sens 5:19–41. https://doi.org/10.3390/rs5010019

Waldner F, Fritz S, Di Gregorio A, et al (2016) A unified cropland layer at 250 m for global agriculture monitoring. Data 1. https://doi.org/10.3390/data1010003

Xiong J, Prasad T, James T et al (2017) Nominal 30-m cropland extent map of continental africa by integrating pixel-based and object-based algorithms using sentinel-2 and landsat-8 data on google earth engine. Remote Sens 9:1065

Yadav K, Congalton RG (2018) Accuracy assessment of Global Food Security-Support Analysis Data (GFSAD) cropland extent maps produced at three different spatial resolutions. Remote Sens 10. https://doi.org/10.3390/rs10111800

Yu L, Wang J, Clinton N et al (2013) FROM-GC: 30 m global cropland extent derived through multisource data integration. Int J Digit Earth 6:521–533. https://doi.org/10.1080/17538947.2013.822574

Yu Q, You L, Wood-Sichra U et al (2020) A cultivated planet in 2010 – Part 2: The global gridded agricultural-production maps. Earth Syst Sci Data 12:3545–3572. https://doi.org/10.5194/essd-12-3545-2020

Global Thematic Land Use Cover Datasets Characterizing Artificial Covers

David García-Álvarez, Javier Lara Hinojosa, and Francisco José Jurado Pérez

Abstract

The mapping of artificial covers at a global scale has received increasing attention in recent years. Numerous thematic global Land Use Cover (LUC) datasets focusing on artificial surfaces have been produced at increasingly high spatial resolutions and using methods that ensure improved levels of accuracy. In fact, there are several long time series of maps showing the evolution of artificial surfaces from the 1980s to the present. Most of them allow for change detection over time, which is possible, thanks to the high level of accuracy at which artificial surfaces can be mapped and because transitions from artificial to non-artificial covers are very rare. Global thematic LUC datasets characterizing artificial covers usually map the extent or percentage of artificial or urban areas across the world. They do not provide thematic detail on the different uses or covers that make up artificial or urban surfaces. Unlike other general or thematic LUC datasets, those focusing on artificial covers make extensive use of radar data. In several cases, optical and radar imagery have been used together, as each source provides complementary information. Global Urban Expansion 1992–2016 and ISA, which were produced at a spatial resolution of 1 km, are the coarsest of the nine datasets reviewed in this chapter. ISA provides information on the percentage of impervious surface area per pixel. The GHSL edition of 2014 and the GMIS at 30 m also provide sub-pixel information, whereas all the other datasets reviewed here only map the extent of artificial/impervious/urban areas. Most of the datasets reviewed in this chapter were produced at a spatial resolution of 30 m. This is due to the extensive use of Landsat imagery in the production of these datasets. Landsat provides a long, high-resolution series of satellite imagery that enables effective mapping of the evolution of impervious surfaces at detailed scales. Of the datasets produced at 30 m, Global Urban Land maps artificial covers for seven different dates between 1980 and 2015, while GHSL does the same for five different dates between 1987 and 2016, although the map for the last date was produced at 20 m. GUB maps the extent of urban land for seven dates between 1990 and 2018 and was produced together with GAIA, which provides an annual series of maps for the period 1985–2018. HBASE, GMIS and GISM, also at 30 m, are only available for one reference year. The same is true of GUF and WSF, which were produced as part of the same effort to map global artificial surfaces as accurately as possible. They provide the most detailed datasets up to date, with spatial resolutions of 12 m (GUF) and 10 m (WSF). Future updates of WSF will produce a consistent time series of global LC maps of artificial areas from the 1980s to the present. It aims to be the longest, most detailed, most accurate dataset ever produced on this subject.

Keywords

Artificial areas • Impervious surfaces • Global Urban Land • GAIA • GUB • GHSL • Global Urban Expansion 1992–2016 • ISA • HBASE • GMIS • GUF • WSF • GISM

D. García-Álvarez (✉)
Departamento de Geología, Geografía y Medio Ambiente, Universidad de Alcalá, Alcalá de Henares, Spain
e-mail: David.garcia@uah.es

J. Lara Hinojosa · F. J. Jurado Pérez
Departamento de Análisis Geográfico Regional y Geografía Física, Universidad de Granada, Granada, Spain

1 Global Urban Land

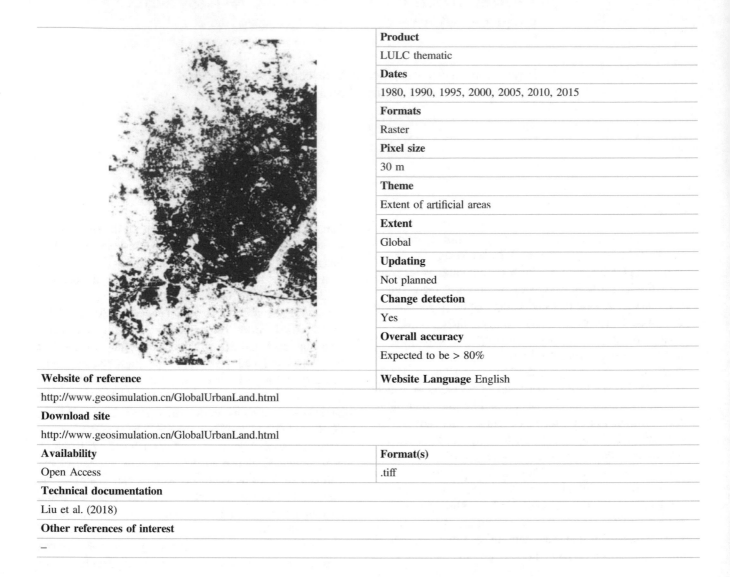

Product	
LULC thematic	
Dates	
1980, 1990, 1995, 2000, 2005, 2010, 2015	
Formats	
Raster	
Pixel size	
30 m	
Theme	
Extent of artificial areas	
Extent	
Global	
Updating	
Not planned	
Change detection	
Yes	
Overall accuracy	
Expected to be > 80%	

Website of reference	**Website Language** English
http://www.geosimulation.cn/GlobalUrbanLand.html	

Download site	
http://www.geosimulation.cn/GlobalUrbanLand.html	

Availability	**Format(s)**
Open Access	.tiff

Technical documentation	
Liu et al. (2018)	

Other references of interest	
–	

Project

Global Urban Land, also referred to as Multi-temporal Global Impervious Surface (MGIS), is a project developed by researchers from different Chinese universities (Sun Yat-sen, East China Normal, Guangzhou and Jiangsu Normal) to create a high-resolution multi-temporal urban land dataset. They aimed to provide high-resolution data about urban areas at multiple dates, which could be useful for those studying urbanization and the impact of artificial surfaces and human activities on the environment.

In this dataset, urban land is understood as an impervious surface. It can therefore be assimilated to all the datasets mapping artificial or impervious surfaces, such as GAIA. Initially, the dataset was produced for the period 1990–2010, with maps every 5 years. However, it has since been updated, with new data for the years 1980 and 2015.

Production method

Global Urban Land is obtained through an index-based method that automatically predicts urban land: the Normalized Urban Areas Composite Index (NUACI). The index, implemented through the Google Earth Engine (GEE) platform, uses Landsat imagery and DMSP-OLS nighttime lights images as inputs.

To calibrate the index, the world was stratified into different urban ecoregion categories, according to the particular physical and socioeconomic characteristics of each urban region. Three indexes (NDWI, NDVI and NDBI) were extracted from Landsat imagery to calculate the NUACI. In addition, a binary mask was obtained by segmenting DMSP-OLS nighttime lights images into urban and non-urban by applying a specific threshold. On the basis of these data, the NUACI index was calculated, obtaining a raster showing the percentage of impervious surface area per pixel.

The final Global Urban Land dataset was obtained after applying region-specific segmentation thresholds to the NUACI images showing the degree of imperviousness. After this step, a binary urban/non-urban map was generated.

For the calibration of the NUACI index, as well as for the application of segmentation thresholds, cities were randomly assigned to three equal-sized groups: centroid sites, threshold sites and testing sites. Different criteria for index calibration and threshold segmentation were decided for each type of site.

Product description

The Global Urban Land dataset can be downloaded from three different servers: Baidu Drive, Google Drive and FTP. From them, users will be able to separately download the dataset for each of the available years of reference. For each year, there is a compressed folder (.zip) containing the whole dataset distributed in tiles.

An auxiliary vector file (.shp) is provided to help users identify the number of the files corresponding to their area of interest (field "grid_id"). The scientific paper presenting the dataset is also available for download, together with a text file with relevant technical information about the product and the reference data used to produce the dataset for the initial period 1990–2010.

Downloads

Global Urban Land 2010
– Raster files with the extent of the artificial surfaces for each tile into which the dataset was divided (.tiff)

Legend and codification

Code	Label	Code	Label
0	Non-urban land	1	Urban land

Practical considerations

The authors have identified several uncertainties and limitations in the dataset. The 1990 map has missing data areas due to the lack of Landsat imagery or reference data for these areas. The binary mask used to create the dataset may also introduce some uncertainties, as it was unable to detect some urban infrastructure. In addition, the accuracy of the dataset is relatively low in arid and tropical areas. The authors also described the limitations associated with a binary (urban/non-urban) mapping approach, which oversimplifies the real situation being mapped.

2 GHSL (Global Human Settlement Layer)—Built-up Area

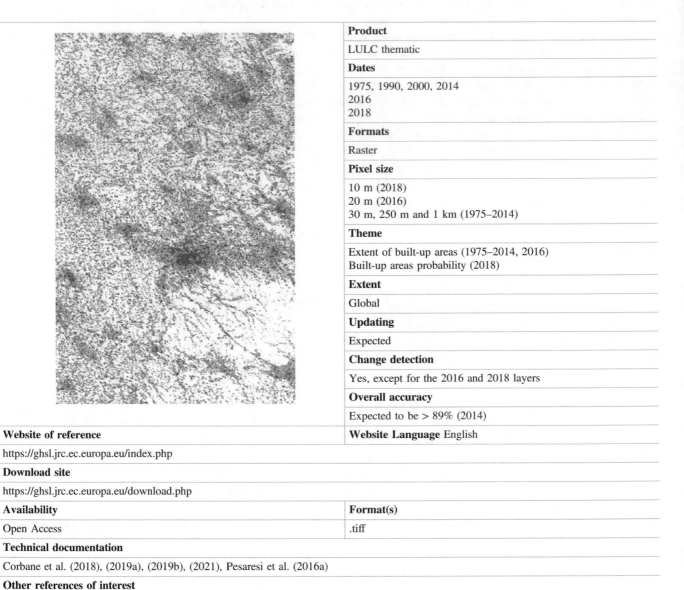

Product
LULC thematic
Dates
1975, 1990, 2000, 2014 2016 2018
Formats
Raster
Pixel size
10 m (2018) 20 m (2016) 30 m, 250 m and 1 km (1975–2014)
Theme
Extent of built-up areas (1975–2014, 2016) Built-up areas probability (2018)
Extent
Global
Updating
Expected
Change detection
Yes, except for the 2016 and 2018 layers
Overall accuracy
Expected to be > 89% (2014)

Website of reference	Website Language English
https://ghsl.jrc.ec.europa.eu/index.php	

Download site
https://ghsl.jrc.ec.europa.eu/download.php

Availability	Format(s)
Open Access	.tiff

Technical documentation
Corbane et al. (2018), (2019a), (2019b), (2021), Pesaresi et al. (2016a)

Other references of interest
Joint Research Centre (2020), Melchiorri et al. (2018), (2019), Pesaresi et al. (2016b)

Project

The GHSL is a project supported by the European Commission through its Joint Research Centre (JRC) and the Directorate General for Regional and Urban Policy (DG REGIO) and for Internal Market, Industry, Entrepreneurship and SMEs (DG GROWTH). The project is part of the Human Planet Initiative of the Group on Earth Observations (GEO). It builds on the research activity carried out by the JRC since 2010.

The project aims to provide high-quality, detailed data that characterize human settlements at a global level over a period of time. The datasets obtained enable us to understand where people live and how human settlements have evolved over time. This provides a useful source of information in support of policy- and decision-making. In this regard, one of the purposes of this project is to contribute to the development of the indicators required to measure different policy objectives.

The project has delivered three main products, one of them referring to the urban footprint of human settlements (GHS-BUILT). This is the product described here, because of its assimilation to a Land Cover product. The other two products include a global grid of population density (GHS-POP) and a spatial layer of urban settlements classified according to their typology (GHS-SMOD). They have been produced for the same three time points and are based on the initial GHS-BUILT layers.

GHS-BUILT was initially produced for the years 1975, 1990, 2000 and 2014, providing a consistent time series of maps. They are available at three spatial resolutions, the finest one (30 m) providing information on the extent of the built-up areas. The aggregated maps (250 m, 1 km) give information on the percent of built-up areas per pixel.

New editions of the GHS-BUILT product have recently been released for the years 2016 and 2018. However, they are based on different imagery (Sentinel-1 and Sentinel-2) and were obtained using different methods. They are therefore not comparable to previous maps.

Production method

The GHS-BUILT maps for the period 1975–2014 were produced by classifying the historical archive of Landsat imagery through a Symbolic Machine Learning (SML) classifier. This is a supervised classifier that builds on a set of learning data. It includes previous information from older versions of the same product and other auxiliary datasets like the GLC30 or a global surface water product.

The classifier helped extract the following earth features from the imagery: clouds, water and built-up. After classifying the imagery mosaics for each of the periods under consideration (1975, 1990, 2000 and 2014), the classifications were then merged, thus ensuring the consistency over time of the historical series of maps.

The 2016 GHS-BUILT was also obtained using the SML classifier. However, the classification was carried out over Sentinel-1 backscatter imagery, so adapting the classifier to the potential and characteristics of this source of imagery. Certain differences can also be identified with regard to the learning data used in the image classification.

The 2018 GHS-BUILT was obtained by classifying a global cloud-free composite of Sentinel-2 imagery through a deep-learning-based framework, which is called the GHS-S2Net approach. A specific model was trained for each UTM grid zone of the global map, which allowed to account for local variability and computational model requirements. The model builds on a convolution neural networks architecture, which calculates the probability of built-up areas per pixel. Each model was trained with data from previous GHS-BUILT datasets, the European Settlement Map (see Sect. 6 in chapter "Supra-National Thematic Land Use Cover Datasets"), Facebook high-resolution settlement data and Microsoft building footprint data.

Product description

GHS-BUILT for the period 1975–2014 can be downloaded in small tiles or as a single global file. It is also provided at three different spatial resolutions (30 m, 250 m and 1 km) and in two different projections (Mollweide and Mercator).

The map at 30 m can only be downloaded as a multi-temporal product, providing information about the urban footprint for the whole period covered by the product (1975–2014). Maps at 250 m and 1 km can also be downloaded for specific years, without reference to built-up areas for other time points.

The dataset for 2016 obtained from Sentinel-1 imagery can only be downloaded for the whole world as a single zipped file. The dataset for 2018 is distributed in tiles corresponding with UTM grid zones. A vector layer representing the UTM grid zones in which the product is split can be downloaded as an auxiliary file, together with the product's metadata.

Downloads

GHS—Built-up 2018 (10 m)
– Raster file with LUC information

GHS—Built-up 2016
– Raster files with LUC information for each of the tiles in which the product is divided (.tiff) (13_OTSU folder) – Global mosaic of the product (.vrt) (V1-0) – Vector file representing the tiles in which the product is distributed (.shp) (V1-0) – PDF with the description of the product

GHS—Built-up 2014 (250 m)
– Raster file with LUC information – PDF with the description of the product

GHS—Built-up multi-temporal (30 m)
– Raster file with LUC information – PDF with the description of the product

Legend and codification

GHS—Built-up 2018 (10 m)

Code	Label
0–100	Probability of being built-up area (1–100)
255	No data

GHS—Built-up 2016 (20 m)

Code	Label
0	No built-up/no data
1	Built-up area

(continued)

GHS—Built-up 2014 (250 m)

Code	Label

GHS—Built-up 2014 (250 m)

Code	Label
0–100	Built-up area density (1–100)
−200	No data

GHS—Built-up multi-temporal (30 m)

Code	Label	Code	Label
0	No data	4	Built-up from 1990 to 2000 epochs
1	Water surface	5	Built-up from 1975 to 1990 epochs
2	Land not built-up in any epoch	6	Built-up to 1975 epoch
3	Built-up from 2000 to 2014 epochs		

Practical considerations

The maps for 2016 and 2018 are a test version of the product obtained with Sentinel-1 and Sentinel-2 imagery. They should not be therefore used together with the other GHS-BUILT maps, as if they were part of the same series of maps.

Users interested in the method used to produce this dataset can find the general workflow for built-up areas extraction in the MASADA (Massive Spatial Automatic Data Analytics) tool.[1]

---

[1] https://ghsl.jrc.ec.europa.eu/tools.php.

3 GAIA—Global Artificial Impervious Areas, GUB—Global Urban Boundaries

Product	
LULC thematic	
Dates	
1985–2018 (GAIA) 1990, 1995, 2000, 2005, 2010, 2015, 2018 (GUB)	
Formats	
Raster (GAIA), Vector (GUB)	
Pixel size	
30 m	
Theme	
Extent of artificial areas Urban boundaries	
Extent	
Global	
Updating	
Not planned	
Change detection	
Yes	
Overall accuracy	
Expected to be > 89% (GAIA)	

Website of reference	**Website Language** English
http://data.ess.tsinghua.edu.cn/	

Download site

http://data.ess.tsinghua.edu.cn/gaia.html (GAIA)
http://data.ess.tsinghua.edu.cn/gub.html (GUB)

Availability	**Format(s)**
Open Access	.tiff, .shp

Technical documentation

Gong et al. (2020), Li et al. (2020)

Other references of interest

Gong et al. (2019), Li et al. (2015), Li and Gong (2016)

Project

This project was led by researchers from Tsinghua University with the collaboration of colleagues from other Chinese and American universities, together with Google and the US Geological Survey. They have produced two different datasets: Global Artificial Impervious Areas (GAIA) and Global Urban Boundaries (GUB). The second was obtained from the first and both were produced to better understand global urbanization and other human socioeconomic activities and their impacts on the environment.

GAIA maps artificial surfaces across the world, whereas GUB maps urban areas. Unlike GAIA, GUB does not include small urban patches. In addition, in the GUB dataset non-artificial areas within cities, such as green areas or water bodies, are considered urban.

The project took advantage of the full Landsat data archive (1985–2018), providing a temporally consistent series of maps in which the only change possible was from non-artificial to artificial surfaces. The project is part of the global LUC mapping efforts carried out by Tsinghua University, such as FROM-GLC or GLC250, which are described in previous chapters of this book.

Production method

GAIA was first produced via the classification of the Landsat imagery archive (1985–2018). The dataset obtained in this way was then used to produce GUB. Google Earth Engine (GEE) was used to create both datasets.

Two different classification methodologies were followed to obtain GAIA: one for non-arid regions and the other for arid ones. This is due to the spectral confusion between impervious areas and bare lands. For classification purposes, the world was split into 583 tiles, of which 155 referred to arid environments.

The classification of non-arid areas was based on previous experiences of the production team in mapping artificial areas at local and national scales. Annual artificial areas were first obtained through an "ExclusionInclusion" algorithm, based on training data from earlier Landsat datasets and Google Earth imagery, and NVDI, MNDWI and SWIR data

from Landsat imagery. The time series of maps obtained in this way was then further refined through a "temporal consistency check" approach.

For arid areas, a primary urban mask was first obtained for the year 2018 based on radar data from Sentinel-1 and VIIRS NTL. The classification of Sentinel-1 data was based on backscatter coefficients and NTL data was classified according to the quantile-based method. In both cases, different parameters were used for each arid biome. Once the two urban masks for 2018 had been obtained, they were mixed. Then, the time series of maps was created using the same "ExclusionInclusion" algorithm and "temporal consistency check" approach applied to the non-arid regions.

The GUB dataset was later obtained on the basis of a combination of two inputs: a kernel density map at a spatial resolution of 1 km obtained from GAIA based on a kernel density estimation (KDE) approach; and an initial urban boundary obtained from a Cellular Automata-based (CA) modelling exercise at 30 m. The results were improved through a morphological approach with dilation and erosion processing. This last step improved the mapped urban boundaries around fringe urban areas. Small holes inside urban areas were removed in a post-processing stage.

Product description

GAIA is distributed in 3.5° × 3.5° tiles, named according to the latitude and longitude of their upper-left coordinates. Users can download a vector file (.shp) drawing all the tiles and providing their names (field "FName_ID").[2] GUB is distributed as a single global file for each of the 7 years available.

Downloads

GAIA
– A raster file with the extent of artificial areas (.tiff)

GUB
– A vector file with urban boundaries (.shp)

---

[2] http://data.ess.tsinghua.edu.cn/data/GAIA/GAIA_shape.zip.

Legend and codification

GAIA

Code	Label[a]	Code	Label	Code	Label	Code	Label	Code	Label	Code	Label
1	2018	8	2011	15	2004	22	1997	29	1990		
2	2017	9	2010	16	2003	23	1996	30	1989		
3	2016	10	2009	17	2002	24	1995	31	1988		
4	2015	11	2008	18	2001	25	1994	32	1987		
5	2014	12	2007	19	2000	26	1993	33	1986		
6	2013	13	2006	20	1999	27	1992	34	1985		
7	2012	14	2005	21	1998	28	1991				

[a]The label refers to the time when the pixel was sealed

Database

GUB

GUB

	ORIG_FID ▼	urbanArea
1	65461	1,67112000000
2	65460	9,33964000000
3	65459	2,31508000000

- Orig_FID: Unique identifier for each polygon
- UrbanArea: area of the delimited urban area

4   Global Urban Expansion 1992–2016

Product
LULC thematic
Dates
1992, 1996, 2000, 2006, 2010, 2016
Formats
Raster
Pixel size
1 km
Theme
Extent of Urban areas
Extent
Global
Updating
Not expected
Change detection
Yes
Overall accuracy
Expected to be > 90%

Website of reference	**Website Language** English
https://doi.pangaea.de/10.1594/PANGAEA.892684	
Download site	
https://doi.pangaea.de/10.1594/PANGAEA.892684	
Availability	**Format(s)**
Open Access	.tiff
Technical documentation	
He et al. (2019)	
Other references of interest	
–	

Project

The dataset on Global Urban Expansion is the result of the work carried out by a group of researchers from the Beijing Normal University, the China University of Geosciences and Murray State University in the USA. Their aim was to create a new dataset on urban expansion using fully convolutional network (FCN)-based methods, which would be able to overcome some of the limitations of previous datasets on the same topic: outdated datasets, low spatial resolutions and low levels of accuracy.

The dataset provides useful information for studies addressing global urbanization and its impacts on the environment. It considers as urban all those built-up areas where human-constructed or artificial elements cover more than half of the area or pixel.

Production method

A specific fully convolutional network (FCN) was developed to map the urban areas in the Global Urban Expansion dataset. FCN are deep learning structures based on convolutional neural networks (CNN) that employ pixel-to-pixel image recognition.

The FCN was fed with different sources of input data: Nighttime Light (NTL) imagery from NOAA and NPP-VIIRS, as well as Normalized Difference Vegetation Index (NDVI) and Land Surface Temperature (LST) data from MODIS. Other auxiliary data sources were also employed to obtain the Global Urban Expansion dataset: urban population statistics, Landsat imagery and the GHS and LC-CCI LUC datasets. LST data is only available for the period 2000–2016 and was not used to map the urban areas in 1992 and 1996.

The FCN was calibrated with data from MODIS Land Cover, differentiating urban from non-urban areas. The calibration provided the weights of the FCN, which were then used to obtain the final Global Urban Expansion dataset.

A post-classification stage using population density data was carried out to ensure the consistency over time of the maps obtained.

Product description

The dataset can be downloaded as a single compressed file (.zip), including the raster files showing the urban expansion for each available year. No auxiliary information is provided with the dataset.

Downloads

Global Urban Expansion
– Raster files with urban expansion data for each mapped year (.tiff)

Legend and codification

Code	Label	Code	Label
0	Non-urban area	1	Urban area

5 ISA—Global Inventory of the Spatial Distribution and Density of Constructed Impervious Surface Area

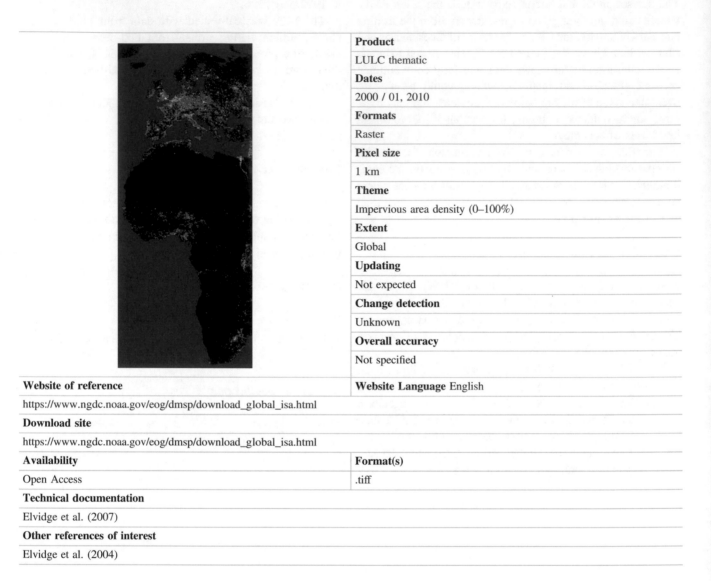

Product	
LULC thematic	
Dates	
2000 / 01, 2010	
Formats	
Raster	
Pixel size	
1 km	
Theme	
Impervious area density (0–100%)	
Extent	
Global	
Updating	
Not expected	
Change detection	
Unknown	
Overall accuracy	
Not specified	

Website of reference	**Website Language** English
https://www.ngdc.noaa.gov/eog/dmsp/download_global_isa.html	

Download site

https://www.ngdc.noaa.gov/eog/dmsp/download_global_isa.html

Availability	**Format(s)**
Open Access	.tiff

Technical documentation

Elvidge et al. (2007)

Other references of interest

Elvidge et al. (2004)

Project

ISA is the result of a project partially funded by NASA's Carbon Cycle research program and is made up of researchers from different American institutions and universities. It builds on a previous attempt to map Impervious Surface Area (ISA) for the USA led by the NOAA (National Oceanic and Atmospheric Administration).

ISA was initially produced for the reference year 2000/01. A new version of the dataset is available for 2010. The dataset is useful for understanding the global distribution of impervious areas and for studies analysing the impact of these covers and their associated uses on the environment.

In addition to the production of an ISA density grid, the project's outputs also include spreadsheets with information about the quantity of ISA per person at a country level and the ISA density per watershed areas. These are classified according to the proportion of ISA in three groups: stressed (1–10% ISA), impacted (10–25%) and degraded (>25%).

Production method

The ISA density grid for the reference year 2000/01 was obtained through a model making use of night-time lights imagery (DMSP OLS) and a population count grid (LandScan). Night lights imagery were captured in 2000–01, whereas the population count grid dates from 2004. A linear regression was defined to estimate the ISA density based on those two inputs. Only cells with a population count of at least 3 were considered in the regression. The model was calibrated with the ISA dataset produced for the USA at 30 m.

There is no accompanying information about the production process of the 2010 map. Therefore, we cannot know if it followed the same method as the previous map or some changes were introduced in the production process.

Product description

The ISA dataset for the reference year 2010 is distributed as a single compressed file (.gz). For the reference year 2000/01, the dataset is distributed in two different projections (GCS, Mollweide) and formats (ENVI, GeoTiff).

Spreadsheets containing ISA information per country and watershed are also available on the project website. This data is distributed together with a text file offering a technical explanation of these results.

Downloads

ISA (GeoTiff 2000–2001)
– Raster file with ISA proportion (.tiff)

Legend and codification

Code	Label
0–100	Impervious area density

Practical considerations

Although there is an ISA map for 2010, no information is available about the way it was produced. If there were important differences between the production methods used in 2000/01 and 2010 editions of ISA, they could not be used for comparison purposes or land change studies.

ISA was obtained from a calibration based on data for the USA. This may make the final result less accurate for countries with different night lights conditions, such as African countries. It is therefore likely that this dataset underestimates ISA densities in many different parts of the world.

6   HBASE and GMIS (Global High Resolution Urban Data from Landsat)

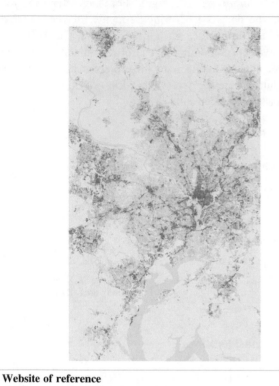

Product
LULC thematic
Dates
2010
Formats
Raster
Pixel size
30 m, 250 m, 1 km
Theme
Extent of urban areas Percentage of impervious areas
Extent Global
Updating
Not expected
Change detection
No (only one date)
Overall accuracy
Not specified

Website of reference	**Website Language** English

https://sedac.ciesin.columbia.edu/data/collection/ulandsat

Download site
https://sedac.ciesin.columbia.edu/data/set/ulandsat-hbase-v1/data-download
https://sedac.ciesin.columbia.edu/data/set/ulandsat-gmis-v1/data-download

Availability	**Format(s)**
Open Access after registration	.tiff

Technical documentation

De Colstoun et al. (2017), Wang et al. (2017)

Other references of interest

–

Project

Researchers from NASA, in collaboration with the University of Maryland and other American institutions, created two datasets to globally map artificial areas across the world: Global Human Built-up and Settlement Extent (HBASE) and Global Man-made Impervious Surface (GMIS). These were created within the context of NASA's Land Cover and Land Use Change (LCLUC) program.

Both datasets used Landsat imagery available through the Landsat Global Land Survey (GLS) archive to consistently map impervious surfaces across the globe at high spatial resolution for the reference year 2010. These datasets aimed to overcome the resolution-related limitations of previous datasets. They can be useful for anyone studying impervious surfaces, their impact on the environment or their relation with other land dynamics. Because of the detail they provide, they can be used for studies and applications at global, supra-national, national and local scales.

HBASE and GMIS are complementary datasets, jointly produced to address the spectral confusion arising from the fact that many impervious areas are sealed with soil, sand, rocks, etc. and can therefore be confused with bare land. The HBASE dataset provides a mask to remove such areas from the GMIS dataset.

Production method

HBASE and GMIS were produced separately, although the first was used as a mask in the production of the second. In both cases, the GLS 2010 Surface Reflectance Dataset from Landsat was the input imagery.

For the production of HBASE, the first stage was to segment the GLC imagery using a Recursive Hierarchical Image Segmentation (RHSeg) software package. This produced a series of objects, from which different textures and other variables were extracted. On the basis of these variables, a random forest (RF) classification was carried out to classify the segmented objects in HBASE/non-HBASE categories. Training data for the classification was obtained from Landsat and Google Earth imagery. In addition, OpenStreetMap was used as an auxiliary dataset in the post-classification process to improve the mapping of the roads, which had not been correctly classified in the previous stages.

GMIS was obtained in two steps, with classifications carried out at the scene level. First, an object-based classification of GLC imagery was performed using the HSeg (Hierarchical Image Segmentation) Learn software to classify all the areas as either impervious or non-impervious. Only pixels effectively classified as HBASE in the previous dataset were considered and pixels with a low-quality classification were discarded. Later, the percentage of impervious area per pixel was calculated for all pixels classified as impervious through a regression-tree algorithm (Cubist). The algorithm was run with reference data from the National Geospatial-Intelligence Agency (NGA) at a spatial resolution of 30 m.

Product description

An online viewer allows users to download HBASE and GMIS: (i) for a specific country, (ii) for the tiles into which the datasets are split[3] or (iii) for user-defined areas of interest (by drawing a polygon or shape or uploading a shapefile file that defines the area). The files can be downloaded at the original resolution (30 m) and resampled at 250 m and 1 km. Users can also choose between two projections: geographic or UTM.

---

[3] The datasets are split into tiles corresponding to the UTM zones.

Other complementary products are also available for download: a layer of standard error for the production of the GMIS dataset and an HBASE probability layer.

Downloads

GMIS

– Raster files with information on the percentage of impervious surface area (.tiff)
– A text document with technical information about the product (.txt)

HBASE

– Raster files with information on the urban extent (.tiff)
– A text document with technical information about the product (.txt)

Legend and codification

Global Man-made Impervious Surface (GMIS)—Percentage

Code	Label
0–100	Percentage of impervious surface area (0–100%)
200	Non-HBASE
255	No data, clouds, shadows

Global Human Built-up and Settlement Extent (HBASE)

Code	Label	Code	Label
200	Non-HBASE	202	Road
201	HBASE	255	No data, clouds, shadows

Practical considerations

Users can explore the different datasets available online,[4] including the complementary layer about the standard error of the Impervious Surface Percentage raster and the HBASE probability layer. Full metadata for GMIS and HBASE is also available online.[5]

GMIS and HBASE have some limitations associated with their production methodology. For example, they may present areas of missing information due to cloud cover or other factors. The technical documents for the product (cited below) provide a detailed description of all these limitations.

As part of the same project, Landsat imagery composites for 66 urban areas are also available for download.[6]

---

[4] https://sedac.ciesin.columbia.edu/mapping/gmis-hbase/explore-view/.
[5] https://sedac.ciesin.columbia.edu/data/set/ulandsat-hbase-v1/metadata .https://sedac.ciesin.columbia.edu/data/set/ulandsat-gmis-v1/metadata.
[6] https://sedac.ciesin.columbia.edu/data/set/ulandsat-cities-from-space.

7 GUF—Global Urban Footprint

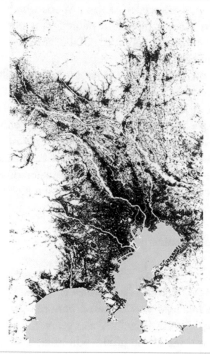

Product	
LULC thematic	
Dates	
2011	
Formats	
Raster	
Pixel size	
0.4 arc seconds ($\sim$ 12 m near the Equator) 2.8 arc seconds ($\sim$ 84 m near the Equator)	
Theme	
Extent of built-up areas	
Extent	
Global	
Updating	
Not expected	
Change detection	
No (only one date)	
Overall accuracy	
Not specified	

Website of reference	**Website Language** English
https://www.dlr.de/eoc/en/desktopdefault.aspx/tabid-9628/16557_read-40454/	
Download site	
https://www.dlr.de/eoc/en/desktopdefault.aspx/tabid-11725/20508_read-47944/	

Availability	**Format(s)**
Open Access on request after filling in a request form	.tiff

Technical documentation

Esch et al. (2010), (2012), (2013), (2017)

Other references of interest

Esch et al. (2011), (2014), (2018a), (2018b), (2020), Marconcini et al. (2014)

Project

Global Urban Footprint (GUF) is a dataset produced by the German Aerospace Center (DLR) from radar imagery at very high spatial resolution: 0.4 arc seconds, which is equivalent to about 12 m at the Equator. The dataset at the highest resolution is envisaged for scientific uses, whereas a coarser resolution of the dataset at 2.8 arc seconds (~ 84 m near the Equator) has also been produced for non-commercial use by the general public.

The dataset aims to facilitate the quantitative and qualitative characterization of urban surfaces (size, form, spatial distribution) at different scales, from local to continental and global. Because of its high resolution, it allows all artificial surfaces to be analysed, in both urban and rural landscapes. This information is useful for researchers investigating the different impacts of the urbanization process, be they environmental, economic, political, societal or cultural.

The dataset was produced to overcome some of the limitations associated with previous global datasets on impervious surfaces, usually produced from demographic data. In this regard, by the time it was produced, high spatial resolution datasets were only available for specific regions, such as North America and Europe.

The project is part of the Urban Thematic Exploitation Platform (U-TEP) of the European Space Agency (ESA), which explores new methods and techniques to understand urban patterns and dynamics across the world. U-TEP is one of the seven Thematic Exploitation Platforms developed by the ESA to help data user communities.

In the context of U-TEP, DLR has also developed the WSF dataset, which outperforms GUF and resolves some of the limitations associated with it. WSF, which is described later on in this chapter, is a natural progression from the work undertaken to produce GUF. The two datasets are closely linked.

Based on GUF, a new layer on global built-up density was produced at a spatial resolution of 30 m for the reference year 2012 (GUF-DenS 2012). It provides information about the percentage of sealed surface or greenness per cell. Other complementary products based on GUF have been also produced, although they have not been made available to the public, namely a layer characterizing settlement properties and patterns (GUF-NetS) and a layer defining the average building height (GUF-3D).

Production method

GUF was produced from radar imagery from the TerraSAR-X/TanDEM-X satellites at a spatial resolution of 3 m. The imagery was captured between 2011 and 2012, except for a few images from the years 2013 to 2014.

The first stage of the production process was to extract a texture feature (speckle divergence) from the input imagery. Then, based on those features, a binary settlement layer differentiating between built-up and non-built-up areas was generated through an automatic unsupervised classifier:

Support Vector Data Description (SVDD) one-class classification. The classification was carried out in $5° \times 5°$ tiles. Once all the tiles had been processed, the obtained layers were mosaicked.

In a post-classification stage, the dataset was assessed against reference data, which confirmed or excluded the presence of built-up surfaces: Open Street Map, GLC30, NLCD, Imperviousness HRL and SRTM DEM.

Seven different layers were finally obtained on the basis of different classification settings: from very conservative settings (version 1) to very relaxed settings (version 7). Version 1 followed very strict criteria for classifying areas as built-up, whereas Version 7 followed much more relaxed, more inclusive criteria.

Product description

Interested users should request the product for their area of interest from the map's producers. Before accessing the dataset, they have to sign a license agreement. Depending on the use they intend to make of the dataset, they can access the fine resolution version of the dataset (0.4 arcsec), which is only available for scientific purposes, or the coarser version (2.8 arc seconds). In both cases, the download only includes the raster file with the LUC information.

Downloads

GUF
– Raster file with built-up areas for the requested area of interest (.tiff)

Legend and codification

Code	Label	Code	Label
0	Non-built-up areas	128	No data
255	Built-up areas		

Practical considerations

The dataset can be consulted online at the two spatial resolutions available.[7] A short document summarizing the technical characteristics of the product and its methodology is also available online.[8]

Many other interesting data sources for characterizing urban areas can be found at the U-TEP Visualisation and Analytics Toolbox.[9] Users can also visualize the GUF-DenS 2012, which is not available for download. This dataset is complementary to GUF and provides information on the percentage of sealed surface for all the areas classified as built-up in GUF.

---

[7] https://geoservice.dlr.de/web/maps/eoc:guf:3857.
[8] https://www.dlr.de/eoc/en/PortalData/60/Resources/dokumente/guf/ GUF_Product_Specifications_GUF_DLR_v01.pdf.
[9] https://urban-tep.eu/puma/tool/?id=567873922.

8 WSF—World Settlement Footprint

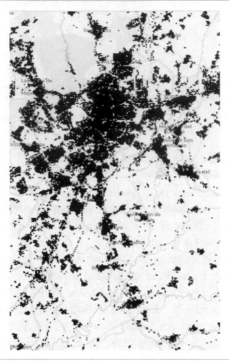

Product	
LULC thematic	
Dates	
1985–2015, 2014 / 15, 2019	
Formats	
Raster	
Pixel size	
10 m, 30 m, 100 m, 250 m, 500 m, 1 km, 10 km	
Theme	
Extent of settlement areas (10 m) Percentage of settlement areas (100 m, 250 m, 500 m, 1 km, 10 km)	
Extent	
Global	
Updating	
Expected	
Change detection	
Not yet (will be available with updates)	
Overall accuracy	
Expected to be > 86%	

Website of reference	Website Language English
https://www.esa.int/Applications/Observing_the_Earth/Mapping_our_global_human_footprint	

Download site	
https://springernature.figshare.com/collections/Outlining_where_humans_live_-_The_World_Settlement_Footprint_2015/4712852/1	

Availability	Format(s)
Open Access	.tiff

Technical documentation	
Marconcini et al. (2020)	

Other references of interest	
Esch et al. (2018a), Esch et al. (2020)	

Project

The World Settlement Footprint (WSF) is a dataset produced by the German Aerospace Center (DLR) within the context of a project (SAR4URBAN) funded by the European Space Agency (ESA) in which Synthetic Aperture Radar (SAR) is used to monitor urbanization. The project aimed to develop a new method to automatically map built-up areas via the joint use of radar and optical data.

The dataset obtained is useful for the characterization and analysis of urban patterns across the world. It overcomes the limitations of previous high spatial resolution datasets mapping impervious surfaces by making use of both radar and optical imagery at the same time. This allows WSF to avoid the misclassifications that can result from using only one of the two types of sensors: optical imagery misclassifies sand and bare soil, whereas radar imagery misclassifies complex topography areas and forested regions.

WSF is produced by the same institution as the Global Urban Footprint (GUF) described earlier in this chapter. In spite of this, it overcomes some of the limitations associated with GUF, such as the misclassifications arising from the use of single-date scenes and the use of commercial imagery, which makes updating more difficult due to the associated costs. Like GUF, WSF was also developed within the framework of the Urban Thematic Exploitation Platform (U-TEP) of the ESA.

The dataset was originally produced at a spatial resolution of 10 m, although resampled versions at 100 m, 250 m, 500 m, 1 km and 10 km are also available for download. The resampled versions show the percent of settlement area in each pixel instead of a binary classification differentiating between settlement and non-settlement areas.

The DLR is currently working with the Google Earth Engine Team on the update of the product, creating a WSF-Evolution dataset that will map the global evolution of built-up surfaces yearly from 1985 to 2015.

Production method

The WSF production methodology was first tested at a range of selected sites and, once validated, was applied to generate the global dataset. It used Sentinel-1 and Landsat 8 data for the reference years 2014 and 2015 as input.

From Sentinel-1 data, key temporal statistics were extracted from the original backscattering value. From Landsat 8 imagery, different spectral indices were extracted: vegetation index, built-up index etc. Based on the extracted information, a binary classification (settlement/non-settlement) was computed through an ensemble of Support Vector Machines (SVM) classifiers for each type of input data: radar and optical. The two results were then combined.

In a post-classification stage, the obtained result was assessed against reference information, following the post-editing object-based approach applied in the production of GUF. The auxiliary datasets were: Open Street Maps, GLC30, SRTM DEM, ASTER DEM, NLCD and the High-Resolution Layer on imperviousness.

Product description

WSF can be downloaded at multiple spatial resolutions. For the original resolution (10 m), the users will download a compressed file (.zip) that includes all the raster files into which the dataset is split (306.tiff files). The download also includes a virtual raster that merges all the tiles in a single mosaic. For all other available resolutions (100 m, 250 m, 500 m, 1 km and 10 km), users can only download a .tiff file with data on the settlement percentage per pixel. No auxiliary information is provided in either of the two cases.

Downloads

WSF 10 m
– Raster files with the settlement extent for the 306 tiles into which the product is divided (.tiff) – Raster file with a mosaic of the WSF tiles (.vrt)

WSF 100 m, 250 m, 500 m, 1 km, 10 km
– Raster files with the settlement percentage (.tiff)

Legend and codification

WSF 10 m

Code	Label	Code	Label
0	Non-settlement	255	Settlement

WSF 100 m, 250 m, 500 m, 1 km, 10 km

Code	Label	Code	Label
0–100	Settlement percent (0–100%)	255	Settlement

Practical considerations

WSF is considered by the authors to be the most accurate dataset of its type. It is part of the U-TEP tool, which also distributes many other datasets for characterizing urban areas that may be of interest to users. Users can access an online visualization of the dataset on the U-TEP tool website.[10]

For more detailed information about the characteristics of the dataset, we recommend interested users to read the scientific paper in which it was presented.

---

[10] https://urban-tep.eu/puma/tool/?id=574795484&lang=en.

9 GISM—Global Impervious Surface Map

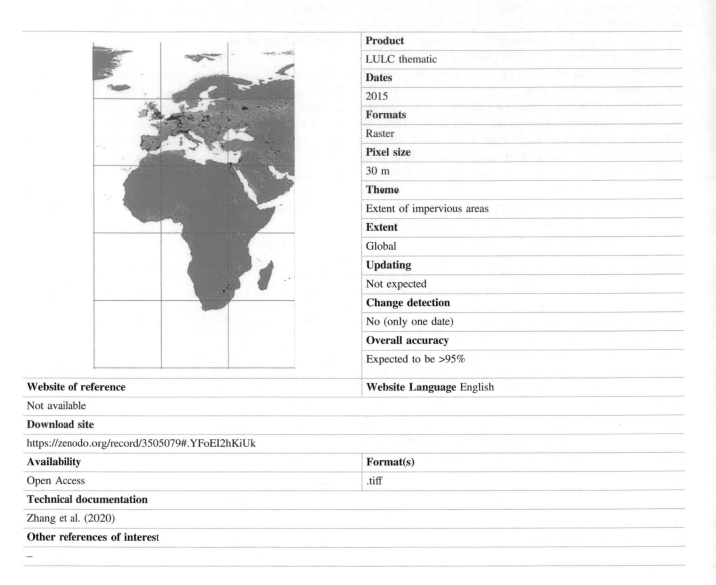

Product
LULC thematic

Dates
2015

Formats
Raster

Pixel size
30 m

Theme
Extent of impervious areas

Extent
Global

Updating
Not expected

Change detection
No (only one date)

Overall accuracy
Expected to be >95%

Website of reference	Website Language English
Not available	

Download site
https://zenodo.org/record/3505079#.YFoEI2hKiUk

Availability	Format(s)
Open Access	.tiff

Technical documentation
Zhang et al. (2020)

Other references of interest
–

Project

A group of researchers from Chinese institutions (Chinese Academy of Sciences, University of Science and Technology) and the University of Wisconsin-Milwaukee produced a Global Impervious Surface Map, which aimed to overcome some of the limitations of previous datasets.

GISM is part of recent efforts to produce a detailed global mapping of artificial or impervious surfaces with a high level of accuracy to provide useful data that can help characterize artificial areas and their associated environmental and socioeconomic impacts. The dataset was produced with that aim, without any further updates being planned.

Production method

GISM was obtained by classifying Landsat and Sentinel-1 data in the Google Earth Engine (GEE) platform, using the MSMT_RF method. First, temporal–spectral–textural features were extracted from Landsat imagery. Then, temporal-SAR features were extracted from Sentinel-1 imagery. On the basis of all these features, a classification was carried out with a random forest classifier in $5° \times 5°$ tiles. Training data for the classification were obtained from GLC30, VIIRS NTL and MODIS EVI imagery. SRTM DEM was used as an auxiliary dataset in the classification process.

Product description

GISM is distributed as a single compressed file (.zip) containing all the raster files into which the product is distributed: 954 5×5 degree tiles. No auxiliary information is provided.

Downloads

GISM
– Raster files mapping impervious areas for each of the tiles into which the dataset was divided (.tiff)

Legend and codification

Code	Label	Code	Label
1	Non-impervious	2	Impervious

Practical considerations

The only other relevant information on the dataset can be found in the scientific paper in which it was presented. Users wishing to find out more about the characteristics of this product should consult this paper.

References

Corbane C, Lemoine G, Pesaresi M et al (2018) Enhanced automatic detection of human settlements using sentinel-1 interferometric coherence. Int J Remote Sens 39:842–853. https://doi.org/10.1080/01431161.2017.1392642

Corbane C, Pesaresi M, Kemper T et al (2019a) Automated global delineation of human settlements from 40 years of Landsat satellite data archives. Big Earth Data 3:140–169. https://doi.org/10.1080/20964471.2019.1625528

Corbane J, Ehrlich C, Freire D, et al (2019b) GHSL Data Package 2019b. https://ghsl.jrc.ec.europa.eu/documents/GHSL_Data_Package_2019b.pdf. Accessed 20 Aug 2020

Corbane C, Syrris V, Sabo F, Politis P, Melchiorri M, Pesaresi M, Soille P, Kemper T (2021) Convolutional neural networks for global human settlements mapping from Sentinel-2 satellite imagery. Neural Comput Appl 33(12):6697–6720. https://doi.org/10.1007/s00521-020-05449-7

De Colstoun ECB, Huang C, Wang P, et al (2017) Documentation for the global man-made impervious surface (GMIS) Dataset from landsat. https://sedac.ciesin.columbia.edu/downloads/docs/ulandsat/ulandsat-gmis-v1-documentation.pdf. Accessed 19 May 2021

Elvidge CD, Milesi C, Dietz JB, Tuttle BT, Sutton PC, Nemani R, Vogelmann JE (2004) U.S. Constructed area approaches the size of Ohio. Eos (Washington. DC). 85, 233. https://doi.org/10.1029/2004EO240001

Elvidge CD, Tuttle BT, Sutton PS, Baugh KE, Howard AT, Milesi C, Bhaduri BL, Nemani R (2007) Global distribution and density of constructed impervious surfaces. Sensors 7:1962–1979. https://doi.org/10.3390/s7091962

Esch T, Thiel M, Schenk A, Roth A, Müller A, Dech S (2010) Delineation of Urban footprints from TerraSAR-X data by analyzing speckle characteristics and intensity information. IEEE Trans Geosci Remote Sens 48:905–916. https://doi.org/10.1109/TGRS.2009.2037144

Esch T, Schenk A, Ullmann T, Thiel M, Roth A, Dech S (2011) Characterization of land cover types in TerraSAR-X images by combined analysis of speckle statistics and intensity information. IEEE Trans Geosci Remote Sens 49:1911–1925. https://doi.org/10.1109/TGRS.2010.2091644

Esch T, Taubenböck H, Roth A, Heldens W, Felbier A, Thiel M, Schmidt M, Müller A, Dech S (2012) TanDEM-X mission—new perspectives for the inventory and monitoring of global settlement patterns. J Appl Remote Sens 6:061702–061711. https://doi.org/10.1117/1.jrs.6.061702

Esch T, Marconcini M, Felbier A, Roth A, Heldens W, Huber M, Schwinger M, Taubenbock H, Muller A, Dech S (2013) Urban footprint processor-fully automated processing chain generating settlement masks from global data of the TanDEM-X mission. IEEE Geosci Remote Sens Lett 10:1617–1621. https://doi.org/10.1109/LGRS.2013.2272953

Esch T, Marconcini M, Marmanis D, Zeidler J, Elsayed S, Metz A, Müller A, Dech S (2014) Dimensioning urbanization - An advanced procedure for characterizing human settlement properties and patterns using spatial network analysis. Appl Geogr 55:212–228. https://doi.org/10.1016/j.apgeog.2014.09.009

Esch T, Heldens W, Hirner A, Keil M, Marconcini M, Roth A, Zeidler J, Dech S, Strano E (2017) Breaking new ground in mapping human settlements from space—the global urban footprint. ISPRS J Photogramm Remote Sens 134:30–42. https://doi.org/10.1016/j.isprsjprs.2017.10.012

Esch T, Asamer H, Bachofer F, Balhar J, Boettcher M, Boissier E, Hirner A, Mathot E, Marconcini M, Metz-Marconcini A, Permana H, Soukup T, Svaton V, Uereyen S, Zeidler J (2018a) New prospects in analysing big data from space—the urban thematic exploitation platform. Int Geosci Remote Sens Symp 2018a-July, 8193–8196. https://doi.org/10.1109/IGARSS.2018.8517493

Esch T, Asamer H, Bachofer F, Balhar J, Boettcher M, Boissier E, d'Angelo P, Gevaert CM, Hirner A, Jupova K, Kurz F, Kwarteng AY, Mathot E, Marconcini M, Marin A, Metz-Marconcini A, Pacini F, Paganini M, Permana H, Soukup T, Uereyen S, Small C, Svaton V, Zeidler JN (2020) Digital world meets urban planet—new prospects for evidence-based urban studies arising from joint exploitation of big earth data, information technology and shared knowledge. Int J Digit Earth 13:136–157.https://doi.org/10.1080/17538947.2018.1548655

Gong P, Li X, Zhang W (2019) 40-year (1978–2017) human settlement changes in China reflected by impervious surfaces from satellite remote sensing. Sci Bull 64:756–763. https://doi.org/10.1016/j.scib.2019.04.024

Gong P, Li X, Wang J, Bai Y, Chen B, Hu T, Liu X, Xu B, Yang J, Zhang W, Zhou Y (2020) Annual maps of global artificial impervious area (GAIA) between 1985 and 2018. Remote Sens Environ 236.https://doi.org/10.1016/j.rse.2019.111510

He C, Liu Z, Gou S, Zhang Q, Zhang J, Xu L (2019) Detecting global urban expansion over the last three decades using a fully convolutional network. Environ Res Lett 14.https://doi.org/10.1088/1748-9326/aaf936

Joint Research Centre (2020) Atlas of the Human Planet 2019. https://ghsl.jrc.ec.europa.eu/documents/Atlas_2019.pdf?t=1584689728. Accessed 20 Aug 2020

Li X, Gong P, Liang L (2015) A 30-year (1984–2013) record of annual urban dynamics of Beijing City derived from Landsat data. Remote Sens Environ 166:78–90. https://doi.org/10.1016/j.rse.2015.06.007

Li X, Gong P (2016) An "exclusion-inclusion" framework for extracting human settlements in rapidly developing regions of China from Landsat images. Remote Sens Environ 186:286–296. https://doi.org/10.1016/j.rse.2016.08.029

Li X, Gong P, Zhou Y, Wang J, Bai Y, Chen B, Hu T, Xiao Y, Xu B, Yang J, Liu X, Cai W, Huang H, Wu T, Wang X, Lin P, Li X, Chen J, He C, Zhu Z et al (2020) Mapping global urban boundaries from the global artificial impervious area (GAIA) data. Environ Res Lett 15(9). https://doi.org/10.1088/1748-9326/ab9be3

Liu X, Hu G, Chen Y, Li X, Xu X, Li S, Pei F, Wang S (2018) High-resolution multi-temporal mapping of global urban land using Landsat images based on the Google Earth Engine Platform. Remote Sens Environ 209:227–239. https://doi.org/10.1016/j.rse.2018.02.055

Marconcini M, Marmanis D, Esch T, Felbier A (2014) A novel method for building height estmation using TanDEM-X data. In: International geoscience and remote sensing symposium (IGARSS), pp 4804–4807. https://doi.org/10.1109/IGARSS.2014.6947569

Marconcini M, Metz-Marconcini A, Üreyen S, Palacios-Lopez D, Hanke W, Bachofer F, Zeidler J, Esch T, Gorelick N, Kakarla A, Paganini M, Strano E (2020) Outlining where humans live, the World Settlement Footprint 2015. Sci Data 7:242.https://doi.org/10.1038/s41597-020-00580-5

Melchiorri M, Florczyk AJ, Freire S, et al (2018) Unveiling 25 years of planetary urbanization with remote sensing: perspectives from the global human settlement layer. Remote Sens 10:https://doi.org/10.3390/rs10050768

Melchiorri M, Pesaresi M, Florczyk AJ, et al (2019) Principles and applications of the global human settlement layer as baseline for the land use efficiency indicator—SDG 11.3.1. ISPRS Int J Geo-Information 8. https://doi.org/10.3390/ijgi8020096

Pesaresi M, Ehrlich D, Ferri S, et al (2016a) Operating procedure for the production of the Global Human Settlement Layer from Landsat data of the epochs 1975, 1990, 2000, and 2014. https://ec.europa.eu/jrc/en/publication/operating-procedure-production-global-human-settlement-layer-landsat-data-epochs-1975-1990-2000-and. Accessed 20 Aug 2020

Pesaresi M, Syrris V, Julea A (2016b) A new method for earth observation data analytics based on symbolic machine learning. Remote Sens 8https://doi.org/10.3390/rs8050399

Wang P, Huang C, de Colstoun ECB, et al (2017) Documentation for the global human built-up and settlement extent (HBASE) Dataset from landsat. https://sedac.ciesin.columbia.edu/downloads/docs/ulandsat/ulandsat-hbase-v1-documentation.pdf. Accessed 19 May 2021

Zhang X, Liu L, Wu C, Chen X, Gao Y, Xie S, Zhang B (2020) Development of a global 30m impervious surface map using multisource and multitemporal remote sensing datasets with the google earth engine platform. Earth Syst Sci Data 12:1625–1648. https://doi.org/10.5194/essd-12-1625-2020

Supra-National Thematic Land Use Cover Datasets

David García-Álvarez, Francisco José Jurado Pérez, and Javier Lara Hinojosa

Abstract

Supra-national thematic Land Use Cover (LUC) datasets are not very common. While there are several general datasets mapping all the land uses or covers in different supra-national areas across the world, LUC datasets with a similar extent that focus on the mapping of specific land covers in greater thematic detail are scarce. In this chapter, we review six different supra-national thematic LUC datasets. Three others were also found in the literature, but are not fully available for download, namely the TREES Vegetation Map of Tropical South America, the Central Africa—Vegetation map and FACET. The Circumpolar Arctic Region Vegetation dataset was also excluded from this review because of its specificity and coarse scale (1:7,500,000). Europe is the continent with the most relevant, most updated and most detailed LUC thematic datasets at supra-national scales. This is due to the work being done by the European Commission through its Joint Research Centre (JRC) and the Copernicus Land Monitoring Programme. The High-Resolution Layers (HRL) provide very detailed information, both thematically and spatially (from 10 m), for five different themes: imperviousness, tree cover, grasslands, water and wet covers, and small woody features. The European Settlement Map also provides information on built-up areas at very detailed scales (from 2.5 m). HRL and ESM are recently launched datasets which, therefore, do not provide a long series of historical data. In addition, ESM is an experimental dataset produced within the framework of a research project funded by the European Commission and no updates are expected. The datasets reviewed in this chapter for other parts of the world focus on vegetation covers of tropical forests and other relevant areas in terms of biodiversity and environmental studies. These datasets were produced within projects funded by the European Commission and the United States Agency for International Development. Unlike the previous datasets for Europe, they are already outdated and are usually produced at coarser spatial resolutions: Insular Southeast Asia—Forest Cover Map (1 km, 1998/00); Continental Southeast Asia—Forest Cover Map (1 km, 1998/02). For its part, the Congo Basin Monitoring dataset, although outdated, provides information at a higher resolution (57 m) for two different dates: 1990, 2000. The Joint Research Centre of the European Commission also produced an African cropland mask as a source of information for policy-makers. Of all the datasets reviewed in this chapter, it is the only one to focus on agricultural covers. It was obtained from data fusion at 250 m. Consequently, it does not show the cropland areas of Africa for a specific date across the whole continent.

Keywords

Supra-National • Forest • Vegetation • Built-up • Insular Southeast Asia—Forest Cover Map • Continental Southeast Asia—Forest Cover Map • Congo Basin Monitoring Maps • MARS Crop Mask Over Africa • High Resolutions Layers • European Settlement Map

D. García-Álvarez (✉)
Departamento de Geología, Geografía y Medio Ambiente, Universidad de Alcalá, Alcalá de Henares, Spain
e-mail: David.garcia@uah.es

F. J. Jurado Pérez · J. Lara Hinojosa
Departamento de Análisis Geográfico Regional y Geografía Física, Universidad de Granada, Granada, Spain

D. García-Álvarez et al. (eds.), *Land Use Cover Datasets and Validation Tools*,
https://doi.org/10.1007/978-3-030-90998-7_22

1   Insular Southeast Asia—Forest Cover Map

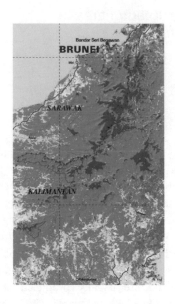

Product	
LULC thematic	
Dates	
1998 / 00	
Formats	
Raster	
Pixel size	
1 km	
Theme	
4 forest classes out of 10	
Extent	
Insular Southeast Asia	
Updating	
Not expected	
Change detection	
No (only one date)	
Overall accuracy	
Not specified	

Website of reference	**Website Language** English
https://forobs.jrc.ec.europa.eu/products/veget_map_insulare-sea/insularSEasia.php	
Download site	
https://forobs.jrc.ec.europa.eu/products/veget_map_insulare-sea/download_forest_cover_map_isea.php	

Availability	**Format(s)**
Open Access	.tiff

Technical documentation

Stibig et al. (2002, 2003a, b)

Other references of interest

–

Project

The Joint Research Centre (JRC) of the European Commission produced a map for Insular Southeast Asia which sought to provide a more accurate characterization of the forest covers in this region. It aimed to overcome the limitations associated with the mapping of vegetation covers in tropical regions, due to the persistence of cloud covers.

The dataset covers Malaysia, Singapore, Indonesia, Brunei, East Timor, the Philippines and Papua New Guinea. It is especially useful for research into deforestation and biodiversity due to the significance of the insular Southeast Asia forest ecosystem for the world as a whole.

The dataset was produced within the context of the TRopical Ecosystem Environment observations by Satellite (TREES) project. The project aimed to produce regularly updated information to monitor forest covers in tropical regions at regional scales.

Production method

The forest map for Insular Southeast Asia was produced through the unsupervised classification (clustering and maximum likelihood classification) of a mosaic of imagery collected by the VEGETATION sensor of the SPOT satellite for the period 1998–2000.

The unsupervised classification identified 60 spectral clusters. They were manually interpreted and labelled on the basis of information provided by other satellite imagery, maps of reference and field data. In addition, the initial set of clusters was regrouped on the basis of information provided by two auxiliary datasets: GTOPO30 DEM and WCMC forest map. After this initial processing, the remaining clusters were finally grouped into 8 LUC categories and a No-Data category.

Product description

The forest map for Insular Southeast Asia can be downloaded as a single compressed file (.zip) containing the raster with the LUC information. No auxiliary information is provided.

Downloads

Insular Southeast Asia—Forest Cover Map
– A raster with the LUC information (.tiff)

Legend and codification

Code	Label	Code	Label
0	No data	5	Cropland
1	Evergreen montane forest	6	Burnt/dry/sparse vegetation
2	Evergreen lowland forest	7	Non-forest vegetation
3	Mangrove forest	8	Water
4	Swamp forest		

Practical considerations

A full characterization of the dataset is provided in the technical report published by the European Commission and in the technical documentation cited above.

The map comes with several limitations: a few seasonal monsoon forests in Sulawesi, New Guinea and Philippines were not mapped as an individual category, while degraded forest cover and mature stages of forest regrowth were sometimes mapped as forest.

2 Continental Southeast Asia—Forest Cover Map

Product	
LULC thematic	
Dates	
1998 / 00	
Formats	
Raster	
Pixel size	
1 km	
Theme	
8 forest / wood classes out of 14	
Extent	
Bangladesh, Myanmar, Thailand, Laos, Cambodia, the Himalayas mountain range, north-eastern India and southern China	
Updating	
Not expected	
Change detection	
No (only one date)	
Overall accuracy	
Not specified	

Website of reference	**Website Language** English
https://forobs.jrc.ec.europa.eu/products/veget_map_continental-sea/continentalSEasia.php	
Download site	
https://forobs.jrc.ec.europa.eu/products/veget_map_continental-sea/download_forest_cover_map_csea.php	
Availability	**Format(s)**
Open Access	.tiff
Technical documentation	
Stibig et al. (2004)	
Other references of interest	
–	

Project

The forest map for Continental Southeast Asia was developed by the Joint Research Centre (JRC) of the European Commission within the context of the TRopical Ecosystem Environment observations by Satellite (TREES) and GLC2000 projects. Other LUC maps on forest covers for Insular Southeast Asia and Central Africa were also developed as part of the TREES project, following similar mapping workflows. They are all reviewed in this chapter.

The project aimed to provide regularly updated LUC information on tropical forests to help monitor activities in these regions. The obtained dataset covers Bangladesh, Myanmar, Thailand, Laos, Cambodia, the Himalaya mountain range and tropical areas of north-eastern India and southern China.

Production method

The dataset was produced through unsupervised classification of a cloud free mosaic of VEGETATION imagery for the period 1998–2000. The classification identified 70 spectral clusters, which were manually labelled and interpreted on the basis of information provided by Landsat imagery, field-collected data and a DEM. For the labelling and interpretation of spectral classes, the mapped area was split into 11 geographic strata, covering the different types of climate, landscape and land cover in the region. Finally, the labelled clusters were grouped together in 12 land cover categories.

Product description

The forest map can be downloaded in a single compressed file (.zip). No additional information is provided.

Downloads

Continental Southeast Asia—Forest Cover Map

– A raster file containing the LUC information (.tiff)

Legend and codification

Code	Label	Code	Label
0	No data	7	Evergreen wood and shrubland and regrowth mosaics
1	Evergreen Mountain forests	8	Deciduous wood and shrubland and regrowth mosaics
2	Evergreen Lowland forests	9	Mosaics of Cropping and Regrowth
3	Fragmented and degraded evergreen forest cover	10	Other lands
4	Deciduous forests	11	Other lands
5	Mangrove forests	12	Rocks
6	Swamp forests and inundated shrubland	13	Water bodies/Sea

Practical considerations

Although a technical report describing the characteristics of the dataset was published, it is not currently available. The available information is therefore limited. In addition, the spatial resolution of the map (1 km) limits its capacity to map gradual local transitions in tree canopies, such as the degradation or fragmentation of forest canopies.

3   Congo Basin Monitoring Maps

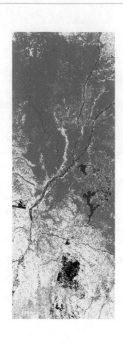

Product
LULC thematic
Dates
1990 / 00
Formats
Raster
Pixel size
57 m
Theme
Forest extent (2000) Forest probability (2000) Forest cover clearing (1990–2000)
Extent
Congo River Basin
Updating
Not expected
Change detection
Information on forest cover clearing for the period 1990–2000
Overall accuracy
Not specified

Website of reference	**Website Language** English
https://glad.umd.edu/congo-basin-monitoring	
Download site	
https://glad.umd.edu/congo-basin-monitoring	
Availability	**Format(s)**
Open Access	.tiff, .img
Technical documentation	
Hansen et al. (2008)	
Other references of interest	
Lindquist et al. (2008)	

Project

Maps of the Congo Basin Monitoring project were developed within the context of the Central African Regional Program for the Environment (CARPE), funded by the United States Agency for International Development (USAID). The program aims to promote sustainable resource management in the Congo Basin region, for which the provision of accurate monitoring data is vital.

The resulting LUC maps provide a useful resource for monitoring humid tropical deforestation at high spatial resolutions. Previous LUC datasets mapping humid tropical regions had insufficient spatial resolution. Central Africa forest covers are not subject to large-scale clearings and instead suffer smaller clearing processes taking place at a local level. This means that monitoring projects at coarse resolution miss many of the key landscape dynamics. Previous attempts to map the humid forests of Central Africa also faced important methodological limitations because of the lack of cloud-free imagery for the area. The Congo Basin Monitoring project aimed to overcome these limitations.

Two maps were produced for the Congo Basin as part of this project: a forest mask and a forest probability map that also offers information on forest clearing for the period 1990–2000. Forest clearing is defined as complete removal of the forest over story.

Production method

A forest mask was first created from a forest percent tree cover layer at 250 m generated after the classification of MODIS imagery (2000–2004) using the Vegetation Continuous Field (VCF) method. 34 metrics from MODIS imagery were extracted to carry out the classification. A threshold of 60% was applied to this layer to generate the forest mask: all pixels with a forest percentage of over 60% were considered forest. All the remaining pixels were considered non-forest. Two other categories were also classified from MODIS imagery based on a classification tree algorithm: water and rural complex. Water pixels were treated as non-land in the forest mask, and rural complex pixels were considered non-forest.

A forest probability layer was obtained from the classification of Landsat imagery at the scene level for two different epochs: pre-1996 (1986–1996) and post-1996 (>1996–2003). The classification was performed on the basis of tree models using the previously obtained forest mask as the dependent variable and the Landsat imagery as the independent variable. Forest cover changes between the two periods were mapped through a multi-date direct classification of change methodology, using training data at the same locations for the two available epochs.

Product description

The forest map can be downloaded as a single compressed file (.zip) in .tiff format. The forest probability layer is available in two different formats (.tif and .img). In both cases, the download includes the raster file with the LUC information and a text file with a technical description of each dataset.

Downloads

Forest probability and forest cover clearing
– Raster file with information on forest probability and forest cover clearing (.tiff)   – A text file with a technical description of the dataset (.txt)

MODIS-based evergreen tropical forest map (forest mask)
– Raster file with information on the forest extent (.tiff)   – A text file with a technical description of the dataset (.txt)

Legend and codification

Forest probability and forest cover clearing

Code	Label
0–100	Forest probability (0–100%)
253	Forest clearing between 1990s and 2000s
250	Water
254, 255	No data

MODIS-based evergreen tropical forest map (forest mask)

Code	Label	Code	Label
0	Non forest	1	Forest

4 MARS Crop Mask Over Africa

Product	
LULC thematic	
Dates	
One-date (varies from one product to the next)	
Formats	
Raster	
Pixel size	
250 m	
Theme	
Cropland extent	
Extent	
Africa	
Updating	
Not expected	
Change detection	
No (only one date)	
Overall accuracy	
Expected to be > 70% for most of the mapped countries	

Website of reference	**Website Language** English
https://ec.europa.eu/jrc/en/mars	
Download site	
Not available	
Availability	**Format(s)**
On request to authors	.tiff
Technical documentation	
Vancutsem et al (2013)	
Other references of interest	
Pérez-Hoyos et al. (2017a, b)	

Project

The Monitoring Agricultural Resources (MARS) unit of the Joint Research Centre (JRC) produced a cropland mask for Africa to assist the unit and Commission's activities with crop and food security monitoring. The mask aimed to provide the most accurate information possible on cropland covers for Africa by merging the best available LUC cropland data sources.

The methodology applied in the production of this dataset has also been used in the development of other cropland masks (ASAP Land Cover Masks) by the same team.

Production method

The MARS crop mask was obtained by merging the best available LUC data sources on cropland covers. To this end, all the input data sources were resampled or rasterized to a common spatial resolution (250 m) and projected with the same parameters. Cropland categories were extracted from each input dataset. LUC categories were considered as cropland when at least 50% of their surface was covered by cropland. LUC categories with a cropland proportion of between 20 and 50% were manually checked by experts, who decided whether to include them as cropland categories at a global level or for just one specific region.

The accuracy of each dataset was assessed against Google Earth imagery. When several datasets were available for the same area, the most accurate one was selected. If several datasets had similar levels of accuracy, the most detailed or recent was selected.

The input datasets were Globcover, SADC, Cropland Use Intensity datasets from USGS, Woody Biomass map of Ethiopia, AFRICOVER, JRC-MARS crop masks, LULC

2000 USGS datasets and national land cover maps of the Democratic Republic of Congo, Mozambique and Senegal.

Product description

The crop mask is available in Google Drive on request to the producers of the map. The download includes a document with a technical description of the product as well as the raster file with the LUC information. Another raster file is provided with information about the data source that was finally selected to create the crop mask in each case.

Downloads

MARS crop mask over Africa
− Raster file with crop extent (.tiff) − Raster file with information on the data source used to map each area (.tiff) − Document with a technical description of the dataset (.doc)

Legend and codification

Code	Label	Code	Label
0	Cropland	1	No cropland

Practical considerations

Users interested in accessing the dataset should apply to the map's authors (Christelle.vancutsem@ec.europa.eu). This map was obtained by merging data from selected data sources. The dataset cannot provide LUC information for any specific reference year as each source had its own.

5   HRL—High Resolution Layers

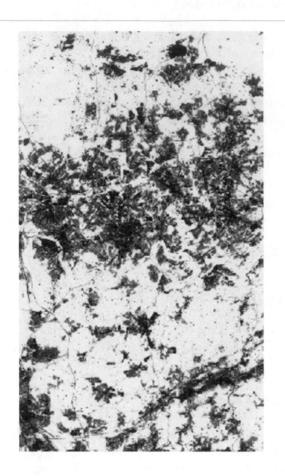

Product
LULC thematic
Dates
2006, 2009, 2012, 2015, 2018 (Imperviousness) 2012, 2015, 2018 (Forests) 2015, 2018 (Grassland, Wetness and Water) 2015 (Small Woody Features)
Formats
Raster
Pixel size
5 m (Small Woody Features) 10 m (Products since 2018) 20 m (Products up to 2015) 100 m (Mosaics)
Themes
Extent and percentage of impervious areas Percentage of tree cover areas, leaf type and forest type Extent of grassland areas Wetness and water covers (5 water/wet classes out of 8) Extent of Small Woody Features
Extent
Europe (39 countries)
Updating
Planned every 3 years
Change detection
Through change layers
Overall accuracy
Imperviousness HRL, Forests: expected to be > 90%Grassland HRL, Wetness and Water HRL: expected to be > 80–80%Wetness and Water: HRL expected to be > 80%

Website of reference	Website **Language** English, German and French

Website of reference

https://land.copernicus.eu/pan-european/high-resolution-layers

Download site

https://land.copernicus.eu/pan-european/high-resolution-layers

Availability	Format(s)
Open Access after registration	.tiff

Technical documentation

Copernicus Land Monitoring Service (2020a, b, c, d), D'amico et al. (2019), Faucqueur et al. (2018), Langangke (2015, 2016), Langangke et al. (2017, 2018a, b, 2019), Pennec et al. (2019a, b), Smith et al. (2019), Weirather et al. (2019a, b)

Other references of interest

Büttner et al. (2016), Manakos et al. (2018), Sannier et al. (2017)

Project

The High-Resolution Layers are produced within the framework of the Copernicus Land Monitoring Programme. They were created as a means of overcoming some of the limitations associated with CORINE Land Cover (CLC), such as lack of detail, the presence of mixed classes and the difficulty of adapting the CLC legend to other common classification schemes, such as the FAO LCSS. Each High-Resolution Layer is associated with one of the CLC Level 1 classes: artificial surfaces (Imperviousness HRL), agricultural areas (Grassland HRL), forest and semi-natural areas (Forests HRL), wetlands and water bodies (Water & Wetness HRL).

The different High-Resolution Layers are separately produced using specific methods. Since 2018, they have been produced at enhanced spatial resolution (10 m) based on Sentinel imagery. This marks a change in the methodology applied in the production of HRL compared to the layers created for previous years of reference.

Some of the HRL layers have been produced for more years than the others, such as the Imperviousness HRL, available since 2006, and the Forests HRL, available since 2012. However, when available, the reference years are almost all the same for all the layers. The only exception is the recently created Small Woody Features HRL. In some cases, when more than one date is available, change layers have been developed.

Production method

Each HRL has its own specific production method, as each theme is characterized in a different way. Nevertheless, all the HRLs are obtained by automatic classification and interactive rule-based classification of high-resolution imagery, mostly from the Sentinel constellation. The Imperviousness HRL and Water and Wetness HRL are obtained from both optical and raster data, while the Forests, Grasslands and Small Woody Features HRLs are obtained exclusively from optical data.

Change layers are obtained by comparing the status layers for two different years of reference. For the changes between 2018 and the previous year of reference, some uncertainties may arise because of the change in the spatial resolution: 10 m vs 20 m. The production teams have implemented various different measures to prevent such uncertainties, including the development of supporting layers that inform about the changes that take place due to technical reasons and the level of confidence of the obtained change layer.

Initial production of the HRL is centralized. Then, each country reviews and verifies the results, so enhancing this initial product. For more detailed information about the production process of all the HRLs, readers are referred to the technical documentation cited above.

Product description

Imperviousness HRL

The Imperviousness HRL can be separately downloaded for each year of reference or for each period of changes. In the latter case, users can choose between an uncategorized file showing the change in the degree of imperviousness and a file that categorizes this change in a series of classes. For the reference year 2018, users can also download the Impervious built-up layer as a separate file. This is a binary map differentiating built-up areas from non-built-up areas.

The layers are disseminated at country level in 100 × 100 km tiles. Users download a single file with all the tiles covering the selected country. A mosaic of all the mapped countries is also available as a single file at two spatial resolutions: 10-20 m (the original resolution) and 100 m.

Different supporting layers are also available for download as part of the Imperviousness HRL. Unlike the previous layers, they are available in the "Expert Products" section as single files covering all of Europe. These supporting layers include (i) a layer indicating the change in the degree of imperviousness between 2015 and 2018 due to technical reasons (IMCS); (ii) a layer showing the confidence level of the Imperviousness density 2018 layer at 10 m (IMDCL); and (iii) an adaptation of the Imperviousness density 2015 layer to a spatial resolution of 10 m, to enable researchers to study changes in the impervious area between 2015 and 2018 (IMDR).

All downloads have the same contents: a raster file containing the LUC information, a file to symbolize it in any GIS software and a metadata file. Files for the pre-2018 editions of Imperviousness HRL also include an Excel file with technical information about the product.

Downloads

Imperviousness built-up 2018 (Status Map)
Imperviousness density 2018 (Status Map)
Imperviousness Change 2015–2018 (Change Map)
Imperviousness Classified Change 2015–2018 (Change Map)

– Raster file with LUC information (.tiff) (*DATA* folder)
– Text file to symbolize the raster in QGIS (.txt) (*Symbology* folder)
– Metadata file (*Metadata* folder)

Legend and codification

Imperviousness built-up (Status Map)

Code	Label	Code	Label
0	Non built-up	255	Outside area
1	Built-up		

Imperviousness density (Status Map)

Code	Label	Code	Label
0	Non-impervious areas	254	Unclassifiable
1–100	Degree of imperviousness (%)	255	Outside area

Imperviousness Change (Change Map)

Code	Label
0–99	Percentage of decreased imperviousness density
100	Unchanged areas with some degree of imperviousness
101–200	Percentage of increased imperviousness density
201	Unchanged areas with no degree of imperviousness
254	Unclassifiable (no satellite image available, or clouds, shadows, or snow)

Imperviousness Classified Change (Change Map)

Code	Label
0	Unchanged areas with Imperviousness Density = 0%
1	New cover (increasing imperviousness density, which was 0% at first reference date)
2	Loss of cover (decreasing imperviousness density, which was 0% at second reference date)
10	Unchanged areas with Imperviousness Density > 0% at both reference dates
11	Increased Imperviousness Density (>0% at both reference dates)
12	Decreased Imperviousness Density (>0% at both reference dates)
254	Unclassifiable
255	Outside area

Forests HRL

For each available year of reference, three different types of layer can be downloaded as part of the Forests HRL: (i) a layer showing the forest density or the degree of tree cover (Tree Cover Density); (ii) a layer informing about the dominant leaf type, distinguishing mainly between broadleaf

and coniferous trees (Dominant Leaf Type); and (iii) a layer informing about the dominant leaf type in treed areas covering more than 0.5 ha and with a tree cover density of over 10%, i.e. those areas considered as forest according to the FAO definition (Forest Type).

Change layers for Tree Cover and Dominant Leaf Type are also provided for each mapped period. A layer of tree cover density changes was initially created for the period 2012–2015. However, it has not been updated for the new mapping periods and is no longer distributed.

In all cases, the layers are distributed at a country level in 100 × 100 km tiles. A single file mosaic of each layer for all the mapped countries is also available at two spatial resolutions: 10–20 m (the original resolution) and 100 m.

Nine additional layers were also produced as supplementary information to the Forests HRL for the year 2018. These can be downloaded from the "Experts products" section. They provide information about the broadleaved and coniferous cover densities at 100 m (BCD, CCD) as well as other relevant technical information about the production of the Forests HRL: level of confidence, data sources, etc. The technical documentation of HRL Forests includes a detailed description of each of these supporting layers.

In all cases, the downloaded files include the raster with LUC information, a file to symbolize it in any GIS software and the product's metadata. Files for the pre-2018 editions of Forests HRL also include an Excel file with technical information about the product.

Downloads

Tree Cover Density 2018
Tree Cover Change Mask 2015–2018
Dominant Leaf Type 2015
Dominant Leaf Type Change 2015–2018
Forest Type 2018

– Raster file with LUC information (.tiff) (*DATA* folder)
– Text file to symbolize the raster in QGIS (.txt) (*Symbology* folder)
– Metadata file (*Metadata* folder)

Legend and codification

Tree Cover Density

Code	Label	Code	Label
0	Non–tree-covered areas	254	Unclassifiable
1–100	Percentage of tree cover density	255	Outside area

Tree Cover Change Mask

Code	Label	Code	Label
0	Unchanged areas with no tree cover	10	Unchanged areas with tree cover
1	New tree cover	254	Unclassifiable in any of parent status layers
2	Loss of tree cover	255	Outside area

Dominant Leaf Type

Code	Label	Code	Label
0	Non–tree-covered areas	254	Unclassifiable
1	Broadleaved trees	255	Outside areas
2	Coniferous trees		

Dominant Leaf Type

Code	Label	Code	Label
0	Unchanged areas with no tree cover	10	Unchanged areas with tree cover
1	New broadleaved cover	12	Potential change among dominant leaf types
2	New coniferous cover	254	Unclassifiable in any of parent status layers
3	Loss of broadleaved cover	255	Outside area
4	Loss of coniferous cover		

Forest Type

Code	Label	Code	Label
0	Non–tree-covered areas	3	Mixed forest (only for aggregated 100 m layer)
1	Broadleaved trees	254	Unclassifiable
2	Coniferous trees	255	Outside areas

Grassland HRL

A status layer for each reference year and a layer of changes for each mapped period can be downloaded separately as part of the Grassland HRL. Moreover, three additional supporting layers are distributed as "Expert products": (i) a layer showing the probability of each pixel being grassland (Grassland Vegetation Probability Index, GRAVPI); (ii) a layer informing about the number of years since the last ploughing (Ploughing Indicator, PLOGH); and (iii) a confidence layer for the Grassland 2018 status map (GRACL).

The status layer and the change layers are distributed at country level in 100 × 100 km tiles. A single file European mosaic is also available at two spatial resolutions: 10-20 m (the original resolution) and 100 m. The three supporting layers can be downloaded as single files covering the whole of the mapped area.

All downloads include the raster with LUC information, a file to symbolize it in GIS and a metadata file. Downloads for the pre-2018 editions of the layers also include an Excel file with technical information about the product.

Downloads

Grassland 2018 (Status Map)
Grassland Change 2015–2018 (Change maps)
– Raster file with LUC information (.tiff) (*DATA* folder) – Text file to symbolize the raster in QGIS (.txt) (*Symbology* folder) – Metadata file (*Metadata* folder)

Legend and codification

Grassland (Status Map)

Code	Label	Code	Label
0	Non–grass areas	254	Unclassifiable
1	Grassy and non–woody vegetation	255	Outside area

Grassland Change (Change maps)

Code	Label	Code	Label
0	All non-grassland areas	11	Unverified grassland gain
1	Grassland gain	22	Unverified grassland loss
2	Grassland loss	254	Unclassifiable in any of parent status layers
10	Unchanged grassland in both years	255	Outside area

Water and Wetness HRL

The Water and Wetness HRL is made up of a main product mapping the different types of water and wetness covers in Europe. Users can also download an additional layer (Expert products) showing the probability of each pixel being water or wetness. Two extra technical layers are also available as expert products: one informs about the confidence of the 2018 status map (WACL) while the other studies the differences in the mapping of water and wetness covers between 2015 and 2018 (WAWCSL).

Different files can be downloaded for each available layer and year. The main layer is distributed at country level in 100 × 100 km tiles. However, a single file mosaic is also available at the original resolution of the product (10–20 m) and at 100 m. The supporting layers are only available at the original resolution as single files covering the whole of Europe.

All downloads include the raster with LUC information, a file to symbolize it in any GIS software and a metadata file. The available layer for 2015 also includes an Excel file with technical information about the product.

Downloads

Water and Wetness 2018–WAW (Status Map)

- Raster file with LUC information (.tif) (*DATA* folder)
- Text file to symbolize the raster in QGIS (.txt) (*Symbology* folder)
- Metadata file (*Metadata* folder)

Legend and codification

Water and Wetness (Status Map)

Code	Label	Code	Label
0	Dry	4	Temporary wet
1	Permanent water	253	Sea water
2	Temporary water	254	Unclassifiable
3	Permanent wet	255	Outside areas

Small Woody Features HRL

The Small Woody Features HRL is available in either vector or raster files. Vector files can be downloaded in two different formats: ESRI Geodatabase and GeoPackage. Raster files can be downloaded at two different spatial resolutions: 5 and 100 m.

The vector and raster files at 5 m are distributed in tiles obtained after splitting each European country into a series of large regions. To find out which tile corresponds to their particular area of interest, users should consult the viewer on the dataset's website.[1] The rasters at 100 m are distributed as single files covering the whole of Europe, without splits into regions.

The raster at 5 m only differentiates between Small Woody Features (SWF) and Additional Woody Features (AWF). The vector file also differentiates between SWF and AWF, although it splits the first category into linear and patchy structures. Three different layers are available at 100 m: (i) the density of small woody features (SWF); (ii) the density of Additional Woody Features (AWF); and (iii) the density of both small and additional woody features (SWFAWF).

Downloads

Small Woody Features 2018 (Geodatabase)

- Vector file with LUC information (*DATA* folder)
- Raster file with information about the accuracy of the product (.tiff)
- PDF with a guide about how to use the ESRI Geodatabase in QGIS (*Documents* folder)
- PDF with information about the product (*Documents* folder)
- Metadata about the product (*Metadata* folder)

Small Woody Features 2018 (Raster 5 m)

- Raster file with LUC information (.tiff) (*Data* folder)
- File to symbolize the raster in GIS (.clr) (*Data* folder)
- PDF with information about the product (*Documents* folder)
- Metadata about the product (*Metadata* folder)

SWF density (Raster 100 m)

AWF density (Raster 100 m)

SWF + AWF density (Raster 100 m)

- Raster file with LUC information (.tiff) (*Documents* folder)
- PDF with information about the product (*Documents* folder)
- Metadata about the product (*Metadata* folder)

---

[1] https://land.copernicus.eu/pan-european/high-resolution-layers/small-woody-features/small-woody-features-2015.

Database

Small Woody Features 2018 (Geodatabase)

gid	code	area	class_name
1	1	5301,36442179704	Linear structures of trees, hedges, bushes and scrub
2	1	5376,72678401566	Linear structures of trees, hedges, bushes and scrub
3	1	3579,55745263859	Linear structures of trees, hedges, bushes and scrub

- Gid: Unique identifier for each polygon
- Code: Thematic code for each polygon
- Area: Area of the polygon, in square meters
- Class_name: Category assigned to each polygon

Legend and codification

Small Woody Features (Geodatabase and GeoPackage)

Code	Label	Code	Label
1	Linear structures of trees, hedges, bushes and scrub	3	Additional woody features
2	Patchy structures of trees, hedges, bushes and scrub		

Small Woody Features (Raster 5 m)

Code	Label	Code	Label
0	Non-SWF area	254	Unclassifiable
1	Patchy structures of trees, hSWF area (Linear or patchy structures of trees, hedges, bushes and scrub)	255	Outside areas
3	Additional woody features		

SWF density (Raster 100 m)

Code	Label	Code	Label
0	Non-SWF area	254	Unclassifiable
0–100	Small Woody Features density	255	Outside areas

AWF density (Raster 100 m)

Code	Label	Code	Label
0	Non-SWF area	254	Unclassifiable
0–100	Additional Woody Features density	255	Outside areas

SWF+ AWF density (Raster 100 m)

Code	Label	Code	Label
0	Non-SWF area	254	Unclassifiable
0–100	Small+ additional Woody Features density	255	Outside areas

Practical considerations

Users can consult the layers via the online viewers available at the product's download website. The technical documents provide useful descriptions of the characteristics of the products and all the layers available for each year of reference, including the expert products, which we have not been reviewed in detail.

6 ESM—European Settlement Map

Product	
LULC Thematic	
Dates	
2012, 2015	
Formats	
Raster	
Pixel size	
2 m, 10 m (2015) 2.5 m, 10 m, 100 m (2012)	
Theme	
Extent of Built-up areas (2015) Extent of Residential areas (2012) 13 built-up categories (2012) Percentage of built-up areas (2012)	
Extent	
Europe	
Updating	
Not planned	
Change detection	
No	
Overall accuracy	
Expected to be > 80% (ESM 2015 - 2 m) Expected to be > 70% (ESM 2015 - 10 m)	

Website of reference	Website Language English
https://land.copernicus.eu/pan-european/GHSL/european-settlement-map	

Download site

https://land.copernicus.eu/pan-european/GHSL/european-settlement-map

Availability	Format(s)
Open access after registration	.tiff

Technical documentation

Ferri et al. (2014, 2016a, 2017), Florczyk et al. (2016), Pafi et al. (2016a), Pesaresi et al. (2013), Sabo et al. (2019), Smith and Sannier (2017)

Other references of interest

Ferri et al. (2016b), Pafi et al. (2016b)

Project

The European Settlement Map (ESM) is part of the Global Human Settlement Layer (GHSL) project, supported by the European Commission through the Joint Research Centre (JRC) and the Directorate General for Regional and Urban Policy (DG REGIO). ESM complements the GHSL global products by providing an urban settlement map for Europe at a very detailed spatial resolution: 2–2.5 m versus 30 m for the GHSL. Both products share similar automatic methods for extracting LUC information from satellite imagery.

ESM was initially released in 2014, with successive updates in 2016, 2017 and 2019. In 2014, a dataset was created for the reference year 2012, showing the percentage of the surface area that was built up. This was revised with a new production methodology in 2016 and again in 2017. The first update improved the accuracy of the product and its consistency with population data. The spatial resolution was also improved: from 100 to 10 m. The second update increased the spatial and thematic detail of the product, at 2.5 m and differentiating between 12 classes. A new dataset at 2 m for the year 2015 was released in 2019, using a different production methodology. Unlike previous editions, this map only shows the extent of built-up areas, without providing further information about the built-up fraction per pixel.

In addition to the base layer delineating built-up areas, the latest edition of the product (2019) includes a classification differentiating residential from non-residential areas at a spatial resolution of 10 m.

Production method

The ESM production method has changed over time, although it has always been fully automatic. The latest edition (2019) was produced at 2 m on the basis of the Copernicus VHR_IMAGE_2015 imagery dataset, made up of images captured by the satellites Pleiades, Deimos-02, WorldView-2, WorldView-3, GeoEye-01 and Spot 6/7. The imagery was classified through a scene-based classification algorithm: Symbolic Machine Learning (SML).

The first three editions of ESM were obtained at 100, 10 and 2.5 m through a textural and morphological technique of unsupervised built-up area detection. Spot 6/7 imagery was used as an input. In the third edition (2017), auxiliary data sources (Open Street Map, Urban Atlas…) were also used to provide more thematic detail, distinguishing between 13 LUC categories, instead of just between built-up and non-built-up areas.

Product description

The ESM for each of the available editions can be downloaded separately as a single file. If more than one spatial resolution is available, users must separately download the specific product for the spatial resolution they require.

The ESM layers at 100 m for the 2014 and 2016 editions are distributed as a single European file. For the 2017 edition of ESM at 100 m, users must download a different file covering the entire mapped area for each of the categories (13 in total). The 2016 edition at 10 m is also distributed in 400 × 400 km tiles. Finally, the ESM layers at 2–2.5 and 10 m are distributed in 100 × 100 km tiles for the 2017 and 2019 editions of the product. In all cases, users can find out which tile or tiles fall within their area of interest by consulting the viewer available on the ESM website.

Downloads

Due to the complexity of this product, with different editions available for the same years of reference at different spatial resolutions, in the following table we present an overview of all the available maps, classified according to the year they were released, their spatial resolution and the year of reference, i.e. the year for which they map the LUC covers. The different files available for download are described below the table.

Available products for download

Product	Edition	Pixel size
ESM 2012	2014	100 m
	2016	100 m
	2016	10 m
	2017	100 m
	2017	10 m
	2017	2.5 m
ESM 2015	2019	10 m
	2019	2 m

ESM 2012 (2014)—100 m

 − Raster file with built-up percentage (*EU_GHSL100m* folder)
 − Raster files with technical information about the product (*EU_GHSL100m_Data_Mask* and *EU_GHSL100m_Data_Processed_Ref_Year* folders)

ESM 2012 (2016)—100 m, 10 m

 − Raster file with built-up percentage
 − Text file with a description of the product (.txt)

ESM 2012 (2017)—100 m

- Raster file with class percentage per pixel for one of the classes mapped in 2nd edition of ESM

ESM 2012 (2017)—10 m

- Raster files with class percentage per pixel for each of the classes mapped in 2nd edition of ESM

ESM 2012 (2017)—2.5 m

- Raster file with LC information
- Layer style file for ArcGIS (.lyr) and QGIS (.qml)
- PDF with technical information about the product

ESM 2015 (2019)—10 m, 2 m

- Raster file with LC information
- TXT files with map legend and copyright information
- File for symbolizing the raster in GIS(.clr)

Legend and codification

ESM 2012 (2014)—100 m

Code	Label	Code	Label
0–1	Built-up percentage (0–100%)	-2	No data

ESM 2012 (2016)—100 m

Code	Label
0–1	Built-up percentage (0–100%)

ESM 2012 (2016)—10 m

Code	Label
0–100	Built-up percentage (0–100%)

ESM 2012 (2017)—100 m-Class 50 (Buildings)

Code	Label
0–1	Percentage (0–100%) of the selected class (50)

ESM 2012 (2017)—10 m-Class 50 (Buildings)

Code	Label
0–100	Percentage (0–100%) of the selected class (50)

ESM 2012 (2017)—2.5 m

Code	Label	Code	Label
50	BU Buildings	20	NBU Area-Green NDVI
45	BU Area-Street Green NDVI	15	NBU Area-Streets
41	BU Area-Green UA	10	NBU Area-Open Space
40	BU Area-Green NDVI	2	Railways
35	BU Area-Streets	1	Water
30	BU Area-Open Space	0	No Data
25	NBU Area-Street Green NDVI		

ESM 2015 (2019)—10 m

Code	Label	Code	Label
0	No data	250	Non-residential built-up area
1	Land	255	Residential built-up area

ESM 2015 (2019)—2 m

Code	Label	Code	Label
0	No data	2	Water
1	Land	255	Built-up area

Practical considerations

All editions of ESM are available for download at the Copernicus Land programme website.[2] The ESM 2015 can be consulted through an online viewer as part of the GHSL framework.[3] It can also be downloaded from the same website in tiles.[4]

The 2016 ESM edition at 10 m is distributed in 237 400 × 400 km tiles. However, of the 237 tiles available for download, only 86 fall within areas with impervious surfaces. Therefore, only 86 out of the 237 tiles include LUC information.

---

[2] https://land.copernicus.eu/pan-european/GHSL/european-settlement-map.
[3] https://ghsl.jrc.ec.europa.eu/ESMVisualisation.php.
[4] https://ghsl.jrc.ec.europa.eu/download.php?ds=ESM.

References

Büttner G, Maucha G, Kosztra B (2016) High-resolution layers. In: Feranec J, Soukup T, Hazeu G, Jaffrain G (eds) European landscape dynamics. CORINE land cover data. CRC Press, Boca Raton, pp 61–69

Copernicus Land Monitoring Service (2020a) Copernicus land monitoring service high resolution land cover characteristics. Lot1: imperviousness 2018, imperviousness change 2015–2018 and built-up 2018. https://land.copernicus.eu/user-corner/technical-library/hrl-imperviousness-2018-user-manual. Accessed 30 Sept 2020

Copernicus Land Monitoring Service (2020b) Copernicus land monitoring service high resolution land cover characteristics. Tree-cover/forest and change 2015–2018. https://land.copernicus.eu/user-corner/technical-library/forest-2018-user-manual-v1-0.pdf. Accessed 30 Sept 2020

Copernicus Land Monitoring Service (2020c) Copernicus land monitoring service high resolution land cover characteristics. Grassland 2018 and grassland change 2015–2018. https://land.copernicus.eu/user-corner/technical-library/hrl-grassland-2018-user-manual. Accessed 30 Sept 2020

Copernicus Land Monitoring Service (2020d) Copernicus land monitoring service high resolution land cover characteristics. Lot4: water & wetness 2018. https://land.copernicus.eu/user-corner/technical-library/hrl-water-and-wetness-2018-user-manual. Accessed 30 Sept 2020

D'amico Q, Corsini N, Vandeputte R, Pennec A (2019) HRL water and wetness 2015 validation report. https://land.copernicus.eu/user-corner/technical-library/hrl-water-and-wetness-2015-validation-report. Accessed 30 Sept 2020

Faucqueur L, Desclée B, Pennec A et al (2018) HRL small woody features validation concept. https://land.copernicus.eu/user-corner/technical-library/clms_hrl_swf_validation_concept_sc03_1_1-2.pdf. Accessed 30 Sept 2020

Ferri S, Syrris V, Florczyk A et al (2014) A new map of the European settlements by automatic classification of 2.5m resolution SPOT data. Int Geosci Remote Sens Symp 1160–1163. https://doi.org/10.1109/IGARSS.2014.6946636

Ferri S, Siragusa A, Halkia M (2016a) The ESM green components. A dedicated focus on the production of the green in the European settlement map's workflow. https://publications.jrc.ec.europa.eu/repository/handle/JRC102521. Accessed 28 Sept 2020

Ferri S, Siragusa A, Pafi M, Halkia S (2016b) How green are the European Cities? Exploring the green european settlement map 2016. https://ec.europa.eu/jrc/en/publication/how-green-are-european-cities-exploring-green-european-settlement-map-2016-systematic-comparison. Accessed 28 Sept 2020

Ferri S, Siragusa A, Sabo F et al (2017) The European settlement map 2017 release. https://publications.jrc.ec.europa.eu/repository/bitstream/JRC105679/kjna28644enn.pdf. Accessed 28 Sept 2020

Florczyk AJ, Ferri S, Syrris V et al (2016) A new European settlement map from optical remotely sensed data. IEEE J Sel Top Appl Earth Obs Remote Sens 9:1978–1992. https://doi.org/10.1109/JSTARS.2015.2485662

Hansen MC, Roy DP, Lindquist E et al (2008) A method for integrating MODIS and landsat data for systematic monitoring of forest cover and change in the Congo Basin. Remote Sens Environ 112:2495–2513. https://doi.org/10.1016/j.rse.2007.11.012

Langangke T (2015) GIO land (GMES/Copernicus initial operations land) High Resolution Layers (HRLs)–summary of product specifications

Langangke T (2016) Copernicus land monitoring service–high resolution layer imperviousness. https://land.copernicus.eu/user-corner/technical-library/hrl-imperviousness-technical-document-prod-2015. Accessed 30 Sept 2020

Langangke T, Desclee B, Faucqueur L et al (2019) Copernicus Land monitoring service–high resolution layer small woody features—2015 reference year. Product specifications & user guidelines. https://land.copernicus.eu/user-corner/technical-library/hrl_lot5_d5-1_product-specification-document_i3-4_public-1.pdf. Accessed 30 Sept 2020

Langangke T, Hermann D, Ramminger G et al (2017) Copernicus land monitoring service–high resolution layer forest: product specifications document. https://land.copernicus.eu/user-corner/technical-library/hrl-forest. Accessed 30 Sept 2020

Langangke T, Richter R, Sandow C et al (2018a) Copernicus land monitoring service–high resolution layer grassland: product specifications document. https://land.copernicus.eu/user-corner/technical-library/hrl-grassland-technical-document-prod-2015. Accessed 30 Sept 2020

Langangke T, Moran A, Dulleck B, Schleicher C (2018b) Copernicus land monitoring service–high resolution layer water and wetness: product specifications document. https://land.copernicus.eu/user-corner/technical-library/hrl-water-wetness-technical-document-prod-2015. Accessed 30 Sept 2020

Lindquist EJ, Hansen MC, Roy DP, Justice CO (2008) The suitability of decadal image data sets for mapping tropical forest cover change in the Democratic Republic of Congo: implications for the global land survey. Int J Remote Sens 29:7269–7275. https://doi.org/10.1080/01431160802275890

Manakos I, Tomaszewska M, Gkinis I, et al (2018) Comparison of global and continental land cover products for selected study areas in South Central and Eastern European Region. Remote Sens 10. https://doi.org/10.3390/rs10121967

Pafi M, Ferri S, Siragusa A, Halkia M (2016a) Visual assessment of the green European settlement map. https://ec.europa.eu/jrc/en/publication/visual-assessment-green-european-settlement-map-exploration-and-quality-performance-measures-green. Accessed 28 Sept 2020

Pafi M, Siragusa A, Ferri S, Halkia S (2016b) Measuring the accessibility of urban green areas. https://ec.europa.eu/jrc/en/publication/measuring-accessibility-urban-green-areas-comparison-green-esm-other-datasets-four-european-cities. Accessed 28 Sept 2020

Pennec A, Deutscher J, Pennec A, Dufourmont H (2019a) HRL forest 2015 validation report. https://land.copernicus.eu/user-corner/technical-library/hrl-forest-2015-final-validation-report. Accessed 30 Sept 2020

Pennec A, Vandeputte R, Sannier C et al (2019b) Analysis of the mixed forest in the 20 m HRL forest type validation report. https://land.copernicus.eu/user-corner/technical-library/analysis-of-the-mixed-forest-in-the-20-m-hrl-forest-type-validation-report. Accessed 30 Sept 2020

Pérez-Hoyos A, Rembold F, Gallego J et al (2017a) Development of a new harmonized land cover/land use dataset for agricultural monitoring in Africa. In: ESA world cover conference, 14–17 Mar 2017. Frascati, Rome (Italy)

Pérez-Hoyos A, Rembold F, Kerdiles H, Gallego J (2017b) Comparison of global land cover datasets for cropland monitoring. Remote Sens 9. https://doi.org/10.3390/rs9111118

Pesaresi M, Huadong G, Blaes X et al (2013) A global human settlement layer from optical HR/VHR RS data: concept and first results. IEEE J Sel Top Appl Earth Obs Remote Sens 6:2102–2131. https://doi.org/10.1109/JSTARS.2013.2271445

Sabo F, Corbane C, Politis P, Kemper T (2019) The European settlement map 2019 release. https://ec.europa.eu/jrc/en/publication/european-settlement-map-2019-release. Accessed 28 Sept 2020

Sannier C, Pennec A, Dufourmont H (2017) Comparative validation of HRL tree cover density and University of Maryland global forest change products. https://land.copernicus.eu/user-corner/technical-library/comparative-validation. Accessed 30 Sept 2020

Smith G, Pennec A, Sannier C, Dufourmont H (2019) HRL Imperviousness degree 2015 validation report. https://land.copernicus.eu/user-corner/technical-library/hrl-imperviousness-2015-validation-report. Accessed 30 Sept 2020

Smith G, Sannier C (2017) European settlement map 2016, validation report. https://land.copernicus.eu/user-corner/technical-library/the-european-settlement-map-validation-report/view. Accessed 30 Sept 2020

Stibig H-J, Achard F, Fritz S (2004) A new forest cover map of continental Southeast Asia derived from satellite imagery of coarse spatial resolution. Appl Veg Sci 7:153–162

Stibig HJ, Beuchle R, Janvier P (2002) Forest cover map of insular Southeast Asia at 1:5 500 000 derived from SPOT-vegetation satellite images. https://land.copernicus.eu/user-corner/technical-library/analysis-of-the-mixed-forest-in-the-20-m-hrl-forest-type-validation-report. Accessed 20 Apr 2021

Stibig HJ, Beuchle R, Achard F (2003a) Mapping of the tropical forest cover of insular Southeast Asia from SPOT4-vegetation images. Int J Remote Sens 24:3651–3662. https://doi.org/10.1080/0143116021000024113

Stibig HJ, Malingreau JP (2003b) Forest cover of insular Southeast Asia mapped from recent satellite images of coarse spatial resolution. AMBIO A J Hum Environ 32:469. https://doi.org/10.1639/0044-7447(2003)032[0469:fcoisa]2.0.co;2

Vancutsem C, Marinho E, Kayitakire F et al (2013) Harmonizing and combining existing land cover/land use datasets for cropland area monitoring at the African continental scale. Remote Sens 5:19–41. https://doi.org/10.3390/rs5010019

Weirather M, Zeug G, Pennec A et al (2019a) HRL Grassland 2015 validation report. https://land.copernicus.eu/user-corner/technical-library/analysis-of-the-mixed-forest-in-the-20-m-hrl-forest-type-validation-report. Accessed 30 Sept 2020

Weirather M, Zeug G, Pennec A et al (2019b) HRL Grassland 2015 validation report. https://land.copernicus.eu/user-corner/technical-library/hrl-grassland-2015-validation-report. Accessed 30 Sept 2020

Correction to: About This Book

David García-Álvarez, María Teresa Camacho Olmedo,
Martin Paegelow, and Jean-François Mas

Correction to:
Chapter "About This Book" in: D. García-Álvarez et al. (eds.), *Land Use Cover Datasets*
and Validation Tools, **https://doi.org/10.1007/978-3-030-90998-7_1**

In the original version of this chapter, the URL link on page 11 in section "5.5 Data" was wrongly pointed to https://dx.doi.org/10.1007/978-3-030-90998-7_16 saying "DOI Not Found". This has been corrected in the updated version.

The updated version of this chapter can be found at
https://doi.org/10.1007/978-3-030-90998-7_1

Printed in the United States
by Baker & Taylor Publisher Services